The Privatized Art of War
Private Military and Security Companies and State Responsibility for Their Unlawful Conduct in Conflict Areas

SCHOOL OF HUMAN RIGHTS RESEARCH SERIES, Volume 67

The titles published in this series are listed at the end of this volume.

The Privatized Art of War

Private Military and Security Companies and State Responsibility for Their Unlawful Conduct in Conflict Areas

Evgeni Moyakine

intersentia

Cambridge – Antwerp – Portland

Intersentia Ltd
Sheraton House | Castle Park
Cambridge | CB3 0AX | United Kingdom
Tel.: +44 1223 370 170 | Email: mail@intersentia.co.uk

This publication is based on a doctoral dissertation defended at Tilburg University in November 2014. The research was supported by the Netherlands Organization for Scientific Research (NWO) under project number 017.007.045.

Evgeni Moyakine
The Privatized Art of War. Private Military and Security Companies and State Responsibility for Their Unlawful Conduct in Conflict Areas

ISBN 978-1-78068-281-5
D/2015/7849/19
NUR 828

British Library Cataloguing in Publication Data. A catalogue record for this book is available from the British Library.

TO MY FAMILY

ACKNOWLEDGMENTS

Writing a doctoral dissertation might seem to some as a lonely enterprise, and this process – at certain points – definitely involves days, weeks, and even months of hard work isolated from civilization. As a whole, however, it can better be described as a collaborative project or joint effort in which you courageously and persistently collect pieces of the puzzle trying to create the complete picture and to enjoy it in its full and undisturbed glory.

Undoubtedly, I am blessed that on my unforgettable journey I was assisted by my supervisors who were always there when I needed them the most. Snežana Trifunovska from Radboud University Nijmegen deserves special thanks as my first mentor, who supervised my Master's thesis five years ago. She took over this role during my PhD research and was always on my side like a caring mother and my greatest role model. Thank you. The words of thanks and appreciation also go to my Tilburg-based supervisors, Willem van Genugten and Nicola Jägers, with whom I spent four fruitful years of research at Tilburg Law School and who have also been my inexhaustible sources of inspiration, constantly offering me assistance and support. I owe my sincere gratitude to all of you.

Furthermore, I am indebted to my friends, colleagues, and professors at Tilburg University, with whom I had incredible time participating in stimulating discussions, sharing nice moments with a warm cup of tea in the hands, and pumping iron in the gym late at night. Bert van Roermund opened my eyes during his fascinating Principles Classes and our conversations and I will never be blind again: "ieder tsèn". Péter Cserne, Angelos Dimopoulos, and Matteo Negrinotti, it is a pleasure to consider you my friends – I have learnt from you more than you would believe. I cannot omit mentioning my two paranymphs, Dražan Djukic and Zlatina Georgieva, who were watching my back as my own private security guards and trustworthy allies during the PhD defense. I am glad that such wonderful friends and roommates still exist and that our paths have crossed. Thank you, Melissa Lewis. Without your birds-eye view and your help that seemed just as a duck takes to water I would have probably laid an egg. Thanks are also due to Ian Curry-Sumner who played a crucial role as my advisor in the tough competition for the NWO Mosaic grant and contributed to my success in obtaining it. Benjamin Biebl, our long-lasting friendship flavored by endless debates and unforgettable adventures in the Netherlands and abroad made me realize what having a real friend actually means – thank you.

I am especially grateful to the members of the PhD Committee – Cedric Ryngaert, Jonathan Verschuuren, Sorcha MacLeod, and Antenor Hallo de Wolf – for reading my work and providing their invaluable insights. My appreciation further extends to Steve Lambley from Steve Lambley Information Design who took up the challenge of editing and typesetting my manuscript and has delivered outstanding work. I also thank his colleague Max Denyer-Green from Penguin Proofing and Tom Scheirs and Isabelle van Dongen from Intersentia for their assistance.

Above all, my deepest gratitude goes out to my family. Without pure and infinite love of my thoughtful and gentle mother, wise and talented father, always cheerful and industrious brother and his wife, I would have never been able to be who I am and where I am now. Finally, I am thankful beyond words to Evgenia, the true love of my life, who has been very supportive of all my endeavors and keeps inspiring me all over again and giving a meaning to my existence. This book is dedicated to you… my family.

Evgeni Moyakine
Tilburg, November 2014

CONTENTS

List of Abbreviations

ACHR	American Convention on Human Rights
ACHPR	African Charter on Human and Peoples' Rights
AECA	Arms Export Control Act
ANSI	American National Standards Institute
AU	African Union
BAPSC	British Association of Private Security Companies
CAT	UN Convention Against Torture
CEDAW	UN Convention on the Elimination of All Forms of Discrimination Against Women
CERD	UN Convention on the Elimination of All Forms of Racial Discrimination
CIA	Central Intelligence Agency
CRC	UN Convention on the Rights of the Child
DARIO	Draft Articles on the Responsibility of International Organizations
DARS	Draft Articles on Responsibility of States for Internationally Wrongful Acts
DCAF	Geneva Centre for the Democratic Control of Armed Forces
DSS	US Department of State's Diplomatic Security Service
ECHR	European Convention on Human Rights
ECtHR	European Court of Human Rights
EU	European Union
FAIR Act	Federal Activities Inventory Reform Act
FAR	Federal Acquisition Regulation
FBI	Federal Bureau of Investigation
IACommHR	Inter-American Commission on Human Rights
IACtHR	Inter-American Court of Human Rights
ICC	International Criminal Court
ICCPR	International Covenant on Civil and Political Rights
ICJ	International Court of Justice
ICoC	International Code of Conduct for Private Security Service Providers

ICoCA	International Code of Conduct Association
ICRC	International Committee of the Red Cross
ICTY	International Criminal Tribunal for the former Yugoslavia
ILC	International Law Commission
ISOA	International Stability Operations Association
ITAR	International Traffic in Arms Regulation
MEJA	US Military Extraterritorial Jurisdiction Act
NATO	North Atlantic Treaty Organization
OAS	Organization of American States
OAU	Organization of African Unity
OHCHR	Office of the United Nations High Commissioner for Human Rights
OMB Circular	Office of Management and Budget Circular A-76
OSCE	Organization for Security and Co-operation in Europe
PCIJ	Permanent Court of International Justice
PMC	Private military company
PMF	Private military firm
PMSC	Private military and security company
PSC	Private security company
SCEG	Security in Complex Environments Group
SOFA	Status of forces agreement
SSR	Security Sector Reform
TNC	Transnational Corporation
TRNC	Turkish Republic of Northern Cyprus
UCMJ	Uniform Code of Military Justice
UDHR	Universal Declaration of Human Rights
UN	United Nations
UN Doc.	United Nations document
US DoD	US Department of Defense
VPs	Voluntary Principles on Security and Human Rights

CHAPTER I
GENERAL INTRODUCTION

States must recognize the possibility that, in certain circumstances, they have State Responsibility for the acts of otherwise private groups, such as PMFs,[1] *under international law.*

Oliver R. Jones[2]

1. INTRODUCTION

The 1990s set in motion the era of "corporate globalization",[3] while the end of the Cold War that took place with the fall of the Soviet Union in 1991 led to the emergence of so-called private military and security companies (hereinafter: PMSCs). Currently, these PMSCs are actively hired by a wide spectrum of clients ranging from States to international organizations and multinational corporations and this reliance on private military and security contractors has increased dramatically over the last two decades. They play an undeniably significant global role and are contracted to provide various military and security services in horrific times of war and blessed times of peace. Performing these essential tasks as professionally as possible, private contractors strive to gain more financial profit and compete with other PMSCs on the expanding market of military and security expertise. Many questions with regard to the functioning of these firms and its consequences, however, remain largely unanswered. Who is responsible for the misconduct of PMSCs and their employees in case of human rights violations and breaches of the law of armed conflict and what is the extent of this responsibility? Can States be held responsible for the illegal activities of PMSCs and their personnel and, if they can, under what circumstances? Which States should face international responsibility in this regard and what is the legal basis for this responsibility?

---

1 PMFs – private military firms.

2 Oliver R. Jones, (2009) "Implausible Deniability: State Responsibility for the Actions of Private Military Firms", *Connecticut Journal of International Law*, Volume 24, Issue 2, p. 243.

3 John G. Ruggie, (2013) *Just Business: Multinational Corporations and Human Rights*, New York & London: W. W. Norton & Company, p. xv.

Some might be inclined to argue that international law primarily focuses on the conduct of States and State actors and does not extensively deal with non-State actors, such as PMSCs active in many conflict areas.[4] There is, for instance, no elaborate legal framework governing international responsibility of non-State actors and thus there is a certain deficit of responsibility of private actors operating on the international plane.[5] Therefore, it is frequently asserted that PMSCs operate in "a legal vacuum", "the grey area of law", "beyond the law", and even "in the shadows".[6] They might create an image of typical guns for hire often unrestricted by international law and domestic legislation who manage to stay below the radar of criminal and civil accountability mechanisms. There are indeed still significant gaps perceivable at both international and national levels in the regulation of PMSCs' activities.[7] These PMSCs seem to operate in conflict areas around the globe without an appropriate level of supervision and accountability.[8] As will be observed, employees of PMSCs usually enjoy immunity and are exempted from local prosecution by States on which territory they are active and can simply escape criminal responsibility for violating international law, more specifically international humanitarian and human rights law. This practice creates a picture of impunity of PMSC personnel. Their international criminal responsibility appears to be a complicated matter in this regard and has little practical relevance. In addition, it is the case that currently PMSCs acting transnationally cannot effectively be held directly accountable for the misconduct in question.

When the lines of responsibility are blurred, the role of States involved in the practice of using this particular type of company is crucial. State responsibility or international responsibility for the conduct of PMSCs and their employees breaching international law offers an important tool in tackling these abuses, attaining justice,

---

4 Stephanie Farrior, (1998) "State Responsibility for Human Rights Abuses by Non-state Actors", *American Society of International Law Proceedings*, Volume 92, p. 300.

5 Robert P. Barnidge, Jr., (2006) "The Due Diligence Principle Under International Law", *International Community Law Review*, Volume 8, p. 81.

6 Chia Lehnardt, "Private Military Companies and State Responsibility", in Simon Chesterman and Chia Lehnardt, (Eds.), (2007) *From Mercenaries to Market: The Rise and Regulation of Private Military Companies*, Oxford: Oxford University Press, p. 140; Kateri Carmola, (2010) *Private Security Contractors and New Wars: Risk, Law, and Ethics*, Oxon and New York: Routledge, p. 99.

7 See, for instance, the recognition of this fact in UN Human Rights Council, *Draft of a Possible Convention on Private Military and Security Companies (PMSCs) for Consideration and Action by the Human Rights Council*, Prepared by the Working Group on the use of mercenaries as a means of violating human rights and impeding the exercise of the right of peoples to self-determination, May 13, 2011, First Session, Geneva, May 23-27, 2011, UN Doc. A/HRC/WG.10/1/2, Preamble, p. 3; Stephanie Brown, (2010) "Bottom-up Lawmaking and the Regulation of Private Military and Security Companies", *Cuadernos de Derecho Transnacional*, Volume 2, Issue 1, pp. 46-47.

8 José L. Gómez del Prado, (2009) "Private Military and Security Companies and the UN Working Group on the Use of Mercenaries", *Journal of Conflict & Security Law*, Volume 13, Issue 3, p. 429.

and ensuring that the firms and individuals in question respect international law.[9] States can be held responsible if this misconduct can be attributed to them or if they fail to comply with an array of their positive obligations under the law of armed conflict and human rights law. In this regard, the concept of accountability mismatch[10] – when States deny any connection to PMSCs and their employees acting in breach of international law – severely diminishes the possibility of holding States responsible for the unlawful conduct of these entities and persons. This notion thus relates to the State practice of supporting the efforts in which PMSCs are engaged as long as it takes place in accordance with rules and principles of international law. When members of PMSC personnel deviate from the right path and commit crimes, States tend to distance themselves from them, to point at the corporate nature of these entities, and to emphasize that the issue of international responsibility is out of the question. From the point of view of international law, there is a need to more closely investigate this problem and to look into the responsibility of States, having the potential to narrow the responsibility gap and, most importantly, to ensure that PMSCs and their personnel comply with international law and do not breach its rules and principles. The present book's focus is on the pertinent question of responsibility of States for the misconduct of PMSCs. An elaborate discussion of other related types of responsibility, such as individual criminal responsibility of PMSC personnel, corporate responsibility of PMSCs, and institutional responsibility of non-State clients of these companies will be left aside.

2. Research Aims and Methodology

The current study aims at answering the main research question in a coherent and systemized manner:

> "To what extent should and can States be held responsible for the misconduct of PMSCs, taking place in conflict areas and constituting breaches of international humanitarian and human rights law?"

---

9 Nicola Jägers, (2002) *Corporate Human Rights Obligations. In Search of Accountability*, School of Human Rights Research Series, Volume 17, Antwerp: Intersentia, p. 257; Smita Narula, "International Financial Institutions, Transnational Corporations and Duties of States", in Malcolm Langford, Wouter Vandenhole, Martin Scheinin, and Willem van Genugten, (2014) *Global Justice, State Duties: The Extra-territorial Scope of Economic, Social and Cultural Rights in International Law*, Cambridge: Cambridge University Press, p. 136; Michael Cottier, (2006) "Elements for Contracting and Regulating Private Security and Military Companies", *International Review of the Red Cross*, Volume 88, Issue 863, p. 639.

10 Mariana M. Prado, "Regulatory Choices in the Privatization of Infrastructure", in Simon Chesterman and Angelina Fisher (Eds.), (2009) *Private Security, Public Order: The Outsourcing of Public Services and Its Limits*, Oxford: Oxford University Press, p. 130.

In the first place, it must be noted that the study looks into the misconduct of both PMSCs and their employees that might engage in such activities, which can result in violations of international law. In this regard, references are made to not only PMSCs, but also the personnel of these companies who have the potential of breaching international humanitarian and human rights law. For the purpose of discussing the issue of international responsibility, it is not the first priority to distinguish between the acts and omissions of PMSCs as legal entities and their employees. In principle, both private corporations and individuals employed by them can violate international law, more specifically various human rights.[11] Hence, the conduct of both actors, PMSCs and members of their staff, can – under certain conditions – lead to the invocation of State responsibility. Moreover, it is important to observe that the analysis of the issue of State responsibility will focus on the circumstances and extent of attribution of the unlawful conduct of PMSCs to States under public international law. Additionally, the circumstances and extent of application of positive State obligations under international humanitarian and human rights law will be examined to establish whether international responsibility of States is incurred when PMSCs and their employees commit human rights violations and breach the law of armed conflict.

By analyzing the relevant rules and principles of international law, this book adopts the hypothesis that States should and can be held responsible for the violations of international law committed by PMSCs under public international law and for breaching existing positive State obligations under international humanitarian and human rights law as two particular branches of public international law. State responsibility arises and should arise under a number of circumstances and to a certain extent identified below. The current study not only assesses international law as it stands, *lex lata*, but adopts a perspective incorporating rules and principles of international law as it should be, *lex ferenda*. The element of progressive development of international law is to be found in the application of due diligence doctrine from both the law of armed conflict and human rights law to States and their division in categories with varying degrees of responsibility for the unlawful conduct of PMSCs and their personnel. Furthermore, progressive development of international law and possibly some traces of emerging customary international law can be observed in certain non-binding international legal instruments that will be discussed and used throughout the study. In the context of State responsibility for the activities of private contractors, an attempt will also be made to clarify some relevant concepts, such as the issue of control, and to argue that a lower threshold of control must be used in relation to the operations of PMSCs employed by States. Additionally, it will be proposed to use a broad interpretation of the concept of internal law authorizing

---

[11] UN Human Rights Council, *Corporations and Human Rights: A Survey of the Scope and Patterns of Alleged Corporate-related Human Rights Abuse*, Report of the Special Representative of the Secretary-General on the Issue of Human Rights and Transnational Corporations and Other Business Enterprises, May 23, 2008, UN Doc. A/HRC/8/5/Add.2.

PMSCs to exercise elements of governmental authority. This necessary step must be made due to the fact that, although many rules on State responsibility still need to be crystallized through State practice and jurisprudence, some answers to pertinent questions need to be given now.

By attributing the actions of PMSCs to States and invoking their positive obligations in relation to these private corporations often employed to perform the most vital State functions, it is possible to make the necessary linkages between the breaches of international law committed on the ground and States being held responsible and facing the legal consequences. The research has its main focus on the circumstances and scope of invoking the doctrine of State responsibility with regard to the State use of PMSCs. In order to tackle the core research question, it additionally addresses such aspects as the phenomenon of privatization of war and security, the existing legal framework of operation of PMSCs, the legal status of these corporations and their personnel, and the application of the State responsibility doctrine. In general, this book presents an assessment of the question of international responsibility for the misconduct of PMSCs and members of their staff by presenting an analysis of States' obligations with regard to private contractors and addressing the circumstances and extent of State responsibility for the actions or omissions of these specific private actors breaching international law.

The present study is mostly based on the library-based research. For the major part, it involves an extensive discussion of international legal norms concerning State responsibility and their application to the misconduct of PMSCs and their employees. In addition, it includes a less extensive assessment of national and transnational legal frameworks focusing on these corporations and other regulatory tools that play a role in their functioning. As such, the study builds on an analysis of treaty law, case law of international, regional, and national judicial bodies, documents of international and regional organizations, reports, relevant academic legal literature, and other publications. Some insights presented in this book, more specifically those relating to self-regulation of PMSCs, have been drawn from the participation in the research project of the Hague Institute for the Internationalization of Law (HiiL) and the qualitative empirical research with private contractors and other actors carried out as a part of the case study on transnational private regulation in the field of PMSCs.[12] The study as a whole does not have the ambition to be an interdisciplinary one. It does, however, contain some elements of interdisciplinarity given that it also takes into consideration materials from other disciplines, such as political sciences and history, especially when dealing with the phenomenon of privatization of war and security and historical developments in this field.

---

12 Nicola Jägers, Evgeni Moyakine, and Willem van Genugten, "Transnational Private Regulation, Private Security Companies and Human Rights: Promise of Effectiveness?", Case Study of the Center for Transboundary Legal Development (CTLD), Tilburg University, for the Hague Institute for the Internationalization of Law (HiiL), September 2012 (on file with the author).

In attempting to answer its research questions, the study seeks to map main types of unlawful conduct in which PMSCs and their staff were implicated in the past and thus can be implicated in the practice. For the purpose of investigating the issue of State responsibility for the violations of international law committed by PMSCs and their employees, it is useful to commence the research by analyzing facts of this misconduct in order to understand the scope of respective illegalities and to establish – at a later stage – whether they constitute breaches of international law. The instances of misconduct are scrutinized on the basis of the analysis of case law, legal and other literature, reports, and media coverage. In order to connect the theoretical elaboration on the doctrine of State responsibility to its general application to PMSCs as non-State business actors and to bring the research findings closer to the day-to-day reality, three particular case studies are included that are useful in clarifying certain concepts presented in the study. In the subsequent Chapters, references will regularly be made to these case studies in the process of applying essential findings of the doctrinal part of the research and indicating how rules and principles of the law of State responsibility can or cannot trigger the responsibility of respective States.

The first case study involves the PMSC called Sandline International that was employed by the government of Papua New Guinea in 1997 and was accused of being a mercenary organization. The second case study concerns torture and other related types of abuse that took place at the Abu Ghraib detention facility in Baghdad, Iraq, between October and December 2003 and were committed by the employees of Titan Corp. (hereinafter: Titan) and CACI International Inc. (hereinafter: CACI). The third case study concerns the shooting incident, in which one of the most prominent US-based PMSCs at the beginning of the twenty-first century – Blackwater, later renamed Xe Services and then, at the end of 2011, Academi[13] – was involved in Baghdad, Iraq, in September 2007. All the above-mentioned cases are illustrative for the illegal conduct of PMSCs and their personnel that can bring about State responsibility. There is a significant amount of information available on these incidents: they received wide coverage in the press, literature, and reports and were addressed in the course of investigations and even legal proceedings that were initiated in the Abu Ghraib and Blackwater scandals. These case studies will be referred to in order to explain a variety of rules and principles of State responsibility in relation to the activities of PMSCs and their employees. More specifically, these incidents will be used to clarify the possibility of attributing acts and omissions of PMSCs and their personnel to States in practice under the law of State responsibility. Furthermore, they will be considered when assessing the question of whether the misconduct of PMSCs and their personnel constitutes breaches of international law. In addition to

---

13 David Isenberg, "Pedagogic Military Contractors", *The Huffington Post*, December 18, 2011, <http://www.huffingtonpost.com/david-isenberg/military-contractors-blackwater_b_1156080.html>, accessed on October 10, 2013.

these incidents, a number of other instances of unlawful conduct that have not been widely covered in the literature and are not well documented will also be discussed.

As follows from the wording of the main research question, the scope of the research is limited to the misconduct of PMSCs and their personnel that takes place in conflict areas. Certainly, these corporations are also active in other environments around the globe, but in order to make the study feasible it has been decided to narrow down the research problem and to focus on the operations of PMSCs in zones of conflict. In addition, it appears to be undeniable that in these scenarios the probability of the involvement of PMSCs and their employees in violations of international human rights law is higher than during their functioning in non-conflict situations. More importantly, the deployment of PMSCs can only lead to breaches of international humanitarian law in places affected by armed conflict. Finally, it comes to the fore from the examination of literature and media coverage that a large variety of abuses were committed by PMSCs and their personnel in this specific context under analysis. With this in mind, it should be emphasized that PMSCs do not form bands of criminals baying for blood, who do nothing but inflict pain and destruction.[14] First and foremost, they are private corporations striving to make a financial profit, providing military and security services, and operating in various conflict zones and other areas. While carrying out their missions, they can, however, breach international law and have done so in the past. The current study focuses on such PMSCs and their personnel that violate rules and principles of international law – as these violations pose the relevant questions of responsibility – and does not deal with private contractors and their employees that comply with them.

3. MISCONDUCT OF PMSCS AND THEIR PERSONNEL

It seems reasonable to assume that PMSCs operating on behalf of States are often in a position of breaching the interests protected by international law. In this regard, it has been rightly pointed out that:

> "While there is no empirical evidence that PMCs are more likely to engage in misconduct than their public counterparts, PMCs can, of course, violate interests protected by international law."[15]

Some documents obtained by the Associated Press indicate that more than 200 private contractors working for the US government committed various previously

14 Peter W. Singer, (2008) *Corporate Warriors: The Rise of the Privatized Military Industry*, Ithaca and London: Cornell University Press, p. 217.

15 Chia Lehnardt, "Private Military Companies and State Responsibility", in Simon Chesterman and Chia Lehnardt (Eds.), (2007) *From Mercenaries to Market: The Rise and Regulation of Private Military Companies*, Oxford: Oxford University Press, p. 139.

undisclosed offenses in Afghanistan, Iraq, and other countries between 2004 and 2008. Those include excessive drinking, drug use, sexual misconduct, weapons violations, and other possible types of misconduct that can easily go unnoticed by general public. There are, however, other offenses, such as indiscriminate shootings, cases of sex slavery, and torture that belong to the most serious crimes, which could also be committed by employees of those companies.[16]

Indiscriminate shootings and civilian casualties

In the first place, employees of PMSCs are often condemned for indiscriminate shootings and civilian casualties caused by their actions and are well known for incidents of escalation of force in the areas where they perform a wide array of their services. Various statements by the local population of the countries where those firms function and reports of the businesses themselves strongly reinforce the view that private contractors often use force indiscriminately and fire their weapons at innocent local nationals.[17] Descriptions of the operation of security guards working for Blackwater – later renamed Xe Services and then Academi – are without any doubt applicable to numerous other firms active in the field of provision of private military and security services: they often appear to "have very quick trigger fingers", to "shoot first and ask questions later", and even to "act like cowboys".[18] It must, however, be pointed out that these firms are usually employed to merely engage in the defensive use of force in order to protect their clients, themselves and others,[19] but can go beyond the limits of their contracts by using excessive violence.

Perhaps, the most illustrative case, which is used in the current study for the purposes of legal analysis, is and remains that of Blackwater.[20] It relates to a violent shooting incident with the involvement of five employees working for Blackwater in Iraq.[21] In essence, the shooting was initiated by the members of Blackwater personnel

---

16 Richard Lardner, "DynCorp, Other Private Contractors in Afghanistan Behaving Badly", *The Huffington Post*, December 19, 2010, <http://www.huffingtonpost.com/2010/12/19/dyncorp-afghanistan-private-contractors_n_798753.html>, accessed on October 10, 2013.

17 Human Rights First, *Private Security Contractors at War: Ending the Culture of Impunity*, Report, 2008, New York, pp. 5-9.

18 US House of Representatives, Committee on Oversight and Government Reform, "Additional Information About Blackwater USA", Memorandum, October 1, 2007, p. 6, <http://graphics8.nytimes.com/packages/pdf/national/20071001121609.pdf>, accessed on September 16, 2014.

19 Ibidem.

20 Sorcha MacLeod, "The Role of International Regulatory Initiatives on Business and Human Rights for Holding Private Military and Security Contractors to Account", in Francesco Francioni and Natalino Ronzitti (Eds.), (2011) *War by Contract: Human Rights, Humanitarian Law, and Private Contractors*, Oxford: Oxford University Press, pp. 350-351.

21 Michael Walzer, "Mercenary Impulse: Is There an Ethics That Justifies Blackwater?", *The New Republic*, March 12, 2008, <http://www.tnr.com/article/mercenary-impulse>, accessed on October 10, 2013; Kateri Carmola, (2010) *Private Security Contractors and New Wars: Risk, Law, and Ethics*, Oxon

and resulted in the death of 17 Iraqi civilians, among which were women and children, and injury of 24 individuals.[22] After the incident, the US Department of State and Federal Bureau of Investigation (hereinafter: FBI) commenced their investigations into the matter. Detailed information on the factual background of the event can be distilled from the criminal case brought before the US District Court for the District of Columbia against the security guards involved in the shooting.[23] In these criminal proceedings, the five defendants – employees of Blackwater – were charged with multiple voluntary manslaughter counts and firearms violations due to their participation in the shooting. All these individuals who discharged their weapons and killed and injured Iraqi civilians were employed by Blackwater, which provided security services to the US employees in Iraq on a contractual basis.[24]

On September 16, 2007, a Blackwater Tactical Support Team called "Raven 23" had as its main task provision of back-up fire support for other Blackwater personnel occupied with private security services in Baghdad. Before noon, Raven 23 team, comprising four vehicles in a convoy, received a message concerning a vehicle-borne improvised explosive device that had been detonated near a compound where US officials were meeting their Iraqi colleagues.[25] Reacting rapidly to this incident, the Raven 23 convoy moved to Nisour Square in order to make sure that an evacuation route would be secured for the US officials and another Blackwater team serving as their security guards. Nisour Square is a traffic circle outside the International Zone in Baghdad and it was considered a suitable location for the members of the Raven 23 team to take up their positions, approximately one mile away from the explosion site. Not long after the convoy entered the square, the Blackwater employees stationed at the square were engaged in a shooting. The other team protecting the US officials did not discharge their weapons.

The media started investigating the incident, contacting possible witnesses, and quoting official investigation reports and, quite importantly, the testimony of Blackwater's CEO Erik Prince prepared for the hearing of the House Committee on Oversight and Government Reform entitled "Blackwater USA: Private Military

---

and New York: Routledge, p. 100; Pratap Chatterjee, "Iraq War Logs: Military Privatisation Run Amok", *The Guardian*, October 23, 2010, <http://www.guardian.co.uk/commentisfree/cifamerica/2010/oct/23/iraq-war-logs-us-military/>, accessed on October 10, 2013; Human Rights First, *Private Security Contractors at War: Ending the Culture of Impunity*, Report, 2008, New York, p. 1.

22 Human Rights First, *Private Security Contractors at War: Ending the Culture of Impunity*, Report, 2008, New York, p. 1; Katherin J. Chapman, (2010) "The Untouchables: Private Military Contractors' Criminal Accountability under the UCMJ", *Vanderbilt Law Review*, Volume 63, Issue 4, pp. 1048, 1050-1051.

23 US District Court for the District of Columbia, *United States of America v. Paul A. Slough et al.*, Case No. CR-08-360, filed on December 4, 2008.

24 US District Court for the District of Columbia, *United States of America v. Paul A. Slough et al.*, Criminal Action No. 08-0360(RMU), Memorandum Opinion, September 8, 2009, p. 2.

25 Ibidem, p. 4.

Contractor Activity in Iraq and Afghanistan".[26] This testimony contains an elaborate description of what happened on September 16, 2007. According to the Chairman of the company, the Raven 23 team came under fire at Nisour Square and instructed the other team protecting the US officials to take a different route. While the Blackwater guards located at the intersection kept receiving fire, they also returned fire at several targets: men with AK-47s shooting at the convoy and approaching suicide car bombers. When one of the Blackwater vehicles was disabled by enemy fire, the rest of the Blackwater employees could not easily exit the area. After having towed the disabled truck out of the square, they left and returned to the International Zone. While leaving, the Raven 23 team was constantly under enemy fire from various individuals, including those who appeared to wear Iraqi National Police uniforms. It appeared that, of 20 members of the team, only five employees discharged their weapons. Mr. Prince added:

> "Despite the valiant missions our people conduct each day with great success, in this September 16 instance, Blackwater and its people have been the subject of negative and baseless allegations reported as truth. There has been a rush to judgment based on inaccurate information, and many public reports have wrongly pronounced Blackwater's guilt for the death of varying numbers of civilians."[27]

The media reports, however, made clear that according to US military officials the members of the Raven 23 team had not been ambushed, but had started the shooting without provocation.[28] It appears that after driving approximately 150 yards away from Nisour Square, the guards opened fire shooting their way out and hit various cars, which resulted in a number of Iraqi civilians being killed and injured. A few moments later, the convoy disappeared. With regard to the use of force by

---

[26] Statement of Erik D. Prince, Chairman and CEO, Blackwater, for the House Committee on Oversight and Government Reform, October 2, 2007, <http://oversight-archive.waxman.house.gov/documents/20071003153621.pdf>, accessed on October 10, 2013; Sudarsan Raghavan, Joshua Partlow, and Karen DeYoung, "Blackwater Faulted in Military Reports From Shooting Scene", *The Washington Post*, October 5, 2007, <http://www.washingtonpost.com/wp-dyn/content/article/2007/10/04/AR2007100402654.html>, accessed on September 16, 2014; US House of Representatives, Committee on Oversight and Government Reform, "Additional Information About Blackwater USA", Memorandum, October 1, 2007, p. 1, <http://graphics8.nytimes.com/packages/pdf/national/20071001121609.pdf>, accessed on September 16, 2014.

[27] Statement of Erik D. Prince, Chairman and CEO, Blackwater, for the House Committee on Oversight and Government Reform, October 2, 2007, p. 5, <http://oversight-archive.waxman.house.gov/documents/20071003153621.pdf>, accessed on October 10, 2013.

[28] Sudarsan Raghavan and Thomas E. Ricks, "Private Security Puts Diplomats, Military at Odds", *The Washington Post*, September 26, 2007, <http://www.washingtonpost.com/wp-dyn/content/article/2007/09/25/AR2007092502675.html>, accessed on September 16, 2014; Sudarsan Raghavan, Joshua Partlow, and Karen DeYoung, "Blackwater Faulted in Military Reports From Shooting Scene", *The Washington Post*, October 5, 2007, <http://www.washingtonpost.com/wp-dyn/content/article/2007/10/04/AR2007100402654.html>, accessed on September 16, 2014.

the Blackwater team, one of the US military officials pointed out: "It was obviously excessive, it was obviously wrong."[29] Then he made the point that the Blackwater guards had used not only machine guns, but also grenade launchers and added:

> "The civilians that were fired upon, they didn't have any weapons to fire back at them. And none of the IP [the Iraqi National Police] or any of the local security forces fired back at them."[30]

Thirty minutes after the shooting, US Army Colonel David Boslego examined the location where the shooting incident took place and collected some evidence.[31] He indicated that no evidence of insurgent activity could be found. Another US officer Colonel Michael Tarsa who inspected the scene one hour after the shooting came to the same conclusion. A large number of US shell casings, a used M203 grenade cartridge, and two spent magazines bearing the name "Liberty" were found. The Iraqi National Police also immediately reacted to the incident and its officers initiated an investigation after the Raven 23 convoy left Nisour Square.[32] The police collected physical evidence and interviewed witnesses and victims.

Hours after the incident, the US Department of State's Diplomatic Security Service (hereinafter: DSS) interviewed all members of the Raven 23 team. During these interviews, four of the members acknowledged discharging their weapons at the intersection. In October 2007, a team of FBI investigators arrived in Baghdad to look into the incident. They interviewed various Iraqi witnesses and victims, members of the Blackwater personnel, US military police, and Iraqi National Police. On September 18, 2007, all members of the Blackwater team submitted the so-called Hunter Memorandum form containing their written statements of the events of September 16, 2007 that expanded oral statements given on the day of the incident. The DSS investigators also interviewed 14 Iraqi witnesses. During the following days, four Blackwater employees who had fired their weapons were again subjected to interviews and questioning by DSS concerning their actions and specific details of the incident. On September 20, 2007, DSS agents investigated Nisour Square, collected photographic evidence, conducted additional witness interviews, and found some physical evidence suitable for analysis. Over the following days, DSS investigators re-interviewed those Blackwater guards who had been involved in the shooting.

On September 26, 2007, FBI agents and the Department of Justice Criminal Division met with representatives of the State Department in order to make an assessment

---

29 Sudarsan Raghavan, Joshua Partlow, and Karen DeYoung, "Blackwater Faulted in Military Reports From Shooting Scene", *The Washington Post*, October 5, 2007, <http://www.washingtonpost.com/wp-dyn/content/article/2007/10/04/AR2007100402654.html>, accessed on September 16, 2014.

30 Ibidem.

31 US District Court for the District of Columbia, *United States of America v. Paul A. Slough et al.*, Criminal Action No. 08-0360(RMU), Memorandum Opinion, December 31, 2009, p. 12.

32 Ibidem, p. 11.

of the DSS investigation findings concerning the shooting incident. Two days later, the State Department's Office of Legal Counsel contacted the Department of Justice Criminal Division and expressed its concerns regarding the report that might have been based on compelled statements given by the Blackwater guards. Therefore, a team of prosecutors and investigators other than those who took part in the meeting of September 26 was assigned by the government. Eventually, on December 4, 2008 five members of the Raven 23 team were indicted on charges relating to the Nisour Square shooting. Those guards were Paul A. Slough, Nicholas A. Slatten, Evan S. Liberty, Dustin L. Heard, and Donald W. Ball.[33] The sixth guard was Jeremy P. Ridgeway who made a deal with the prosecutors and pleaded guilty.[34] The charges of five employees of Blackwater concerned voluntary manslaughter, attempting to commit manslaughter, and using and discharging a firearm during and in relation to a crime of violence.[35] In 2009, US District Judge Ricardo Urbina dismissed the case due to the fact that the protected statements given by the Blackwater guards were given under promise of immunity[36] and these were improperly used by the US government.[37] In April 2011, the case was revived by the US Court of Appeals for the D.C. Circuit.[38] On January 6, 2012, the case was finally settled by Blackwater on behalf of its guards.

---

[33] US District Court for the District of Columbia, *United States of America v. Paul A. Slough et al.*, Criminal Action No. 08-0360(RMU), Indictment, December 4, 2008, <http://www.justice.gov/opa/documents/grandjury.pdf>, accessed on October 15, 2013.

[34] US District Court for the District of Columbia, *United States of America v. Jeremy P. Ridgeway*, Criminal Action No. 08-341-01(RMU), Indictment, December 4, 2008, <http://www.justice.gov/opa/documents/us-v-ridgeway2.pdf>, accessed on October 1, 2013; see also the factual proffer in support of guilty plea by Jeremy P. Ridgeway, <http://www.justice.gov/opa/documents/us-v-ridgeway.pdf>, accessed on October 15, 2013.

[35] US District Court for the District of Columbia, *United States of America v. Paul A. Slough et al.*, Criminal Action No. 08-0360(RMU), Indictment, December 4, 2008, <http://www.justice.gov/opa/documents/grandjury.pdf>, accessed on October 15, 2013.

[36] David Johnston, "Immunity Deals Offered to Blackwater Guards", *The New York Times*, October 30, 2007, <http://www.nytimes.com/2007/10/30/washington/30blackwater.html>, accessed on October 15, 2013.

[37] Del Quentin Wilber, "Judge Dismisses Charges Against Blackwater Guards in Iraq Deaths", *The Washington Post*, January 1, 2010, <http://www.washingtonpost.com/wp-dyn/content/article/2010/01/01/AR2010010100562.html>, accessed on September 16, 2014.

[38] US Court of Appeals for the District of Columbia Circuit, *United States of America v. Paul Alvin Slough, et al.*, Case No. 10-3006, Appeal from the US District Court for the District of Columbia (No. 1:08-cr-00360), April 22, 2011, <http://www.justice.gov/usao/dc/programs/vw/pdf/nisursquare_docs/nisursquare_update_april_22_2011_english.pdf>, accessed on October 15, 2013; James Risen, "Ex-Blackwater Guards Face Renewed Charges", *The New York Times*, April 22, 2011, <http://www.nytimes.com/2011/04/23/us/23blackwater.html>, accessed on October 15, 2013; Del Quentin Wilber and Spencer S. Hsu, "Judge Ordered to Revisit Blackwater Guard Shooting Case", *The Washington Post*, April 22, 2011, <http://www.washingtonpost.com/local/judge-ordered-to-revisit-blackwater-guard-shooting-case/2011/04/22/AF4flCRE_story.html>, accessed on September 16, 2014; Evan Perez, "Appeals Court Revives Blackwater Case", *The Wall Street Journal*, April 22, 2011, <http://online.wsj.com/news/articles/SB10001424052748703521304576278921617509908>, accessed on October 15, 2013.

In October 2013, the case was reopened again with the Justice Department filing new charges against four Blackwater employees.[39] The new trial against them commenced on June 11, 2014 and is under way at the time of writing (June 2014).

This incident was judged by a senior US military official to be a horrific action that "is going to hurt us badly" and that "may be worse than Abu Ghraib",[40] a case of detainee abuse by US soldiers and private guards, which will be addressed below. At that time, Blackwater was a major player in the field of provision of private military and security services. As emphasized by legal experts, "Blackwater had become the Sandline of the US: a large, privately financed, well-established firm whose close ties with the US government and the Republican Party made it look like they were the paramilitary wing of the Bush Administration".[41] After the shooting, the United States Congress decided to research the activities in which that company was involved in the past years and to list possible problems with its use by the State Department. In 2007, Chairman of the House Committee on Government Oversight and Reform Henry Waxman made clear the following:

> "New documents indicate that there have been a total of 195 shooting incidents involving Blackwater forces since 2005. Blackwater's contract says the company is hired to provide defensive services, but in most of these incidents it was Blackwater forces who fired first. We have also learned that 122 Blackwater employees, one seventh of the company's current work force in Iraq, have been terminated for improper conduct."[42]

Furthermore, a representative in Congress from the State of Virginia Tom Davis added:

---

39 US District Court for the District of Columbia, *United States of America v. Paul Alvin Slough et al.*, Criminal Action No. CR-08-360(RCL), Indictment, October 17, 2013, <http://www.justice.gov/usao/dc/programs/vw/pdf/nisursquare_docs/nisursquare_Superseding_Indictment_October)17_2013.pdf>, accessed on June 10, 2014; Noel Brinkerhoff, "Blackwater Guards Finally to Be Tried for Killing 14 Iraqi Civilians", *AllGov*, April 13, 2014, <http://www.allgov.com/news/us-and-the-world/blackwater-guards-finally-to-be-tried-for-killing-14-iraqi-civilians-140413?news=852903>, accessed on May 10, 2014.

40 Sudarsan Raghavan and Thomas E. Ricks, "Private Security Puts Diplomats, Military at Odds", *The Washington Post*, September 26, 2007, <http://www.washingtonpost.com/wp-dyn/content/article/2007/09/25/AR2007092502675.html>, accessed on September 16, 2014; US House of Representatives, Committee on Oversight and Government Reform, "Additional Information About Blackwater USA", Memorandum, October 1, 2007, p. 6, <http://graphics8.nytimes.com/packages/pdf/national/20071001121609.pdf>, accessed on September 16, 2014.

41 Kateri Carmola, (2010) *Private Security Contractors and New Wars: Risk, Law, and Ethics*, Oxon and New York: Routledge, p. 101.

42 Ibidem, p. 100; see also Steve Fainaru, "Guards in Iraq Cite Frequent Shootings", *The Washington Post*, October 3, 2007, <http://www.washingtonpost.com/wp-dyn/content/article/2007/10/02/AR2007100202456.html>, accessed on September 16, 2014; US House of Representatives, Committee on Oversight and Government Reform, "Additional Information About Blackwater USA", Memorandum, October 1, 2007, p. 6, <http://graphics8.nytimes.com/packages/pdf/national/20071001121609.pdf>, accessed on September 16, 2014.

> "Incidents of erratic and dangerous behavior by security personnel from all the companies involved, not just Blackwater, are handled with little or no regard to Iraqi law. Usually, the bad actor is simply whisked out of the country, whether the offence is a civilian casualty, negligent discharge or a weapon, alcohol or drug abuse, or destruction of property. To date, there has not been a single successful prosecution of a security provider in Iraq for criminal misconduct."[43]

The above-described events are, however, not isolated incidents: in the past, Blackwater – like other PMSCs – was involved in a variety of other shootings, as that of May 14, 2005.[44] A Blackwater convoy was riding down "Route Irish" – the road to the Baghdad airport – when its private contractors started to fire gunshots at an Iraqi civilian vehicle. Approaching the 69th Regiment of the United States Army, Blackwater "cowboys" opened fire over the heads of the US soldiers and quickly disappeared in their armored truck. The result of this reckless behavior was the death of the Iraqi driver and many injuries to his wife and daughter.

According to the reporter who analyzed information made public by WikiLeaks, this incident is merely one example of a major problem. Having concerns over the whole idea of privatization of war, he pointed out that:

> "The incident is one of several dozen involving private security companies in Iraq – almost all of which have never been previously reported – that led to an '*escalation of force*', which is military parlance for a series of actions that begins with nonlethal measures (such as visual signals with flags, spotlights or flares), but may graduate to potentially lethal force, with warning shots, disabling shots (to vehicle tyres) or, if all other measures have failed, deadly shots, depending on the perceived level of threat. Blackwater, the company from Moyock, North Carolina, is responsible for about half of the incidents, closely followed by Erinys, a British private security company registered in the Virgin Islands, which seems to have an unusually high number of vehicle crashes."[45]

On May 28, 2005, 16 employees of another US PMSC, Zapata Security, many of whom were former US Marines, were driving in Fallujah and at a certain moment with no justification started to shoot at the US military posts and innocent Iraqi citizens. After this incident, the contractors were taken into custody by a US Marine

---

[43] US House of Representatives, Committee on Oversight and Government Reform, "Blackwater USA. Hearing before the House Committee on Oversight and Government Reform", Serial No. 110-89, October 2, 2007, <http://www.gpo.gov/fdsys/pkg/CHRG-110hhrg45219/html/CHRG-110hhrg45219.htm>, accessed on October 16, 2013.

[44] Pratap Chatterjee, "Iraq War Logs: Military Privatisation Run Amok", *The Guardian*, October 23, 2010, <http://www.guardian.co.uk/commentisfree/cifamerica/2010/oct/23/iraq-war-logs-us-military/>, accessed on October 10, 2013.

[45] Ibidem.

unit.[46] Eventually, they were accused of "repeatedly firing weapons at civilians and Marines, erratic driving, and possession of illegal weapons", but also forming a "direct threat to Marine personnel".[47]

Starting with videos showing private contractors of Aegis Defence firing at unarmed Iraqi civilians,[48] many other examples of the questionable use of force by private military and security contractors can be provided. In its report on the private security providers, Human Rights First published the story of an Iraqi civilian, Ali Ismael.[49] He described that he had pulled his car onto a Baghdad highway on July 14, 2005. It was not far away from a private security convoy consisting of four vehicles. At a certain moment, the backdoor of the vehicle opened and a man wearing sunglasses aimed his rifle and fired a shot. Mr. Ismael received a serious head wound, but fortunately survived the incident.

A list of that type of misconduct committed by other PMSCs can be made as extensive as necessary. Private contractors of another firm, Erynis, with its head offices in Dubai and subsidiary businesses operating in South Africa and the United Kingdom, were also implicated in the commission of serious crimes. They killed a US soldier in Iraq on October 25, 2005 and, escorting a convoy of vehicles with the personnel of the US Army Corps of Engineers, opened fire on a taxi in Kirkuk on October 18, 2007, injuring three Iraqi civilians.[50]

Employees of another major PMSC Triple Canopy have been implicated in the commission of crimes related to causing civilian casualties and damaging property of local nationals. On July 8, 2006, its security guards fired several times at the vehicles of Iraqi civilians, damaging them and possibly killing or wounding some

---

46 Kateri Carmola, (2010) *Private Security Contractors and New Wars: Risk, Law, and Ethics*, Oxon and New York: Routledge, p. 87.

47 Josh White and Griff Witte, "Tension, Confusion Between Troops, Contractors in Iraq", *The Washington Post*, July 10, 2005, <http://www.washingtonpost.com/wp-dyn/content/article/2005/07/09/AR2005070901175.html>, accessed on September 16, 2014; Human Rights First, *Private Security Contractors at War: Ending the Culture of Impunity*, Report, 2008, New York, p. 2.

48 Human Rights First, *Private Security Contractors at War: Ending the Culture of Impunity*, Report, 2008, New York, pp. 8, 49; Pratap Chatterjee, "Iraq War Logs: Military Privatisation Run Amok", *The Guardian*, October 23, 2010, <http://www.guardian.co.uk/commentisfree/cifamerica/2010/oct/23/iraq-war-logs-us-military/>, accessed on October 10, 2013.

49 Kateri Carmola, (2010) *Private Security Contractors and New Wars: Risk, Law, and Ethics*, Oxon and New York: Routledge, p. 3; Human Rights First, *Private Security Contractors at War: Ending the Culture of Impunity*, Report, 2008, New York, p. 7.

50 Suzanne Goldenberg, "US Soldier's Family Brings Legal Action Against British Private Security Firm", *The Guardian*, October 30, 2007, <http://www.guardian.co.uk/world/2007/oct/30/iraq.usa1>, accessed on October 17, 2013; Luke Baker, "British Security Co. Sued over Death of U.S. Soldier", *Reuters.com*, October 26, 2007, <http://www.reuters.com/article/2007/10/26/us-iraq-britain-lawsuit-idUSL2666274820071026>, accessed on September 16, 2014; "Licence to Kill for Private Armies", *WarOnWant.org*, October 19, 2009, <http://www.waronwant.org/news/press-releases/16692-licence-to-kill-for-private-armies>, accessed on October 18, 2013.

individuals.[51] A number of members of the security team referred to at least one of those incidents as "unwarranted" with no situations posing threat to the private contractors. According to the guards working for Triple Canopy, one of the shooters told his colleagues that on that day he just "want[ed] to kill somebody".[52] The general public acquired knowledge of the incidents only because of a wrongful termination suit filed by the two of the five Triple Canopy employees. The US State Department appeared to know nothing about the incidents and no investigations have been carried out by the US government.

Elaborating on the issue of excessive use of force by the PMSC personnel, discussion of the following cases cannot be omitted. One of the employees working for the PMSC Blackwater, Andrew Moonen, allegedly shot and killed one of the members of the Iraqi Vice President Adil Abdul-Mahdi's security personnel, Raheem Khalif Hulaichi, on Christmas Eve 2006.[53] According to the reports, after drinking heavily on that day, the private contractor decided to pass through the gate of the Iraqi Prime Minister's compound. After being confronted by Mr. Hulaichi, he fired three shots at him, left the man dying, and fled the crime scene. The next violent shooting of Blackwater took place in Baghdad on February 7, 2007 when private contractors shot three Iraqi guards of the government-funded Al-Iraqiya television network.[54] Several months later, on September 9, 2007, Blackwater employees killed five Iraqi civilians and injured 10 individuals not far away from the Baghdad municipality building; three days later Blackwater guards severely wounded five more people in east Baghdad.[55] Another shooting incident, in which its employees were implicated, happened more recently in Kabul, Afghanistan, on May 5, 2009. On that day, four members of Blackwater personnel shot and killed two Afghan civilians.[56]

The above-described shooting incidents involving Blackwater, Zapata Security, Triple Canopy, and Erynis are just a few examples of misconduct by PMSCs

---

[51] Human Rights First, *Private Security Contractors at War: Ending the Culture of Impunity*, Report, 2008, New York, pp. 2, 48.

[52] Ibidem, p. 48.

[53] US House of Representatives, Committee on Oversight and Government Reform, "Additional Information About Blackwater USA", Memorandum, October 1, 2007, pp. 9-10, <http://graphics8.nytimes.com/packages/pdf/national/20071001121609.pdf>, accessed on September 16, 2014; Human Rights First, *Private Security Contractors at War: Ending the Culture of Impunity*, Report, 2008, New York, p. 2.

[54] Joshua Partlow and Sudarsan Raghavan, "Iraq Probe of U.S. Security Firm Grows: Blackwater, Accused of Killing 11 on Sunday, Cited in Earlier Deaths", *The Washington Post*, September 22, 2007, <http://www.washingtonpost.com/wp-dyn/content/article/2007/09/21/AR2007092102271.html?sid=ST2007092200176>, accessed on September 16, 2014.

[55] Ibidem.

[56] Lillian Tan, "Private Military and Security Companies Wanted for Hire by CIA as 'Hitmen'", *The Huffington Post*, August 27, 2009, <http://www.huffingtonpost.com/amnesty-international/private-military-and-secu_b_270734.html>, accessed on October 17, 2013; David Zucchino, "Deadly Contractor Incident Sours Afghans", *The Los Angeles Times*, August 13, 2009, <http://www.latimes.com/la-fg-afghan-contractors13-2009aug13-story.html#page=1>, accessed on September 16, 2014.

operating in conflict zones. It is the case that governments usually do not keep record of illegal activities of the companies they hire and their personnel, their possible victims, or numbers of contractors killed and wounded.[57] While the general public has no easy access to all violations of international and national laws committed by private contractors, the representatives of the industry itself remain reluctant to elaborate on the issue, downplaying any real concern by making statements similar to the comment made by one ArmorGroup official: "After 25 years in the business, I have not yet seen a case of improper conduct."[58] It does not mean, however, that PMSCs do not engage in the excessive use of force resulting in indiscriminate shooting and civilian casualties.

Indeed, Blackwater, as many other firms from the same industry, do not fight their own private war, but function on behalf of legitimate governments, yet often appear to participate in "firefights on their own, for their own reasons", which are not always acknowledged and approved by the US military.[59] Many of more than 100 PMSCs operating in Iraq seem to open fire more frequently than has been acknowledged before.[60] The findings made by Blackwater itself and the US State Department pointed out that the use of – often lethal – force by Blackwater guards is "frequent and extensive, resulting in significant casualties and property damage". According to the statements made by the US officials and former and current contractors working for those businesses, incidents involving the use of force, shootings at civilian population, and property damage are rarely reported to the US or Iraqi authorities, while this lack of reporting undermines official statistics the regular military of the States needs to comply with. Even the death toll from the Nisour Square incident involving Blackwater, in which 17 people were killed and 24 severely wounded, was incorrectly reported in governmental records as nine deaths among the Iraqi citizens.[61]

It is often highlighted that PMSCs make frequent use of firearms, considering the fact that they operate in war zones and other unstable and insecure environments with many dangers and risks for the safety of their clients and personnel. David Horner, one of the contractors working for Crescent Security Group, a firm having its

---

57 Michael Walzer, "Mercenary Impulse: Is There an Ethics That Justifies Blackwater?", *The New Republic*, March 12, 2008, <http://www.tnr.com/article/mercenary-impulse>, accessed on October 10, 2013.

58 Kateri Carmola, (2010) *Private Security Contractors and New Wars: Risk, Law, and Ethics*, Oxon and New York: Routledge, p. 101.

59 Michael Walzer, "Mercenary Impulse: Is There an Ethics That Justifies Blackwater?", *The New Republic*, March 12, 2008, <http://www.tnr.com/article/mercenary-impulse>, accessed on October 10, 2013.

60 Steve Fainaru, "Guards in Iraq Cite Frequent Shootings", *The Washington Post*, October 3, 2007, <http://www.washingtonpost.com/wp-dyn/content/article/2007/10/02/AR2007100202456.html?>, accessed on September 16, 2014.

61 Pratap Chatterjee, "Iraq War Logs: Military Privatisation Run Amok", *The Guardian*, October 23, 2010, <http://www.guardian.co.uk/commentisfree/cifamerica/2010/oct/23/iraq-war-logs-us-military/>, accessed on October 10, 2013.

headquarters in Kuwait City, said that his colleagues carrying out their tasks in Baghdad started to fire automatic weapons pre-emptively every time they had to pass the town. The security contractors often open fire at suspicious individuals and vehicles without being attacked first. The same operational policy seems to have been adopted and realized on a daily basis by the employees of Blackwater,[62] but possibly also by a large number of other PMSCs. According to the private guards of Crescent Security Group, this policy was used as a response to an attack by a roadside bomb carried out against one of their convoys.[63] Mr. Horner made the following statement with regard to the use of firearms by his fellow contractors during their service in Iraq:

> "I know that I personally never saw anyone shoot at us, but we blazed through that town all the time… Personally I did not take aim at one person. But I do not know what everybody else did. We'd come back at the end of the day, and a lot of times we were out of ammo."[64]

David Horner also said that, in one particular instance, an employee of Crescent Security Group was firing a machine gun from the bed of Horner's truck and hit two members of the Iraqi National Guard. This type of incident is often witnessed during the operation of other PMSCs as well. One might find it obvious that such reckless behavior and serious misconduct can easily result not only in property damage caused by indiscriminate pre-emptive shootings of the contractors, but also massive numbers of killed and injured individuals among innocent civilians, which is, unfortunately, also frequently the case. Notably enough, in three years – from 2005 until 2007 – PMSC Blackwater reported to have caused 16 civilian casualties in Iraq and to have been involved in 162 incidents of property damage.[65] It must be emphasized that in more than 80% of those incidents Blackwater guards where the ones to fire the first shot.[66] DynCorp International and Triple Canopy are also reported to have shot first in more than a half of all shooting incidents.[67]

---

62 US House of Representatives, Committee on Oversight and Government Reform, "Additional Information About Blackwater USA", Memorandum, October 1, 2007, p. 6, <http://graphics8.nytimes.com/packages/pdf/national/20071001121609.pdf>, accessed on September 16, 2014.

63 Human Rights First, *Private Security Contractors at War: Ending the Culture of Impunity*, Report, 2008, New York, p. 8.

64 Steve Fainaru, "Guards in Iraq Cite Frequent Shootings", *The Washington Post*, October 3, 2007, <http://www.washingtonpost.com/wp-dyn/content/article/2007/10/02/AR2007100202456.html>, accessed on September 16, 2014.

65 US House of Representatives, Committee on Oversight and Government Reform, "Additional Information About Blackwater USA", Memorandum, October 1, 2007, p. 7, <http://graphics8.nytimes.com/packages/pdf/national/20071001121609.pdf>, accessed on September 16, 2014.

66 Ibidem, p. 12.

67 Ibidem, p. 2.

Intentional killings

After indiscriminate shootings, the second type of misconduct is the hiring of PMSCs to eliminate certain individuals on behalf of governments. There are great obstacles to conducting a solid investigation into the practice of State employment of PMSCs, such as by various governmental agencies of the United States and the United Kingdom, for intentional killings. There are no contracts available and these cases are extremely hard to prove, cannot easily be identified and researched. One might state, however, that certain governments could violate rules and principles of international law by hiring PMSCs in order to intentionally deprive someone of his life. According to some reports, the Central Intelligence Agency (hereinafter: the CIA) might have contracted the US PMSC Blackwater as a part of a secret program to locate and to assassinate top members of the terrorist organization al-Qaeda.[68] This program, which was not even known to the US Congress, was characterized as "an attempt to carry out a 2001 presidential authorization to capture or kill al Qaeda operatives".[69] *The New York Times* reported:

> "It is unclear whether the C.I.A. had planned to use the contractors to actually capture or kill Qaeda operatives, or just to help with training and surveillance in the program. American spy agencies have in recent years outsourced some highly controversial work, including the interrogation of prisoners. But government officials said that bringing outsiders into a program with lethal authority raised deep concerns about accountability in covert operations."[70]

Some sources familiar with the program made clear that the team consisting of Blackwater contractors and the CIA agents was appointed to go to Germany without notifying "their own station" or the German government in order to trace and to follow a naturalized German citizen from Syria, who was an al-Qaeda financier. Another target for assassination was a Pakistani scientist who was tracked down in Dubai. However, at the final stage of the operations, the authorities in Washington decided in both cases not to kill the above-mentioned individuals. As a result, the Blackwater and CIA team had to cancel its assassination plans.[71]

---

68 Mark Mazzetti, "C.I.A. Sought Blackwater's Help to Kill Jihadists", *The New York Times*, August 19, 2009, <http://www.nytimes.com/2009/08/20/us/20intel.html?_r=1&>, accessed on September 16, 2014; Lillian Tan, "Private Military and Security Companies Wanted for Hire by CIA as 'Hitmen'", *The Huffington Post*, August 27, 2009, <http://www.huffingtonpost.com/amnesty-international/private-military-and-secu_b_270734.html>, accessed on October 17, 2013.

69 Siobhan Gorman, "CIA Had Secret Al Qaeda Plan", *The Wall Street Journal*, July 13, 2009, <http://online.wsj.com/article/SB124736381913627661.html>, accessed on October 19, 2013.

70 Mark Mazzetti, "C.I.A. Sought Blackwater's Help to Kill Jihadists", *The New York Times*, August 19, 2009, <http://www.nytimes.com/2009/08/20/us/20intel.html?_r=1&>, accessed on September 16, 2014.

71 Adam Ciralsky, "Tycoon, Contractor, Soldier, Spy", *Vanity Fair*, January 2010, <http://www.vanityfair.com/politics/features/2010/01/blackwater-201001?currentPage=all>, accessed on October 18, 2013.

According to US State officials, the above-mentioned CIA program of using teams of private contractors to kill terrorists was associated with a number of "logistical, legal and diplomatic hurdles almost from the outset",[72] given the fact that killing terrorists who are hiding on the territory of a foreign country is an issue other than lawful killing of armed soldiers in a war. It might appear obvious that when the governmental agencies belonging to the State apparatus hire outside private entities, such as PMSCs, with an aim to eliminate certain individuals or groups of people in foreign countries without any form of trial before the court, it could be considered not only a serious crime, but also a violation of international law. In this regard, such a construction would lead to various issues associated with the accountability and criminal responsibility of private contractors and governmental officials and may lead to the responsibility of the respective State.

Property damage

The third type of misconduct can be derived from the analysis made above and constitutes property damage caused by the actions of PMSCs. One can conclude that the functioning of PMSCs in danger zones generally goes hand in hand with various shooting incidents, frequently initiated by the contractors themselves, that lead to third party deaths, injuries, and property damage, which are all closely linked to the improper behavior of private contractors. According to the internal reports of the PMSCs, such as Blackwater, property damage is very common, given the existence of pre-emptive shooting policies used on various occasions.[73]

Reports indicate that Blackwater has repeatedly caused damage to the property of local nationals of the countries where they perform their duties. For instance, on November 28, 2005, a Blackwater motorcade traveling to and from the Ministry of Oil for official meetings collided with 18 different vehicles during the round trip causing great damage.[74] This and other similar incidents seem to take place all too often during the carrying out of military and security services by PMSCs.

Torture and other cruel, inhuman, or degrading treatment or punishment

Fourthly, PMSCs have been involved in the commission of torture and other cruel, inhuman, or degrading treatment or punishment. The NGO Human Rights First claims that besides of the use of excessive and lethal force used by PMSCs there have been

72 Mark Mazzetti, "C.I.A. Sought Blackwater's Help to Kill Jihadists", *The New York Times*, August 19, 2009, <http://www.nytimes.com/2009/08/20/us/20intel.html?_r=1&>, accessed on September 16, 2014.
73 US House of Representatives, Committee on Oversight and Government Reform, "Additional Information About Blackwater USA", Memorandum, October 1, 2007, p. 7, <http://graphics8.nytimes.com/packages/pdf/national/20071001121609.pdf>, accessed on September 16, 2014.
74 Ibidem, p. 8.

numerous abuses of detainees committed by private contractors in Iraq and Afghanistan in the context of interrogations of those suspected of belonging to terrorist groups.[75] In February 2006 alone, 20 cases of various abuses were reported and referred to the US Justice Department. However, only one civilian contractor working for the CIA, David Passaro, has ever been prosecuted for violence towards local Afghani citizens. In the course of interrogation, Passaro beat a local national of Afghanistan, who died in custody the next day. In August 2006, Passaro faced a trial and was convicted of multiple assault charges and sentenced to more than eight years in prison.[76]

As stated, the CIA frequently makes use of private contractors. In some instances, those contractors perform specific tasks for the US agency that can qualify as torture or cruel, inhuman, and degrading treatment. Reporters have indicated that employees of PMSCs can engage in using horrific interrogation techniques, such as "waterboarding", in order to make suspects of, for example, terrorist attacks testify.[77] Private contractors working for PMSCs have not only been implicated in the actual commission of torture and comparable treatments, but also advised the US governmental agencies in these methods of gathering valuable information from captured individuals. One of these private businesses, Mitchell Jessen & Associates, has advised the CIA in torture techniques and their proper implementation.[78] Another PMSC, Global Risks Solution Inc., was alleged to have trained the Policía Federal in León, Mexico, in various torture techniques.

One of the scandals associated with private contractors that received a lot of publicity in the past years was without doubt that of the Abu Ghraib detention facility.[79]

---

[75] Human Rights First, *Private Security Contractors at War: Ending the Culture of Impunity*, Report, 2008, New York, p. 2.

[76] US District Court for the Eastern District of North Carolina, Western Division, *United States of America v. David A. Passaro*, Case No. 5:04-CR-211-BO(1); US Department of Justice, United States Attorney's Office, Eastern District of North Carolina, "David Passaro Sentenced to 100 Months Imprisonment: First American Civilian Convicted of Detainee Abuse During the Wars in Iraq and Afghanistan", Press Release, February 13, 2007, <http://www.justice.gov/usao/nce/press/2007/2007-Feb-13.html>, accessed on July 10, 2014; Human Rights First, *Private Security Contractors at War: Ending the Culture of Impunity*, Report, 2008, New York, p. 2.

[77] Siobhan Gorman, "US: CIA Likely Let Contractors Perform Waterboarding", *The Wall Street Journal*, February 8, 2008, <http://www.corpwatch.org/article.php?id=14926>, accessed on October 19, 2013.

[78] Scott Shane, "2 U.S. Architects of Harsh Tactics in 9/11's Wake", *The New York Times*, August 11, 2009, <http://www.nytimes.com/2009/08/12/us/12psychs.html?pagewanted=all>, accessed on October 17, 2013; "The Story of Mitchell Jessen & Associates: How a Team of Psychologists in Spokane, WA, Helped Develop the CIA's Torture Techniques", *Democracy Now!: The War and Peace Report*, April 21, 2009, <http://www.democracynow.org/2009/4/21/the_story_of_mitchell_jessen_associates>, accessed on October 17, 2013.

[79] Human Rights First, *Private Security Contractors at War: Ending the Culture of Impunity*, Report, 2008, New York, pp. 1-2, 15-16; Kateri Carmola, (2010) *Private Security Contractors and New Wars: Risk, Law, and Ethics*, Oxon and New York: Routledge, p. 27; Moshe Schwartz, *Department of Defense Contractors in Iraq and Afghanistan: Background and Analysis*, Congressional Research Service Report

It is considered by some to be an international embarrassment[80] for the United States and has led to a firestorm in the media.[81] The Abu Ghraib prison, or the Baghdad Central Detention Center, is a detention facility located in a city called Abu Ghraib 32 km to the west of Baghdad. The prison complex was built in the 1960s by Western contractors.[82] Under the regime of Saddam Hussein, prisoners residing here in poor conditions were frequently subjected to torture and executions.[83] In May 2003, the Coalition Provisional Authority decided to use the prison as "a criminal detention facility"[84] and – not long after that decision was taken – the US took over the prison. In June 2003, first prisoners started to arrive and the facility was extensively used to detain up to 5,000 Iraqis for committing various offenses. Not many could have imagined what horrific acts would take place inside the walls of the Abu Ghraib detention facility. US President George W. Bush noted:

> "Under the dictator, prisons like Abu Ghraib were symbols of death and torture. That same prison became a symbol of disgraceful conduct by a few American troops who dishonored our country and disregarded our values."[85]

Indeed, investigations into the incident revealed that American soldiers and a number of employees of two PMSCs, Titan, which is now L-3 Services as part of L-3

---

for Congress, July 2, 2010, p. 16, <http://www.fas.org/sgp/crs/natsec/R40764.pdf>, accessed on October 18, 2013; Amol Mehra, (2010) "Bridging Accountability Gaps – The Proliferation of Private Military and Security Companies and Ensuring Accountability for Human Rights Violations", *The Pacific McGeorge Global Business & Development Law Journal*, Volume 22, p. 325.

80 Greg Stohr, "Abu Ghraib Contractor Suit Draws U.S. Supreme Court Interest", *Bloomberg*, October 4, 2010, <http://www.bloomberg.com/news/2010-10-04/abu-ghraib-case-involving-private-contractors-draws-top-court-s-interest.html>, accessed on September 10, 2013.

81 Human Rights First, *Private Security Contractors at War: Ending the Culture of Impunity*, Report, 2008, New York, pp. 1-2, 15-16; Kateri Carmola, (2010) *Private Security Contractors and New Wars: Risk, Law, and Ethics*, Oxon and New York: Routledge, p. 27; Moshe Schwartz, *Department of Defense Contractors in Iraq and Afghanistan: Background and Analysis*, Congressional Research Service Report for Congress, July 2, 2010, p. 16, <http://www.fas.org/sgp/crs/natsec/R40764.pdf>, accessed on October 18, 2013; Amol Mehra, (2010) "Bridging Accountability Gaps – The Proliferation of Private Military and Security Companies and Ensuring Accountability for Human Rights Violations", *The Pacific McGeorge Global Business & Development Law Journal*, Volume 22, p. 325.

82 Martin Asser, "Abu Ghraib: Dark Stain on Iraq's Past", *BBC News*, May 25, 2004, <http://news.bbc.co.uk/2/hi/americas/3747005.stm>, accessed on October 15, 2013.

83 Seymour M. Hersh, "Torture at Abu Ghraib: American Soldiers Brutalized Iraqis. How Far up Does the Responsibility Go?", *The New Yorker*, May 10, 2004, <http://www.newyorker.com/archive/2004/05/10/040510fa_fact>, accessed on October 15, 2013.

84 Major General George R. Fay, Investigating Officer, Article 15-6 Investigation of the Abu Ghraib Prison and 205th Military Intelligence Brigade, August 2004, p. 33, <news.findlaw.com/hdocs/docs/dod/fay82504rpt.pdf>, accessed on October 15, 2013.

85 Dana Milbank, "Bush Seeks to Reassure Nation on Iraq", *The Washington Post*, May 25, 2004, <http://www.washingtonpost.com/wp-dyn/articles/A52711-2004May24.html>, accessed on October 15, 2013.

Communications Holdings Inc., and CACI, had been implicated in detainee abuse between October and December 2003. CACI and Titan, working alongside the US troops had been employed by the US Department of Defense (hereinafter: US DoD) to carry out several activities, such as interrogation and translation at the Abu Ghraib prison facilities in Iraq during 2003.[86] The contract between CACI and the US army was administered by the US Interior Department.[87] The PMSCs focussed on two activities: CACI helped with the interrogation of the prisoners at the detention facility while the employees of Titan provided translation services.[88]

After the release of the confidential report of the International Committee of the Red Cross (hereinafter: ICRC) on the abuses committed against prisoners of war and other protected individuals under the law of armed conflict,[89] the formal US army investigations commenced led by Major General Antonio Taguba. In March 2004, Taguba identified various abuses of detainees at Abu Ghraib and pointed out a number of problems, such as those of oversight and inadequate training of the implicated personnel.[90] In the summer 2004, the Fay Report appeared. Investigation of the abuses was conducted under the lead of Major General George R. Fay and indicated

---

[86] Laura A. Dickinson, "Contract as a Tool for Regulating Private Military Companies", in Simon Chesterman and Chia Lehnardt (Eds.), (2007) *From Mercenaries to Market: The Rise and Regulation of Private Military Companies*, Oxford: Oxford University Press, p. 221; Renae Merle and Ellen McCarthy, "6 Employees From CACI International, Titan Referred for Prosecution", *The Washington Post*, August 26, 2004, <http://www.washingtonpost.com/wp-dyn/articles/A33834-2004Aug25.html>, accessed on November 10, 2013; Ellen McCarthy, "CACI Contract: From Supplies to Interrogation", *The Washington Post*, May 17, 2004, <http://www.washingtonpost.com/wp-dyn/articles/A31611-2004May16.html>, accessed on November 10, 2013; Edmond Lococo, "Titan Competing with Northrop, L-3 to Keep Its Largest Contract", *Bloomberg*, June 7, 2004, <http://www.bloomberg.com/apps/news?pid=newsarchive&sid=afR4zWgUnpDo>, accessed on October 17, 2013.

[87] Ellen McCarthy, "CACI Contract: From Supplies to Interrogation", *The Washington Post*, May 17, 2004, <http://www.washingtonpost.com/wp-dyn/articles/A31611-2004May16.html>, accessed on November 10, 2013.

[88] Major General George R. Fay, Investigating Officer, Article 15-6 Investigation of the Abu Ghraib Prison and 205th Military Intelligence Brigade, August 2004, p. 9, <news.findlaw.com/hdocs/docs/dod/fay82504rpt.pdf>, accessed on October 15, 2013; Antenor Hallo de Wolf, (2012) *Reconciling Privatization with Human Rights*, School of Human Rights Research Series, Volume 49, Cambridge, Antwerp and Portland: Intersentia, p. 498; Katja Nieminen, (2004) "The Rules of Attribution and the Private Military Contractors at Abu Ghraib: Private Acts or Public Wrongs?", *Finnish Yearbook of International Law*, Volume 15, p. 299; Greg Stohr, "Abu Ghraib Contractor Suit Draws U.S. Supreme Court Interest", *Bloomberg*, October 4, 2010, <http://www.bloomberg.com/news/2010-10-04/abu-ghraib-case-involving-private-contractors-draws-top-court-s-interest.html>, accessed on September 10, 2013.

[89] ICRC, Report of the International Committee of the Red Cross (ICRC) on the Treatment by the Coalition Forces of Prisoners of War and Other Protected Persons by the Geneva Conventions in Iraq During Arrest, Internment and Interrogation, February 2004, <http://www.informationclearinghouse.info/pdf/icrc_iraq.pdf>, accessed on November 11, 2013.

[90] Major General Antonio N. Taguba, Report of the Article 15-6 Investigation of the 800th Military Police Brigade, March 2004, <http://www.dod.mil/pubs/foi/operation_and_plans/Detainee/taguba/TAGUBA_REPORT_CERTIFICATIONS.pdf>, accessed on October 15, 2013.

that there had been 44 instances of detainee abuse carried out by the US soldiers and personnel of PMSCs. It came to light that the employees of the latter were involved in 10 of those incidents.[91] The incidents included physical and sexual abuse, such as rape, threats of rape, sexual assaults, and electric shock beatings; improper use of military working dogs; humiliating and degrading treatment; and improper use of isolation. It was revealed that at least five contractors of the companies mentioned above and US soldiers appeared to be involved in the commission of torture and other cruel, inhuman, or degrading treatment or punishment.[92] Furthermore, the Fay Report noted that:

> "... abuse is defined as treatment of detainees that violated U.S. criminal law or international law or treatment that was inhumane or coercive without lawful justification. Whether the Soldier or contractor knew, at the time of the acts, that the conduct violated any law or standard, is not an element of the definition."[93]

The abuses at Abu Ghraib primarily fall into two categories: "intentional violent or sexual abuses" and "abusive actions taken based on misinterpretations of or confusion regarding law or policy".[94] The US Army investigations showed that there were six employees of Titan and CACI implicated in 36% of all abuses at Abu Ghraib.[95] While the soldiers implicated in the unlawful acts were charged under the Uniform Code of Military Justice (hereinafter: UCMJ), the PMSC employees managed to escape responsibility.[96] On June 9, 2004, a large group of Iraqis who were detainees

---

91 Major General George R. Fay, Investigating Officer, Article 15-6 Investigation of the Abu Ghraib Prison and 205th Military Intelligence Brigade, August 2004, pp. 68-95, <news.findlaw.com/hdocs/docs/dod/fay82504rpt.pdf>, accessed on October 15, 2013.

92 Lieutenant General Anthony R. Jones, Article 15-6 Investigation of the Abu Ghraib Prison and 205th Military Intelligence Brigade, August 2004, <news.findlaw.com/hdocs/docs/dod/fay82504rpt.pdf>, accessed on October 15, 2013; Major General George R. Fay, Investigating Officer, Article 15-6 Investigation of the Abu Ghraib Prison and 205th Military Intelligence Brigade, August 2004, <news.findlaw.com/hdocs/docs/dod/fay82504rpt.pdf>, accessed on October 15, 2013; Major General Antonio Taguba also named two contractors in his investigation. See Major General Antonio N. Taguba, Report of the Article 15-6 Investigation of the 800th Military Police Brigade, March 2004, <http://www.dod.mil/pubs/foi/operation_and_plans/Detainee/taguba/TAGUBA_REPORT_CERTIFICATIONS.pdf>, accessed on October 15, 2013; Human Rights First, *Private Security Contractors at War: Ending the Culture of Impunity*, Report, 2008, New York, p. 2.

93 See Executive Summary: Investigation of Intelligence Activities At Abu Ghraib, p. 3, in Major General George R. Fay, Investigating Officer, Article 15-6 Investigation of the Abu Ghraib Prison and 205th Military Intelligence Brigade, August 2004, <news.findlaw.com/hdocs/docs/dod/fay82504rpt.pdf>, accessed on 15 October 15, 2013.

94 Major General George R. Fay, Investigating Officer, Article 15-6 Investigation of the Abu Ghraib Prison and 205th Military Intelligence Brigade, August 2004, p. 4, <news.findlaw.com/hdocs/docs/dod/fay82504rpt.pdf>, accessed on 15 October 15, 2013.

95 Peter W. Singer, (2005) "Outsourcing War", *Foreign Affairs*, Volume 84, Issue 2, pp. 127-128.

96 Ibidem, p. 128.

at the Abu Ghraib prison brought a federal lawsuit against both CACI and Titan.[97] The case was brought under the US Alien Tort Statute and charged the defendants with various violations of state, federal, and international law, including torture and sexual assault. In November 2007, the case against L-3 was dismissed by the district judge on the basis of "government contractor defense". In September 2009, the Court of Appeals gave its ruling in favor of both defendants and agreed with the dismissal of the claims against L-3 by ruling that CACI also had government contractor immunity. On June 27, 2011, the Supreme Court made clear that it would not hear the case and ended it. In 2008, some Iraqi inmates sued CACI under the US Alien Tort Statute for torturing them.[98] In August 2008, the plaintiffs' claims against L-3 were dismissed and after the appeal of the court's decision in March 2009, the plaintiffs and L-3 reached an out-of-court settlement the terms of which remain confidential. The case against CACI was dismissed by the District Court in June 2013 in accordance with the reasoning of the Supreme Court in the *Kiobel v. Shell* lawsuit[99] concerning the lack of jurisdiction. In August 2013, CACI decided to seek payment from the former Abu Ghraib inmates for its legal expenses.[100] On October 29, 2013, the plaintiffs appealed the decision challenging both the payment of legal fees and the dismissal of the case.[101] On July 30, the Fourth Circuit Court of Appeals vacated the judgment of the District Court and remanded all claims made by the plaintiffs for further legal proceedings.[102]

---

97 US District Court for the District of Columbia, *Saleh et al. v. CACI et al.*, Case No. 05-cv-1165; Center for Constitutional Rights, "Saleh et al v. Titan et al", <http://ccrjustice.org/Saleh-v-Titan>, accessed on October 17, 2013; Business & Human Rights Resourse Center, "Case Profile: Abu Ghraib Lawsuits Against CACI, Titan (now L-3)", <http://www.business-humanrights.org/Categories/Lawlaw suits/Lawsuitsregulatoryaction/LawsuitsSelectedcases/AbuGhraiblawsuitsagainstCACITitannowL-3>, accessed on October 16, 2013.

98 US District Court for the Eastern District of Virginia, Alexandria Division, *Suhail Najim Abdullah al Shimari v. CACI et al.*, Case No. 08-cv-0827 GBL-JFA; "Abu Ghraib Inmates Sue US Firms", *BBC News*, July 1, 2008, <http://news.bbc.co.uk/2/hi/americas/7482617.stm>, accessed on October 15, 2013.

99 US Supreme Court, *Kiobel v. Royal Dutch Petroleum Co.*, Opinion of the Court, April 17, 2013, 133 S. Ct. 1659 (2013).

100 Marjorie Censer, "CACI Seeking Payment from Former Abu Ghraib Prisoners", *The Washington Post*, August 12, 2013, <http://www.washingtonpost.com/blogs/capital-business/post/caci-seeking-payment-from-former-abu-ghraib-prisoners/2013/08/12/4c573550-035d-11e3-a07f-49ddc7417125_blog.html>, accessed on October 17, 2013.

101 Laura Raymond, "Abu Ghraib Torture Survivors Fight on Against Military Contractors", *The Huffington Post*, November 6, 2013, <http://www.huffingtonpost.com/the-center-for-constitutional-rights/abu-ghraib-torture-surviv_b_4227182.html>, accessed on November 13, 2013; Center for Constitutional Rights, "Al Shimari et al v. CACI et al", <http://ccrjustice.org/ourcases/current-cases/al-shimari-v-caci-et-al>, accessed on November 13, 2013.

102 US Court of Appeals for the Fourth Circuit, *Suhail Najim Abdullah al Shimari et al. v. CACI et al.*, No. 13-1937 and No. 13-2162.

Sex trafficking, sex slavery, and running prostitution networks

Fifthly, PMSC employees have been found to be active in the area of sex trafficking, sex slavery, and running prostitution networks. Men fighting in wars can and often do carry out appalling activities. One of the most terrifying faces of war is that of the blood, pain, and suffering of women and children. The Texas-based company, DynCorp, was hired by the US government to train the US police officers functioning as part of a UN Police Task Force contingent in Bosnia set up in 1995.[103] The personnel of this firm were operating under the US contract but under the command of the UN.[104] Some employees of DynCorp hit the news headlines because of their participation in sex trafficking and prostitution networks in the late 1990s while working on the police training contract.[105] They were accused of "engaging in perverse, illegal and inhumane behavior... purchasing illegal weapons, women, forged passports and participating in other immoral acts".[106] Those men were also reported as regularly "buying and selling women for their own personal enjoyment, and employees would brag about the various ages and talents of the individual slaves they had purchased".[107] Among these women were also girls who were 12 years old.[108] These facts were summed up by the former DynCorp aircraft mechanic who decided to file a lawsuit in Texas against the perpetrators and the company.[109] Eventually, the scandal created a negative image of the UN and led to general distrust in its peacekeeping operations.[110] The most notable aspect of this incident is the fact that none of the individuals involved in the misconduct has been prosecuted and that the US

---

103 Colum Lynch, "Misconduct, Corruption by U.S. Police Mar Bosnia Mission", *The Washington Post*, May 29, 2001, <http://dailyuw.com/archive/2001/05/29/imported/misconduct-corruption-us-police-mar-bosnia-mission#.VBlpSfmSyiw>, accessed on September 16, 2014.

104 Lou Pingeot, (2012) *Dangerous Partnership: Private Military & Security Companies and the UN*, Report, June 2012, Global Policy Forum and Rosa Luxemburg Foundation: New York and Berlin, p. 22.

105 Human Rights Watch, *Hopes Betrayed: Trafficking of Women and Girls to Post-Conflict Bosnia and Herzegovina for Forced Prostitution*, Report, November 2002, pp. 65-67, <http://www.hrw.org/reports/2002/bosnia/Bosnia1102.pdf>, accessed on September 16, 2014; Kathryn Bolkovac and Cari Lynn, (2011) *The Whistleblower: Sex Trafficking, Military Contractors, and One Woman's Fight for Justice*, New York: Palgrave Macmillan.

106 Kelly P. O'Meara, "US: DynCorp Disgrace", *Insight*, January 14, 2002, <http://www.corpwatch.org/article.php?id=11119>, accessed on September 16, 2014.

107 Ibidem.

108 David Isenberg, "PMC Sexual Violence: It's Still a Problem", *The Huffington Post*, January 30, 2012, <http://www.huffingtonpost.com/david-isenberg/pmc-sexual-violence-its-s_b_1240751.html?view=screen>, accessed on February 20, 2014.

109 Kelly P. O'Meara, "US: DynCorp Disgrace", *Insight*, January 14, 2002, <http://www.corpwatch.org/article.php?id=11119>, accessed on September 16, 2014.

110 Lou Pingeot, (2012) *Dangerous Partnership: Private Military & Security Companies and the UN*, Report, June 2012, Global Policy Forum and Rosa Luxemburg Foundation: New York and Berlin, p. 22.

still employs the corporation in question.[111] Those men were merely sent back home with no further consequences. The scandal in Bosnia was not an isolated event of this nature. Security guards operating in Afghanistan were accused of running a bordello and requiring that their security personnel engaged in the lewd conduct in order to get a promotion.[112]

Drug running, profiteering in transnational criminal networks, and weapons violations

Sixthly, one can also consider drug running, profiteering in transnational criminal networks, and weapons violations as the types of misconduct in which PMSCs have been involved. It appears that the misconduct involving excessive drinking and drugs has been one of the most commonly committed offenses by the US contractors in Iraq, Afghanistan, and other countries.[113] Besides the commission of that type of "less serious" offense, there have been reports on a number of PMSCs, such as DynCorp, that were implicated in the drugs business, selling and buying large amounts of various drugs. In Colombia, DynCorp contractors were possibly involved in smuggling drugs from the country.[114] The Colombian authorities appear to have found heroin worth $100,000 in a package sent from Colombian offices of DynCorp to its airbase in Florida.

Recently, the Pentagon issued new rules with regard to the process of hiring private contractors in Afghanistan.[115] Some concern and debate seem to center on the

---

111 David Isenberg, "PMC Sexual Violence: It's Still a Problem", *The Huffington Post*, January 30, 2012, <http://www.huffingtonpost.com/david-isenberg/pmc-sexual-violence-its-s_b_1240751.html?view=screen>, accessed on February 20, 2014.

112 Ginger Thompson, "2 Say They Reported Abuses at Embassy", *The New York Times*, September 10, 2009, <http://www.nytimes.com/2009/09/11/world/asia/11contractor.html>, accessed on November 13, 2013; Scott Horton, "Security Contractors Immune from Torture Charges, Judges Rule", *Harpers*, September 14, 2009, <http://harpers.org/blog/2009/09/security-contractors-immune-from-torture-charges-judges-rule/>, accessed on September 16, 2014.

113 "Booze, Drugs and Hookers: Hundreds of Offences by Afghan Contractors Uncovered", *The Associated Press*, December 19, 2010, <http://www.rawstory.com/rs/2010/12/hundreds-offenses-afghan-contractors-uncovered/>, accessed on November 15, 2013; Kelly P. O'Meara, "US: DynCorp Disgrace", *Insight*, January 14, 2002, <http://www.corpwatch.org/article.php?id=11119>, accessed on September 16, 2014.

114 Robert Lawson, "DynCorp: Beyond the Rule of Law", *Colombia Journal*, August 27, 2001, <http://colombiajournal.org/colombia78.htm>, accessed on November 14, 2013.

115 NATO/ISAF, *COMISAF's Counterinsurgency (COIN) Contracting Guidance*, September 8, 2010, <http://www.isaf.nato.int/images/stories/File/100908-NUI-COMISAF%20COIN%20GUIDANCE.pdf>, accessed on November 14, 2013; Deb Riechmann, "Petraeus Issues Afghan Contracting Guidance", *The Associated Press*, September 13, 2010, <http://www.airforcetimes.com/article/20100913/NEWS/9130308/Petraeus-issues-Afghan-contracting-guidance>, accessed on September 16, 2014; "NATO Issues New Guidelines for Afghanistan Contracts", *BBC News*, September 13, 2010, <http://www.bbc.co.uk/news/world-south-asia-11287074>, accessed on November 15, 2013.

fact that PMSCs can cooperate with criminal networks and benefit in financial and other senses without those linkages to the criminals being detected, considering the large number of sub-contracts usually concluded between firms. The participation of private contractors in criminal networks appears to be a matter of serious concern for States hiring them and an issue to be aware of.

In recent years, PMSCs have been involved in a number of violations with regard to arms embargoes, illegal arms shipments, and illegal possession of weapons. One of the most famous cases is the so-called "Arms-to-Africa scandal" in Sierra Leone involving a UK-based firm, Sandline International, which was hired in 1997 by the proponents of the President Ahmed Tejan Kabbah in order to provide "logistical support" for a counter-coup.[116] The company was accused of breaching UN Resolution 1132[117] by shipping 35 tons of Bulgarian-made AK-47 rifles to Sierra Leone in early 1998. The statements made by the representatives of Sandline included those relating to the approval by the US and UK governments of the preparations for the coup and the arms shipment to supply the Kabbah militia. The State representatives repeatedly denied that any approval had been given.

UN Resolution 1132 was passed on October 8, 1997 and contained provisions aimed to impose an arms embargo on Sierra Leone where according to the UN Security Council, the situation constituted a threat to international peace and security. Furthermore, the Security Council decided that all States must prevent the sale or supply to Sierra Leone of not only petroleum and petroleum products, but also arms and related material of all types, such as weapons, ammunition, military vehicles, and equipment. Allegedly, weapons were brought to Sierra Leone by Lieutenant-Colonel Tim Spicer, head of Sandline International, and were flown from Bulgaria to the Lungi airport in Sierra Leone. It happened in breach of the imposed UN embargo[118] and with British consent, as stated by Mr. Spicer.[119] In the course of investigations of the case, the statements made by the head of Sandline International proved to be true. It also came to light that the British Foreign Ministry appeared to have had knowledge of the firm's operations.[120] Arms meant to be delivered to Sierra Leone did not

---

116 Ed O'Loughlin, "Sandline Scandal Arms Shipment Reaches Forces", *The Independent*, May 22, 2000; "Q&A: Arms-to-Africa Scandal", *BBC News*, May 10, 1998, <http://news.bbc.co.uk/2/hi/uk_news/90526.stm>, accessed on November 14, 2013; Lieutenant-Colonel Tim Spicer Obe, (2000) *An Unorthodox Soldier: Peace and War and the Sandline Affair*, Edinburgh and London: Mainstream Publishing, pp. 190, 192, 196-198.

117 UN Security Council Resolution 1132 of October 8, 1997, UN Doc. S/Res/1132.

118 Andrew Buncombe, Paul Routledge, and Fran Abrams, "Inquiry Finds Sandline Did Breach Arms Embargo", *The Independent*, May 17, 1998, <http://www.independent.co.uk/news/inquiry-finds-sandline-did-breach-arms-embargo-1159472.html>, accessed on February 18, 2014.

119 Peter W. Singer, (2008) *Corporate Warriors: The Rise of the Privatized Military Industry*, Ithaca and London: Cornell University Press, p. 115.

120 Ibidem.

reach their destination. The weapon shipment fell into the hands of Nigerian forces and weapons were not distributed among the Kabbah militia.[121]

Apparently, some PMSCs seem not merely to violate arms embargoes and to provide weapons to embargoed States, but even to make use of illegal weapons in the course of carrying out their operations under a contract. The firm Executive Outcomes, a joint UK and South African company,[122] met much criticism because of their use of cluster bombs, napalm, and fuel air explosives during their operations in order to gain advantage against the enemy.[123] The aforementioned PMSC Crescent Security Group lost its license to work for the US government because of a number of violations committed by its employees, which included the possession of illegal weapons that may not be used by private security contractors, such as rocket launchers and grenades.[124] In the living quarters of the employees, the US military police also found 143 cans of beer and illegal steroids. One month later, more prohibited weapons were found: more rocket launchers, machine guns, and ammunition.[125] Further, the PMSC Zapata Security was also found to be in the possession of illegal weapons and after the shooting incident, which had occurred in Iraq on May 28, 2005, private contractors employed by that firm were taken into custody by the US Marine Unit.[126] Very probably, more private contractors working for private military and security businesses are involved in the commission of these offenses, but remain in the shadows, out of the sight of the public eye.

Mercenary activities

Finally, it is to be noted that PMSCs are often associated with mercenary activities. In recent decades, some PMSCs have been accused of participating in coups on the territory of a number of States and received negative publicity by being described as mercenary organizations. One of the most colorful examples of such companies is the PMSC Executive Outcomes, which according to Peter W. Singer "has become

---

121 "Q&A: Arms-to-Africa Scandal", *BBC News*, May 10, 1998, <http://news.bbc.co.uk/2/hi/uk_news/90526.stm>, accessed on November 14, 2013.

122 Kateri Carmola, (2010) *Private Security Contractors and New Wars: Risk, Law, and Ethics*, Oxon and New York: Routledge, p. 2.

123 Peter W. Singer, (2008) *Corporate Warriors: The Rise of the Privatized Military Industry*, Ithaca and London: Cornell University Press, p. 116; Amol Mehra, (2010) "Bridging Accountability Gaps – The Proliferation of Private Military and Security Companies and Ensuring Accountability for Human Rights Violations", *The Pacific McGeorge Global Business & Development Law Journal*, Volume 22, p. 325.

124 Steve Fainaru, "Cutting Costs, Bending Rules, and a Trail of Broken Lives", *The Washington Post*, July 29, 2007, <http://www.pulitzer.org/archives/7688>, accessed on February 18, 2014.

125 Robert Y. Pelton, "Crescent Goes Down Hard in Washington Post: Violations and Kidnapping Take Down Media-Friendly Security Company", *Privateforces.com*, July 29, 2007, <http://privateforces.com/index.php?option=com_content&task=view&id=1894&Itemid=99999999>, accessed on February 20, 2012.

126 Human Rights First, *Private Security Contractors at War: Ending the Culture of Impunity*, Report, 2008, New York, p. 2.

emblematic of the overall phenomenon of corporate armies".[127] This company caused a great military impact in various wars due to its organizational structure and the scope of its political connections, business links, and incorporation into a larger economic holding.[128] The company consisted of contractors having their origins in the elite forces of the apartheid-era South African Defence Force and was frequently accused of being "a mercenary army of racist killers".[129]

The general public often refers to PMSCs as "mercenaries", "hired guns", or "soldiers of fortune" because of the confusion regarding the actual nature of those companies and the qualification and legal status of their employees in international law.[130] It might appear very confusing to make a clear distinction between mercenaries and firms performing military and security services on behalf of States or other entities, because of the fact that no clear consensus has been reached so far on how those soldiers "paid to serve in a foreign army or other military organization"[131] should be defined.[132]

Many PMSCs clearly distinguish themselves from commonly known mercenaries by the corporatization of their service provision, as pointed out by Peter W. Singer,[133] representing an exception to the rule that mercenary armies are the sole source of force and security that had been changed after the Westphalian Peace of 1648.[134] However, some PMSCs do not try to legitimize their image even creating a stronger impression of their mercenary character. The operation of two firms, Executive Outcomes and Sandline International, known by their public face in person of Lieutenant-Colonel Tim Spicer, led to many concerns among legal and other scholars with regard to their "mercenary activities".[135]

The first company, Executive Outcomes, was hired by Angola in 1993 to train the Angolan army and then to fight the UNITA rebel guerrilla group controlling some oil fields.[136] The firm consisted of members of a former South African battalion that had

---

[127] Peter W. Singer, (2008) *Corporate Warriors: The Rise of the Privatized Military Industry*, Ithaca and London: Cornell University Press, p. 101.

[128] Ibidem.

[129] Ibidem.

[130] Nils Rosemann, (2005) "The Privatization of Human Rights Violations – Business' Impunity or Corporate Responsibility? The Case of Human Rights Abuses and Torture in Iraq", *Non-State Actors and International Law*, Volume 5, p. 82.

[131] Definition provided by the Oxford English Dictionary, OED Online version June 2014, <http://www.oed.com>.

[132] Peter W. Singer, (2008) *Corporate Warriors: The Rise of the Privatized Military Industry*, Ithaca and London: Cornell University Press, p. 40.

[133] Ibidem, p. 45.

[134] Ibidem, pp. 19-39.

[135] Kateri Carmola, (2010) *Private Security Contractors and New Wars: Risk, Law, and Ethics*, Oxon and New York: Routledge, p. 2.

[136] Peter W. Singer, (2008) *Corporate Warriors: The Rise of the Privatized Military Industry*, Ithaca and London: Cornell University Press, pp. 107-110.

been used to attack African National Congress rebels under the apartheid regime. The company acquired bad reputation as a group of mercenaries due to their employees being ex-South African Defense Forces, was automatically brought in connection with a scandal involving Tim Spicer, and had to disband in 1999. Another company, Sandline International, established by Spicer, was a large PMSC that in January 1997 signed an agreement with the Deputy Prime Minister of Papua New Guinea for carrying out a variety of tasks, such as training of the Special Forces of the Papua New Guinea Defense Force, participation in offensive operations, and gathering intelligence.[137] Sandline had to actively engage in these activities in order to rout a rebel group called the Bougainville Revolutionary Army that had captured one of the largest copper mines in the world on the island of Bougainville. Most of these tasks were sub-contracted to Executive Outcomes.[138] The contract, worth $36,000,000, concluded between the government and Sandline, was later approved by the National Security Council of Papua New Guinea allowing the company to start providing its services.[139] The work that Sandline was about to commence led, however, to various protests and demonstrations in the country against it and the corrupt government of Papua New Guinea. Subsequently, employees of Sandline were arrested and their equipment was confiscated. The Prime Minister of Papua New Guinea suspended the contract with the company and promised a judicial inquiry to investigate the matter. After Spicer and his firm had quickly left the country, the firm was finally closed in 2004. For the purposes of legal analysis, this case study represents a very valuable addition given that the contract concluded between the company and Papua New Guinea is publicly available and can be thoroughly scrutinized.

Executive Outcomes and Sandline International are not the only businesses active in the field of private military and security service provision that are associated with mercenaries, but all of them face the same criticisms. Very recently, claims have been made of the involvement of the US PMSC Greystone Limited in the conflict in Ukraine.[140] Some have stressed that the US government, avoiding direct military

---

137 Antenor Hallo de Wolf, (2006) "Modern Condottieri in Iraq: Privatizing War from the Perspective of International Human Rights Law", *Indiana Journal of Global Legal Studies*, Volume 13, Issue 2, p. 326; Peter W. Singer, (2008) *Corporate Warriors: The Rise of the Privatized Military Industry*, Ithaca and London: Cornell University Press, pp. 192-193, 263-264.

138 Hannah Tonkin, (2011) *State Control over Private Military and Security Companies in Armed Conflict*, Cambridge: Cambridge University Press, p. 44.

139 Peter W. Singer, (2008) *Corporate Warriors: The Rise of the Privatized Military Industry*, Ithaca and London: Cornell University Press, p. 193.

140 Damien Gayle, "Has Blackwater Been Deployed to Ukraine? Notorious U.S. Mercenaries 'Seen on the Streets of Flashpoint City' as Russia Claims 300 Hired Guns Have Arrived in Country", *MailOnline*, March 8, 2014, <http://www.dailymail.co.uk/news/article-2576490/Are-Blackwater-active-Ukraine-Videos-spark-talk-U-S-mercenary-outfit-deployed-Donetsk.html>, accessed on May 15, 2014; "Kämpfen US-Söldner in der Ukraine?", *Bild*, March 9, 2014, <http://www.bild.de/politik/ausland/ukraine/us-soeldner-von-blackwater-im-einsatz-34992896.bild.html>, accessed on May 15, 2014; "400 US Commandos Help Kiev in Its Military Offensive in East Ukraine – Reports", *The Voice of Russia*, May

intervention and responsibility for their actions, used this company to invade the country.[141] The US authorities have fiercely denied these accusations.[142] It is frequently noted that some PMSCs do not want to be put on the same list with Executive Outcomes, Sandline International, or a new firm established by Lieutenant-Colonel Tim Spicer called Aegis.[143] Logically, a large number of other legitimate private military and security contractors that operate in foreign countries and perform their quite ambiguous tasks would also like to avoid this stigma.

Interim conclusion

To sum up, it is to be emphasized that in the last two decades PMSCs have been accused of the excessive use of force, indiscriminate shootings, and civilian casualties caused by those incidents; murder of individuals performed on behalf of States; property damage; torture and other cruel, inhuman, or degrading treatment or punishment; sex trafficking, sex slavery, and running of prostitution networks; drug running; profiteering in transnational criminal networks; weapons violations; and mercenary activities and interference in internal affairs of States. Considering their operational environment, it is not entirely impossible that private contractors have committed other serious crimes, such as a murder of a specific population with an aim of ethnic cleansing.

Not all the above-listed illegal acts can be determined to be violations of international law, more specifically international humanitarian law and human rights law. Some of these types of misconduct, however, can amount not only to the breaches of the law of armed conflict and human rights abuses, but also be seen as the most serious international crimes that could trigger international criminal responsibility of the responsible individuals. The possibility to consider them as such will be discussed further in the present study. The excessive use of force often employed by PMSCs and their indiscriminate shootings leading to civilian casualties and property damage

---

11, 2014, <http://voiceofrussia.com/news/2014_05_11/400-US-commandos-help-Kiev-in-its-military-offensive-in-east-Ukraine-reports-8880/>, accessed on May 15, 2014; Stephen Lendman, "Blackwater Killers in Eastern Ukraine", *thepeoplesvoice.org*, May 12, 2014, <http://www.thepeoplesvoice.org/TPV3/Voices.php/2014/05/12/blackwater-killers-in-eastern-ukraine>, accessed on May 15, 2014; "Einsatz Gegen Separatisten: Ukrainische Armee Bekommt Offenbar Unterstützung von US-Söldnern", *Der Spiegel*, May 11, 2014, <http://www.spiegel.de/politik/ausland/ukraine-krise-400-us-soeldner-von-academi-kaempfen-gegen-separatisten-a-968745.html>, accessed on May 15, 2014.

[141] "US Mercenaries Pose Threat to Ukrainians Being Ultra-Violent Individuals – US Military Vet", *The Voice of Russia*, April 11, 2014, <http://voiceofrussia.com/2014_04_11/US-mercenaries-pose-threat-to-Ukrainians-being-ultra-violent-individuals-US-military-vet-2839/>, accessed on May 15, 2014.

[142] Michael Birnbaum, "Russia to Ukraine: Your Military Actions are 'Criminal'", *The Washington Post*, May 2, 2014, <http://www.washingtonpost.com/world/europe/russia-to-ukraine-your-military-actions-are-criminal/2014/05/02/bd009b57-0ff4-403d-83a3-67f3f9db8908_story.html>, accessed on May 15, 2014.

[143] Kateri Carmola, (2010) *Private Security Contractors and New Wars: Risk, Law, and Ethics*, Oxon and New York: Routledge, p. 3.

constitute, in fact, the most common form of misconduct in which private contractors are implicated. Other forms of misconduct are less common, but nevertheless usually have the same or even higher degree of gravity, such as the commission of torture and other cruel, inhuman, or degrading treatment or punishment. These actions may result in the application of the law of armed conflict, which they violate, and be defined as international crimes and human rights abuses – in case of successful attribution – leading to the acceptance of the responsibility of hiring States. Additionally, other States might be held responsible for their non-compliance with positive obligations under international humanitarian and human rights law.

4. Moral implications

Apart from the legal aspects involved in the question of State responsibility for the misconduct of PMSCs and their personnel violating international law, there are also some moral implications and dilemmas present in the outsourcing of crucial governmental activities to private actors that deserve attention. The position that States should bear international responsibility for breaches of international law committed by PMSCs and their employees can be defended given that the State use of PMSCs in armed conflicts is morally problematic for two main reasons.

The first main moral objection that can be raised against the presence of PMSCs on the battlefield is their lack of attachment to a cause.[144] Causes justifying participation in wars could be said to vary over time and place, but appear to include the pursuit and protection of national or religious interests and interests of sovereign States.[145] Thomas Hobbes understood the core of the problem and stated in his *Leviathan* that individuals striving to achieve their own goals and providing their own security in protecting own interests will create an unstable society resulting in chaos and conflicts.[146] According to him, "competition of riches, honour, command, or other power inclineth to contention, enmity, and war, because the way of one competitor to the attaining of his desire is to kill, subdue, supplant, or repel the other".[147] Just as many scholars consider mercenaries to be immoral,[148] Nicolò Machiavelli was

---

144 Sarah Percy, "Morality and Regulation", in Simon Chesterman and Chia Lehnardt, (Eds.), (2007) *From Mercenaries to Market: The Rise and Regulation of Private Military Companies*, Oxford: Oxford University Press, p. 14; Dimitrios Machairas, (2014) "The Ethical Implications of the Use of Private Military Force: Regulatable or Irreconcilable?", *Journal of Military Ethics*, Volume 13, Issue 1, pp. 50-53.

145 Sarah Percy, (2007) *Mercenaries: The History of a Norm in International Relations*, Oxford: Oxford University Press, p. 54.

146 Adam White, (2010) *The Politics of Private Security: Regulation, Reform and Re-Legitimation*, Houndmills, Basingstoke, New Hampshire: Palgrave Macmillan, p. 23.

147 Thomas Hobbes, (1651) *Leviathan*, Revised Edition, Edited by A.P. Martinich, and Brian Battiste, 2011, Peterborough [etc.]: Broadview Press, p. 105, Chapter XI, para. 3.

148 Kim R. Nossal, (1998) "Roland Goes Corporate: Mercenaries and Transnational Security Corporations in the Post-Cold War Era", *Civil Wars*, Volume 1, Issue 1, p. 18.

fiercely against the use of mercenaries who do not fight for the public good and can cause significant damage to the republic.[149] He wrote:

> "Mercenaries and auxiliaries are useless and dangerous; and if one holds his state based on these arms, he will stand neither firm nor safe; for they are disunited, ambitious and without discipline, unfaithful, valiant before friends, cowardly before enemies; they have neither the fear of God nor fidelity to men, and destruction is deferred only so long as the attack is; for in peace one is robbed by them, and in war by the enemy. The fact is, they have no other attraction or reason for keeping the field than a trifle of stipend, which is not sufficient to make them willing to die for you. They are ready enough to be your soldiers whilst you do not make war, but if war comes they take themselves off or run from the foe..."[150]

It is widely known that private corporations – just as other business enterprises – engaging in hostilities and providing security perform their tasks for the sole sake of financial gain and not abstract ideals.[151] The problem is that the public good and the interests of these firms are not always in line with one another.[152] The whole business model is based on the fact that the PMSC employees are often in the position to use force and therefore capable of killing people. Killing or injuring individuals for the purpose of earning money is widely regarded as unacceptable on moral grounds.[153] Throughout the course of human history, the same type of criticism has often been expressed and is still expressed with regard to mercenaries.[154] They appear to be devoted to war and fighting and not the military profession of an official soldier. In the first place, they participate in armed conflicts in order to receive financial and

---

149 Sarah Percy, (2007) *Mercenaries: The History of a Norm in International Relations*, Oxford: Oxford University Press, p. 76.

150 Nicolò Machiavelli, (1513) *The Prince*, Translated by W.K. Marriott, in Robert Maynard Hutchins (Ed. in Chief), (1952) *Great Books of the Western World: Machiavelli, Hobbes*, Volume 23, Chicago [etc.]: Encyclopaedia Britannica, Inc., p. 18, Chapter XII.

151 Sebastian Mallaby, "Paid to Make Peace: Mercenaries Are No Altruists, But They Can Do Good", *The Washington Post*, June 4, 2001, <http://www.sandline.com/hotlinks/article9.html>, accessed on September 16, 2014; Peter W. Singer, (2008) *Corporate Warriors: The Rise of the Privatized Military Industry*, Ithaca and London: Cornell University Press, pp. 217-218; Adam White, (2010) *The Politics of Private Security: Regulation, Reform and Re-Legitimation*, Houndmills, Basingstoke, New Hampshire: Palgrave Macmillan, p. 19; Doug Brooks and Hussein Solomon, (2000) "From the Editor's Desk", *Conflict Trends*, Issue 1, p. 2, <http://www.accord.org.za/downloads/ct/ct_2000_1.pdf?phpMyAdmin=ceeda2df659e6d3e35a63d69e93228f1>, accessed on December 20, 2011.

152 Peter W. Singer, (2008) *Corporate Warriors: The Rise of the Privatized Military Industry*, Ithaca and London: Cornell University Press, p. 217.

153 Sarah Percy, "Morality and Regulation", in Simon Chesterman and Chia Lehnardt, (Eds.), (2007) *From Mercenaries to Market: The Rise and Regulation of Private Military Companies*, Oxford: Oxford University Press, p. 14.

154 Benedict Sheehy and Jackson N. Maogoto, (2008) "The Private Military Company – Unravelling the Theoretical, Legal & Regulatory Mosaic", *ILSA Journal of International & Comparative Law*, Volume 15, Issue 1, p. 152.

other remuneration from their "clients".[155] Both PMSCs and mercenaries are able to share the causes set out above when taking part in hostilities, but do not necessarily do so: they typically engage in armed conflicts, also on behalf of governments, to make money or to receive other forms of remuneration without being loyal to the hiring States or taking care of the governmental priorities.[156]

Before the French Revolution and Napoleonic wars, mercenary activities did not have an obviously negative connotation and were not considered to be dishonorable.[157] However, after sovereign States had emerged in the seventeenth century and the Napoleonic wars were over, nationalism became the main driving force behind wars. Armed conflicts were fought by people for national ideals, which did not mean much to mercenaries aiming at earning money by inflicting human suffering.[158] The use of these persons fighting for the "love of war" and not for their homeland started to be seen as morally problematic and as violating the principle of State neutrality and was eventually outlawed by many nations.[159] PMSCs do not always share in the legitimate causes used to justify engagement in armed conflict. However, in contrast to adventurers seeking to earn money in armed struggles, private contractors do not participate in wars for the mere sake of fighting and killing: they are to be found in the conflict areas in order to perform specific tasks and to achieve the goals stated in the contracts concluded between them and their clients.[160] Their activities include a spectrum of not only military, but also security services.

Two decades ago, PMSCs entered the scene and often carry out various activities on behalf of States, with which they did not have any formal or informal connections besides the instrument of a contract. They also did not necessarily share common interests with the citizens of those countries. Apparently, an idea becoming more prevalent in the Middle Ages that killing people in the absence of an appropriate

---

155 Anthony Mockler, (1985) *The New Mercenaries: The History of the Mercenary from the Congo to the Seychelles*, Letchworth: The Garden City Press, p. 17.

156 Robert Mandel, (2002) *Armies Without States: the Privatization of Security*, Boulder: Lynne Rienner Publishers, p. 23.

157 Tony Lynch and A. J. Walsh, (2000) "The Good Mercenary?", *The Journal of Political Philosophy*, Volume 8, Issue 2, p. 133.

158 Benedict Sheehy and Jackson N. Maogoto, (2008) "The Private Military Company – Unravelling the Theoretical, Legal & Regulatory Mosaic", *ILSA Journal of International & Comparative Law*, Volume 15, Issue 1, p. 149; Peter W. Singer, (2008) *Corporate Warriors: The Rise of the Privatized Military Industry*, Ithaca and London: Cornell University Press, p. 216.

159 Sarah Percy, "Morality and Regulation", in Simon Chesterman and Chia Lehnardt, (Eds.), (2007) *From Mercenaries to Market: The Rise and Regulation of Private Military Companies*, Oxford: Oxford University Press, p. 15; Janice E. Thomson, (1994) *Mercenaries, Pirates and Sovereigns: State-Building and Extraterritorial Violence in Early Modern Europe*, Princeton and Chichester: Princeton University Press, pp. 79, 84; Anthony Mockler, (1985) *The New Mercenaries: The History of the Mercenary from the Congo to the Seychelles*, Letchworth: The Garden City Press, p. 17; Sarah Percy, (2007) *Mercenaries: The History of a Norm in International Relations*, Oxford: Oxford University Press, pp. 77, 96-97.

160 Peter W. Singer, (2008) *Corporate Warriors: The Rise of the Privatized Military Industry*, Ithaca and London: Cornell University Press, p. 217.

cause is morally wrong and unjustified is still relevant nowadays.[161] That is also the reason why many PMSCs distance themselves from purely private military enterprises and emphasize their sole expertise in the provision of security to a variety of customers. It is also true that PMSCs are not motivated by legitimate causes of participating in armed conflicts as stated above and often engage in a wide range of operations to earn money. Nevertheless, they can also be hired to promote those legitimate causes and to provide training and to give advice to armed forces of specific States regarding their compliance with rules and principles of international law and protection of State interests justifying participation in a war.[162]

The second main moral objection to the use of PMSCs – which is also applicable to the situations of employing regular mercenaries – is based on the notion of State monopoly on violence and the democratic control over national military and security matters.[163] This moral concern can be derived from the relationships between the State and its citizens. Thomas Hobbes believed that the latter conclude a social contract with the Leviathan or the modern State and transfer their resources for collective security arrangements as the public good.[164] Acting on their own free will, citizens give to the State apparatus their power, strength, and means[165] ensuring the legitimacy and long-term existence of the State.[166] From a more practical point of view, citizens of a certain State are in the position to exercise a specific degree of democratic oversight over the citizens' army and public security arrangements. With regard to armed conflicts and peace operations, public opinion ideally influences a set of decisions of a country to get involved in conflicts with one or more other countries or groups, or to participate in peacekeeping missions. The degree of legitimate control, under which regular armed forces and other private actors operate, varies significantly.[167] State troops are usually allowed to exercise violence under the strict control of the State apparatus.[168] The use of PMSCs reduces the level of democratic control over

---

161 Sarah Percy, "Morality and Regulation", in Simon Chesterman and Chia Lehnardt, (Eds.), (2007) *From Mercenaries to Market: The Rise and Regulation of Private Military Companies*, Oxford: Oxford University Press, p. 17.

162 Peter W. Singer, (2008) *Corporate Warriors: The Rise of the Privatized Military Industry*, Ithaca and London: Cornell University Press, p. 217.

163 Dimitrios Machairas, (2014) "The Ethical Implications of the Use of Private Military Force: Regulatable or Irreconcilable?", *Journal of Military Ethics*, Volume 13, Issue 1, p. 53.

164 Adam White, (2010) *The Politics of Private Security: Regulation, Reform and Re-Legitimation*, Houndmills, Basingstoke, New Hampshire: Palgrave Macmillan, p. 23.

165 Thomas Hobbes, (1651) *Leviathan*, Revised Edition, Edited by A.P. Martinich, and Brian Battiste, 2011, Peterborough [etc.]: Broadview Press, pp.160-161, Chapter XVII, para. 13.

166 Adam White, (2010) *The Politics of Private Security: Regulation, Reform and Re-Legitimation*, Houndmills, Basingstoke, New Hampshire: Palgrave Macmillan, p. 23.

167 Sarah Percy, (2007) *Mercenaries: The History of a Norm in International Relations*, Oxford: Oxford University Press, p. 57.

168 Deborah D. Avant, (2005) *The Market for Force: The Consequences of Privatizing Security*, Cambridge: Cambridge University Press, pp. 45-46.

military and security affairs and allows governments to more easily engage in armed conflicts and similar activities.[169] It is obvious that the general public is usually more familiar with the losses suffered by the members of the regular armed forces rather than private individuals employed by PMSCs and will not be aware of or concerned with the latter dying on the battlefield. In the end, such a policy of employing private firms and individuals might endanger the democracy of a State as a whole. Wealthy corporate managers who run PMSCs can influence democratic decision-making processes disturbing the normal political process and undermining transparency of the use of violence and accountability of involved entities and individuals.[170]

In addition, citizens can be said to have a moral duty to their nations to fight in wars on their behalf with the aim of protecting common interests. In a survey held among the US population, it was demonstrated that respondents seemed to regard US soldiers as motivated by patriotism, while the main motive for private contractors was according to them material gain.[171] It is obvious that third parties, such as PMSCs, do not necessarily share the same level of loyalty to the community as citizens. Therefore, nationals of a State are not protected by mercenary groups or private firms, but can find themselves in a morally weaker position when their most essential rights might be violated and their public life might be dominated by a tyrant acting in breach of the people's will.[172] It is not entirely correct to say that public forces always act in the interests of a given State protecting the public good or that citizens are always loyal to their States.[173] Historical evidence suggests that official State troops can also take over control of a State and refuse to transfer their power to the legitimate government. Even if such instances of misbehavior are not unthinkable, most representatives of State armed forces often do act in accordance with the public will and surrender to it, as their legitimacy is derived from the idea of public

---

[169] Sarah Percy, (2007) *Mercenaries: The History of a Norm in International Relations*, Oxford: Oxford University Press, p. 23; Benedict Sheehy and Jackson N. Maogoto, (2008) "The Private Military Company – Unravelling the Theoretical, Legal & Regulatory Mosaic", *ILSA Journal of International & Comparative Law*, Volume 15, Issue 1, p. 150; Zoe Salzman, (2008) "Private Military Contractors and the Taint of a Mercenary Reputation", *New York University Journal of International Law and Politics*, Volume 40, pp. 866-867.

[170] Benedict Sheehy and Jackson N. Maogoto, (2008) "The Private Military Company – Unravelling the Theoretical, Legal & Regulatory Mosaic", *ILSA Journal of International & Comparative Law*, Volume 15, Issue 1, p. 151.

[171] Deborah D. Avant and Lee Sigelman, (2008) "What Does Private Security in Iraq Mean for US Democracy at Home?", Research Paper, p. 31, <http://bc.sas.upenn.edu/system/files/Avant_08.pdf>, accessed on July 15, 2014.

[172] Sarah Percy, "Morality and Regulation", in Simon Chesterman and Chia Lehnardt, (Eds.), (2007) *From Mercenaries to Market: The Rise and Regulation of Private Military Companies*, Oxford: Oxford University Press, pp. 18-19.

[173] Benedict Sheehy and Jackson N. Maogoto, (2008) "The Private Military Company – Unravelling the Theoretical, Legal & Regulatory Mosaic", *ILSA Journal of International & Comparative Law*, Volume 15, Issue 1, p. 170; Dimitrios Machairas, (2014) "The Ethical Implications of the Use of Private Military Force: Regulatable or Irreconcilable?", *Journal of Military Ethics*, Volume 13, Issue 1, p. 56.

accountability and respect for the rule of law. PMSCs are, however, known for disregarding the public interests of States and going "rogue" in pursuit of financial and other interests.[174] Therefore, the employment of PMSCs as non-State actors decreases the level of democratic control over those entities and may endanger the existence of States themselves.[175]

It follows from the foregoing that there are a number of moral implications in the use of mercenaries and the employment of PMSCs by States. The extensive privatization of military and security matters can be associated with the lack of appropriate causes for the participation of private corporations in armed conflicts and the reduced democratic control over their engagement. This process appears to undermine the State's monopoly on the use of force belonging to sovereign States.[176] In their capacity of private providers of security and force, PMSCs are central actors in an emerging phenomenon of privatization of war and security that cannot easily be reconciled with the idea of the existence of monopoly on the use of force.[177] In addition, this privatization and outsourcing of crucial State tasks to PMSCs threaten a wide range of public law values, such as norms with regard to the law of armed conflict, human rights, and democratic values.[178] Some of these norms and values might be both norms against corruption, abuse of taxpayer money, cronyism and excessive government secrecy, and norms from the military field, such as loyalty to a country, units, fellow soldiers, and devotion to a mission.[179] In general, priorities in armed conflicts and peace operations shift from the public good to the private good.[180] As a result, voices have been raised not only calling for the regulation of the private military and security industry, but also for the abolition of the whole phenomenon and absolute prohibition of the industry.[181] One should, however, understand that the State as a concept introduced in the seventeenth century is a relatively new construct.

---

174 Ibidem; ibidem, p. 59.

175 Laura A. Dickinson, (2006) "Public Law Values in a Privatized World", *The Yale Journal of International Law*, Volume 31, Issue 2, p. 390; Kathleen M. Jennings, (2006) *Armed Security: Regulating the Private Military Industry*, New Security Programme, Fafo-report 532, p. 45, <http://www.fafo.no/pub/rapp/532/532.pdf>, accessed on May 8, 2011.

176 Zoe Salzman, (2008) "Private Military Contractors and the Taint of a Mercenary Reputation", *New York University Journal of International Law and Politics*, Volume 40, pp. 860-866.

177 Benedict Sheehy and Jackson N. Maogoto, (2008) "The Private Military Company – Unravelling the Theoretical, Legal & Regulatory Mosaic", *ILSA Journal of International & Comparative Law*, Volume 15, Issue 1, pp. 157-158.

178 Laura A. Dickinson, (2006) "Public Law Values in a Privatized World", *The Yale Journal of International Law*, Volume 31, Issue 2, p. 400.

179 Kathleen M. Jennings, (2006) *Armed Security: Regulating the Private Military Industry*, New Security Programme, Fafo-report 532, p. 33, <http://www.fafo.no/pub/rapp/532/532.pdf>, accessed on May 8, 2011.

180 Zoe Salzman, (2008) "Private Military Contractors and the Taint of a Mercenary Reputation", *New York University Journal of International Law and Politics*, Volume 40, p. 872.

181 Peter W. Singer, (2004) "War, Profits and the Vacuum of Law: Privatized Military Firms and International Law", *Columbia Journal of International Law*, Volume 42, Issue 2, p. 544.

As will be seen later, for centuries mercenaries have been used by States and other entities and persons. They were able to effectively protect interests of their employers and to decide many wars to their advantage. This and other reasons still play an important role in the decision of governments to consider the use of private actors capable of exercising violence on their behalf as a substitute for or a complement to the regular armed forces.

It has been argued that, with the spread of Enlightenment ideas, the world turned in the direction of significant reduction of mercenary activities and acknowledgment of their immoral nature in favor of the use of permanent State armies. Soldiers serving in those official State armies were regarded as true protectors of their homeland, fighting for the public good of their country. The employment of these troops was placed under the high degree of democratic control with the consequence that possible perpetrators of war crimes and other illegalities could easily be held accountable for their misconduct.[182] As was argued by Hugo Grotius, soldiers and other persons under the command of State authorities committing serious crimes in a war without a just cause are obliged to make satisfaction for the incurred damages.[183] If there is just cause, those individuals may be held accountable and be answerable to their own State. In addition, States could be held responsible for the activities of their military personnel if certain violations of international or national law occurred. This approach is probably less strict when it comes to similar illegal activities of mercenary groups and PMSCs having no official connection to States employing them.

One might maintain that the monopoly on the legitimate use of force is the main prerogative of the nation State, which must be able to exercise control over violence.[184] This is often believed to be the only way of guiding a particular society in the right direction of prosperity and avoiding anarchy.[185] From the point of view of transaction cost economics, State bureaucracies are to be preferred over private corporations under certain circumstances.[186] The practice of States contracting with private firms to perform military-related services results in the costliness of identifying possible contingencies and specifying required responses in a contract or even makes

---

182 Kevin A. O'Brien, "What Should and What Should Not Be Regulated", in Simon Chesterman and Chia Lehnardt, (Eds.), (2007) *From Mercenaries to Market: The Rise and Regulation of Private Military Companies*, Oxford: Oxford University Press, p. 32.

183 Hugo Grotius, (1625) *The Rights of War and Peace*, Book III, Edited and with an Introduction by Richard Tuck, in Jean Barbeyrac (Ed.), (2005) *Major Legal and Political Works of Hugo Grotius*, Indianapolis: Liberty Fund, p. 1532.

184 Damian Lilly, (2000) *The Privatization of Security and Peacebuilding: A Framework for Action*, International Alert, p. 22.

185 Deborah D. Avant, (2005) *The Market for Force: The Consequences of Privatizing Security*, Cambridge: Cambridge University Press, p. 46.

186 Oliver E. Williamson, (1999) "Public and Private Bureaucracies: A Transaction Cost Economics Perspective", *Journal of Law, Economics and Organization*, Volume 15, Issue 1.

such identification and specification impossible.[187] From that perspective, contracts can also better be avoided when the market for private military and security service provision is saturated with a large number of competing businesses striving to make a profit in various conflict areas. When PMSCs engage in armed conflicts and peace operations, their activities are often not restricted to the mere logistical support and other related non-military tasks, but cover provision of military training and advice, operational support in combat zones, and security-related missions. In these scenarios, difficulties may arise with respect to the loyalty of those private entities and their personnel, the lack of attachment to the causes of participation in hostilities, and the degree of democratic control exercised over them. As purported by Deborah Avant, the privatization of such tasks decreases functional control leading to rising costs and inappropriate adaptations. Furthermore, it changes political control connected with the allocation of power causing imbalance and opportunism.[188] In considering the question of PMSCs exercising core State functions, Oliver Williamson even believes that "replication of a public bureau by a private firm, with or without the support of regulation, is impossible".[189]

Nowadays, governments rely on the expertise of PMSCs having the main focus on their own business interests. Often perceived as modern-day mercenaries,[190] PMSCs represent a type of business entity appearing on the international scene as a result of substantial developments in the modern warfare that came into being because of the dissolution of the world order and the end of the Cold War. Driven by eagerness to gain more profit, PMSCs can go beyond the limits of the contracts concluded with their customers and get involved in various illegal practices on the territory of a State in which they are stationed.[191] They might turn their back on legitimate governments and cooperate with other States or criminal organizations and individuals creating a situation of instability and insecurity.[192] It is also logical to maintain that private actors performing their activities in conflict-affected zones easily escape the public eye and

---

187 Deborah D. Avant, (2005) *The Market for Force: The Consequences of Privatizing Security*, Cambridge: Cambridge University Press, p. 46.

188 Ibidem, p. 49.

189 Oliver E. Williamson, (1999) "Public and Private Bureaucracies: A Transaction Cost Economics Perspective", *Journal of Law, Economics and Organization*, Volume 15, Issue 1, p. 311.

190 Sarah Percy, "Morality and Regulation", in Simon Chesterman and Chia Lehnardt, (Eds.), (2007) *From Mercenaries to Market: The Rise and Regulation of Private Military Companies*, Oxford: Oxford University Press, p. 11.

191 Peter W. Singer, (2008) *Corporate Warriors: The Rise of the Privatized Military Industry*, Ithaca and London: Cornell University Press, p. 219; Kathleen M. Jennings, (2006) *Armed Security: Regulating the Private Military Industry*, New Security Programme, Fafo-report 532, p. 39, <http://www.fafo.no/pub/rapp/532/532.pdf>, accessed on May 8, 2011.

192 Kathleen M. Jennings, (2006) *Armed Security: Regulating the Private Military Industry*, New Security Programme, Fafo-report 532, p. 44, <http://www.fafo.no/pub/rapp/532/532.pdf>, accessed on May 8, 2011.

may go unpunished if serious crimes are committed.[193] The claim can be made that such a broad reliance on private expertise can be explained by the willingness of legitimate governments to escape democratic control and procedural requirements for using force[194] and possibly to avoid being held responsible for the breaches of international law that might be committed by private warriors and security guards.[195] So far, States have managed to escape accountability for the misconduct of PMSCs and their personnel, which is a prominent threat to international law and, more specifically, international humanitarian and human rights law.[196] While States are responsible for acts and omissions of their armed forces breaching international law, the same cannot be said for situations when they employ PMSCs engaging in the similar type of misconduct. These circumstances create much "normative uncertainty"[197] originating from the fact that PMSCs do not share just causes of wars and are not placed under the same degree of democratic control as official State troops. In this respect, a significant question of who should be held responsible for possible violations of international law committed by the personnel of private contractors arises.

It stands to reason that not only private military and security experts working for PMSCs or companies themselves are to be held accountable: States taking part in armed conflicts and peace operations, using services provided by those firms in conflict areas and having close ties with them should also bear a certain responsibility. From an ethical perspective, one might acknowledge the fact that States employing PMSCs cannot distance themselves from the illegal activities of those businesses and their personnel that, for instance, carry out inherently governmental functions or operate under a certain degree of State control. In general, States should not be allowed to escape their obligations under international law, more specifically international humanitarian and human rights law, by outsourcing some of their core capacities to private corporations. Certain links between the State apparatus and PMSCs

---

193 Ibidem, p. 34.

194 Mariana M. Prado, "Regulatory Choices in the Privatization of Infrastructure", in Simon Chesterman and Angelina Fisher (Eds.), (2009) *Private Security, Public Order: The Outsourcing of Public Services and Its Limits*, Oxford: Oxford University Press, p. 109-110; Kathleen M. Jennings, (2006) *Armed Security: Regulating the Private Military Industry*, New Security Programme, Fafo-report 532, p. 41, <http://www.fafo.no/pub/rapp/532/532.pdf>, accessed on May 8, 2011.

195 Carlos Ortiz, "The Private Military Company: An Entity at the Center of Overlapping of Overlapping Spheres of Commercial Activity and Responsibility", in Thomas Jäger and Gerhard Kümmel (Eds.), (2007) *Private Military and Security Companies: Chances, Problems, Pitfalls and Prospects*, Wiesbaden: VS Verlag für Sozialwissenschaften, p. 61; Nigel D. White, "Institutional Responsibility for Private Military and Security Companies", in Francesco Francioni and Natalino Ronzitti (Eds.), (2011) *War by Contract: Human Rights, Humanitarian Law, and Private Contractors*, Oxford: Oxford University Press, p. 383.

196 José L. Gómez del Prado, (2009) "Private Military and Security Companies and the UN Working Group on the Use of Mercenaries", *Journal of Conflict & Security Law*, Volume 13, Issue 3, p. 439.

197 Peter W. Singer, (2008) *Corporate Warriors: The Rise of the Privatized Military Industry*, Ithaca and London: Cornell University Press, p. 221.

involved in the violations of international law and other relevant circumstances might indicate both the possibility and necessity of holding States responsible for the misconduct of those private corporations and their employees. In the present study, it will be argued that States that should face international responsibility are not only those employing PMSCs, but also those in other ways directly and indirectly involved in their functioning on the ground.

5. Structure of the Study

This study is divided into eight Chapters including the introductory one and will unfold as follows. *Chapter II* will essentially focus on the phenomenon of privatization of war and security posing many challenges to the current international legal order and constituting both an issue of theoretical and practical relevance. It aims at discussing historical developments in the field of private military and security service provision and addressing PMSCs as non-State business actors. This Chapter will set up a general framework within which the analysis will be performed for the purposes of understanding the industry of PMSCs with its characteristics and specific features, which is required for finding an answer to the main research question. Among other things, it is intended to cover such important matters as the definition of PMSCs, their legal status under international and national law, and the range of clients they work for.

Chapter III will set up a legal framework for the further analysis of the issue of State responsibility by introducing a variety of different forms of regulation applicable to PMSCs active in conflict zones. In the first instance, this part of the study will have one of its focal points on international legal order. The discussion will center on general rules and principles of public international law, international human rights law, international humanitarian law, and other forms of international legal regulation that apply to PMSC operations and need to be taken into account in the examination of international responsibility. Some attention will also be devoted to the domestic legal order and the legal frameworks of some States will be elaborated upon, given that they clarify the set of obligations of PMSCs and the nature of their activities. The next subject will be self-regulation and transnational private regulatory frameworks in the context of the PMSC industry influencing the State use of private contractors in conflicts areas and the functioning of PMSCs. In addition, a few words will be said about the hiring contracts concluded between States and PMSCs, standard operating procedures, rules of engagement, and status of forces agreements, which form part of the legal framework of PMSC operations and account for some of their particularities. Finally and more importantly, this Chapter will identify regulatory gaps perceivable at different levels of regulation.

Chapter IV will then consider the legal status of PMSC employees in armed conflicts that might have consequences for international responsibility of States associated with the deployment of these companies in situations of armed conflict. It will provide a definition of the members of PMSC personnel and clarify their actual position in both international and non-international armed conflicts. The different types of status that PMSC employees can have will be considered and their rights and duties will be identified.

Chapter V of this book will constitute the first part of the core of the study specifically examining the doctrine of State responsibility as laid down in the Draft Articles on Responsibility of States for Internationally Wrongful Acts (hereinafter: the Draft Articles or DARS). In general terms, it will explore the history, framework, and legal status of the Draft Articles and, *inter alia*, specify conditions for engaging State responsibility, its implementation, and legal consequences. More significantly, this particular Chapter will elaborate on the possibility of attributing misconduct of PMSCs and their personnel to States associated with their employment by discussing different attribution grounds and their relevance for the State use of private contractors. Finally, it will establish whether different types of misconduct, in which PMSCs have been or can be engaged, fall under the qualification of breaches of international obligations of States.

In *Chapter VI*, the second element of the main focal point of this study will be found. This part will concern the discussion centered on a range of positive State obligations under international humanitarian and human rights law, the breach of which can lead to their international responsibility. It will cover the notion of indirect responsibility and set out a spectrum of positive duties of result and diligent conduct that can be distilled from not only human rights law, but also the law of armed conflict. These obligations will be assessed on the basis of their application to three types of States: hiring, host, and home countries of PMSCs.

Finally, *Chapter VII* will introduce the summary of the study, overall conclusions, and final observations. The summaries of the book will also be provided in two additional languages: Dutch and Russian.

CHAPTER II
PRIVATIZATION OF WAR AND SECURITY – WAR IS BUSINESS AND VICE VERSA

Force should be right; or, rather, right and wrong –
Between whose endless jar justice resides –
Should lose their names, and so should justice too.
Then everything includes itself in power,
Power into will, will into appetite;
And appetite, an universal wolf,
So doubly seconded with will and power,
Must make perforce an universal prey,
And last eat up himself.

William Shakespeare[1]

War ... always has been a matter of profit and spoils.
What is changing is how the profits are distributed.

Kathleen M. Jennings[2]

1. INTRODUCTION

Without doubt, in the past two decades, private military and security companies hired by a large number of entities have become important actors in both the international arena and domestic affairs of a large number of States. Many scholars agree that there are signs of a "revolution in military affairs":[3] nowadays, PMSCs freely operate around the world and take an active part in not merely peace support operations, but also international and non-international armed conflicts. They are frequently

1 William Shakespeare, (1602) *The History of Troilus and Cressida*, Stephen Orgel and A.R. Braunmuller (Eds.), (2002) *William Shakespeare: The Complete Works*, New York: Penguin Group, Act I, Scene 3, p. 489.

2 Kathleen M. Jennings, (2006) *Armed Security: Regulating the Private Military Industry*, New Security Programme, Fafo-report 532, p. 7, <http://www.fafo.no/pub/rapp/532/532.pdf>, accessed on May 8, 2011.

3 Michael N. Schmitt, (2005) "Humanitarian Law and Direct Participation in Hostilities by Private Contractors or Civilian Employees", *Chicago Journal of International Law*, Volume 5, Issue 2, p. 511.

employed by States in the situations of armed conflict, by non-State entities participating in hostilities in internal conflicts, by international organizations in the course of conducting peacekeeping missions, and even by private companies paying large amounts of money for their military and security expertise. It is no secret that the concept of PMSCs as private corporations exercising core State functions has been around for longer than 20 years. Nowadays, the industry of private military and security specialists still shows signs of significant development and unprecedented growth.[4] Currently, the world has to deal with many challenges posed by the emergence and proliferation of these companies. There are regulatory issues associated with the PMSC industry, transparency and legitimacy concerns raised against its representatives, problems of individual accountability and State responsibility for the actions of private warriors, who carry weapons, fight wars, and protect persons and assets of legitimate governments and other actors. Before going into the substance of the discussion focusing on the responsibility of States for the illegal conduct of those corporations, some attention should be devoted to the historical developments in the field of the use of force by private persons and entities. Those historical aspects can be helpful in comparing PMSCs to other private actors exercising violence and building a foundation of understanding of their roots, connections to governments increasingly relying on their services, legal status, and the possibility of attributing their crimes to respective States that often hide behind private corporations and the general concept of privatization. Additionally, it is crucial to consider PMSCs themselves as non-State business entities operating on behalf of States for the coherent understanding of the companies under examination for the purpose of the legal analysis included in the subsequent Chapters.

This Chapter will basically deal with historical origins of PMSCs and offer valuable insights into these modern providers of military and security services. From ancient times to the present day, it will address the main problems arising from the use of mercenaries and PMSCs by States. In addition, it will explain reasons for the emergence of private corporations performing military and security companies and compare the phenomenon of privatizing war and security matters with the concept of mercenaries and mercenary organizations employed by States and non-State actors for many centuries. It will then examine the nature of PMSCs, addressing the legal status of those private businesses under both national and international law

---

[4] Peter W. Singer, (2008) *Corporate Warriors: The Rise of the Privatized Military Industry*, Ithaca and London: Cornell University Press, p. 40; Rebecca R. Vernon, (2003-2004) "Battlefield Contractors: Facing the Tough Issues", *Public Contract Law Journal*, Volume 33, Issue 2, p. 371; Simon Chesterman and Chia Lehnardt, "Introduction", in Simon Chesterman and Chia Lehnardt (Eds.), (2007) *From Mercenaries to Market: The Rise and Regulation of Private Military Companies*, Oxford: Oxford University Press; Adam Ebrahim, (2010) "Going to War with the Army You Can Afford: The United States, International Law and the Private Military Industry", *Boston University International Law Journal*, Volume 28, pp. 184-185.

and discussing the validity of the term "mercenaries" often used in literature and academic debates to refer to them. In addition, it will elaborate on the legal nature of the military and security firms as for-profit business entities, their relations with numerous clients, the broad scale of services performed, their main features, and the industry as a whole. Consequently, an attempt will be made to offer a comprehensive definition of PMSCs, which can cover the most significant aspects of those private contractors. Finally, a number of conclusions will be drawn.

2. HISTORICAL ORIGINS OF PMSCS

As claimed by some experts, the process of hiring outside warriors to participate in armed conflicts is nothing unusual and is "as old as war itself".[5] It is appropriate, therefore, to commence the historical analysis of the phenomenon of "mercenary" – understood to mean an individual hired to exercise violence on behalf of its client(s) in return for financial gain – and "private military and security company" with a journey into the roots of human society and a brief description of the first division of labor. This significant milestone in the history of humankind likely appears to be at the core of the emergence of individuals specialized in warfare.[6] While the first division of labor probably started with the specialization of not only men and women, but also adults and children in certain tasks, the invention and further development of agriculture triggered and accelerated the division of activity and tasks between other groups of people. Certain specialized functions could be carried out more efficiently and more effectively by those with more skills and experience in the field, who became indispensable for human society. Even Plato argued in *The Republic* that the division of labor among different classes of individuals is necessary to fulfill the needs of a certain society transforming itself into a State.[7] According to Adam Smith, the main effects of the division of labor seem to be "the greatest improvements in the productive powers of labor, and the greater part of the skill, dexterity, and

---

5 Sarah Percy, (2007) *Mercenaries: The History of a Norm in International Relations*, Oxford: Oxford University Press, p. 1; Peter W. Singer, (2008) *Corporate Warriors: The Rise of the Privatized Military Industry*, Ithaca and London: Cornell University Press, p. 19; Ken Silverstein, (2000) *Private Warriors*, New York and London: Verso, p. 145; Fred Schreier and Marina Caparini, (2005) *Privatising Security: Law, Practice and Governance of Private Military and Security Companies*, Occasional Paper No. 6, DCAF, p. 1; Laura A. Dickinson, (2006) "Public Law Values in a Privatized World", *The Yale Journal of International Law*, Volume 31, Issue 2, p. 394.

6 Peter W. Singer, (2008) *Corporate Warriors: The Rise of the Privatized Military Industry*, Ithaca and London: Cornell University Press, p. 20; Frederic C. Lane, (1979) *Profits from Power: Readings in Protection Rent and Violence-Controlling Enterprises*, Albany: SUNY Press.

7 Plato, *The Republic*, as found in (1973) *The Republic of Plato*, Translated with Introduction and Notes by Francis MacDonald Cornford, London, Oxford and New York: Oxford University Press, pp. 53-63, Chapters VI and VII.

judgment, with which it is anywhere directed, or applied".[8] While the whole process undoubtedly had its negative effects on the individual as forming part of the society, it also presented new possibilities to humanity striving to achieve greater production of goods and led to the rise of new professions and experts.

As war appears to be a constant in human history – there have been only roughly 230 years of peace in the last 3,500 years – and armed conflicts form and have always formed "an integral part of the human condition",[9] people specialized in carrying out military activities could sell them to a large variety of clients and make living out of so doing. Therefore, individuals fighting in wars for remuneration – mercenaries – have always formed an integral part of the social and political history of warfare.[10] Their profession appears to be often regarded as the second oldest profession in the world.[11] An important realization in this context must be that the non-State provision of violence was a prevailing standard in international relations before the twentieth century. In fact, States appearing in the international arena four centuries ago relied on the system of using private force provided by mercenaries in order to construct their public apparatus. Those private individuals capable of effectively exercising violence against their opponents could not only sell their expertise to local rulers and other wealthy persons, but also provide their services to foreign customers.[12] This was and still remains one of the remarkable characteristics of the most old-fashioned soldiers of fortune and private contractors, who can be said to be operating in the same area of private military and security service provision.

The main reason for soldiers hired for pay to organize and to form groups can be found in the advantages of scale that is required for the effective use of violence.[13] Most rulers and princes incapable of controlling their territories, creating administrative structures, and setting up regular armed forces were dependent on the services

---

[8] Adam Smith, (1776) *An Inquiry into the Nature and Causes of the Wealth of Nations*, the Edition Published in 2007, Petersfield: Harriman House, Book 1, Chapter 1.

[9] David Turns, "The Law of Armed Conflict (International Humanitarian Law)", in Malcolm D. Evans (Ed.), (2010) *International Law*, Third Edition, Oxford: Oxford University Press, p. 814.

[10] Sarah Percy, (2007) *Mercenaries: The History of a Norm in International Relations*, Oxford: Oxford University Press, p. 1; James L. Taulbee, (1998) "Reflections on the Mercenary Option", *Small Wars and Insurgencies*, Volume 9, Issue 2, pp. 145-163; Peter W. Singer, (2008) *Corporate Warriors: The Rise of the Privatized Military Industry*, Ithaca and London: Cornell University Press, p. 20; Deborah D. Avant, (2004) "The Privatization of Security and Change in the Control of Force", *International Studies Perspectives*, Volume 5, p. 153; Uwe Steinhoff, "What are Mercenaries?", in Andrew Alexandra, Deane-Peter Baker, and Marina Caparini (Eds.), (2008) *Private Military and Security Companies: Ethics, Policies and Civil-Military Relations*, London: Routledge, p. 28; Robert Mandel, (2002) *Armies Without States: the Privatization of Security*, Boulder: Lynne Rienner Publishers, p. 9; Damian Lilly, (2000) *The Privatization of Security and Peacebuilding: A Framework for Action*, International Alert, p. 8.

[11] Peter Tickler, (1987) *The Modern Mercenary: Dog of War, or Soldier of Honour?*, Wellingborough: Stephens; Ken Silverstein, (2000) *Private Warriors*, New York and London: Verso, p. 145.

[12] Peter W. Singer, (2008) *Corporate Warriors: The Rise of the Privatized Military Industry*, Ithaca and London: Cornell University Press, p. 20.

[13] Ibidem.

provided by trained soldiers, which were hired into ancient armies or functioned as ad-hoc units. Nowadays, experiencing a mixture of horror and admiration, military and security experts with extensive skills and experience follow similar trends. They set up, register, and join legally established firms capable of contracting with States and non-State entities around the world and employ their activities intrinsically connected with the use of force.

2.1 Ancient Times

Peter W. Singer claims that the practice of employing private warriors was significantly widespread among ancient rulers in the Ancient Times (ca. 3000-500 B.C.) and it can hardly be argued that he misinterprets historical facts. According to this expert, the first official use of mercenaries has been documented at the time of Mesopotamian King Shulgi of Ur who ruled during the Third Dynasty of Ur on the territory of modern southern Iraq (ca. 2094-2047 B.C.).[14] A thorough reading of the Bible reveals stories about the hired troops used by the Pharaoh to remove the Israelites from Egypt. In the battle of Kadesh that took place on the territory of the modern Syrian Arab Republic in 1274 B.C., the army of Egypt led by Pharaoh Ramses II containing Numidian units fought the forces of the Hittite Empire under Muwatalli II.[15]

The ancient Greek city-States were well-known for their extensive use of outside soldiers:[16] troops of Cretan slingers, Syracusan hoplites, and Thessalian cavalry are the most colorful examples of such units.[17] In the Persian wars, Athens used to rely on a large number of private soldiers forming naval units.[18] Xenophon, the renowned Greek writer and soldier, describes in his book *Anabasis* – Greek for "going up" – his military adventures with the expedition of the "Ten Thousand" during the Persian civil war (401-400 B.C.).[19] "Ten Thousand" was a mercenary unit of Greek soldiers

---

14 Peter W. Singer, (2008) *Corporate Warriors: The Rise of the Privatized Military Industry*, Ithaca and London: Cornell University Press, p. 20.

15 James L. Taulbee, (1998) "Reflections on the Mercenary Option", *Small Wars and Insurgencies*, Volume 9, Issue 2, p. 145.

16 Uwe Steinhoff, "What are Mercenaries?", in Andrew Alexandra, Deane-Peter Baker, and Marina Caparini (Eds.), (2008) *Private Military and Security Companies: Ethics, Policies and Civil-Military Relations*, London: Routledge, p. 20.

17 Guy T. Griffith, (1968) *The Mercenaries of the Hellenistic World*, Groningen: Boom's Boekhuis B.V., p. 4; Peter W. Singer, (2008) *Corporate Warriors: The Rise of the Privatized Military Industry*, Ithaca and London: Cornell University Press, p. 21.

18 Peter W. Singer, (2008) *Corporate Warriors: The Rise of the Privatized Military Industry*, Ithaca and London: Cornell University Press, p. 21.

19 Xenophon, *Anabasis*, translated by Henry G. Dakyns, in Henry G. Dakyns, (1897) *The Works of Xenophon*, London, New York: Macmillan, <http://www.gutenberg.org/files/1170/1170-h/1170-h.htm>, accessed on August 2, 2012; Anthony Mockler, (1985) *The New Mercenaries: The History of the Mercenary from the Congo to the Seychelles*, Letchworth: The Garden City Press, p. 9.

hired by Cyrus the Younger who revolted against his brother King Artaxerxes II Mnemon of the Persian Empire in order to conquer the throne.[20]

Alexander the Great (356-323 B.C.) used to primarily hire mercenary troops into his armies.[21] In general, Macedonian soldiers fought on behalf of Greek city-States in the Peloponnesian War (431-404 B.C.) and then against them in the wars of King Philip II of Macedon. Hellenic empires were highly dependent on the recruitment of foreign soldiers to fulfill essential security needs.[22] Rulers of the Carthaginian Empire had a long history of employing mercenaries as well.[23] After the First Punic War (264-241 B.C.), provoked by mercenary activities,[24] the so-called Mercenary War (ca. 240 B.C.) broke out when those soldiers turned against their clients in the uprising because of the latter's refusal to pay them. In the Second Punic War (218-202 B.C.), the army of Hannibal, consisting of foreign troops, destroyed the Roman army.[25] The latter maintained a citizen army, but could not survive without the use of mercenaries. It frequently hired experts of certain specialties, such as archers and cavalry, from the poorer areas of the ancient world. This practice went so far that, at the end of the third century A.D., the whole army consisted of more Germanic rather than Roman soldiers.[26] The same strategy of reliance on mercenary skills also played a prominent role in the Byzantine Empire and other empires around the globe.[27]

2.2 The Middle Ages

The period of the Middle Ages (ca. 500-1500) was largely characterized by feudalism. It constituted a system of political organization with a ruler lending lands to individuals in exchange for paying rent in form of, originally, military service, but also other services. At the time, it was the main instrument for creating armies[28] that

---

20 Anthony Mockler, (1985) *The New Mercenaries: The History of the Mercenary from the Congo to the Seychelles*, Letchworth: The Garden City Press, p. 9; Peter W. Singer, (2008) *Corporate Warriors: The Rise of the Privatized Military Industry*, Ithaca and London: Cornell University Press, p. 21.

21 Ken Silverstein, (2000) *Private Warriors*, New York and London: Verso, p. 145.

22 Peter W. Singer, (2008) *Corporate Warriors: The Rise of the Privatized Military Industry*, Ithaca and London: Cornell University Press, p. 21.

23 Ken Silverstein, (2000) *Private Warriors*, New York and London: Verso, p. 145.

24 Kim R. Nossal, (1998) "Roland Goes Corporate: Mercenaries and Transnational Security Corporations in the Post-Cold War Era", *Civil Wars*, Volume 1, Issue 1, p. 19.

25 Peter W. Singer, (2008) *Corporate Warriors: The Rise of the Privatized Military Industry*, Ithaca and London: Cornell University Press, p. 21.

26 Hans Delbrück, (1975) *History of the Art of War: Within the Framework of Political History*, Westport: Greenwood Press, Volume II: The Germans, Walter J. Renfroe Jr. tr., Westport and London: Greenwood Press, p. 250; Peter W. Singer, (2008) *Corporate Warriors: The Rise of the Privatized Military Industry*, Ithaca and London: Cornell University Press, p. 21.

27 Peter W. Singer, (2008) *Corporate Warriors: The Rise of the Privatized Military Industry*, Ithaca and London: Cornell University Press, pp. 21-22.

28 Ibidem, p. 22.

undoubtedly presented restrictions on the process of delivering military services and conducting offensive wars on a massive scale.[29] The whole military system of feudalism was inherently intertwined with the main idea of defense. Knights were obliged to serve locally for a certain amount of time: they had no obligation to perform their military and other services abroad.[30] These are some of the most important reasons why hired troops found their way into the feudal system of mutual relationships between lords and vassals and formed a crucial element of practically every army.[31] As one class of warriors prevailing in the Middle Ages – the other two being the people's levy and vassalage – they came out the strongest and the most vital.[32] The use of mercenaries increased over the years and made these hired soldiers the most dominant type of provider of violence. They could perform more specific technical tasks that regular feudal troops were unable to deliver.[33] Working for the highest bidder, hired warriors specialized in some weapons or techniques requiring much knowledge, skill, and practice to master, which could not easily be acquired by non-experts. Therefore, these soldiers actively engaged in wars as their essential "profit making enterprise"[34] and were undoubtedly valued by their customers.

With the emergence of the first trading companies in Italy in the thirteenth century, the system of contract was introduced. This century was characterized by the perceivable growth of banking and private investments. The system of contract influenced the provision of military services that were in many instances performed by private units on the basis of a contract or *condotta*.[35] Mercenaries were usually hired by large business guilds seeking to protect their interests without mobilizing all society, which could not easily be realized. In addition, the aristocracy also seemed to favor the use of private warriors rather than soldiers from the masses. Many Italian cities contracted with mercenary units: for instance, Venice had an established practice of

---

29 Janice E. Thomson, (1994) *Mercenaries, Pirates and Sovereigns: State-Building and Extraterritorial Violence in Early Modern Europe*, Princeton and Chichester: Princeton University Press, p. 27.

30 Ibidem.

31 Shantanu Chakrabarti, (2009) *Privatisation of Security in the Post Cold-War Period: An Overview of its Nature and Implications*, Institute for Defence Studies and Analyses Monograph Series, No. 2, New Delhi: Magnum Custom Publishing, p. 3.

32 Hans Delbrück, (1975) *History of the Art of War: Within the Framework of Political History*, Volume III: The Middle Ages, Walter J. Renfroe Jr. tr., Westport and London: Greenwood Press, p. 505.

33 Peter W. Singer, (2008) *Corporate Warriors: The Rise of the Privatized Military Industry*, Ithaca and London: Cornell University Press, p. 22.

34 Christopher Coker, (1999) "Outsourcing War", *Cambridge Review of International Affairs*, Volume 13, Issue 1, p. 96.

35 Jurgen Brauer, (1999) "An Economic Perspective on Mercenaries, Military Companies, and the Privatisation of Force", *Cambridge Review of International Affairs*, Volume 13, Issue 1, p. 132; Anthony Mockler, (1985) *The New Mercenaries: The History of the Mercenary from the Congo to the Seychelles*, Letchworth: The Garden City Press, pp. 7-8; Sarah Percy, (2007) *Mercenaries: The History of a Norm in International Relations*, Oxford: Oxford University Press, pp. 74-75; Kim R. Nossal, (1998) "Roland Goes Corporate: Mercenaries and Transnational Security Corporations in the Post-Cold War Era", *Civil Wars*, Volume 1, Issue 1, p. 19.

employing mercenaries to fight during the Crusades (1095-1270). Other European regions did not lag far behind and likewise actively took advantage of the variety of military services provided by private military units.[36] In this period, the qualitative characteristics of the troops – their skills and experience – started to matter more than the quantities of soldiers sent to battle. Several kings in Europe began charging costs of hiring mercenaries – *scutagium* – and employed those individuals to fight on their behalf.[37] The end of the fourteenth century witnessed the domination of mercenary units over their feudal counterparts in warfare matters. More hired soldiers became available and the market started to grow considerably. As a consequence, wars influencing essential aspects of human life could be waged more often and more effectively.[38]

In the following centuries, the Hundred Years War (1337-1453) introduced new challenges to the political order of Europe. Some authors claim that, in the first place, the emerging mercenary system gained a clearly visible influence and domination in England. Notably, English mercenaries operated mostly in France, where the English kings were involved in the Hundred Years War with their rivals. This development urged those rivals from France to expand their mercenary forces as well. German hired soldiers also fought in wars, but only in those taking place outside of Germany: in the Anglo-French wars and in Italy.[39] Most soldiers involved in wars in Italy already belonged to the class of mercenaries.[40] In sum, this period is typified by growing instability and saw a further proliferation of mercenary units, which started to evaluate and to change their structure and managed to win great military victories in the battles of the Hundred Years War.

Since there was no centralized control, mercenaries spread further across Europe. Rather than taking part in wars as individual soldiers and remaining on their own in times of peace, private warriors became aware that it would be beneficial for them to form "companies", which is derived from *cum* and *panis* and means "bread comrades".[41] These organizations were created to ensure the employment of mercenaries as a group and to provide protection and other forms of support for their members.[42]

---

36 Peter W. Singer, (2008) *Corporate Warriors: The Rise of the Privatized Military Industry*, Ithaca and London: Cornell University Press, pp. 22-23.

37 Ibidem, p. 23.

38 Ibidem.

39 Hans Delbrück, (1975) *History of the Art of War: Within the Framework of Political History*, Volume III: The Middle Ages, Walter J. Renfroe Jr. tr., Westport and London: Greenwood Press, p. 505.

40 Ibidem, p. 506.

41 Peter W. Singer, (2008) *Corporate Warriors: The Rise of the Privatized Military Industry*, Ithaca and London: Cornell University Press, p. 24; Hans Delbrück, (1975) *History of the Art of War: Within the Framework of Political History*, Volume III: The Middle Ages, Walter J. Renfroe Jr. tr., Westport and London: Greenwood Press, p. 511.

42 Peter W. Singer, (2008) *Corporate Warriors: The Rise of the Privatized Military Industry*, Ithaca and London: Cornell University Press, p. 24.

While many of these companies did not serve a king, but came into existence in accordance with the medieval right of feudal lords,[43] their further proliferation was one of the reasons for the decline of the feudal system. In addition to safeguarding the interests of the highest bidder, hired soldiers also appeared to frequently participate in all possible forms of misconduct. There have been reports of mercenary companies that, on the one hand, protected their clients, but, on the other, blackmailed entire cities and villages, forcing the inhabitants of various areas to pay for their campaigns, plundering, and torturing them.[44] They engaged in the intimidation of local authorities for money, spread chaos and destruction, or moved on, if blackmail practices were successful and they received their pay. As independent entities, they kept growing and living off the areas where they settled. One such company was the famous Great Company or "la Gran Compagna" with Duke Werner von Urslinger as commander that terrorized many Italian regions and seemed to run a widespread "protection racket".[45]

As time went by, companies initially emerging as temporary enterprises started to evolve: their nature slowly shifted to permanent organizations providing military services for remuneration offered by those willing to pay.[46] Also, the contracts signed between companies and their clients were set up by persons with knowledge of these matters and became more sophisticated and very detailed. In this period, the free companies led by determined leaders and consisting of loyal fighters – loyal to their organization and its chief, but not to a country or a client – came to realize the importance of promotion of their activities. They set up "marketing strategies"[47] in order to create an image of furious warriors capable of dominating their enemy by spreading tales about numerous adventures aimed at attracting potential customers and deterring possible opponents on the battlefield.

The proliferation of free companies could have gone on for a very long time, but it started to significantly reduce in scope in France, the region where many of these mercenary organizations were founded or managed to find several employment possibilities. France was confronted with a myriad of problems that originated from mercenary companies misbehaving during the times of peace and causing troubles

---

43 Hans Delbrück, (1975) *History of the Art of War: Within the Framework of Political History*, Volume III: The Middle Ages, Walter J. Renfroe Jr. tr., Westport and London: Greenwood Press, p. 507.

44 Peter W. Singer, (2008) *Corporate Warriors: The Rise of the Privatized Military Industry*, Ithaca and London: Cornell University Press, pp. 24-25; Hans Delbrück, (1975) *History of the Art of War: Within the Framework of Political History*, Volume III: The Middle Ages, Walter J. Renfroe Jr. tr., Westport and London: Greenwood Press, pp. 506-507.

45 Ibidem, p. 25; ibidem, pp. 506-507; Anthony Mockler, (1985) *The New Mercenaries: The History of the Mercenary from the Congo to the Seychelles*, Letchworth: The Garden City Press, p. 11.

46 Peter W. Singer, (2008) *Corporate Warriors: The Rise of the Privatized Military Industry*, Ithaca and London: Cornell University Press, p. 24.

47 Ibidem.

to local merchants.[48] Due to the efforts of King Charles VII in the fifteenth century, funds were raised among the bourgeois class to employ some free companies on a regular basis and to destroy other competing companies in France. Subsequently, those mercenary groups hired by the king were then kept in service. In fact, they formed the first standing army in Europe. The main rival of the French king, Duke Charles the Bold of Burgundy, scared by such a massive regular army decided to follow the same route of protecting his own interests: the process of emerging national armies was set in motion.

Another interesting phenomenon that is worth noting for the current historical analysis and comparison between mercenaries of the past and PMSCs of today was the emergence of Swiss mercenary units originating from the Swiss Confederation formed in 1291. Each town of the confederation provided citizens united in units of pikemen to grant protection from foreign powers.[49] Well-trained, coordinated, and powerful Swiss units[50] – some of which were organized into companies – proved to be very effective on the battlefield, achieving victory in various battles.[51] They were capable of creating a new industry for the Swiss Confederation, using not only strong tactics of warfare, but also fascinating skills of selling and exporting their military expertise. This model seemed to be so attractive – just as it is nowadays – that Swiss neighbors from south Germany and Austria decided to copy it and set up their own mercenary troops, so-called *landsknechts* or *lansquenets*.[52] Remarkably, the *landsknechts* had no connection with their domestic social institutions. Accepting the skills of conducting war as a sole aim of the business enterprise, they finally started to take the dominating position on the battlefields from the hands of their Swiss adversaries.[53] The *landsknechts* could more easily adapt to the changing climate of warfare and were known for employing different types of units and weapons rather than remaining strictly specialized as were the Swiss pikemen. Finally, at the battle of Bicocca that took place in 1522 many Swiss mercenaries were skillfully defeated by well-trained *landsknechts*. After that, Swiss pikemen could not regain the same position of overshadowing other units in wars and their success story was over.

---

48 Ibidem, p. 26.

49 Anthony Mockler, (1985) *The New Mercenaries: The History of the Mercenary from the Congo to the Seychelles*, Letchworth: The Garden City Press, p. 8.

50 Kim R. Nossal, (1998) "Roland Goes Corporate: Mercenaries and Transnational Security Corporations in the Post-Cold War Era", *Civil Wars*, Volume 1, Issue 1, p. 19.

51 Peter W. Singer, (2008) *Corporate Warriors: The Rise of the Privatized Military Industry*, Ithaca and London: Cornell University Press, p. 26; John McCormack, (1993) *On Million Mercenaries: Swiss Soldiers in the Armies of the World*, London: Leo Cooper, p. 38; Sarah Percy, (2007) *Mercenaries: The History of a Norm in International Relations*, Oxford: Oxford University Press, p. 73.

52 Anthony Mockler, (1985) *The New Mercenaries: The History of the Mercenary from the Congo to the Seychelles*, Letchworth: The Garden City Press, p. 20.

53 Peter W. Singer, (2008) *Corporate Warriors: The Rise of the Privatized Military Industry*, Ithaca and London: Cornell University Press, p. 27.

2.3 The Peace of Westphalia and the Age of Reason

In the following centuries, war and activities associated with it unmistakably formed one of the most significant industries of Europe. By the seventeenth century, individuals possessing enough wealth were able to successfully wage wars and to achieve compelling advantages over their enemies by using private armies seeking to earn money. To the large extent, most European troops of that period consisted of hired soldiers from different regions, which did not fight to protect their homeland or rulers, but to earn a living and to make a profit. In this period, a new class of military businessmen emerged: persons who had enough resources to recruit military experts and to hire them out to those willing to pay. This practice enabled many European rulers to engage in wars without the need for introducing changes in their internal governance systems.[54] One of the most prominent traders in the field of military affairs was without doubt Count Albrecht von Wallenstein, who managed to create one of the biggest and most impressive private companies of Europe before the twentieth century and became the wealthiest man of Europe of his time.[55] The officers serving in the army of Von Wallenstein invested heavily in the military operations conducted and strived to safeguard the business interests of their company, in a similar way to modern corporate managers.

During the Thirty Years War fought between 1618 and 1648, most military troops caught in battles consisted of mercenaries. The end of this major war and the conclusion of the Peace of Westphalia in 1648 signified the emergence of a new concept of State intrinsically connected with the idea of sovereignty over internal affairs. This major development in human history led to the dominating scheme based on the use of official State armies consisting of State nationals loyal to a particular nation and not to some local rulers or other influential individuals. In the coming centuries, the idea of States controlling, sanctioning, and using violence would become a common sense notion prevailing in the international community.[56]

With the Napoleonic wars, the character of war slowly began to change: they were now not wars of kings and other rulers, but wars of people.[57] These were political

---

[54] Ibidem, p. 28.

[55] Ibidem.

[56] Deborah D. Avant, (2005) *The Market for Force: The Consequences of Privatizing Security*, Cambridge: Cambridge University Press, p. 1; Benedict Sheehy and Jackson N. Maogoto, (2008) "The Private Military Company – Unravelling the Theoretical, Legal & Regulatory Mosaic", *ILSA Journal of International & Comparative Law*, Volume 15, Issue 1, p. 167.

[57] Deborah D. Avant, (2000) "From Mercenary to Citizen Armies: Explaining Change in the Practice of War", *International Organization*, Volume 54, Issue 1, p. 43; Anthony Mockler, (1985) *The New Mercenaries: The History of the Mercenary from the Congo to the Seychelles*, Letchworth: The Garden City Press, pp. 6, 21.

wars involving entire societies.[58] The State armies involved in conflicts increased in size and relied mostly on the musket as their effective firearms. The skills of military experts were now less important than numbers of soldiers. It was much more cost-effective for State governments to train musketeers over short periods of time and to gather massive troops in order to be advantageous on the new type of battlefield.[59] States kept growing and expanding territorially, developing their State machinery and taxation system, maintaining and advancing their military potential.

As a new relationship between States and citizens came into being, supported by the Enlightenment ideas of the eighteenth century,[60] many economic and ethical advantages in employing permanent State armies that are loyal to their governments rather than temporary bands of mercenaries were discovered. Those hired troops often disregarded interests of their clients and moral obligations towards nations. Guided in their endeavors by the political and social ideals of this period, many European States commenced to actively rely on the new monopoly model for shaping national law and order.[61] The implementation of a citizen army was in the first place achieved in France and then this idea was supported and further implemented by Prussia.[62] Following this trend on a more global scale, national troops consisting of citizens protecting their home countries evolved into both a new international norm and an international model that was used more extensively by emerging States claiming their sovereignty.[63] The essential actions of State authorities in the field of military and security affairs were now carried out by their citizens and those in power had to take responsibility for the use of violence employed by them.[64] Gradually, the State developed into an instrument governing and controlling many areas of human life.

Although mercenaries could still find employment during the Age of Reason (ca. 1650-1800),[65] after the emergence of the nation-State, their truly glorious years when

---

58 Christopher Coker, (1999) "Outsourcing War", *Cambridge Review of International Affairs*, Volume 13, Issue 1, p. 96.

59 Peter W. Singer, (2008) *Corporate Warriors: The Rise of the Privatized Military Industry*, Ithaca and London: Cornell University Press, pp. 29-30.

60 Deborah D. Avant, (2000) "From Mercenary to Citizen Armies: Explaining Change in the Practice of War", *International Organization*, Volume 54, Issue 1, p. 42.

61 Adam White, (2010) *The Politics of Private Security: Regulation, Reform and Re-Legitimation*, Houndmills, Basingstoke, New Hampshire: Palgrave Macmillan, p. 22.

62 Deborah D. Avant, (2000) "From Mercenary to Citizen Armies: Explaining Change in the Practice of War", *International Organization*, Volume 54, Issue 1, pp. 47-48, 52, 59-60; Anthony Mockler, (1985) *The New Mercenaries: The History of the Mercenary from the Congo to the Seychelles*, Letchworth: The Garden City Press, p. 6.

63 Deborah D. Avant, (2000) "From Mercenary to Citizen Armies: Explaining Change in the Practice of War", *International Organization*, Volume 54, Issue 1, pp. 42-43.

64 Shantanu Chakrabarti, (2009) *Privatisation of Security in the Post Cold-War Period: An Overview of its Nature and Implications*, Institute for Defence Studies and Analyses Monograph Series, No. 2, New Delhi: Magnum Custom Publishing, p. 3.

65 Ibidem.

they were regarded as a norm[66] were practically over for at least two centuries.[67] In general, they were not considered more effective or economical than citizens' armies, but were frequently used by some nations for political reasons to supplement regular troops: they were seen as an instrument of diplomacy capable of avoiding tensions between States and their subjects.[68] The Prussian army in the eighteenth century consisted for the most part of hired soldiers, the British employed German and Swiss mercenaries in the Crimean War in 1853, and mercenary units played an important part in the wars during the breakdown of Chinese Imperial rule.[69]

Besides scenarios when foreign volunteers join armed forces of various nations guided by political and other considerations, there were also cases of States hiring out their own State troops to other States in order for the latter to achieve certain goals. British governments have a long history of using mercenaries in their numerous endeavors.[70] For the American Revolutionary War (1775-1783), the British Army was able to mobilize merely 30,000 soldiers, while American rebels numbered 50,000 militia who could be sent into battle.[71] British hopes were immediately pinned on hiring 20,000 Russian mercenaries to stabilize the situation and the British were willing to pay any price for the Russian manpower.[72] Regardless of the generous monetary offer of George III, Catherine the Great refused to hire out the Russian might out of principle, as it was intended "simply to calm a rebellion which is not supported by any foreign power". Following the refusal of the Russian Queen, the British attempted to hire the Dutch troops. They were particularly interested in the Scots Brigade in the service of the United Provinces, which formed part of the permanent military establishment of the Dutch Republic and consisted of mercenaries from all over Europe. The Dutch States General agreed to hire out the Scots Brigade to Great Britain on the condition that its personnel should not be used outside Europe:

---

66 Laura A. Dickinson, (2006) "Public Law Values in a Privatized World", *The Yale Journal of International Law*, Volume 31, Issue 2, p. 394.

67 Sarah Percy, (2007) *Mercenaries: The History of a Norm in International Relations*, Oxford: Oxford University Press, p. 167; Anthony Mockler, (1985) *The New Mercenaries: The History of the Mercenary from the Congo to the Seychelles*, Letchworth: The Garden City Press, p. 7; Kim R. Nossal, (1998) "Roland Goes Corporate: Mercenaries and Transnational Security Corporations in the Post-Cold War Era", *Civil Wars*, Volume 1, Issue 1, p. 19.

68 Deborah D. Avant, (2000) "From Mercenary to Citizen Armies: Explaining Change in the Practice of War", *International Organization*, Volume 54, Issue 1, pp. 45-46.

69 Peter W. Singer, (2008) *Corporate Warriors: The Rise of the Privatized Military Industry*, Ithaca and London: Cornell University Press, p. 32; Anthony Mockler, (1985) *The New Mercenaries: The History of the Mercenary from the Congo to the Seychelles*, Letchworth: The Garden City Press, p. 7; Deborah D. Avant, (2000) "From Mercenary to Citizen Armies: Explaining Change in the Practice of War", *International Organization*, Volume 54, Issue 1, p. 45.

70 Anthony Mockler, (1985) *The New Mercenaries: The History of the Mercenary from the Congo to the Seychelles*, Letchworth: The Garden City Press, p. 1.

71 Ken Silverstein, (2000) *Private Warriors*, New York and London: Verso, p. 146.

72 Anthony Mockler, (1985) *The New Mercenaries: The History of the Mercenary from the Congo to the Seychelles*, Letchworth: The Garden City Press, p. 2.

a clear refusal that made the British government look for alternatives. In Germany, it finally managed to hire troops from a number of German princes, such as the highly effective Hessian soldiers.[73] In addition to other possible examples, the United States is known for using the troops of Australia, New Zealand, South Korea, Thailand, and the Philippines made available by the respective governments during the Vietnam war.[74] Pakistan also supplied large numbers of troops to Saudi Arabia in the 1980s.[75]

It is worth observing that, in addition to the continuing but not so widespread practice of hiring and employing mercenaries and mercenary units by governments,[76] in the seventeenth and eighteenth centuries, private businesses took the lead in carrying out a scale of military tasks on their own. As the two most illustrative examples, the English East India Company and the Dutch East India Company can be mentioned. These enterprises, based on the charter company system, controlled trade between the East and West and had great powers to conduct war and to protect their own interests.[77] They operated in colonial areas with weak or absent governance structures, exercised a variety of military functions, and dominated political affairs of the region. It goes without saying that both companies having monopoly power and striving to increase their profits often crossed interests of their home countries. When the political situation on the Indian sub-continent started to stabilize, both business entities disappeared: the Dutch East India Company was dissolved as a legal entity in 1799 and the British East India Company in 1873.[78]

2.4 Modern times and the rise of PMSCs

The system of mercenarism can be said to go hand in hand with such notions as democratization, marketization, and internationalization of military service.[79] In the twentieth century, the international system dominated by States, which relied heavily upon the concept of sovereignty, became the reality. Even nowadays, State-centric and government-related approaches seem to prevail in the evaluation of global military

---

73 Ibidem, pp. 3-4; Jurgen Brauer, (1999) "An Economic Perspective on Mercenaries, Military Companies, and the Privatisation of Force", *Cambridge Review of International Affairs*, Volume 13, Issue 1, p. 132; Shantanu Chakrabarti, (2009) *Privatisation of Security in the Post Cold-War Period: An Overview of its Nature and Implications*, Institute for Defence Studies and Analyses Monograph Series, No. 2, New Delhi: Magnum Custom Publishing, p. 3.

74 Janice E. Thomson, (1994) *Mercenaries, Pirates and Sovereigns: State-Building and Extraterritorial Violence in Early Modern Europe*, Princeton and Chichester: Princeton University Press, p. 94.

75 Ibidem, p. 90.

76 Kim R. Nossal, (1998) "Roland Goes Corporate: Mercenaries and Transnational Security Corporations in the Post-Cold War Era", *Civil Wars*, Volume 1, Issue 1, p. 19.

77 Peter W. Singer, (2008) *Corporate Warriors: The Rise of the Privatized Military Industry*, Ithaca and London: Cornell University Press, p. 34.

78 Ibidem, pp. 36-37.

79 Janice E. Thomson, (1994) *Mercenaries, Pirates and Sovereigns: State-Building and Extraterritorial Violence in Early Modern Europe*, Princeton and Chichester: Princeton University Press, p. 58.

and security challenges.[80] Since the end of the nineteenth century, the employment of mercenaries could be said to be substantially insignificant. Soldiers of fortune started to prominently appear on the international stage only after the 1960s.[81] At this particular time, norms against the use of mercenaries began to manifest themselves and international law was confronted with particular challenges of regulating and limiting the scope of the problem in this and the coming decades.[82]

Now, those hired soldiers were mainly individual ex-soldiers who sold their expertise to States, rebel movements, or companies functioning in danger areas of the world, such as Latin America and Africa.[83] Typically, they were both those cruel mercenaries and for many fascinating individuals involved in the African civil wars, such as the war in the Congo in the 1960s.[84] In the years to come, many governments continued to use services provided by soldiers of fortune.[85] Russian mercenaries are known to have been engaged in various military operations in the former Soviet Union and the former Yugoslavia.[86] Even nowadays, the French government uses hired guns in its Foreign Legion. The authorities of the United Arab Emirates have a practice of an almost exclusive reliance on the use of hired soldiers from Great Britain, Pakistan, Oman, Yemen, and Jordan.

Since the end of the Cold War in the early 1990s, the world has witnessed the massive expansion of the private military market and the emergence of the privatized military and security industry with its main actors: private military and security companies.[87] Before the fall of the Berlin Wall, the world politics were dominated by

80 Robert Mandel, (2002) *Armies Without States: the Privatization of Security*, Boulder: Lynne Rienner Publishers, p. 7.

81 Sarah Percy, (2007) *Mercenaries: The History of a Norm in International Relations*, Oxford: Oxford University Press, p. 167.

82 Ibidem, pp. 26, 167.

83 Peter W. Singer, (2008) *Corporate Warriors: The Rise of the Privatized Military Industry*, Ithaca and London: Cornell University Press, p. 37.

84 Ibidem; Sarah Percy, (2007) *Mercenaries: The History of a Norm in International Relations*, Oxford: Oxford University Press, p. 167; Sarah Percy, "Morality and Regulation", in Simon Chesterman and Chia Lehnardt (Eds.), (2007) *From Mercenaries to Market: The Rise and Regulation of Private Military Companies*, Oxford: Oxford University Press, p. 12; Kim R. Nossal, (1998) "Roland Goes Corporate: Mercenaries and Transnational Security Corporations in the Post-Cold War Era", *Civil Wars*, Volume 1, Issue 1, p. 17.

85 Peter Tickler, (1987) *The Modern Mercenary: Dog of War, or Soldier of Honour?*, Wellingborough: Stephens; Ken Silverstein, (2000) *Private Warriors*, New York and London: Verso, p. 124.

86 Peter W. Singer, (2008) *Corporate Warriors: The Rise of the Privatized Military Industry*, Ithaca and London: Cornell University Press, p. 38.

87 Ibidem, p. 49; Peter W. Singer, (2004) "War, Profits and the Vacuum of Law: Privatized Military Firms and International Law", *Columbia Journal of International Law*, Volume 42, Issue 2, p. 522; Andrew Alexandra, Deane-Peter Baker, and Marina Caparini, "Introduction: The Ethics and Governance of Private Military and Security Companies", in Andrew Alexandra, Deane-Peter Baker, and Marina Caparini (Eds.), (2008) *Private Military and Security Companies: Ethics, Policies and Civil-Military Relations*, London and New York: Routledge, pp. 1-2; Shantanu Chakrabarti, (2009) *Privatisation of*

two major superpowers, the United States of America and the Soviet Union. After the collapse of the latter, the global order has dramatically changed. Peter W. Singer underlines three main reasons for the rapid proliferation of private military and security companies in the post-Cold War era, which will now be covered in more detail.[88]

The first main factor that provided stimulus for the emergence of new private military and security firms according to Peter W. Singer is the fact that the collapse of the Soviet Union led to the dissolution of the established global order. This dissolution had major consequences for the global market of force with its supply and demand of military and security services.[89] First, when two superpowers ceased to provide order and to exercise control over the regions of instability, the world experienced an increase in the levels of global violence and the occurrence of many conflicts.[90] Secondly, a large number of non-State conflict groups, such as terrorist organizations, rebel movements, and drug cartels, emerged around the globe, while the international community was unable to regulate and to suppress them.[91] Thirdly, the end of the Cold War led to the massive downsizing of State armies and oversupply of military and security expertise, meaning that many ex-soldiers found themselves unemployed and looking for new employment possibilities outside the State apparatus.[92] In the former Soviet Union alone, tens of thousands of demobilized soldiers joined various private military and security corporations. In 10 years, there were more than 12,000 PMSCs formally registered with the Russian government.[93] In addition, the

---

Security in the Post Cold-War Period: An Overview of its Nature and Implications, Institute for Defence Studies and Analyses Monograph Series, No. 2, New Delhi: Magnum Custom Publishing, p. 5; Carlos Ortiz, "The Private Military Company: An Entity at the Center of Overlapping of Overlapping Spheres of Commercial Activity and Responsibility", in Thomas Jäger and Gerhard Kümmel (Eds.), (2007) *Private Military and Security Companies: Chances, Problems, Pitfalls and Prospects*, Wiesbaden: VS Verlag für Sozialwissenschaften, pp. 64-65; David Shearer, (1999) "Private Military Force and Challenges for the Future", *Cambridge Review of International Affairs*, Volume 13, Issue 1, p. 81.

88 Peter W. Singer, (2008) *Corporate Warriors: The Rise of the Privatized Military Industry*, Ithaca and London: Cornell University Press, p. 49.

89 Ibidem.

90 Deborah D. Avant, (2004) "The Privatization of Security and Change in the Control of Force", *International Studies Perspectives*, Volume 5, p. 154; Deborah D. Avant, (2005) *The Market for Force: The Consequences of Privatizing Security*, Cambridge: Cambridge University Press, p. 36; Ken Silverstein, (2000) *Private Warriors*, New York and London: Verso, p. 203; Kim R. Nossal, (1998) "Roland Goes Corporate: Mercenaries and Transnational Security Corporations in the Post-Cold War Era", *Civil Wars*, Volume 1, Issue 1, p. 30.

91 Kim R. Nossal, (1998) "Roland Goes Corporate: Mercenaries and Transnational Security Corporations in the Post-Cold War Era", *Civil Wars*, Volume 1, Issue 1, pp. 30-31; M.J. Peterson, (1992) "Transnational Activity, International Society and World Politics", *Millennium: Journal of International Studies*, Volume 21, Issue 3, pp. 382-383.

92 Kim R. Nossal, (1998) "Roland Goes Corporate: Mercenaries and Transnational Security Corporations in the Post-Cold War Era", *Civil Wars*, Volume 1, Issue 1, pp. 29-30.

93 Kevin A. O'Brien, "What Should and What Should Not Be Regulated", in Simon Chesterman and Chia Lehnardt (Eds.), (2007) *From Mercenaries to Market: The Rise and Regulation of Private Military Companies*, Oxford: Oxford University Press, p. 31.

open market was flooded with weapons and other military equipment, which became available as a result of the military downsizing. Finally, without support of two major world powers and other countries, many States – faced with the lack of action on the part of the United Nations and regional organizations[94] – remained without proper military response instruments needed to guarantee their security and to protect their citizens from serious dangers emanating from transnational criminal networks, rebels, warlords, and other elements.[95] They experienced the decline of State governance and were unable to effectively exercise control over their sovereign territories and to secure their borders. Therefore, those weak nations frequently sought help from privately run organizations providing military and security services. Many State functions were often outsourced to those firms.[96] In general, some criticism has been expressed with respect to this practice of hiring PMSCs by weak States, which is occasionally referred to as "armed corporate imperialism".[97] Currently, the world has arrived at the stage of its development when State governments increasingly rely on a variety of services offered by non-State actors, such as PMSCs. According to Elke Krahmann, these corporations are capable of supporting States in their efforts to face various challenges of national and international security. She perceptively notes:

> "They [non-State actors] can operate more easily across national borders, they have particular expertise in dealing with non-traditional security issues and they can draw on formal and informal networks with other actors."[98]

The second reason for the proliferation of PMSCs can be found in the many changes introduced in modern warfare itself. Those changes and innovations led to the increased need in using private military and security experts.[99] After the end of the Cold War, the military power possessed by States came into hands of various non-State actors,

---

94 Deborah D. Avant, (2005) *The Market for Force: The Consequences of Privatizing Security*, Cambridge: Cambridge University Press, pp. 37-38.

95 Ibidem, p. 36; Didier Bigo, (2003) "Les Entreprises de Coercition Para-Privées: De Nouveaux Mercenaires?", *Cultures & Conflits*, Volume 52, Issue 4, pp. 5-10; Kim R. Nossal, (1998) "Roland Goes Corporate: Mercenaries and Transnational Security Corporations in the Post-Cold War Era", *Civil Wars*, Volume 1, Issue 1, p. 30.

96 Vadim Volkov, (2002) *Violent Entrepreneurs: the Use of Force in the Making of Russian Capitalism*, Ithaca: Cornell University Press; Deborah D. Avant, (2005) *The Market for Force: The Consequences of Privatizing Security*, Cambridge: Cambridge University Press, p. 36.

97 Uwe Steinhoff, "What are Mercenaries?", in Andrew Alexandra, Deane-Peter Baker, and Marina Caparini (Eds.), (2008) *Private Military and Security Companies: Ethics, Policies and Civil-Military Relations*, London and New York: Routledge, p. 19.

98 Elke Krahmann, "The Rise of Non-State Actors in Security Governance", in Patricia Kennett (Ed.), (2008) *Governance, Globalization and Public Policy*, Cheltenham and Northampton: Edward Elgar Publishing, p. 198.

99 Peter W. Singer, (2008) *Corporate Warriors: The Rise of the Privatized Military Industry*, Ithaca and London: Cornell University Press, pp. 49, 60-66.

such as terrorist groups, and this circumstance has certainly changed the dynamics of conventional warfare. In addition, a new type of criminalized war involving criminal organizations and other groups and governments emerged in contrast to politically motivated armed conflicts between States or organized groups. This development attracted profit-driven entities, such as PMSCs, and led them to take part in hostilities. Also importantly, two of the most essential elements of modern warfare have become information dominance and technical expertise in many fields that are developed, managed, and supported by private entities and civilian personnel. These corporations and their employees proved to be better in many respects than their military counterparts in performing similar tasks.[100] Former CIA director George Tenet once stated in his speech addressed to intelligence professionals in 2003 that the "private sector has stepped forward for over half a century to offer the best technology and expertise".[101] His belief in private corporations is reflected in the words clearly indicating the daily reality the US government and other governments around the world are facing with regard to the heavy reliance on private corporations and their expertise:

> "They are part of our intelligence family because they share our same passion for our mission – to safeguard the freedoms that make America great. Throughout our history, American industry has given every DCI [Director of Central Intelligence] the tools needed to solve our greatest challenges – and we need them now more than ever."[102]

Finally, the "privatization revolution", set in motion in the 1980s, created fertile ground for privatizing many governmental functions and establishment of private enterprises capable of carrying them out.[103] In various countries, it meant the sale of parts of industries owned by the State representing the public sector to the private sector, consisting of a variety of private businesses and organizations.[104] The

---

[100] Deborah D. Avant, (2004) "The Privatization of Security and Change in the Control of Force", *International Studies Perspectives*, Volume 5, p. 154.

[101] Tim Shorrock, (2008) *Spies for Hire: The Secret World of Intelligence Outsourcing*, New York: Simon & Schuster, p. 74.

[102] Ibidem.

[103] Kim R. Nossal, (1998) "Roland Goes Corporate: Mercenaries and Transnational Security Corporations in the Post-Cold War Era", *Civil Wars*, Volume 1, Issue 1, p. 31; Peter W. Singer, (2008) *Corporate Warriors: The Rise of the Privatized Military Industry*, Ithaca and London: Cornell University Press, pp. 49, 66-70; Kevin A. O'Brien, "What Should and What Should Not Be Regulated", in Simon Chesterman and Chia Lehnardt (Eds.), (2007) *From Mercenaries to Market: The Rise and Regulation of Private Military Companies*, Oxford: Oxford University Press, p. 33; Deborah D. Avant, (2005) *The Market for Force: The Consequences of Privatizing Security*, Cambridge: Cambridge University Press, p. 35; Tim Shorrock, (2008) *Spies for Hire: The Secret World of Intelligence Outsourcing*, New York: Simon & Schuster, p. 73.

[104] Deborah D. Avant, (2005) *The Market for Force: The Consequences of Privatizing Security*, Cambridge: Cambridge University Press, p. 24.

phenomenon of privatization is currently a matter of fact[105] and involves an array of policies introduced with the aim to reduce governmental involvement in economic and social life.[106] To some extent, this means the transfer of all or some responsibilities from governments to the private sector.[107] Privatization is justified by both many failures on the part of governments and successes of private actors. First of all, a shift from public to private provision of services and goods can make it possible to limit governmental costs, as the employment of private actors may often be cheaper.[108] Furthermore, privatization is often more efficient and easier for States. Finally, governments may choose to privatize some activities when there is not enough material or technical capacity to complete them by relying on the public apparatus.

According to James Wilson, there are no governmental activities that have never been or cannot be outsourced to private firms.[109] Whereas it might appear obvious to some, the process of privatization was launched in many different fields and is often considered an important domestic and international phenomenon.[110] It took place not only in the area of education, healthcare, and telecommunications, but also in the field of military affairs and security provision, where the group of outsourcers included States and various non-State entities, such as international organizations, NGOs, multinational corporations, and even individuals.[111] The business activities of private companies operating in the field of military and security service provision pertain to the phenomenon of privatization of war and security – also called privatization of war or privatization of violence[112] – that attracted attention of many

---

105 Jurgen Brauer, (1999) "An Economic Perspective on Mercenaries, Military Companies, and Privatisation of Force", *Cambridge Review of International Affairs*, Volume 13, Issue 1, p. 133.

106 Daphne Barak-Erez, "The Privatization Continuum", in Simon Chesterman and Angelina Fisher (Eds.), (2009) *Private Security, Public Order: The Outsourcing of Public Services and Its Limits*, Oxford: Oxford University Press, p. 72.

107 Robert W. Poole Jr., "Privatization", in D.R. Henderson (Ed.), *The Concise Encyclopedia of Economics*, <http://www.econlib.org/library/Enc/Privatization.html>, accessed on March 1, 2012; Ken Silverstein, (2000) *Private Warriors*, New York and London: Verso, p. 143.

108 Mariana M. Prado, "Regulatory Choices in the Privatization of Infrastructure", in Simon Chesterman and Angelina Fisher (Eds.), (2009) *Private Security, Public Order: The Outsourcing of Public Services and Its Limits*, Oxford: Oxford University Press, pp. 109-110.

109 James Q. Wilson, (1989) *Bureaucracy*, New York: Basic Books, p. 346.

110 Laura A. Dickinson, (2006) "Public Law Values in a Privatized World", *The Yale Journal of International Law*, Volume 31, Issue 2, p. 384.

111 Deborah D. Avant, (2005) *The Market for Force: The Consequences of Privatizing Security*, Cambridge: Cambridge University Press, p. 7; Deborah D. Avant, (2004) "The Privatization of Security and Change in the Control of Force", *International Studies Perspectives*, Volume 5, p. 154; Laura A. Dickinson, (2006) "Public Law Values in a Privatized World", *The Yale Journal of International Law*, Volume 31, Issue 2, pp. 384-385, 389; Anna Leander, (2005) "The Power to Construct International Security: On the Significance of Private Military Companies", *Millennium: Journal of International Studies*, Volume 33, Issue 3, pp. 805-806.

112 Kevin A. O'Brien, "What Should and What Should Not Be Regulated", in Simon Chesterman and Chia Lehnardt (Eds.), (2007) *From Mercenaries to Market: The Rise and Regulation of Private Military*

scholars.[113] It indicates "the public-private partnership based foundations of violence, organized through a particular set of institutional, regulatory, and contractual means" that is used to exercise force or provide security.[114] This phenomenon is often seen as global[115] and is without doubt essential to the foreign and military policy of a large number of States.[116]

The phenomenon of privatization of war and security appears to have had a substantial impact on the essence of the State sovereignty concept with States as sole providers of legitimate violence.[117] The result is the availability of a variety of military capacities in the hands of citizens rather than governments, creating a new picture of the current world order.[118] Surveying the activities of PMSCs appearing on the international plane, Jessica T. Matthews of the Carnegie Endowment for International Peace expressed this idea by saying that "the steady concentration of power in the hands of States, which began in 1648 with the Peace of Westphalia, is over, at least for a while".[119] The emergence of PMSCs in the global arena is the result of the above-mentioned developments taking place in the international and domestic

---

Companies, Oxford: Oxford University Press, p. 33; Kim R. Nossal, (1998) "Roland Goes Corporate: Mercenaries and Transnational Security Corporations in the Post-Cold War Era", *Civil Wars*, Volume 1, Issue 1, p. 26; Benedict Sheehy and Jackson N. Maogoto, (2008) "The Private Military Company – Unravelling the Theoretical, Legal & Regulatory Mosaic", *ILSA Journal of International & Comparative Law*, Volume 15, Issue 1, p. 149; Michael Likosky, "The Privatization of Violence", in Simon Chesterman and Angelina Fisher (Eds.), (2009) *Private Security, Public Order: The Outsourcing of Public Services and Its Limits*, Oxford: Oxford University Press, p. 11.

113 Michael Likosky, "The Privatization of Violence", in Simon Chesterman and Angelina Fisher (Eds.), (2009) *Private Security, Public Order: The Outsourcing of Public Services and Its Limits*, Oxford: Oxford University Press, p. 12; Simon Chesterman and Chia Lehnardt, "Introduction", in Simon Chesterman and Chia Lehnardt (Eds.), (2007) *From Mercenaries to Market: The Rise and Regulation of Private Military Companies*, Oxford: Oxford University Press, p. 1; Deborah D. Avant, (2005) *The Market for Force: The Consequences of Privatizing Security*, Cambridge: Cambridge University Press, pp. 3-5; Daphne Barak-Erez, "The Privatization Continuum", in Simon Chesterman and Angelina Fisher (Eds.), (2009) *Private Security, Public Order: The Outsourcing of Public Services and Its Limits*, Oxford: Oxford University Press, p. 71.

114 Michael Likosky, "The Privatization of Violence", in Simon Chesterman and Angelina Fisher (Eds.), (2009) *Private Security, Public Order: The Outsourcing of Public Services and Its Limits*, Oxford: Oxford University Press, p. 11.

115 Robert Mandel, (2002) *Armies Without States: the Privatization of Security*, Boulder: Lynne Rienner Publishers, p. 2; Laura A. Dickinson, (2006) "Public Law Values in a Privatized World", *The Yale Journal of International Law*, Volume 31, Issue 2, p. 384.

116 Ken Silverstein, (2000) *Private Warriors*, New York and London: Verso, p. 187.

117 Daphne Barak-Erez, "The Privatization of Continuum", in Simon Chesterman and Angelina Fisher (Eds.), (2009) *Private Security, Public Order: The Outsourcing of Public Services and Its Limits*, Oxford: Oxford University Press, p. 71.

118 Robert Mandel, (2002) *Armies Without States: the Privatization of Security*, Boulder: Lynne Rienner Publishers, p. 7.

119 Paul Lewis, "It's Not Just Governments That Make War and Peace Now", *The New York Times*, November 28, 1998, <http://www.nytimes.com/1998/11/28/arts/not-just-governments-make-war-or-peace.html>, accessed on July 18, 2012.

spheres, which represent a process that cannot be reversed.[120] It is quite shocking to see that the private armies of military and security experts often outnumber the national armies and are in a position to influence military and security affairs to a great extent.[121] For example, Pakistani leaders have often tried to deny that one of the biggest companies in the PMSC business called Academi, formerly known as Blackwater and Xe Services, had thousands of employees operating in the capital of Pakistan, Islamabad. Claims have been made, however, that approximately 9,000 Academi employees were active in the capital, which is significant when compared to 7,000 members of the personnel of local police.[122]

2.5 PMSCs and mercenaries of the past

A possible legal qualification of PMSC employees as mercenaries will be addressed at a later stage in the study. What is important to realize now, is that PMSCs as main characters playing a crucial role in the manifestation of the phenomenon of privatization of war and security emerged naturally in accordance with the historical developments the world went through.[123] Juan Carlos Zarate even referred to a private corporation providing military and security services as "a recycled form of past mercenary organizations, like the condottieri and the free companies".[124] The question of origins of PMSCs and their connection with typical mercenaries, however, still might be a subject for debate for some academics.[125]

Since time immemorial, mercenaries were capable of performing a wide array of required functions more effectively and efficiently, just as the modern corporations

---

120 Eugenio Cusumano, "Policy Prospects for Regulating Private Military and Security Companies", in Francesco Francioni and Natalino Ronzitti (Eds.), (2011) *War by Contract: Human Rights, Humanitarian Law, and Private Contractors*, Oxford: Oxford University Press, p. 13.

121 T. C. Miller, "Private Contractors Outnumber U.S. Troops in Iraq", *The Los Angeles Times*, July 4, 2007, <http://articles.latimes.com/2007/jul/04/nation/na-private4>, accessed on July 18, 2012; Robert Mandel, (2002) *Armies Without States: the Privatization of Security*, Boulder: Lynne Rienner Publishers, p. 8; Joseph P. Vasquez III, (2010) "Blackwater's Rise and the Draft's Demise", *Yale Journal of International Affairs*, Volume 5, Issue 2, pp. 164-165; Fred Schreier and Marina Caparini, (2005) *Privatising Security: Law, Practice and Governance of Private Military and Security Companies*, Occasional Paper No. 6, DCAF, p. 2.

122 Noel Brinkerhoff and David Wallechinsky, "Blackwater Mercenaries Rumored to Outnumber Local Police in Pakistani Capital", *AllGov*, February 10, 2010, <http://www.allgov.com/news/us-and-the-world/blackwater-mercenaries-rumored-to-outnumber-local-police-in-pakistani-capital?news=840337>, accessed on August 5, 2012.

123 Juan C. Zarate, (1998) "The Emergence of a New Dog of War: Private International Security Companies, International Law, and the New World Disorder", *Stanford Journal of International Law*, Volume 34, Issue 1, pp. 81-82.

124 Ibidem, p. 91.

125 Simon Chesterman and Chia Lehnardt, "Introduction", in Simon Chesterman and Chia Lehnardt (Eds.), (2007) *From Mercenaries to Market: The Rise and Regulation of Private Military Companies*, Oxford: Oxford University Press, p. 1.

supporting States.[126] However, the massive growth of the market for military and security service provision after the 1990s cannot be said to be unavoidable or natural.[127] Many developments and issues the world order was faced with after the end of the Cold War led to more extensive reliance on the expertise of PMSCs by States and other actors. Loren Thompson, a Senior Fellow of the Alexis de Tocqueville Institution, stated with regard to PMSCs that "the only difference between what these firms do and what mercenaries do is that the companies have gained the imprimatur of government for their actions".[128] This statement indicates the many similarities between both types of private force in conflict areas – PMSCs and mercenaries – that can provide some clarity regarding the general functioning of those corporations. It cannot be denied that PMSCs should be distinguished from typical Swiss or English mercenaries operating centuries ago, but there are undoubtedly some features common to both.

Nowadays, States seem to value the quality of services skillfully provided by PMSCs in the era of modern warfare when the quantities of military and security experts are far less important and when the use of regular armed forces can be avoided for a variety of reasons. The military and security services provided by PMSCs are well-known for their colorful diversity, just as were those of *landsknechts* who managed to dominate the battlefield adapting to the reality of conducting warfare in the sixteenth century. In addition, the skills and military and security expertise of private contractors using sophisticated marketing strategies are carefully promoted. This trend is very similar to what happened during the Middle Ages with the first companies of mercenaries emerging in the thirteenth century in Italy and spreading all over Europe.[129] While Swiss units of pikemen created a whole new industry in the Swiss Confederation and were able to successfully sell and to export their military expertise, the daily activities of modern PMSCs also include the same type of operation. As war appears to be a constant in human history, it is apparent that PMSCs, just as mercenary organizations under the leadership of some wealthy individuals functioning many centuries ago, carry out their services for significant amounts of money and manage to become very prosperous even in the times of economic uncertainty.

The main features of the modern private military and security corporations having more power than some countries and becoming separate powerful entities within

---

126 Kathleen M. Jennings, (2006) *Armed Security: Regulating the Private Military Industry*, New Security Programme, Fafo-report 532, pp. 35-38, <http://www.fafo.no/pub/rapp/532/532.pdf>, accessed on May 8, 2011.

127 Kim R. Nossal, (1998) "Roland Goes Corporate: Mercenaries and Transnational Security Corporations in the Post-Cold War Era", *Civil Wars*, Volume 1, Issue 1, p. 31; Deborah D. Avant, (2005) *The Market for Force: The Consequences of Privatizing Security*, Cambridge: Cambridge University Press, p. 38.

128 Ken Silverstein, (2000) *Private Warriors*, New York and London: Verso, p. 187.

129 Anna Leander, (2005) "The Power to Construct International Security: On the Significance of Private Military Companies", *Millennium: Journal of International Studies*, Volume 33, Issue 3, pp. 822-823.

them can be traced back many centuries, for instance, to the business model of Count Albrecht von Wallenstein. Currently, significant financial resources enable PMSCs to employ the most qualified personnel, to equip them with the most advanced weaponry and technology, to provide the most sophisticated training, and thus to effectively conduct military operations, or to handle security matters. Observing some historical facts on the functioning of the English East India Company and the Dutch East India Company, one may conclude that modern corporations can also be deployed on territories with weak or absent governance structures. By exercising a wide spectrum of military and security functions and responding not only to market forces, but also to political concerns as regular mercenaries,[130] those companies can in theory dominate political affairs of countries and become influential actors in other areas.

As with the first companies working in accordance with a *condotta*, modern private contractors engage in armed conflicts and peace operations on the basis of a contract concluded between them and a State or other non-State entity. These documents were drafted by experts many centuries ago and were quite detailed. Currently, the state of affairs remains the same. Contracts show a great degree of detail, contain various provisions on the tasks to be provided by the contractors and limits of their operations. The essence of a contract as a significant link between PMSCs and mercenary companies, on the one hand, and States and other persons and entities, on the other, remains unchanged and constitutes one of the most essential features of the operation of all private military and security industry representatives that will be elaborated upon in the following Chapters.

Just like old-fashioned companies consisting of mercenaries from the Middle Ages, PMSCs fall in the first instance under the category of for-profit enterprises having the main motivation of essentially maximizing their financial gain, which is the main goal of all private firms.[131] They might be concerned with compliance with rules and principles of international law and essential interests of their customers, but in the first place strive to profit financially and might be unable to resist the temptation of earning more money by working for less credible persons or entities and engaging in some forms of unlawful conduct. It has happened in the past that mercenary organizations hired by the kings, States, and other wealthy individuals and entities committed various crimes during or after the fulfillment of their obligations. The main issue at stake and at the same time "the essential problem posed by

[130] Janice E. Thomson, (1994) *Mercenaries, Pirates and Sovereigns: State-Building and Extraterritorial Violence in Early Modern Europe*, Princeton and Chichester: Princeton University Press, p. 57.

[131] Adam White, (2010) *The Politics of Private Security: Regulation, Reform and Re-Legitimation*, Basingstoke: Palgrave Macmillan, p. 175; Michael J. Trebilcock and Edward M. Iacobucci, (2003) "Privatization and Accountability", *Harvard Law Review*, Volume 116, p. 1424; Anna Leander, (2005) "The Power to Construct International Security: On the Significance of Private Military Companies", *Millennium: Journal of International Studies*, Volume 33, Issue 3, p. 808; Benedict Sheehy and Jackson N. Maogoto, (2008) "The Private Military Company – Unravelling the Theoretical, Legal & Regulatory Mosaic", *ILSA Journal of International & Comparative Law*, Volume 15, Issue 1, p. 150.

the mercenary" concerned the responsibility of those hiring mercenaries, especially the responsibility of States for the unlawful actions of their citizens comprising mercenary armies.[132] It is not unthinkable, and indeed apparent from various reports, that PMSCs and their employees can also be involved in illegal activities and violate rules and principles of international humanitarian law and human rights. This fact makes more plausible the call for assessing the question of responsibility with regard to private contractors and respective States using their services.

The rise of private corporations, such as Academi and Triple Canopy, obviously constitutes a new stage in the development of private actors carrying out military and security tasks.[133] There are also a number of striking differences that can be found between PMSCs and regular mercenary organizations. According to Peter W. Singer, the concept of corporatization of military and security activities forms the main distinguishable feature of the new non-State warriors.[134] PMSCs remain legally established businesses to be distinguished from the loose bands of mercenaries that are not created in accordance with national laws of a State. They have an obvious corporate structure, are tied to corporate holdings and financial markets, and have the main motive of maximizing business rather than individual profit.[135] While legitimate States prefer not to be specific about their contractual relationships with PMSCs, in practice they cannot be held accountable for employing those legitimate businesses. It is also the case that in the past not only individual mercenaries could have been employed into the armies of a certain State, but also members of the State armed forces could have been hired out to other States to fight in their conflicts as mercenaries.[136] Nowadays, it is perhaps more a theoretical possibility rather than a matter of practical relevance. While it is highly arguable whether PMSCs can potentially form part of regular armed forces, they cannot be contracted out by one State to another State. It is, however, not uncommon that private corporations sub-contract their personnel to other firms, which engage in contractual relationships with States. As it will be explained in the following Chapters, this practice of sub-contracting is

---

[132] Janice E. Thomson, (1994) *Mercenaries, Pirates and Sovereigns: State-Building and Extraterritorial Violence in Early Modern Europe*, Princeton and Chichester: Princeton University Press, p. 55.

[133] Peter W. Singer, (2008) *Corporate Warriors: The Rise of the Privatized Military Industry*, Ithaca and London: Cornell University Press, p. 45.

[134] Kim R. Nossal, (1998) "Roland Goes Corporate: Mercenaries and Transnational Security Corporations in the Post-Cold War Era", *Civil Wars*, Volume 1, Issue 1, p. 26; Peter W. Singer, (2008) *Corporate Warriors: The Rise of the Privatized Military Industry*, Ithaca and London: Cornell University Press, p. 45; Robert Mandel, (2002) *Armies Without States: the Privatization of Security*, Boulder: Lynne Rienner Publishers, p. 9.

[135] Peter W. Singer, (2008) *Corporate Warriors: The Rise of the Privatized Military Industry*, Ithaca and London: Cornell University Press, p. 47.

[136] Juan C. Zarate, (1998) "The Emergence of a New Dog of War: Private International Security Companies, International Law, and the New World Disorder", *Stanford Journal of International Law*, Volume 34, Issue 1, pp. 86-87.

an unavoidable aspect of the whole industry, which makes it more difficult to hold States responsible for PMSC misconduct.

3. Private Military and Security Companies as Non-State Business Entities

In order to build a system of coherent understanding of the complex issue of State responsibility for the misconduct of private military and security contractors, it is useful to examine PMSCs as non-State actors often operating in the shadows. This should be done to differentiate between PMSCs and other non-State actors, such as other corporations and mercenaries, and to enable a discussion on the legal status of PMSCs, their employees, and eventually the issue of international responsibility.

This sub-Chapter will deal with a number of significant matters. In practical terms, for the purposes of the analysis of the question of State responsibility for acts and omissions of PMSCs and their personnel some light will be shed on the actual industry of private military and security service providers. Special attention will also be given to the nature of the industry, its main developments, and a variety of issues associated with it, which are of relevance for the understanding of activities of these firms, their legal status, relationships with States, and application of the doctrine of State responsibility. Furthermore, it will be discussed whether there is an accepted definition of a PMSC and what such a definition used within the framework of this study could be. The type of actors PMSCs are and their fundamental distinguishing features setting them apart from the plethora of other corporations will be addressed. To deal adequately with this essential issue of defining the above-mentioned businesses and assessing their characteristics, it may be necessary to shift the focus to their relationships with a variety of clients and the legal basis of operation. The emerging fact that private contractors are frequently hired by States will require an investigation into the links between these two types of actors and categories of States, which can be distinguished in the process of contracting PMSCs. Analysis of the State practice of using PMSCs will also focus on the tasks performed by private contractors and should provide more information regarding the broad spectrum of services offered by those corporations and falling within the domain of their daily operation. These services and their actual nature as a common denominator for the array of PMSC functions will be used as a starting point in the analysis of the State responsibility doctrine as triggered by the misconduct of PMSCs and their employees, which, for instance, perform elements of governmental authority mentioned in one of the core provisions of the Draft Articles on Responsibility of States for Internationally Wrongful Acts. Essential efforts will be devoted to providing a more understandable image of those firms and their legal position not only in the international arena, but also at the State level, given the fact that the present study aims at examining violations of international law committed by PMSCs and their employees working on a contractual basis and performing their tasks as members of those firms under various circumstances. The set of aims of the current research project clearly justify the need for assessing PMSCs as corporate non-State entities engaging in military and security activities on behalf of States: a logical step, which cannot be omitted and will be carried out below.

3.1 Industry

With the emergence of a new corporate entity in the international arena involved in the provision of military and security-related tasks, the modern industry of private military and security service providers and the global market for private force came into existence. In the examination of the industry of PMSCs as a whole, one should in the first place focus on the environment in which these firms operate.[137] As observed above, the world order changed dramatically after the introduction of the Westphalian notion of State sovereignty in the seventeenth century. The dissolution of the notion that the monopoly on violence rests in the hands of sovereign States became more visible when the Cold War ended. Developments taking place after the end of this struggle for supremacy between the United States and the Soviet Union had massive effects on the process of conducting war and providing security formerly regulated by State authorities. Currently, not only States can use violence, but also other non-State actors that have emerged into the world scene. PMSCs constitute entities capable of providing safety for many populations and effectively fighting terrorists and other criminals by resorting to the use of force. Therefore, they were and still are frequently employed in armed conflicts – both international and internal – and are used to perform their activities during peace operations.

In a globalizing world, the PMSC industry continues to maintain a significant growth.[138] Generating annually around $100 billion,[139] it has developed a clear transnational character,[140] as the industry representatives are established in one or more

---

[137] Kevin A. O'Brien, "What Should and What Should Not Be Regulated", in Simon Chesterman and Chia Lehnardt (Eds.), (2007) *From Mercenaries to Market: The Rise and Regulation of Private Military Companies*, Oxford: Oxford University Press, p. 33.

[138] Zoe Salzman, (2008) "Private Military Contractors and the Taint of a Mercenary Reputation", *New York University Journal of International Law and Politics*, Volume 40, pp. 858-860; David Isenberg, "How May We Service You?", *The Huffington Post*, August 22, 2012, <http://www.huffingtonpost.com/david-isenberg/how-may-we-service-you_b_1814311.html?utm_source=Alert-blogger>, accessed on August 23, 2012.

[139] Adam Ebrahim, (2010) "Going to War with the Army You Can Afford: The United States, International Law and the Private Military Industry", *Boston University International Law Journal*, Volume 28, p. 184.

[140] Kim R. Nossal, (1998) "Roland Goes Corporate: Mercenaries and Transnational Security Corporations in the Post-Cold War Era", *Civil Wars*, Volume 1, Issue 1, p. 34; Michael Likosky, "The Privatization of Violence", in Simon Chesterman and Angelina Fisher (Eds.), (2009) *Private Security, Public Order: The Outsourcing of Public Services and Its Limits*, Oxford: Oxford University Press, p. 23; Angelina

States and operate on the territory of the same countries or abroad. Many States actively outsource State capacities to PMSCs, but international organizations, NGOs, corporations, and wealthy individuals are also often interested in hiring those corporations carrying out numerous tasks on the basis of a contract.[141] The industry is growing and PMSCs establish their offices around the world, easily moving from one jurisdiction to another in the pursuit of financial gain. It is impressive that during the 1990s PMSCs trained military personnel in more than 42 countries around the globe.[142] In general, these companies specialize in the provision of military and security services. Some scholars, however, argue that the vast majority of PMSCs essentially have "a military rather than a security profile" given that their employees have a great deal of military experience and experience in using different types of weapons that they carry, are usually dressed in uniforms similar to those of regular State troops, and often operate under a hierarchy comparable to military command structures.[143] In reality, most of these companies have a tendency to emphasize an array of security tasks and logistical services as the main type of activity and position themselves as sole private security service providers. This business model is used by PMSCs to remove a negative stigma of their mercenary heritage and to disconnect them from private military companies (hereinafter: PMCs) that are known for the active engagement in hostilities and various types of misconduct.

Despite significant growth in the number and activities of PMSCs, the international community has not yet succeeded in providing an adequate and sufficient legal framework regulating important aspects of this industry. In that sense, it would be correct to say that one of the characteristics of the private military and security industry is the low level of effective regulation in both international and national legal dimensions specifically aimed at addressing the issues of the industry.[144] Some

---

Fisher, "Accountability to Whom?", in Simon Chesterman and Angelina Fisher (Eds.), (2009) *Private Security, Public Order: The Outsourcing of Public Services and Its Limits*, Oxford: Oxford University Press, p. 47.

141 Deborah D. Avant, (2005) *The Market for Force: The Consequences of Privatizing Security*, Cambridge: Cambridge University Press, p. 7.

142 Ibidem, p. 9.

143 Vesna Crnić-Grotić, "Non-State Actors and International Humanitarian Law: The Status of Private Military Companies", in Budislav Vukas and Trpimir M. Šošić (Eds.), (2010) *International Law: New Actors, New Concepts – Continuing Dilemmas; Liber Amicorum Božidar Bakotić*, Leiden [etc.]: Brill, p. 457.

144 Peter W. Singer, (2004) "War, Profits and the Vacuum of Law: Privatized Military Firms and International Law", *Columbia Journal of International Law*, Volume 42, Issue 2, p. 525; Eugenio Cusumano, "Policy Prospects for Regulating Private Military and Security Companies", in Francesco Francioni and Natalino Ronzitti (Eds.), (2011) *War by Contract: Human Rights, Humanitarian Law, and Private Contractors*, Oxford: Oxford University Press, p. 11; Doug Brooks and Hussein Solomon, (2000) "From the Editor's Desk", *Conflict Trends*, Issue 1, p. 1, <http://www.accord.org.za/downloads/ct/ct_2000_1.pdf?phpMyAdmin=ceeda2df659e6d3e35a63d69e93228f1>, accessed on December 20, 2011.

authors emphasize that the PMSC industry is still in its infancy[145] and appears to be "less regulated than the cheese industry"[146] or even the toy industry.[147] While this might be true, there are certainly some regulatory mechanisms applicable to PMSCs and their personnel on the national level.[148] In contrast, however, the position of those companies in international law still remains unclear.[149] Such lack of clarity goes so far that the representatives of the whole industry are sometimes regarded as operating in a legal vacuum with no laws of international or national character applicable to them.[150] There are, however, certain transnational private regulatory regimes often established by several actors – such as most PMSCs concerned about their image,[151] States, and NGOs – that aim at tackling some prominent problems of the industry by means of self-regulation. It is worth mentioning that a significant number of companies representing the private military and security industry work with governments and NGOs towards the establishment of self-regulatory mechanisms meant to ensure compliance of those firms with rules and principles of international law. These and

---

[145] Doug Brooks, (2000) "Messiahs or Mercenaries? The Future of International Private Military Services", *International Peacekeeping*, Volume 7, Issue 4, p. 133.

[146] Peter W. Singer, (2004) *The Private Military Industry and Iraq: What Have We Learned and Where to Next*, DCAF Policy Paper, p. 14.

[147] Office of the United Nations High Commissioner for Human Rights (OHCHR), "Guns for Hire", April 29, 2010, <http://www.ohchr.org/EN/NewsEvents/Pages/Gunsforhire.aspx>, accessed on January 10, 2012.

[148] Anna Leander, "Regulating the Role of Private Military Companies in Shaping Security and Politics", in Simon Chesterman and Chia Lehnardt (Eds.), (2007) *From Mercenaries to Market: The Rise and Regulation of Private Military Companies*, Oxford: Oxford University Press, p. 49; for the analysis of existing regulation at the domestic level of various EU countries see Christine Bakker and Mirko Sossai (Eds.), (2012) *Multilevel Regulation of Military and Security Contractors: The Interplay Between International, European and Domestic Norms*, Oxford and Portland, Oregon: Hart Publishing.

[149] Chia Lehnardt, "Private Military Companies and State Responsibility", in Simon Chesterman and Chia Lehnardt (Eds.), (2007) *From Mercenaries to Market: The Rise and Regulation of Private Military Companies*, Oxford: Oxford University Press, p. 142; Doug Brooks and Hussein Solomon, (2000) "From the Editor's Desk", *Conflict Trends*, Issue 1, p. 1, <http://www.accord.org.za/downloads/ct/ct_2000_1.pdf?phpMyAdmin=ceeda2df659e6d3e35a63d69e93228f1>, accessed on December 20, 2011.

[150] Hin-Yan Liu, (2010) "Leashing the Corporate Dogs of War: The Legal Implications of the Modern Private Military Company", *Journal of Conflict & Security Law*, Volume 15, Issue 1, p. 142; Chia Lehnardt, "Private Military Companies and State Responsibility", in Simon Chesterman and Chia Lehnardt (Eds.), (2007) *From Mercenaries to Market: The Rise and Regulation of Private Military Companies*, Oxford: Oxford University Press, p. 142; Sarah Percy, (2006) *Regulating the Private Security Industry*, Adelphi Series, Volume 46, Issue 384, Oxon: Routledge, p. 7; Peter W. Singer, (2004) "War, Profits and the Vacuum of Law: Privatized Military Firms and International Law", *Columbia Journal of International Law*, Volume 42, Issue 2; Angelina Fisher, "Accountability to Whom?", in Simon Chesterman and Angelina Fisher (Eds.), (2009) *Private Security, Public Order: The Outsourcing of Public Services and Its Limits*, Oxford: Oxford University Press, p. 46; Thomas Catan, "Private Armies March into a Legal Vacuum", *Financial Times*, February 10, 2005, <http://www.ft.com/cms/s/0/66d5a032-7b0e-11d9-a3ea-00000e2511c8.html#axzz1p06MGP1k>, accessed on January 15, 2012.

[151] Robert Mandel, (2002) *Armies Without States: the Privatization of Security*, Boulder: Lynne Rienner Publishers, p. 10.

other regulatory instruments relevant for the purposes of the current study will be discussed in Chapter III.

In relation to the issues of the PMSC industry, it can be seen that in the first instance the industry is well-known for its lack of transparency and a great degree of secrecy when it comes to the functioning on behalf of States or other business-related matters.[152] As stressed by the head of one of the industry associations, "we can do things on short notice and keep our mouths shut".[153] It is apparent that much information about the work of PMSCs with governments and other entities is unavailable.[154] Little is known about the actual substance of PMSC operations, activities, their partnerships with States, and their financing.[155] In the words of Lou Pingeot, these companies "actively manage their image through secrecy, lobbying, public relations campaigns and non-disclosure".[156] The contracts, for instance, can specify that private contractors and their employees are not allowed to give details of their operations. The agreement between government of Papua New Guinea and the PMSC Sandline included the following requirement:

> "... Sandline and its personnel are well versed in the requirement to maintain absolute secrecy with regard to all aspects of its activities in order to guard against compromising operations and will apply the necessary safeguards."[157]

---

152 Doug Brooks, (2000) "Messiahs or Mercenaries? The Future of International Private Military Services", *International Peacekeeping*, Volume 7, Issue 4, p. 134; Benedict Sheehy and Jackson N. Maogoto, (2008) "The Private Military Company – Unravelling the Theoretical, Legal & Regulatory Mosaic", *ILSA Journal of International & Comparative Law*, Volume 15, Issue 1, p. 170; Anna Leander, (2005) "The Power to Construct International Security: On the Significance of Private Military Companies", *Millennium: Journal of International Studies*, Volume 33, Issue 3, pp. 819-820; Kathleen M. Jennings, (2006) *Armed Security: Regulating the Private Military Industry*, New Security Programme, Fafo-report 532, p. 43, <http://www.fafo.no/pub/rapp/532/532.pdf>, accessed on May 8, 2011.

153 Leslie Wayne, "America's For-Profit Secret Army", *The New York Times*, October 13, 2002, <http://www.nytimes.com/2002/10/13/business/america-s-for-profit-secret-army.html?pagewanted=all&src=pm>, accessed on February 10, 2014.

154 Deborah D. Avant and Lee Sigelman, (2008) "What Does Private Security in Iraq Mean for US Democracy at Home?", Research Paper, p. 21, <http://bc.sas.upenn.edu/system/files/Avant_08.pdf>, accessed on July 15, 2014.

155 Carlos Ortiz, "The Private Military Company: An Entity at the Center of Overlapping of Overlapping Spheres of Commercial Activity and Responsibility", in Thomas Jäger and Gerhard Kümmel (Eds.), (2007) *Private Military and Security Companies: Chances, Problems, Pitfalls and Prospects*, Wiesbaden: VS Verlag für Sozialwissenschaften, p. 68.

156 Lou Pingeot, (2012) *Dangerous Partnership: Private Military & Security Companies and the UN*, Report, June 2012, Global Policy Forum and Rosa Luxemburg Foundation: New York and Berlin, p. 10.

157 See the contract concluded between the government of Papua New Guinea and Sandline obtained through the Project on Papua New Guinea at Australian National University in Peter W. Singer, (2008) *Corporate Warriors: The Rise of the Privatized Military Industry*, Ithaca and London: Cornell University Press, Appendix 2, p. 266.

This lack of information creates an obvious "empirical deficit"[158] concerning the PMSC industry and its representatives. While the secrecy surrounding PMSCs and their activities might explain the rapid prosperity of the industry,[159] it makes it very difficult if not impossible to conduct a thorough research of the business activities of those PMSCs hired by States, which might be necessary to assess the relationships between PMSCs and their customers and to establish the degree of connections between those firms and States.[160] The contracts concluded between PMSCs and States could serve as a good starting point for such examination, but these are also not freely available or are hidden from the public eye behind a chain of sub-contracts. Some contracts specifically contain non-disclosure clauses making their terms secret. Such was the above-mentioned agreement concluded between the government of Papua New Guinea and Sandline that was included in the work of Peter W. Singer as an appendix.[161] It provided the following with regard to the company's responsibilities:

> "Sandline will ensure that the contents of this agreement shall remain strictly confidential and will not be disclosed to any third party. Sandline will not acknowledge the existence of this contract prior to the State issuing notifications in accordance with paragraph 4.11 below and will not take credit for any successful action unless this is mutually agreed by the parties."[162]

It also appears to be the case with the client lists and funding sources of PMSCs, which are regularly confidential.[163] The processing of requests to release information

---

[158] Michael Likosky, "The Privatization of Violence", in Simon Chesterman and Angelina Fisher (Eds.), (2009) *Private Security, Public Order: The Outsourcing of Public Services and Its Limits*, Oxford: Oxford University Press, pp. 23-24.

[159] Peter W. Singer, (2008) *Corporate Warriors: The Rise of the Privatized Military Industry*, Ithaca and London: Cornell University Press, p. x.

[160] Martha Minow, (2005) "Outsourcing Power: How Privatizing Military Efforts Challenges Accountability, Professionalism and Democracy", *Boston College Law Review*, Volume 46, p. 999; Deborah D. Avant and Lee Sigelman, (2008) "What Does Private Security in Iraq Mean for US Democracy at Home?", Research Paper, p. 22, <http://bc.sas.upenn.edu/system/files/Avant_08.pdf>, accessed on July 15, 2014.

[161] Peter W. Singer, (2008) *Corporate Warriors: The Rise of the Privatized Military Industry*, Ithaca and London: Cornell University Press, pp. 263-270; as noted by Antenor Hallo de Wolf, Sandline ceased its activities and the text of the contract that had been provided on its website became unavailable. It is, however, on file with the author himself. See Antenor Hallo de Wolf, (2012) *Reconciling Privatization with Human Rights*, School of Human Rights Research Series, Volume 49, Cambridge, Antwerp and Portland: Intersentia, p. 481, footnote 103.

[162] Peter W. Singer, (2008) *Corporate Warriors: The Rise of the Privatized Military Industry*, Ithaca and London: Cornell University Press, p. 266.

[163] Doug Brooks and Hussein Solomon, (2000) "From the Editor's Desk", *Conflict Trends*, Issue 1, p. 2, <http://www.accord.org.za/downloads/ct/ct_2000_1.pdf?phpMyAdmin=ceeda2df659e6d3e35a63d69e93228f1>, accessed on December 20, 2011.

on contract documents and performance reports by governments is often very slow. An expert in the field of private military and security service provision, David Isenberg, had requested some written communications from the US State Department, but the information made available appeared to be poor and incomplete.[164] Although many companies around the world use the same strategy of staying in the shadows in order to protect their business interests, concerns can be raised with regard to the activities of PMSCs often entrusted with governmental tasks and exercising essentially a government's power over legitimate violence on behalf of States.[165]

Together with the greater reliance on PMSCs in recent years, their involvement in breaches of international humanitarian law and human rights has increased. While it is not correct to refer to PMSCs as bands of criminals, their broader deployment in conflict areas does increase the number of incidents they may be involved in. Without doubt, private contractors can potentially violate rules and principles of international law[166] when they are in a position to exercise lethal force. Some might point to the similarities between PMSCs and mercenaries and argue that employees of these corporations "are more susceptible to commit war crimes".[167] Sadly, experience teaches us that PMSCs have frequently infringed human rights and international humanitarian law in the past and engaged in the commission of severe crimes, such as torture.[168] These instances of illegal misconduct shaped a negative image for the whole industry and led to much criticism by academics, civil society actors, and others. They emphasized an apparent lack of oversight and monitoring that must be exercised by the hiring States over PMSCs,[169] which are very important, as they can be used to prevent

---

164 David Isenberg, "The State Department Tells Us How They Really Felt: Part 1", *The Huffington Post*, February 17, 2012, <http://www.huffingtonpost.com/david-isenberg/state-department-blackwater_b_1280847.html>, accessed on March 28, 2012.

165 Michael Likosky, "The Privatization of Violence", in Simon Chesterman and Angelina Fisher (Eds.), (2009) *Private Security, Public Order: The Outsourcing of Public Services and Its Limits*, Oxford: Oxford University Press, p. 12.

166 Daphne Barak-Erez, "The Privatization of Continuum", in Simon Chesterman and Angelina Fisher (Eds.), (2009) *Private Security, Public Order: The Outsourcing of Public Services and Its Limits*, Oxford: Oxford University Press, p. 71.

167 Peter W. Singer, (2008) *Corporate Warriors: The Rise of the Privatized Military Industry*, Ithaca and London: Cornell University Press, p. 217.

168 Peter W. Singer, (2004) "War, Profits and the Vacuum of Law: Privatized Military Firms and International Law", *Columbia Journal of International Law*, Volume 42, Issue 2, p. 523; Mervyn Frost, "The Ethics of PMCs in Global Civil Society", in Andrew Alexandra, Deane-Peter Baker, and Marina Caparini (Eds.), (2008) *Private Military and Security Companies: Ethics, Policies and Civil-Military Relations*, London: Routledge, p. 43; Katherine Fallah, (2006) "Corporate Actors: The Legal Status of Mercenaries in Armed Conflict", *International Review of the Red Cross*, Volume 88, Issue 863, p. 600.

169 Jan Stöber, "Contracting in the Fog of War... Private Security Providers in Iraq: A Principal-Agent Analysis", in Thomas Jäger and Gerhard Kümmel (Eds.), (2007) *Private Military and Security Companies: Chances, Problems, Pitfalls and Prospects*, Wiesbaden: VS Verlag für Sozialwissenschaften, p. 133.

the commission of a large number of abuses by the members of PMSCs deployed in various operations and to address them.[170]

The assessment of types of misconduct made in the previous Chapter is merely an indication of the broad spectrum of breaches of international and municipal law involving PMSCs. More interesting for the current study are the consequences of those illegal activities often carried out in unstable regions of the world where the issue of accountability cannot be addressed in practical terms or is legally impossible.[171] They illustrate the lack of responsibility of PMSCs, their employees, States, and other involved actors for the acts and omissions of these particular firms and their personnel. It might be argued that "privatization of security functions and proliferation of PMSCs have taken place without a corresponding outsourcing of accountability regimes".[172] There are not only uncertainties regarding the possibility and measures of accountability, but also the question of to whom that accountability is owed and should be owed. This fact creates a picture of impunity of PMSC personnel[173] and even of the States outsourcing their functions to private military and security firms. Often members of the personnel of those corporations managed to escape any form of individual responsibility and continued their activities for the same employer. Companies appearing in the media as a result of various scandals rebranded themselves, but kept working for legitimate governments. Finally, until now, States have denied any connection to the perpetrators of the crimes and have not been held responsible for their misconduct. This "accountability mismatch" and the possibility of States escaping responsibility raise serious concerns about the whole industry and its representatives.

The above-mentioned accountability problems originate from the specific nature of the PMSC industry and many legal and regulatory problems associated with this new type of military and security actors active in the international arena.[174] These companies are private corporations performing the variety of essential military and

---

170 Martha Minow, (2005) "Outsourcing Power: How Privatizing Military Efforts Challenges Accountability, Professionalism and Democracy", *Boston College Law Review*, Volume 46, pp. 1005-1012.

171 Simon Chesterman, (2008) "We Can't Spy ... If We Can't Buy!: The Privatization of Intelligence and the Limits of Outsourcing 'Inherently Governmental Functions'", *The European Journal of International Law*, Volume 19, Issue 5, p. 1057.

172 Angelina Fisher, "Accountability to Whom?", in Simon Chesterman and Angelina Fisher (Eds.), (2009) *Private Security, Public Order: The Outsourcing of Public Services and Its Limits*, Oxford: Oxford University Press, p. 46.

173 Peter W. Singer, (2008) *Corporate Warriors: The Rise of the Privatized Military Industry*, Ithaca and London: Cornell University Press, pp. 220-221; Stephen Lendman, "Outsourcing War: The Rise of Private Military Contractors", *The Intelligence Daily*, January 20, 2010, <http://dissidentvoice.org/2010/01/outsourcing-war-the-rise-of-private-military-contractors-pmcs/>, accessed on March 12, 2012.

174 Peter W. Singer, (2004) "War, Profits and the Vacuum of Law: Privatized Military Firms and International Law", *Columbia Journal of International Law*, Volume 42, Issue 2, p. 524; Adam Ebrahim, (2010) "Going to War with the Army You Can Afford: The United States, International Law and the Private Military Industry", *Boston University International Law Journal*, Volume 28, pp. 183-184.

security-related tasks on behalf of governments and other actors. As stated above, the whole industry of PMSCs is often regarded as under-regulated.[175] In addition, there are no generally accepted definitions of companies representing the PMSC industry and their qualification on the battlefield.[176] Although PMSCs do not operate in a regulatory void, the applicability of international law to them and their employers, which will be examined further in the following part of the present study, poses significant questions[177] and is not as straightforward as it might be the case with the regular armed forces representing a certain State.[178] The international legal framework appears to be incomplete, incoherent, vague, and even contradictory.[179] In that regard, a logical comparison can be made between PMSCs and typical mercenaries, the use of which is not entirely prohibited by international law, but can result in international responsibility. At a later stage in the study, a legal assessment will be made of the possibility to classify PMSCs as mercenary organizations and their employees as mercenaries. In addition, in many countries there are no legal rules specifically designed to deal with PMSCs and to ensure their compliance with international humanitarian law and human rights.[180] In general, there appears to be a lack of clear understanding regarding the applicability of international or national legal rules and their enforcement with regard to the activities of private military and security contractors. In their operations, PMSCs seem to be mainly guided not by the rule of law, but by economics and the drive to make a profit.[181]

There are also legitimacy concerns[182] with respect to the role played by PMSCs in military and security operations and the actual role of States in the use of force

---

175 Eugenio Cusumano, "Policy Prospects for Regulating Private Military and Security Companies", in Francesco Francioni and Natalino Ronzitti (Eds.), (2011) *War by Contract: Human Rights, Humanitarian Law, and Private Contractors*, Oxford: Oxford University Press, p. 11; Doug Brooks, (2000) "Messiahs or Mercenaries? The Future of International Private Military Services", *International Peacekeeping*, Volume 7, Issue 4, p. 136.

176 Doug Brooks and Hussein Solomon, (2000) "From the Editor's Desk", *Conflict Trends*, Issue 1, p. 2, <http://www.accord.org.za/downloads/ct/ct_2000_1.pdf?phpMyAdmin=ceeda2df659e6d-3e35a63d69e93228f1>, accessed on December 20, 2011.

177 Laura A. Dickinson, (2006) "Public Law Values in a Privatized World", *The Yale Journal of International Law*, Volume 31, Issue 2, p. 398.

178 Peter W. Singer, (2008) *Corporate Warriors: The Rise of the Privatized Military Industry*, Ithaca and London: Cornell University Press, p. 220.

179 Katherine Fallah, (2006) "Corporate Actors: The Legal Status of Mercenaries in Armed Conflict", *International Review of the Red Cross*, Volume 88, Issue 863, p. 601.

180 See, for instance, Guido den Dekker, "The Netherlands", in Christine Bakker and Mirko Sossai (Eds.), (2012) *Multilevel Regulation of Military and Security Contractors: The Interplay Between International, European and Domestic Norms*, Oxford and Portland, Oregon: Hart Publishing, p. 253.

181 Peter W. Singer, (2004) "War, Profits and the Vacuum of Law: Privatized Military Firms and International Law", *Columbia Journal of International Law*, Volume 42, Issue 2, pp. 524, 543.

182 Kathleen M. Jennings, (2006) *Armed Security: Regulating the Private Military Industry*, New Security Programme, Fafo-report 532, pp. 43-44, <http://www.fafo.no/pub/rapp/532/532.pdf>, accessed on May 8, 2011.

and provision of those core services. Due to the many issues touched upon above, many PMSCs seem to struggle to maintain a legitimate face and keep changing their names,[183] promoting the legitimacy of operations in different forums and cooperating with other businesses within the framework of self-regulatory instruments. As entities capable of exercising violence on behalf of legitimate States and even non-State actors, PMSCs should be approached with utmost care when they are involved in conflicts around the globe. It could be assumed that those private corporations performing military and security services are more likely to engage in violations of international law and analysis of historical evidence may prove this. More importantly, members of the personnel of PMSCs violating international humanitarian law and human rights, just as firms themselves and States, are often in a position to escape accountability for the atrocities which they might have committed during a mission.[184] Time and again, this implies a need to address the question of responsibility of States contracting with PMSCs on a more frequent basis and strengthens this need even further in cases when States opt to outsource essential governmental prerogatives to private companies. The subsequent Chapters will deal with this issue in more detail.

3.2 Definition of a PMSC

After many years of extensive debates on the phenomenon of privatization of war and security associated with the outsourcing of military and security services to private companies, scholars still keep fiercely elaborating on the concept of private military and security companies.[185] Nevertheless, it is very difficult to speak about them in general terms.[186] To a certain extent, PMSCs are immune to any form of categorization or typology, which makes it nearly impossible for academics to get a grip on

---

[183] Jeffrey Benzing, "Rebranded Academi Looks to Shed Blackwater Baggage", *Just Anti-Corruption*, March 27, 2012, <http://www.mainjustice.com/justanticorruption/2012/03/27/rebranded-academi-looks-to-shed-blackwater-baggage>, accessed on March 28, 2012.

[184] Kathleen M. Jennings, (2006) *Armed Security: Regulating the Private Military Industry*, New Security Programme, Fafo-report 532, p. 42, <http://www.fafo.no/pub/rapp/532/532.pdf>, accessed on May 8, 2011.

[185] David Isenberg, (2009) *Shadow Force: Private Security Contractors in Iraq*, Westport and London: Praeger Security International, p. 25; Eugenio Cusumano, "Policy Prospects for Regulating Private Military and Security Companies", in Francesco Francioni and Natalino Ronzitti (Eds.), (2011) *War by Contract: Human Rights, Humanitarian Law, and Private Contractors*, Oxford: Oxford University Press, p. 14; Fred Schreier and Marina Caparini, (2005) *Privatising Security: Law, Practice and Governance of Private Military and Security Companies*, Occasional Paper No. 6, DCAF, pp. 17-18.

[186] David Isenberg, (2009) *Shadow Force: Private Security Contractors in Iraq*, Westport and London: Praeger Security International, p. 157; Emanuela-Chiara Gillard, (2006) "Business Goes to War: Private Military / Security Companies and International Humanitarian Law", *International Review of the Red Cross*, Volume 88, Issue 863, pp. 529-530.

their actual nature and functioning.[187] David Isenberg believes that "although they have attracted growing attention from analysts, scholars, governments, and the general public in the past decade, there are still no agreements on how to define them, let alone categorize them".[188] It should be stressed, however, that in many publications the term PMSCs is often cumulatively used to refer to two types of corporate entities.[189] In the first instance, there are private military companies that perform tasks of a military nature. One can think of services related to the offensive combat and military operations carried out on the ground, on the sea, and in the air. In addition, there are also private security companies, which largely carry out security services, such as guarding prisons and providing personal security. This distinction is based on the general level of the PMSC activity in order to distinguish "passive" and "active" PMSCs.[190] Using this classification, analysts point out that active companies often engage in offensive military operations, while passive firms have as their main task to defend their clients, such as State officials and members of corporate staff, or to provide training and advice. Similarly, some authors identify two main variants of PMSCs: combat private military companies and non-combat PMCs.[191] This dichotomy is exactly what both terms imply: some private companies engage in activities having a structured military character, while other firms are employed to carry out non-combat related tasks, such as offering protection and logistic services. Generally speaking, PMSCs offering services of a clear military nature are more likely to

---

187 Shantanu Chakrabarti, (2009) *Privatisation of Security in the Post Cold-War Period: An Overview of its Nature and Implications*, Institute for Defence Studies and Analyses Monograph Series, No. 2, New Delhi: Magnum Custom Publishing, p. 7.

188 David Isenberg, (2009) *Shadow Force: Private Security Contractors in Iraq*, Westport and London: Praeger Security International, p. 157.

189 Carlos Ortiz, (2009) "Private Military and Security Companies: Misconceptions, Reconceptualisations, and Regulation", Working Paper No. 9, The Centre for Global Political Economy, University of Sussex, p. 2; Eugenio Cusumano, "Policy Prospects for Regulating Private Military and Security Companies", in Francesco Francioni and Natalino Ronzitti (Eds.), (2011) *War by Contract: Human Rights, Humanitarian Law, and Private Contractors*, Oxford: Oxford University Press, p. 15; Damian Lilly, (2000) "The Privatization of Peacekeeping: Prospects and Realities", *Disarmament Forum*, Issue 3, p. 53; Benedict Sheehy and Jackson N. Maogoto, (2008) "The Private Military Company – Unravelling the Theoretical, Legal & Regulatory Mosaic", *ILSA Journal of International & Comparative Law*, Volume 15, Issue 1, p. 148.

190 Peter W. Singer, (2008) *Corporate Warriors: The Rise of the Privatized Military Industry*, Ithaca and London: Cornell University Press, p. 89; Eugenio Cusumano, "Policy Prospects for Regulating Private Military and Security Companies", in Francesco Francioni and Natalino Ronzitti (Eds.), (2011) *War by Contract: Human Rights, Humanitarian Law, and Private Contractors*, Oxford: Oxford University Press, p. 15.

191 Sarah Percy, "Morality and Regulation", in Simon Chesterman and Chia Lehnardt (Eds.), (2007) *From Mercenaries to Market: The Rise and Regulation of Private Military Companies*, Oxford: Oxford University Press, pp. 12-13; Zoe Salzman, (2007-2008) "Private Military Contractors and a Taint of a Mercenary Reputation", *New York University Journal of International Law and Politics*, Volume 40, Issue 3, p. 857.

engage in violations of most essential rules and norms of international law, specifically international humanitarian law and human rights law, than their counterparts that merely perform security tasks and other non-combat operations.

Christopher Kinsey is of the opinion that PMSCs should be classified as three main categories of businesses: (a) private combat companies, (b) proxy military companies, and (c) private security companies. Freelance operators belong to the third class of firms.[192] Peter W. Singer appears to employ his "tip-of-the-spear typology" dividing all PMSCs into three basic groups: (a) military provider firms, (b) military consultant firms, and (c) military support firms.[193] The first group of businesses in Singer's typology – military provider firms – are those companies that provide direct military assistance, for instance, by engaging in actual combat operations. Military consultant firms specialize in services in the sphere of training and advice given to their potential clients, but do not participate in the actual military operations. Finally, military support firms concentrate on the provision of supplementary military services, such as technical support, transportation, intelligence, and logistics. According to Singer, this particular type of classification can be very useful in exploring the variation within the PMSC industry and the variation of the organization, operation, and impact of those firms.[194]

Doug Brooks, president emeritus of the International Stability Operations Association (hereinafter: ISOA), points out that there are three distinguished groups identified within the spectrum of all PMSCs: (a) private military companies, (b) private security companies, and (c) private support companies.[195] According to him, private military firms are of particular importance nowadays, getting more and more contracts with a variety of clients.[196] It is, however, worth noting that the artificial distinction between these three types of companies or the above-made division in relation to private military and security companies can hardly be maintained in the most cases. PMSCs are usually involved in activities having a very broad spectrum and

---

[192] Christopher Kinsey, (2006) *Corporate Soldiers and International Security: The Rise of Private Military Companies*, London: Routledge, pp. 13-21.

[193] Peter W. Singer, (2008) *Corporate Warriors: The Rise of the Privatized Military Industry*, Ithaca and London: Cornell University Press, p. 91; Peter W. Singer, (2005) "Outsourcing War", *Foreign Affairs*, Volume 84, Issue 2, pp. 120-121; Shantanu Chakrabarti, (2009) *Privatisation of Security in the Post Cold-War Period: An Overview of its Nature and Implications*, Institute for Defence Studies and Analyses Monograph Series, No. 2, New Delhi: Magnum Custom Publishing, p. 8; Eugenio Cusumano, "Policy Prospects for Regulating Private Military and Security Companies", in Francesco Francioni and Natalino Ronzitti (Eds.), (2011) *War by Contract: Human Rights, Humanitarian Law, and Private Contractors*, Oxford: Oxford University Press, p. 15.

[194] Peter W. Singer, (2008) *Corporate Warriors: The Rise of the Privatized Military Industry*, Ithaca and London: Cornell University Press, pp. 91-92.

[195] Shantanu Chakrabarti, (2009) *Privatisation of Security in the Post Cold-War Period: An Overview of its Nature and Implications*, Institute for Defence Studies and Analyses Monograph Series, No. 2, New Delhi: Magnum Custom Publishing, p. 9.

[196] Ibidem.

overlapping each other. One single similarity between the main actors – private military and security companies – is particularly striking: they both focus on exchanging military and security services for payment, while often also performing other related support tasks.[197] Therefore, further in the text the term PMSCs will be used as understood, accepted, and used by a large number of experts[198] to refer to two main types of companies: private military and security firms.

It has been rightly pointed out by Peter W. Singer that there is "a vacuum of established facts" with regard to PMSCs and, more significantly, the absence of consensus among the experts in the field and general understanding regarding those firms and the PMSC industry as a whole.[199] Doug Brooks and Hussein Solomon add to the discussion on the issue that "there are no universally accepted definitions of even the most widely used terms."[200] David Isenberg indicates that many uncertainties with regard to PMSCs and their employees – such as many questions about the numbers of their contingent operating abroad – arise from the general lack of agreement as to what those entities and persons actually are.[201] It is, however, obvious that formulating a definition of PMSCs is not merely a purely theoretical issue with some taxonomic connotations. It is an essential operation necessary to accumulate knowledge about the whole concept of these private entities offering military and security expertise which is necessary for understanding and regulating the industry.[202] As the longest journey starts with a single step, the academic research into the possibility of holding States responsible for the violations of international law committed by the firms providing military

---

[197] Sarah Percy, "Morality and Regulation", in Simon Chesterman and Chia Lehnardt (Eds.), (2007) *From Mercenaries to Market: The Rise and Regulation of Private Military Companies*, Oxford: Oxford University Press, p. 13.

[198] Francesco Francioni and Natalino Ronzitti, "Introduction", in Francesco Francioni and Natalino Ronzitti (Eds.), (2011) *War by Contract: Human Rights, Humanitarian Law, and Private Contractors*, Oxford: Oxford University Press, p. 1; Emanuela-Chiara Gillard, (2006) "Business Goes to War: Private Military / Security Companies and International Humanitarian Law", *International Review of the Red Cross*, Volume 88, Issue 863; Amol Mehra, (2010) "Bridging Accountability Gaps – The Proliferation of Private Military and Security Companies and Ensuring Accountability for Human Rights Violations", *Pacific McGeorge Global Business & Development Legal Journal*, Volume 22, Issue 2, and other publications.

[199] Peter W. Singer, (2008) *Corporate Warriors: The Rise of the Privatized Military Industry*, Ithaca and London: Cornell University Press, p. viii.

[200] Doug Brooks and Hussein Solomon, (2000) "From the Editor's Desk", *Conflict Trends*, Issue 1, p. 2, <http://www.accord.org.za/downloads/ct/ct_2000_1.pdf?phpMyAdmin=ceeda2df659e6d3e35a63d69e93228f1>, accessed on December 20, 2011.

[201] Won Kidane, (2010) "The Status of Private Military Contractors under International Humanitarian Law", *Denver Journal of International Law and Policy*, Volume 38, Issue 3, p. 391; David Isenberg, (2009) *Shadow Force: Private Security Contractors in Iraq*, Westport and London: Praeger Security International, p. 27.

[202] Eugenio Cusumano, "Policy Prospects for Regulating Private Military and Security Companies", in Francesco Francioni and Natalino Ronzitti (Eds.), (2011) *War by Contract: Human Rights, Humanitarian Law, and Private Contractors*, Oxford: Oxford University Press, p. 14.

and security services should also commence with the inquiry into the nature of those corporations and their functional definition. In order to work through the "definitional morass"[203] with regard to PMSCs, it seems intuitively right to determine the core characteristics of both private military and private security companies.

In the first instance, private military and private security companies are non-State actors sharing the same organizational business structure in their capacity of for-profit companies. In this respect, non-State actors can be defined as "all those actors that are not (representatives of) states, yet that operate at the international level and are potentially relevant to international relations".[204] Thus, private contractors belong to the group of individuals and entities which do not act under the lawful authority of States in performing their activities falling under the scope of international law.[205] Being private businesses, PMSCs all serve the various needs of their clients by providing a large number of professional services on the basis of a contract. One of the main features of these entities is their private character, as they remain the property of non-State actors and do not belong to the official State machinery.[206] However, one should understand that PMSCs cannot be said to be easily categorized and made comparable to each other. Also, they often do not serve the same markets.[207] Finally, not all PMSCs are established in the same manner: some of them are divisions of large firms, while others are just small businesses acting on a less prominent level as "tiny niche players".[208]

A general definition of a PMSC was used by the UN Working Group on the use of mercenaries as a means of violating human rights and impeding the exercise of the right of peoples to self-determination (hereinafter: the UN Working Group on the Use of Mercenaries) in its 2010 Draft International Convention on PMSCs (hereinafter: the Draft International Convention on PMSCs). It took the view that this type of corporation must be seen as "a corporate entity which provides on a compensatory basis military and/or security services by physical persons and/or legal entities".[209] Being

---

[203] David Isenberg, (2009) *Shadow Force: Private Security Contractors in Iraq*, Westport and London: Praeger Security International, p. 14.

[204] Bas Arts, (2003-2004) *Non-State Actors in Global Governance: Three Faces of Power*, Max Planck Project Group on Common Goods, Preprint, Bonn, <http://www.coll.mpg.de/pdf_dat/2003_04online.pdf>, accessed on November 10, 2011.

[205] Vesna Crnić-Grotić, "Non-State Actors and International Humanitarian Law: The Status of Private Military Companies", in Budislav Vukas and Trpimir M. Šošić (Eds.), (2010) *International Law: New Actors, New Concepts – Continuing Dilemmas; Liber Amicorum Božidar Bakotić*, Leiden [etc.]: Brill, p. 455; see also UN Security Council Resolution 1540 of April 28, 2004, UN Doc. S/RES/1540.

[206] Stephanie Brown, (2010) "Bottom-up Lawmaking and the Regulation of Private Military and Security Companies", *Cuadernos de Derecho Transnacional*, Volume 2, Issue 1, p. 46.

[207] Ibidem.

[208] David Isenberg, (2009) *Shadow Force: Private Security Contractors in Iraq*, Westport and London: Praeger Security International, p. 67.

[209] UN Human Rights Council, *Draft of a Possible Convention on Private Military and Security Companies (PMSCs) for Consideration and Action by the Human Rights Council*, Prepared by the Working

aware of a great variety of possible definitions of PMSCs, the Geneva Centre for the Democratic Control of Armed Forces (hereinafter: DCAF) frames its own description of PMSCs in fairly broad terms referring to them as to the "businesses that offer specialized services related to war and conflict, including combat operations, strategic planning, intelligence collection, operational and logistical support, training, procurement and maintenance".[210] This is, in fact, a general definition of PMSCs, which can be used to identify a large number of businesses as companies of the global private military and security industry. Undoubtedly, there could be good reasons to adopt such a broad definition of the firms offering military and security expertise in order not to exclude any of their representatives from its application. In addition, the backgrounder published by the DCAF stipulates that the organizational structure of PMSCs providing their services for the main reason of making a profit rather than for political reasons is that of registered businesses with corporate structures.

Another clear and fairly precise description is provided by Francesco Francioni and Natalino Ronzitti in the latest book written collaboratively by a large number of experts within the framework of the PRIV-WAR research project. The legal scholars note that PMSCs form part of a modern phenomenon of privatization of services lying in the sphere of the military and security services provision that have usually been seen and are still usually regarded as typical governmental functions. They define PMSCs as "corporations offering security, defense and/or military services to states, international organizations, non-governmental organizations, and private companies and/or armed groups".[211]

Carlos Ortiz, a visiting research fellow at the Centre for Global Political Economy at the University of Sussex, who has given his own definition of PMSCs, puts into his description the core nature of PMSCs as international enterprises and underlines the idea that those companies can often resort to the use of force that has long been regarded as the main domain of sovereign States. The author also expressly recognizes the fact of legal establishment of those businesses on the territory of some States, such as the United States and the United Kingdom. According to him, PMSCs can be defined as:

> "… legally established international firms offering services that incorporate the 'potential' to exercise force in a systematic way and by military or paramilitary means, and/or the transfer or enhancement of that potential to clients. The potential can materialize

---

Group on the use of mercenaries as a means of violating human rights and impeding the exercise of the right of peoples to self-determination, May 13, 2011, First Session, Geneva, May 23-27, 2011, UN Doc. A/HRC/WG.10/1/2, Article 2(a).

210 DCAF, (2006) *Private Military Companies*, DCAF Backgrounder, <http://www.dcaf.ch/publications/kms/details.cfm?lng=en&id=18412&nav1=5>, accessed on May 5, 2011.

211 Francesco Francioni and Natalino Ronzitti, "Introduction", in Francesco Francioni and Natalino Ronzitti (Eds.), (2011) *War by Contract: Human Rights, Humanitarian Law, and Private Contractors*, Oxford: Oxford University Press, p. 1.

when delivering protection services to public and private entities in climates of instability, such as armed conflicts and humanitarian crises."[212]

Following the reasoning of another expert working for the institute for Defence Studies and Analyses in New Delhi, Shantanu Chakrabarti, one can define PMSCs as "legitimate national corporations organized in accordance with the legal codes of their respective home countries".[213] The core distinctive feature of PMSCs, which sets them apart from mercenaries, is, according to many authors, the idea of corporatization of the provision of military and security services.[214] In the first instance, those firms remain legally established commercial enterprises created and registered in accordance with the domestic legislation of respective States. These entities have a clear hierarchical structure and offer a plethora of services associated with a wide range of military and security skills.

Doug Brooks predicted in 2000 that in the future the offer and provision of services by PMSCs would broaden to include not only security-related tasks, but also military services.[215] The services offered by PMSCs and provided to a broad spectrum of clients often include, but are not limited to:

> "… services that involve the potential to exercise force in a systematic way and by military means and/or the transfer of enhancement of that potential to clients… (PMSCs) also offer services that do not involve the potential to exercise force. These include training in non-conflict environments, and management and/or maintenance of strategic facilities, transport, and communications infrastructures. Still, the delivery of such services involves civilian rendering of expertise that, nonetheless, directly enhances the recipient's military and security capabilities. All the services on offer by the private military industry have a direct impact on and contribute to the management of the State's monopoly of legitimate violence."[216]

In general, the services realized by PMSCs can be divided into two main categories, encompassing armed and unarmed tasks, which are associated with a wide scope of

---

212 Carlos Ortiz, (2010) "The New Public Management of Security: the Contracting and Managerial State and the Private Military Industry", *Public Money & Management*, Volume 30, Issue 1, p. 35.

213 Shantanu Chakrabarti, (2009) *Privatisation of Security in the Post Cold-War Period: An Overview of its Nature and Implications*, Institute for Defence Studies and Analyses Monograph Series, No. 2, New Delhi: Magnum Custom Publishing, p. 6.

214 Peter W. Singer, (2008) *Corporate Warriors: The Rise of the Privatized Military Industry*, Ithaca and London: Cornell University Press, p. 45.

215 Doug Brooks, (2000) "Messiahs or Mercenaries? The Future of International Private Military Services", *International Peacekeeping*, Volume 7, Issue 4, p. 140.

216 Carlos Ortiz, "The Private Military Company: An Entity at the Center of Overlapping of Overlapping Spheres of Commercial Activity and Responsibility", in Thomas Jäger and Gerhard Kümmel (Eds.), (2007) *Private Military and Security Companies: Chances, Problems, Pitfalls and Prospects*, Wiesbaden: VS Verlag für Sozialwissenschaften, pp. 60-61.

critical jobs.[217] While the provision of armed services is often considered as closely related to the direct participation of private contractors in armed conflicts, it can also include the provision of static security in the form of protection of fixed or static sites, such as government buildings, convoy security, security escorts, and personal security to high-ranking individuals. Unarmed security services include security training, operational coordination, and intelligence analysis. In this regard, experts make a division between three principal areas of PMSCs specialization on the ground, which can vary significantly among various representatives of the industry.[218] In the first place, those companies focus on the provision of protection, including personal protection details, convoy protection, and site protection. In addition, they provide consulting services, such as security and intelligence reports and security management advice. Finally, PMSCs specialize in training for military and security missions.[219]

Francesco Francioni and Natalino Ronzitti state that PMSCs carry out various services that are of particular importance to their clients operating in the globalizing world. The point is that "these services include armed guarding and protection of persons and objects or buildings, maintenance and operation of weapons systems, prisoner detention and interrogation, intelligence, risk assessment and military research analysis, as well as advice to or training of local forces and security personnel".[220] Private military and security contractors appear to be operating in many situations around the globe and find themselves participating in actual hostilities and supporting "armed forces in prolonged military occupation", taking care of "security in peace-keeping/peace-enforcing operations", "international administration in post-conflict institution building and intelligence gathering". As a matter of fact, only a very small number of contracts require PMSCs to exercise core military functions and to deploy their personnel on the battlefield. More often these companies offer advice and training services to military forces, provide operational support, such as maintenance of weapons systems, transport, and command and control, or provide

---

217 US Congressional Research Service, *Private Security Contractors in Iraq: Background, Legal Status, and Other Issues*, Report for Congress, August 25, 2008, pp. 2-3; Peter W. Singer, (2005) "Outsourcing War", *Foreign Affairs*, Volume 84, Issue 2, pp. 122-123; David Isenberg, (2009) *Shadow Force: Private Security Contractors in Iraq*, Westport and London: Praeger Security International, p. 161; Todd S. Milliard, (2003) "Overcoming Post-Colonial Myopia. A Call to Recognize and Regulate PMCs", *Military Law Review*, Volume 176, Issue 1, pp. 11-14; Alexis P. Kontos, (2004) "'Private' Security Guards: Privatized Force and State Responsibility under International Human Rights Law", *Non-State Actors and International Law*, Volume 4, pp. 202-203.

218 Jan Stöber, "Contracting in the Fog of War... Private Security Providers in Iraq: A Principal-Agent Analysis", in Thomas Jäger and Gerhard Kümmel (Eds.), (2007) *Private Military and Security Companies: Chances, Problems, Pitfalls and Prospects*, Wiesbaden: VS Verlag für Sozialwissenschaften, p. 127.

219 Ibidem.

220 Francesco Francioni and Natalino Ronzitti, "Introduction", in Francesco Francioni and Natalino Ronzitti (Eds.), (2011) *War by Contract: Human Rights, Humanitarian Law, and Private Contractors*, Oxford: Oxford University Press, p. 1.

internal security, acting not only as international civilian police and site security, but also offering police training, crime prevention, and intelligence services.[221] In essence, most private firms working in this sphere focus on the provision of logistics, while, for instance, security services are offered merely by five per cent of all industry representatives that are members of the International Stability Operations Association, the largest standards-based PMSC association.[222] For example, after the invasion of Iraq by a multinational force in Operation Iraqi Freedom, PMSCs have focused on three main categories of services on the invaded territory: personal security details for senior civilian officials, non-military site security, and non-military security of convoys.[223] Nevertheless, it is obvious even for a non-expert in the field that armed professionals hired by and working for PMSCs and carrying out military and security tasks can easily switch from a purely passive defensive role to a more active offensive function. The historical evidence seems to suggest that on various occasions PMSC employees crossed this line and offered direct combat support to their clients.[224] It goes without saying that the observed practice of PMSCs taking an active part in armed conflicts can lead to the commission of a large number of violations of both international humanitarian and human rights law.[225]

Seeking to elaborate on the definition of PMSCs, we can refer to the Montreux Document on Pertinent International Legal Obligations and Good Practices for States Related to Operations of Private Military and Security Companies During Armed Conflict (hereinafter: the Montreux Document) from September 2008.[226] This legal

---

[221] Deborah D. Avant, (2005) *The Market for Force: The Consequences of Privatizing Security*, Cambridge: Cambridge University Press, pp. 17-21.

[222] Eugenio Cusumano, "Policy Prospects for Regulating Private Military and Security Companies", in Francesco Francioni and Natalino Ronzitti (Eds.), (2011) *War by Contract: Human Rights, Humanitarian Law, and Private Contractors*, Oxford: Oxford University Press, p. 15.

[223] David Isenberg, (2009) *Shadow Force: Private Security Contractors in Iraq*, Westport and London: Praeger Security International, p. 158.

[224] Terry D. Gill and Dieter Fleck, "Private Contractors and Security Companies", in Terry D. Gill and Dieter Fleck (Eds.), (2010) *The Handbook of the International Law of Military Operations*, Oxford: Oxford University Press, p. 490.

[225] José L. G. del Prado, (2009) "Private Military and Security Companies and the UN Working Group on the Use of Mercenaries", *Journal of Conflict & Security Law*, Volume 13, Issue 3, pp. 435-436; Eugenio Cusumano, "Policy Prospects for Regulating Private Military and Security Companies", in Francesco Francioni and Natalino Ronzitti (Eds.), (2011) *War by Contract: Human Rights, Humanitarian Law, and Private Contractors*, Oxford: Oxford University Press, p. 15.

[226] The Montreux Document is a product of a joint initiative of the Swiss government and the International Committee of the Red Cross that brought together 17 governments, non-governmental organizations, industry representatives, and academic experts in order to regulate the relationships between States and PMSCs. Initially, 17 States endorsed the Montreux Document in 2008: Afghanistan, Angola, Australia, Austria, Canada, China, France, Germany, Iraq, Poland, Sierra Leone, South Africa, Sweden, Switzerland, the United Kingdom of Great Britain and Northern Ireland, Ukraine, and the United States of America. Currently, there are 50 States and three international organizations supporting the initiative. See UN General Assembly, (2008) *Montreux Document on Pertinent International Legal Obligations*

instrument, in the form of an international agreement, provides the following extensive definition of the above-mentioned companies:

> "PMSCs are private business entities that provide military and/or security services, irrespective of how they describe themselves. Military and security services include, in particular, armed guarding and protection of persons and objects, such as convoys, buildings and other places; maintenance and operation of weapons systems; prisoner detention; and advice to or training of local forces and security personnel."[227]

According to the same international document, there are three main types of States playing an important role in the functioning of PMSCs employing their activities on behalf of State governments: contracting States, territorial States, and home States.[228] Contracting or hiring States are those States that directly contract for services offered by PMSCs. It is to be noted that PMSCs sub-contracting their services to other private contractors also fall under the definition of business entities working for contracting States. In addition, territorial or host States are States on which territory PMSCs carry out their activities. Finally, home States are countries where particular PMSCs are registered or incorporated. Home States where PMSCs have been established and host States where PMSCs provide their services can be contracting States while it is also possible that some States are purely contracting States. It is often assumed that PMSCs usually have former members of the State armed forces among their personnel and on many occasions appear to have certain ties with their home States.[229] This means that PMSCs and their personnel will be reluctant to act in contravention of the interests of those specific States, but will strive to protect them. Therefore, relations between PMSCs and home States play an important role in the examination of the PMSC presence on the territory of host States, for many such firms are frequently referred to as "foreign policy by proxy" or "covert wings of governments".[230] But in general, there seems to be a common understanding and agreement among the experts that all three types of States are significant actors in the process of hiring and deploying PMSCs, given the fact that they are capable of legally regulating the operations of those entities to a certain extent. Those are also

---

and Good Practices for States Related to Operations of Private Military and Security Companies During Armed Conflict, October 8, 2008, UN Doc. A/63/467-S/2008/636, <http://www.icrc.org/eng/assets/files/other/icrc_002_0996.pdf>, accessed on July 18, 2014; Swiss Federal Department of Foreign Affairs, "Participating States of the Montreux Document", February 21, 2014, <http://www.eda.admin.ch/eda/en/home/topics/intla/humlaw/pse/parsta.html>, accessed on July 18, 2014.

227 Montreux Document, Preface, p. 9, para. 9(a).

228 Ibidem, Preface, p. 10, paras. 9(c), 9(d), 9(e).

229 Chia Lehnardt, "Private Military Companies and State Responsibility", in Simon Chesterman and Chia Lehnardt (Eds.), (2007) *From Mercenaries to Market: The Rise and Regulation of Private Military Companies*, Oxford: Oxford University Press, p. 140.

230 Ibidem.

the States that in theory can face international responsibility and its numerous legal consequences for the unlawful conduct of private contractors disregarding rules and principles of international law.

There are different types of relationships existing between States and PMSCs, which should be identified by analyzing the State practice of using those business corporations. In the first place, PMSCs can be incorporated and registered on the territory of one State and provide their military and security services to another State. This is normally the case if those firms are hired by a State other that the State of registration and carry out their tasks on the territory of that State or a third State.[231] Additionally, PMSCs can be employed by a State, in which they are incorporated, and provide a great variety of their services to that State on its territory or abroad.[232] There is also a possibility that private military and security entities are hired by non-State actors, such as international organizations or other corporations, that need to protect their assets and personnel requiring better security than the protection offered by States, on the territory of which they employ their activities.[233]

In conclusion, in order to study the concept of a PMSC and implications of its use by States it is necessary to establish a definition coherently incorporating the most important aspects of this rapidly growing industry and its representatives in the form of private entities offering a wide array of military and security tasks. This process of defining a PMSC involves the integration of the most prominent features of this type of business as accentuated by a large number of authors whose works have been assessed above into its description. The following comprehensive definition covering the majority of the elements and features present in different types of PMSCs is established for the purposes of the present study:

> "PMSCs are legally established private for-profit businesses that have a corporate structure often operating as international firms and having the potential to use violence or to exercise force. They offer on a contractual basis to States and non-State entities, such as international organizations and other corporations, a broad spectrum of military and security services related to armed conflicts, which often include, but are not limited to, the provision of security, protection, guarding, interrogation, detention, participation in hostilities, and other activities."

3.3 Clients of PMSCs

As was said above, PMSCs do not exclusively operate on behalf of governments, but perform military and security services for a plethora of actors with a variety of

---

[231] Olivier De Schutter, "The Responsibility of States", in Simon Chesterman and Angelina Fisher (Eds.), (2009) *Private Security, Public Order: The Outsourcing of Public Services and Its Limits*, Oxford: Oxford University Press, p. 25.

[232] Ibidem.

[233] Ibidem.

private and public interests.[234] PMSCs are now to a great extent a part of the military and security landscape. In 2000, Doug Brooks identified some trends in the PMSC industry. He assumed that they would not only play a more prominent role in providing military services to the State militaries, but also be employed by the UN and regional organizations and extensively involved in peace operations.[235] Additionally, multinational corporations and other private entities active in unstable States will more often rely on their expertise.[236] Nowadays, predictions made by Doug Brooks appear to resonate with the truth. States, international organizations, NGOs, and private corporations have clear interests in using private contractors and hiring them regularly to offer a broad array of expertise.[237] In practice, PMSCs often take significant roles in the operations of these actors. While the current study has its main focus on the relationship between States and PMSCs and the possibility of engaging State responsibility by attributing illegal activities performed by the latter to States and the application of positive State obligations, it is necessary for clarification of the phenomenon of privatization of war and security to briefly elaborate upon those three main types of PMSC clients, i.e. private corporations, NGOs and international organizations, and States.

One of the main categories of PMSC employers is comprised by private companies, often large multinationals, conducting their business activities around the world.[238] Due to the fact that these corporations are often active on the territory of weak States that are incapable of maintaining law and order inside their own borders, they usually are in need of protection services that in the most cases cannot be provided by State authorities.[239] The security measures are necessary because of many risks to investment faced by large corporations functioning in the developing coun-

[234] Zoe Salzman, (2007-2008) "Private Military Contractors and a Taint of a Mercenary Reputation", *New York University Journal of International Law and Politics*, Volume 40, Issue 3, p. 862; Nicholas Dorn and Michael Levi, (2007) "European Private Security, Corporate Investigation and Military Services: Collective Security, Market Regulation and Structuring the Public Sphere", *Policing and Society*, Volume 17, Issue 3, p. 214; Jan Grofe, "Human Rights and Private Military Companies: A Double-Edged Sword Too Dangerous to Use?", in Thomas Jäger and Gerhard Kümmel (Eds.), (2007) *Private Military and Security Companies: Chances, Problems, Pitfalls and Prospects*, Wiesbaden: VS Verlag für Sozialwissenschaften, p. 244; Michael Cottier, (2006) "Elements for Contracting and Regulating Private Security and Military Companies", *International Review of the Red Cross*, Volume 88, Issue 863, p. 637.

[235] Doug Brooks, (2000) "Messiahs or Mercenaries? The Future of International Private Military Services", *International Peacekeeping*, Volume 7, Issue 4, p. 140.

[236] Ibidem, pp. 140-141.

[237] DCAF, (2006) *Private Military Companies*, DCAF Backgrounder, p. 2, <http://www.dcaf.ch/publications/kms/details.cfm?lng=en&id=18412&nav1=5>, accessed on May 5, 2011; Eugenio Cusumano, "Policy Prospects for Regulating Private Military and Security Companies", in Francesco Francioni and Natalino Ronzitti (Eds.), (2011) *War by Contract: Human Rights, Humanitarian Law, and Private Contractors*, Oxford: Oxford University Press, p. 14.

[238] Deborah D. Avant, (2004) "The Privatization of Security and Change in the Control of Force", *International Studies Perspectives*, Volume 5, p. 154.

[239] Ibidem, p. 178.

tries.[240] Not only their assets, but also employees frequently find themselves at the center of violence stemming from various groups. Therefore, physical security is one of the main priorities of multinationals and smaller firms employing their business activities in those risky areas.[241] The security tasks aimed at ensuring safe operations of companies abroad can be provided by a large number of PMSCs as "investment enablers"[242] willing to broaden the scope of their PMSC business activities and to deploy their personnel in those often underdeveloped regions of the world.

Multinational corporations allocate large sums of money to hire PMSCs capable of granting them protection and ensuring their safe operation in danger areas. As claimed by scholars, in Algeria – a country experiencing frequent terrorist violence – oil companies spend a large portion of their operational budgets, almost nine per cent, on private security provided by PMSCs. In Colombia, the same security costs of foreign businesses amount to six per cent, which are used to prevent the attacks of guerrillas, paramilitaries, and narco-terrorists.[243] If PMSCs did not offer their services to those companies, it is highly unlikely that multinationals would be able to take a risk of operating in a large number of countries of the developing world.

Searching for new markets, extractable resources, and inexpensive labor, a wide spectrum of foreign companies, such as Shell, Chevron, and BP, are settled in various less developed areas abroad employing their economic activities. Many of these places belong to unstable regions or conflict zones.[244] As an example of the necessity on the part of companies to hire PMSCs, one can consider the case of Shell and Chevron in Nigeria.[245] In this country, one of the world's largest oil exporters, conflict started in the early 1990s between multinationals exploiting natural resources of the Niger Delta and local communities keen to halt this exploitation. Many protests had been organized by local population that were fiercely repressed by the Nigerian governmental forces supported and subsidized by oil companies. Between 1998 and 1999, there was a massive escalation of violence resulting in a major confrontation between the protest movements and multinationals to a certain degree supported by the State. While

---

240 Peter W. Singer, (2008) *Corporate Warriors: The Rise of the Privatized Military Industry*, Ithaca and London: Cornell University Press, p. 80.

241 Ibidem, pp. 80-81.

242 Ibidem, p. 81.

243 Ibidem, pp. 81-82.

244 Deborah D. Avant, (2004) "The Privatization of Security and Change in the Control of Force", *International Studies Perspectives*, Volume 5, p. 154; Deborah D. Avant, (2005) *The Market for Force: The Consequences of Privatizing Security*, Cambridge: Cambridge University Press, pp. 180-181.

245 Angela McIntyre and Taya Weiss, "Weak Governments in Search of Strength: Africa's Experience of Mercenaries and Private Military Companies", in Simon Chesterman and Chia Lehnardt (Eds.), (2007) *From Mercenaries to Market: The Rise and Regulation of Private Military Companies*, Oxford: Oxford University Press, pp. 79-80; Rita Abrahamsen and Michael C. Williams, *The Globalisation of Private Security, Country Report: Nigeria*, Report, January 2005, Aberystwyth: The University of Wales, pp. 13-14, <http://users.aber.ac.uk/rbh/privatesecurity/country%20report-nigeria.pdf>, accessed on January 15, 2014.

kidnapping, attacks on the oil installations, and other activities kept intensifying, Shall was one of the companies seeking to ensure safety of its operations by hiring a private security company, Group4Securitor (now: G4S). This private security provider, operating closely with police and armed forces in Nigeria, carried out an important task of protecting assets and personnel of Shell and did it not without success.[246] Chevron as another corporation concluded an extensive contract with Outsourcing Services Limited, a subsidiary of Group4Securitor.[247] This PMSC also performed a large number of services, including marine security. Currently, G4S is an industry leader and can now probably be considered the most powerful PMSC in the world,[248] at the same time the second largest private employer worldwide with over 620,000 employees active in more than 125 countries.[249] The biggest clients of the company are from the sectors of major corporates and industrials (28%), governments (23%), financial institutions (19%), retail (9%), and private energy and utilities (8%).[250]

In addition to private corporations interested in the expertise of PMSCs, non-governmental organizations and international organizations appear to be willing to frequently make use of their services.[251] Those entities, especially non-governmental organizations and several agencies of the United Nations, hire private military and security contractors to assist them in safeguarding areas of operation and in facing major humanitarian disasters.[252] Taking into account the fact that non-governmental organizations and aid agencies are usually to be found in danger zones, the services

246 Deborah D. Avant, (2005) *The Market for Force: The Consequences of Privatizing Security*, Cambridge: Cambridge University Press, p. 184.

247 Rita Abrahamsen and Michael C. Williams, *The Globalisation of Private Security, Country Report: Nigeria*, Report, January 2005, Aberystwyth: The University of Wales, pp. 13-14, <http://users.aber.ac.uk/rbh/privatesecurity/country%20report-nigeria.pdf>, accessed on January 15, 2014.

248 "30 Most Powerful Private Security Companies in the World", *Security Degree Hub*, January 11, 2014, <http://www.securitydegreehub.com/30-most-powerful-private-security-companies-in-the-world/>, accessed on January 13, 2014.

249 Luke McKenna and Robert Johnson, "A Look at the Largest Private Armies in the World", *Business Insider*, June 3, 2012, <http://www.businessinsider.com/worlds-most-powerful-mercenary-armies-2012-06?op=1>, accessed on January 13, 2014; G4S plc, *Annual Report and Accounts 2012*, pp. 1, 30, <http://www.g4s.com/~/media/C15A0466A2F842258247347391C48379.ashx>, accessed on February 17, 2014.

250 G4S plc, *Annual Report and Accounts 2012*, p. 3, <http://www.g4s.com/~/media/C15A0466A2F842258247347391C48379.ashx>, accessed on February 17, 2014.

251 Peter W. Singer, "Humanitarian Principles, Private Military Agents: Some Implications of the Privatised Military Industry for the Humanitarian Community", in Victoria Wheeler and Adele Harmer (Eds.), *Resetting the Rules of Engagement: Trends and Issues in Military Humanitarian Relations*, Humanitarian Policy Group Report 21, March 2006, London: Overseas Development Institute, p. 71; Shantanu Chakrabarti, (2009) *Privatisation of Security in the Post Cold-War Period: An Overview of its Nature and Implications*, Institute for Defence Studies and Analyses Monograph Series, No. 2, New Delhi: Magnum Custom Publishing, p. 35.

252 Peter W. Singer, (2008) *Corporate Warriors: The Rise of the Privatized Military Industry*, Ithaca and London: Cornell University Press, p. 82.

offered by PMSCs are essential to their proper operation and, in fact, constitute the only viable way to receive protection.[253] As recent studies indicate, more than 40 different contracts have been concluded between PMSCs and humanitarian actors, for instance, the International Rescue Committee, the United Nations Children's Fund, CARE, and others in order to invest in security and stability.[254] A Christian relief, development, and advocacy organization World Vision in Sierra Leone has, for example, always been dependent upon the private security and advisory services offered by private contractors, such as Lifeguard.[255] In addition, international and regional organizations appear to rely on the expertise of PMSCs: not only the UN and the European Union (hereinafter: EU), but also the North Atlantic Treaty Organization (hereinafter: NATO), the African Union (hereinafter: AU), and the Organization for Security and Co-operation in Europe (hereinafter: OSCE).[256] The UK Secretary of State for Foreign and Commonwealth Affairs indicated in the Green Paper on options for regulating PMSCs:

> "A further source of demand for private military services could be international organisations. The private sector is already active and effective in areas that would once have been seen as the preserve of the military – demining for example. And both the UN and international NGOs employ private companies to provide them with security and logistics support."[257]

The UN itself directly employs PMSCs and its various agencies, departments, and funds also regularly hire these corporations.[258] At least seven different UN agencies have extensively been employing PMSCs for such tasks as the provision of armed and unarmed security, consultancy, risk assessment, security training, and logistical support, including such services as the provision of food to refugees.[259] It has been

---

253 Shantanu Chakrabarti, (2009) *Privatisation of Security in the Post Cold-War Period: An Overview of its Nature and Implications*, Institute for Defence Studies and Analyses Monograph Series, No. 2, New Delhi: Magnum Custom Publishing, pp. 32-36.

254 Peter W. Singer, "Humanitarian Principles, Private Military Agents: Some Implications of the Privatised Military Industry for the Humanitarian Community", in Victoria Wheeler and Adele Harmer (Eds.), *Resetting the Rules of Engagement: Trends and Issues in Military Humanitarian Relations*, Humanitarian Policy Group Report 21, March 2006, London: Overseas Development Institute, pp. 69-70.

255 Peter W. Singer, (2008) *Corporate Warriors: The Rise of the Privatized Military Industry*, Ithaca and London: Cornell University Press, p. 82.

256 Herbert Wulf, (2005) *Internationalizing and Privatizing War and Peace*, Basingstoke: Palgrave Macmillan, p. 42.

257 UK Foreign and Commonwealth Office, *Private Military Companies: Options for Regulation 2001-02*, February 2002, HCP 577, London: The Stationary Office, p. 4, <http://www.official-documents.gov.uk/document/hc0102/hc05/0577/0577.pdf>, accessed on February 10, 2014.

258 Åse Gilje Østensen, (2011) *UN Use of Private Military and Security Companies: Practices and Policies*, SSR Paper 3, DCAF, p. 12.

259 Lou Pingeot, (2012) *Dangerous Partnership: Private Military & Security Companies and the UN*, Report, June 2012, Global Policy Forum and Rosa Luxemburg Foundation: New York and Berlin,

established that the United Nations High Commissioner for Refugees on the Afghan border used to rely on PMSC expertise.[260] Demining tasks during almost every UN operation have been contracted out to PMSCs. Police and logistics services have also been carried out by contractors, such as in the UN Missions in Liberia and Sierra Leone.[261] The Africa Peacekeeping Programme that was created to assist African countries in regional peacekeeping operations and crisis management and to enhance these abilities was controlled by the US Department of State and heavily relied on the expertise of PMSCs, such as DynCorp.[262] In general, while data on the UN's contracting of private contractors appears to be incomplete, there are indications of an increase of the number of contracts concluded with PMSCs since 2006-2007.[263] Obviously, the use of PMSCs has significant advantages but can also be problematic. In the words of the UK Foreign Secretary:

> "A strong and reputable private military sector might have a role in enabling the UN to respond more rapidly and more effectively in crises. The cost of employing private military companies for certain functions in UN operations could be much lower than that of national armed forces. Clearly there are many pitfalls in this which need to be considered carefully. There are, for example, important concerns about human rights, sovereignty and accountability."[264]

On July 31, 2013, the UN Working Group on the Use of Mercenaries organized a special panel event to discuss the use of PMSCs by the UN.[265] On that day, the Work-

---

pp. 24-27; Peter W. Singer, "Humanitarian Principles, Private Military Agents: Some Implications of the Privatised Military Industry for the Humanitarian Community", in Victoria Wheeler and Adele Harmer (Eds.), *Resetting the Rules of Engagement: Trends and Issues in Military Humanitarian Relations*, Humanitarian Policy Group Report 21, March 2006, London: Overseas Development Institute, p. 70; Shantanu Chakrabarti, (2009) *Privatisation of Security in the Post Cold-War Period: An Overview of its Nature and Implications*, Institute for Defence Studies and Analyses Monograph Series, No. 2, New Delhi: Magnum Custom Publishing, p. 36.

260 Peter W. Singer, (2008) *Corporate Warriors: The Rise of the Privatized Military Industry*, Ithaca and London: Cornell University Press, p. 82.

261 Kwesi Aning, Thomas Jaye, and Samuel Atuobi, (2008) "The Role of Private Military Companies in US-Africa Policy", *Review of African Political Economy*, Volume 35, Issue 118, p. 623; Peter W. Singer, (2008) *Corporate Warriors: The Rise of the Privatized Military Industry*, Ithaca and London: Cornell University Press, p. 82.

262 Shantanu Chakrabarti, (2009) *Privatisation of Security in the Post Cold-War Period: An Overview of its Nature and Implications*, Institute for Defence Studies and Analyses Monograph Series, No. 2, New Delhi: Magnum Custom Publishing, p. 27.

263 Lou Pingeot, (2012) *Dangerous Partnership: Private Military & Security Companies and the UN*, Report, June 2012, Global Policy Forum and Rosa Luxemburg Foundation: New York and Berlin, p. 23.

264 UK Foreign and Commonwealth Office, *Private Military Companies: Options for Regulation 2001-02*, February 2002, HCP 577, London: The Stationary Office, p. 4, <http://www.official-documents.gov.uk/document/hc0102/hc05/0577/0577.pdf>, accessed on February 10, 2014.

265 The UN Working Group on the Use of Mercenaries, Summary Report of the Expert Panel on the Use of Private Military and Security Companies by the United Nations, July 31, 2013, New York, <http://

ing Group essentially initiated a study into the use of private contractors by different agencies of the UN that will be presented to the UN General Assembly in 2014. The results of this study will contribute to developing a coherent policy of the United Nations on this matter.

The third group of entities contracting PMSCs on a frequent basis is that of States.[266] Since the end of the Cold War, States have been the major clients of private contractors willing to invest substantial financial resources in the provision of not only security services, but also military services by private actors. Some claim that the US government, and more specifically the US Department of Defense, constitutes the main client of the industry offering private military and security services.[267] The United Kingdom and Canada also rely extensively on PMSCs. PMSCs provide hiring governments with professional services associated with warfare and security not only in armed conflicts, but also in peace operations.[268] The phenomenon of the privatization of war and security has become the intense reality of modern warfare.[269] Many authors emphasize the active reliance of the military on PMSCs and their services.[270] Experts believe that those companies make significant contributions to the battlefields and are here to stay as a logical development in the military and security affairs of States after a chain of historical events influencing the course of their development.[271] Hence, it is asserted that the whole concept of corporate entities as effective and experienced providers of force and security will expand further and set its foot more firmly in the field of military and security service provision.

In situations of armed conflict, individual States contact PMSCs and conclude contracts for the realization of the above-mentioned tasks. In practice, the same path is also taken when States are in need of extra manpower, for instance, during

---

www.ohchr.org/Documents/Issues/Mercenaries/WG/StudyPMSC/EventSummary.pdf>, accessed on January 20, 2014.

[266] Chia Lehnardt, "Private Military Companies and State Responsibility", in Simon Chesterman and Chia Lehnardt (Eds.), (2007) *From Mercenaries to Market: The Rise and Regulation of Private Military Companies*, Oxford: Oxford University Press, p. 139; Carlos Ortiz, (2009) *Private Military and Security Companies: Misconceptions, Reconceptualisations, and Regulation*, Working Paper No. 6, Centre for Global Political Economy at the University of Sussex; Peter W. Singer, (2005) "Outsourcing War", *Foreign Affairs*, Volume 84, Issue 2.

[267] Eugenio Cusumano, "Policy Prospects for Regulating Private Military and Security Companies", in Francesco Francioni and Natalino Ronzitti (Eds.), (2011) *War by Contract: Human Rights, Humanitarian Law, and Private Contractors*, Oxford: Oxford University Press, p. 14; Rebecca DeWinter-Schmitt (Ed.), *Montreux Five Years On: An Analysis of State Efforts to Implement Montreux Document Legal Obligations and Good Practices*, Report, 2013, p. 17.

[268] Peter W. Singer, (2005) "Outsourcing War", *Foreign Affairs*, Volume 84, Issue 2, p. 120.

[269] Adam Ebrahim, (2010) "Going to War with the Army You Can Afford: The United States, International Law and the Private Military Industry", *Boston University International Law Journal*, Volume 28, p. 214.

[270] Rebecca R. Vernon, (2003-2004) "Battlefield Contractors: Facing the Tough Issues", *Public Contract Law Journal*, Volume 33, Issue 2, p. 394.

[271] Ibidem, p. 371.

peacekeeping missions. As is known, peacekeeping operations by the United Nations and NATO depend on the willingness of their particular Member States to provide contingents in order to successfully conclude those missions. The inability of States or general lack of enthusiasm among them to provide national troops may render a particular peacekeeping task impossible to fulfill and lead to major humanitarian catastrophes. In this regard, the use of PMSCs to supplement regular armed forces of States and to provide their personnel for peacekeeping operations can be essential to achieving the goals of those missions.[272] The idea of PMSCs providing military and security services in peacekeeping operations seems to be supported not merely by some States, such as the United Kingdom, but also by UN officials. In one of his interviews, the former UN Under Secretary Sir Brian Urquart stated: "In a perfect world, we don't need them or want them. But the world isn't perfect."[273] For these reasons, States providing their troops for a peacekeeping mission often enter into contractual relationships with PMSCs offering essential military and security expertise.[274] Such firms have been supporting the US troops deployed in the Balkans, such as Brown & Root Services and DynCorp, which were hired to train the US police officers operating as part of the UN Police Task Force contingent in Bosnia established in 1995. Additionally, private contractors proved to be very effective on the African continent, supporting the UN peacekeeping forces on the side of the government, such as the PMSC Executive Outcomes hired by Sierra Leone in 1995,[275] and in general have a long history of working alongside UN peacekeepers.

Interestingly enough, all three above-mentioned groups of clients frequently relying on the services provided by PMSCs often engage in contractual relationships with firms that are known for the misbehavior of their employees, involvement in financial scandals, violations of international law and domestic regulations of countries of their operation, and other unlawful activities. Not only do States and private companies strive to hire those corporations capable of performing their tasks in a professional manner and being cost-effective providers of security and other expertise, but also agencies of the United Nations follow the same approach without taking into

---

272 Deborah D. Avant, (2004) "The Privatization of Security and Change in the Control of Force", *International Studies Perspectives*, Volume 5, p. 154; Peter W. Singer, (2003) "Peacekeepers, Inc.", *Policy Review*, Volume 119, p. 60; Shantanu Chakrabarti, (2009) *Privatisation of Security in the Post Cold-War Period: An Overview of its Nature and Implications*, Institute for Defence Studies and Analyses Monograph Series, No. 2, New Delhi: Magnum Custom Publishing, p. 15; Don Mayer, (2010) "Peaceful Warriors: Private Military Security Companies and the Quest for Stable Societies", *Journal of Business Ethics*, Volume 89, p. 389.

273 Peter W. Singer, (2003) "Peacekeepers, Inc.", *Policy Review*, Volume 119, pp. 60-61.

274 Åse Gilje Østensen, (2011) *UN Use of Private Military and Security Companies: Practices and Policies*, SSR Paper 3, DCAF, p. 12.

275 Leslie Hough, (2007) "A Study of Peacekeeping, Peace-enforcement and Private Military Companies in Sierra Leone", *African Security Review*, Volume 16, Issue 4, p. 14; James T. Gathii, (2009-2010) *Commercializing War: Private Military and Security Companies, Mercenaries and International Law*, Legal Studies Research Paper Series, No. 49, Albany Law School, pp. 5-7.

account the background of PMSCs, to the great surprise of researchers.[276] The United Nations has regularly relied on such disreputable firms as DynCorp, Saracen Uganda, G4S, and ArmorGroup, posing threats to its image, credibility, and effectiveness.[277]

3.4 Legal status

After the privatization revolution took off and the growing practice of outsourcing started to spread around the globe, a great variety of PMSCs emerged. These corporations have different sets of characteristics, such as history of establishment and operation, market capitalization and market share, corporate interrelationships, nationality and origin of their personnel, employee experience and expertise, and many other aspects. While it remains true that all PMSCs share the same feature of offering services falling within the scope of the military and security domain,[278] their military and security expertise is variable and is highly dependent on the terms of specific contracts concluded between companies and States and thus the needs and expectations of the latter.[279] These non-State actors can be said to have a particular legal status under national law of various States and under international law. This status characterizes the nature of PMSCs and confers a set of certain rights and duties on them under national and international law. In addition, this status might influence a variety of relationships between States and PMSCs and create certain obligations for the former to take positive measures aimed at the latter.

3.4.1 Legal status under national law

The examination of the concept of PMSCs from the national perspective reveals that they can be said to be a legal form of private business entities,[280] operating as well-organized companies with an evident corporate structure, including boards of directors and share-holdings.[281] The business activities of these firms have a strong focus on the provision of a broad spectrum of military and security services on a for-profit basis.[282]

---

[276] Lou Pingeot, (2012) *Dangerous Partnership: Private Military & Security Companies and the UN*, Report, June 2012, Global Policy Forum and Rosa Luxemburg Foundation: New York and Berlin, pp. 28-30.

[277] Ibidem, p. 28.

[278] Peter W. Singer, (2008) *Corporate Warriors: The Rise of the Privatized Military Industry*, Ithaca and London: Cornell University Press, p. 88.

[279] Carlos Ortiz, (2010) "The New Public Management of Security: the Contracting and Managerial State and the Private Military Industry", *Public Money & Management*, Volume 30, Issue 1, p. 36.

[280] Montreux Document, Preface, p. 9, para. 9(a).

[281] Kevin A. O'Brien, "What Should and What Should not Be Regulated", in Simon Chesterman and Chia Lehnardt (Eds.), (2007) *From Mercenaries to Market: The Rise and Regulation of Private Military Companies*, Oxford: Oxford University Press, p. 38.

[282] Nils Rosemann, (2008) "Code of Conduct: Tool for Self-Regulation for Private Military and Security Companies", DCAF, p. 10; Sarah Percy, "Morality and Regulation", in Simon Chesterman and Chia

Interestingly enough, as argued by some experts working in this sphere, the private contractor industry appears to be familiar with the specific type of PMSCs that do not have strong corporate structures, but are usually established only for a short period of time and, therefore, show many similarities with common mercenary groups.[283] There is a huge variety of PMSCs in terms of size, ranging from small consulting firms to massive transnational corporations simultaneously employing their activities in more than one country.[284] PMSCs can often constitute the following types of business enterprises in the United Kingdom and the United States where the most PMSCs active in the world are headquartered: UK public limited companies (PLC), such as BAE Systems PLC; UK private companies limited by shares (Ltd.), such as Erinys UK Ltd. as a member of the British private security company Erinys International; US companies with limited liability (LLC), such as Aegis LLC as a member of the British security and risk management company Aegis Group, and DynCorp International LLC.

Companies or corporations are legal entities that are established under the laws of a specific home State.[285] PMSCs constitute legally created businesses incorporated under the jurisdiction of home States and there is a close relationship between the two. In this regard, home States might be thought to have certain legal duties to ensure that these companies do not breach international law. Furthermore, in their capacity of corporations and legal persons, PMSCs are, in fact, able to own property, to enter into contracts with other entities, such as States, to take part in legal disputes, or to be sued, while their main goal is and remains to make a profit and to generate more business.[286] As legal persons in the form of private, non-governmental business entities, PMSCs cannot be arrested or imprisoned and do not have voting rights within the States of their establishment in contrast to natural persons. As such, companies are not capable of having a criminal intent and of physically committing crimes, but they can be held liable for various violations of national and international law caused by their personnel in the course of performing their duties, be fined, or denied certain rights and privileges. The concepts of corporate responsibility and accountability can

Lehnardt (Eds.), (2007) *From Mercenaries to Market: The Rise and Regulation of Private Military Companies*, Oxford: Oxford University Press, pp. 12-13; Benedict Sheehy and Jackson N. Maogoto, (2008) "The Private Military Company – Unravelling the Theoretical, Legal & Regulatory Mosaic", *ILSA Journal of International & Comparative Law*, Volume 15, Issue 1, p. 149.

283 Kevin A. O'Brien, "What Should and What Should not Be Regulated", in Simon Chesterman and Chia Lehnardt (Eds.), (2007) *From Mercenaries to Market: The Rise and Regulation of Private Military Companies*, Oxford: Oxford University Press, p. 39.

284 DCAF, (2006) *Private Military Companies*, DCAF Backgrounder, <http://www.dcaf.ch/publications/kms/details.cfm?lng=en&id=18412&nav1=5>, accessed on May 5, 2011.

285 Roger L. Miller and Gaylord A. Jentz, (2008) *Business Law Today: The Essentials*, 8th Edition, Mason: Thomson West, p. 174.

286 Andrew Alexandra, Deane-Peter Baker, and Marina Caparini, "Introduction: The Ethics and Governance of Private Military and Security Companies", in Andrew Alexandra, Deane-Peter Baker, and Marina Caparini (Eds.), (2008) *Private Military and Security Companies: Ethics, Policies and Civil-Military Relations*, London: Routledge, p. 8.

be of major significance for the discussion of breaches of international law committed by private contractors and their employees, but taking into account the fact that they fall outside the scope of this study they will not be considered.

3.4.2 Legal status under international law

In addition to the assessment of the legal position of PMSCs as private corporations under domestic laws, the legal status of PMSCs under international law can be established. However, it should be noted that in academic and public debates this issue still remains largely unclarified. PMSCs constitute legitimate businesses[287] established in accordance with the legislation of a certain State, such as the United States or the United Kingdom, that, in the first instance, derive their legal status and competence to act from national law[288] and can be regarded as being under an obligation to comply with the rules of both domestic and international law.[289]

The actual legal status of PMSCs remains rather indefinite under international law, for there are no clear regimes providing a definition or regulating them.[290] State practice that in conjunction with *opinio juris sive necessitatis* or simply *opinio juris* is necessary to establish customary international law[291] indicates that PMSCs are not prohibited as such and are widely used by States in different scenarios. This might be an argument for their legitimacy and general acceptance of the whole phenomenon of PMSCs.[292] These companies are, in fact, private entities that can qualify as "quasi-State actors in the international arena"[293] and are often treated as "secondary international actors".[294] As non-State actors, PMSCs function at the international level, operate on behalf of different States, and offer a great variety of military- and security-oriented services. Currently, it is, however, not possible to hold these companies

---

[287] Anna Leander, (2005) "The Power to Construct International Security: On the Significance of Private Military Companies", *Millennium: Journal of International Studies*, Volume 33, Issue 3, pp. 807-808.

[288] Willem van Genugten, Kees Homan, and Others, (2006) *The United Nations of the Future: Globalization with a Human Face*, Amsterdam: KIT Publishers, pp. 27-28.

[289] Doug Brooks, (2000) "Write a Cheque, End a War: Using Private Military Companies to End African Conflicts", *Conflict Trends*, Issue 1, p. 34, <http://www.accord.org.za/downloads/ct/ct_2000_1.pdf>, accessed on April 22, 2011.

[290] Peter W. Singer, (2004) "War, Profits and the Vacuum of Law: Privatized Military Firms and International Law", *Columbia Journal of International Law*, Volume 42, Issue 2, p. 534.

[291] International Court of Justice, *North Sea Continental Shelf Case (Federal Republic of Germany v. Netherlands)*, Judgment, February 20, 1969, ICJ Reports 1969.

[292] Peter W. Singer, (2004) "War, Profits and the Vacuum of Law: Privatized Military Firms and International Law", *Columbia Journal of International Law*, Volume 42, Issue 2, p. 533.

[293] Juan C. Zarate, (1998) "The Emergence of a New Dog of War: Private International Security Companies, International Law, and the New World Disorder", *Stanford Journal of International Law*, Volume 34, Issue 1, p. 145.

[294] Anna Leander, (2005) "The Power to Construct International Security: On the Significance of Private Military Companies", *Millennium: Journal of International Studies*, Volume 33, Issue 3, p. 806.

directly accountable at the international level: in case of violations of international law, they can be held accountable before national courts.[295]

From the perspective of international law, one may presume that most PMSCs are to be regarded as transnational corporations – one of the main categories of non-State actors in the field of international law and human rights[296] – registered or incorporated on the territory of specific States that employ their business activities within the borders of those States or freely operate in the international arena. It is true that most operations of private military and security contractors, especially those from the United States and Europe, take place abroad.[297] "Transnational corporation" as a specific type of business entity is to be defined in this context as "an economic entity operating in more than one country or a cluster of economic entities operating in two or more countries – whatever their legal form, whether in their home country or country of activity, and whether taken individually or collectively".[298]

The last 50 years of economic globalization cannot be seen in isolation from the expansion of these corporations.[299] Essentially, most PMSCs can qualify as such transnational or multinational enterprises and private business organizations, which often consist of various legal entities belonging to parent corporations. In the light of the legal status of PMSCs, they can further be defined as legally established multinational commercial enterprises offering a wide array of military and security services to various clients, ranging from States to multinationals. As argued by Malcolm Shaw, these companies are potentially candidates for having international legal personality, which brings a certain degree of controversy.[300] In the era of globalization, when non-State actors play a more important role, transnational corporations operating in more than

---

295 Nicola Jägers, (2002) *Corporate Human Rights Obligations: in Search of Accountability*, School of Human Rights Research Series, Volume 17, Antwerpen, Oxford, New York: Intersentia, p. 178.

296 Philip Alston, "The 'Not-a-Cat' Syndrome: Can the International Human Rights Regime Accommodate Non-State Actors", in Philip Alston (Ed.), (2005) *Non-State Actors and Human Rights*, Oxford: Oxford University Press, p. 3.

297 Francesco Francioni, "The Role of the Home State in Ensuring Compliance with Human Rights by Private Military Contractors", in Francesco Francioni and Natalino Ronzitti (Eds.), (2011) *War by Contract: Human Rights, Humanitarian Law, and Private Contractors*, Oxford: Oxford University Press, p. 104.

298 Sub-Commission on the Promotion and Protection of Human Rights, *Norms on the Responsibilities of Transnational Corporations and Other Business Enterprises with Regard to Human Rights*, August 23, 2003, UN Doc. E/CN.4/Sub.2/2003/12/Rev.2, <http://www.unhchr.ch/huridocda/huridoca.nsf/%28Symbol%29/E.CN.4.Sub.2.2003.12.Rev.2.En>, accessed on November 12, 2011.

299 Smita Narula, "International Financial Institutions, Transnational Corporations and Duties of States", in Malcolm Langford, Wouter Vandenhole, Martin Scheinin, and Willem van Genugten, (2014) *Global Justice, State Duties: The Extraterritorial Scope of Economic, Social and Cultural Rights in International Law*, Cambridge: Cambridge University Press, p. 129.

300 Malcolm N. Shaw, (2008) *International Law*, Sixth Edition, Cambridge: Cambridge University Press, pp. 249-250; Nicola Jägers, (2002) *Corporate Human Rights Obligations: in Search of Accountability*, School of Human Rights Research Series, Volume 17, Antwerpen, Oxford, New York: Intersentia, pp. 27-35.

one country or nation at a time, such as many PMSCs, find themselves at the heart of these highly significant developments that take place in the international sphere.[301] Generally, transnational corporations are not considered as being bound by international human rights law.[302] As often observed, however, the focus of the international human rights machinery and, perhaps, international humanitarian law[303] in future will be on transnational corporations and other similar large business entities.[304] The conduct of transnational corporations is, in fact, regulated to a specific extent by international soft law instruments placing human rights obligations on them. Those soft law sources are: the UN "Guiding Principles on Business and Human Rights: Implementing the United Nations 'Protect, Respect and Remedy' Framework" (hereinafter: the UN Guiding Principles or Guiding Principles), the Global Compact, the ILO Tripartite Declaration of Principles Concerning Multinational Enterprises and Social Policy, the ILO Declaration on Fundamental Principles and Rights at Work, the Organisation for Economic Co-operation and Development Guidelines for Multinational Enterprises, the Global Sullivan Principles, the Global Reporting Initiative, and Social Accountability 8000.[305] In spite of the absence of legally binding obligations laid down in those documents and supported by supervision and enforcement mechanisms, they are often perceived as not entirely ineffective and thus capable of ensuring compliance among private transnational corporations with human rights norms, breaches of which form the core element in the analytical framework of the current study.

There are indications that non-State actors, such as PMSCs, play a significant role in the international arena and challenge the claim that merely sovereign States – and not private actors – are entitled to legally use force.[306] Hence, there are many problems associated with the State's use of PMSCs capable of exercising force and using violence, taking into account the fact that PMSCs might be more likely to violate international humanitarian and human rights law, respect of which is essential in

301 Elena Pariotti, (2009) "International Soft Law, Human Rights and Non-State Actors: Towards the Accountability of Transnational Corporations?", *Human Rights Review*, Volume 10, Issue 2, pp. 140-141.

302 Smita Narula, "International Financial Institutions, Transnational Corporations and Duties of States", in Malcolm Langford, Wouter Vandenhole, Martin Scheinin, and Willem van Genugten, (2014) *Global Justice, State Duties: The Extraterritorial Scope of Economic, Social and Cultural Rights in International Law*, Cambridge: Cambridge University Press, p. 129.

303 Philip Alston, "The 'Not-a-Cat' Syndrome: Can the International Human Rights Regime Accommodate Non-State Actors", in Philip Alston (Ed.), (2005) *Non-State Actors and Human Rights*, Oxford: Oxford University Press, p. 7.

304 Ibidem, p. 6.

305 Elena Pariotti, (2009) "International Soft Law, Human Rights and Non-State Actors: Towards the Accountability of Transnational Corporations?", *Human Rights Review*, Volume 10, Issue 2, p. 145.

306 Simon Chesterman and Chia Lehnardt, "Introduction", in Simon Chesterman and Chia Lehnardt (Eds.), (2007) *From Mercenaries to Market: The Rise and Regulation of Private Military Companies*, Oxford: Oxford University Press, pp. 1-2.

times of war, peace, or both.[307] Furthermore, States often lack sufficient oversight and monitoring mechanisms over PMSCs, while no clear formal control instruments and effective control procedures are available.[308] For the most part, legal and other experts appear to agree that PMSCs still occupy an uncertain position in international law and that there is not much clarity with respect to their actual legal status.[309]

4. Conclusion

The current Chapter was intended to provide greater clarity regarding the general phenomenon of privatization of war and security and to look more specifically at PMSCs as non-State business entities. It addressed the key moments in the historical development of the private provision of violence and security and contained a comparison of main development lines of the century-old concept of mercenary with the modern PMSCs. This part of the present study also discussed PMSCs and the whole industry with its problems, assessed their legal status under national and international law, provided a definition for these entities, and analyzed a variety of their clients and the legal basis of operation. The findings of the current Chapter may be used for the further analysis of the actual legal status of PMSC employees and the doctrine of State responsibility.

When assessing the main historical developments in the field of mercenary activity and private military and security service provision, it becomes clear that the whole industry of PMSCs, just as the phenomenon of privatization of war and security, is not going to disappear in the future.[310] The State control of security and military matters is quite a modern concept that has progressively been developed after the Peace of Westphalia concluded in 1648 and should most probably not be regarded as absolute.[311] PMSCs are currently considered highly important actors in the field

---

307 Claude Voillat, (2004) "Private Military Companies: A Word of Caution", *Humanitarian Exchange Magazine*, Issue 28, <http://www.odihpn.org/report.asp?id=2675>, accessed on April 24, 2011.

308 David Isenberg, "A Government in Search of Cover: Private Military Companies in Iraq", in Simon Chesterman and Chia Lehnardt (Eds.), (2007) *From Mercenaries to Market: The Rise and Regulation of Private Military Companies*, Oxford: Oxford University Press, pp. 85-86; Martha Minow, (2005) "Outsourcing Power: How Privatizing Military Efforts Challenges Accountability, Professionalism, and Democracy", *Boston College Law Review*, Volume 46, Issue 5, p. 1005.

309 Chia Lehnardt, "Private Military Companies and State Responsibility", in Simon Chesterman and Chia Lehnardt (Eds.), (2007) *From Mercenaries to Market: The Rise and Regulation of Private Military Companies*, Oxford: Oxford University Press, p. 142.

310 Sarah Percy, (2006) *Regulating the Private Security Industry*, Adelphi Series, Volume 46, Issue 384, Oxon: Routledge, p. 23; David Shearer, (1999) "Private Military Force and Challenges for the Future", *Cambridge Review of International Affairs*, Volume 13, Issue 1, p. 92.

311 Adam Ebrahim, (2010) "Going to War with the Army You Can Afford: The United States, International Law and the Private Military Industry", *Boston University International Law Journal*, Volume 28, pp. 182-183.

of provision of military and security services.[312] They set their feet firmly on the soil of many countries around the world in a climate of "pluralised security arrangements" and coexist with the official structures of State apparatus, its norms, and principles developed in the period of Enlightenment.[313] Therefore, the industry appears to constitute "a window into both the future and the past".[314] One may reasonably expect that the scope of the industry will become even broader in the years to come. States will more actively rely on the activities performed by PMSCs, making the public-private partnership even stronger.[315] Furthermore, examination of the history of both mercenaries and PMSCs combined with the set of moral objections to their use reveals that there are indeed many similarities between those different types of private force rather than structural differences.[316]

History is known for repeating itself. A claim can be made that PMSCs are nothing more than mercenary organizations with corporate structures hired by legitimate governments and other actors. They essentially strive to safeguard own business interests, can potentially engage in violations of the law of armed conflict and human rights, and often manage to avoid being held accountable. Most importantly, clients of PMSCs, including legitimate States, are frequently in a position to hide behind these representatives of the PMSC industry covered by long chains of contracts and to escape responsibility for a variety of breaches of international law, such as violations of international humanitarian law and human rights. In reality, private companies successfully carry out a broad range of activities originally belonging to the operational field of the official State machinery and its organs. A number of similarities in the historical developments between mercenaries and PMSCs and moral implications of the existence and employment of these firms indicate an evident urge for analyzing normative aspects and the possibility of holding States responsible for breaches of international law. Therefore, when governments opt to outsource the main State prerogatives to private corporations, the use of which is regarded as morally problematic, the issue of international responsibility of respective States becomes undeniably

---

312 Anna Leander, (2005) "The Power to Construct International Security: On the Significance of Private Military Companies", *Millennium: Journal of International Studies*, Volume 33, Issue 3, p. 825; Benedict Sheehy and Jackson N. Maogoto, (2008) "The Private Military Company – Unravelling the Theoretical, Legal & Regulatory Mosaic", *ILSA Journal of International & Comparative Law*, Volume 15, Issue 1, p. 150; Zoe Salzman, (2008) "Private Military Contractors and the Taint of a Mercenary Reputation", *New York University Journal of International Law and Politics*, Volume 40, p. 858.

313 Adam White, (2010) *The Politics of Private Security: Regulation, Reform and Re-Legitimation*, Houndmills, Basingstoke, New Hampshire: Palgrave Macmillan, p. 19.

314 Ibidem.

315 Christopher Coker, (1999) "Outsourcing War", *Cambridge Review of International Affairs*, Volume 13, Issue 1, p. 112.

316 Sarah Percy, "Morality and Regulation", in Simon Chesterman and Chia Lehnardt (Eds.), (2007) *From Mercenaries to Market: The Rise and Regulation of Private Military Companies*, Oxford: Oxford University Press, p. 22.

prominent from the moral point of view and requires a broad elaboration from the perspective of international law.

From the above, the conclusion can also be drawn that the PMSC industry is actually remarkably diverse and fragmented. The discussion of issues associated with the industry of private military and security service provision focused on the growing sector itself and identified the main areas of concern, such as transparency, accountability, and legitimacy. PMSCs as non-State actors were determined to be providing essential services of a military and security character, lacking any uniform qualification and definition. They carry out those significant tasks on a daily basis to a wide array of clients, ranging from States to other businesses. The expertise of private contractors is to be regarded as extremely diversified, covering many areas of human activity. As has been argued, perhaps the most important characteristic of PMSCs is their nature as for-profit businesses with a corporate structure, which often operate on the basis of a contract as transnational enterprises and have the potential to use violence or to exercise force. Under national State law, these firms can have different forms depending on the jurisdiction where they are registered or incorporated. Under international law, most PMSCs are seen as non-State actors and transnational corporations that form the main focus of some soft law instruments established at the international level, such as the UN Guiding Principles.

CHAPTER III
EXISTING LEGAL FRAMEWORK OF PMSC OPERATION

Describing the absence of effective regulation for the private security industry as a vacuum is slightly misleading. Various domestic and international legal instruments may apply. However, even when these legal instruments are specifically intended to deal with today's private security industry – and many of them are not – they are often poorly designed.

Sarah Percy[1]

1. INTRODUCTION

Generally speaking, PMSCs as non-State actors constantly challenging the main conceptions of the international legal system based on the State monopoly of violence require regulation.[2] In legal and other literature, references are often made to the concept of legal void characterizing the functioning of PMSCs in conflict areas. It is, however, not entirely correct to speak of the absence of legal norms concerning PMSCs and States associated with the practice of their employment and applicable to their operations: the industry of private military and security service provision is certainly regulated, but the extent of this regulation at different levels varies considerably. Given that these corporations operate in a complicated legal labyrinth made up of a variety of international rules and principles, domestic legal instruments, self-regulatory mechanisms, and contractual obligations,[3] attention should be devoted to these different forms of regulation in order to understand the legal framework governing their operations and use by governments in zones of conflict. PMSCs are registered

---

1 Sarah Percy, (2006) *Regulating the Private Security Industry*, Adelphi Series, Volume 46, Issue 384, Oxon: Routledge, p. 9.

2 Kevin A. O'Brien, "What Should and What Should Not Be Regulated?", in Simon Chesterman and Chia Lehnardt (Eds.), (2007) *From Mercenaries to Market: The Rise and Regulation of Private Military Companies*, Oxford: Oxford University Press, p. 31.

3 Eugenio Cusumano, "Policy Prospects for Regulating Private Military and Security Companies", in Francesco Francioni and Natalino Ronzitti (Eds.), (2011) *War by Contract: Human Rights, Humanitarian Law, and Private Contractors*, Oxford: Oxford University Press, p. 11.

or incorporated as private businesses on the territory of home States and legal rules governing the establishment of private contractors allow them to operate and to carry out certain services. These rules may specify obligations of PMSCs during their operations and clarify the nature of their activities, but do not provide much guidance in determining the scope and circumstances of State responsibility for possible misconduct of employees of these companies constituting breaches of international law. In this regard, a variety of international and transnational legal instruments applicable to PMSC operations and the hiring, host, and home States are crucial.

The current Chapter seeks to briefly elaborate on national legal rules of some countries regarding PMSCs while also addressing international and transnational levels of regulation. In essence, this Chapter will set up the framework for further analysis performed through the lens of State responsibility and will accordingly serve as the basis of considerations made in subsequent Chapters. As there are still gaps present in both international and national legal regimes of PMSC regulation, it will also identify these gaps existing in the legal frameworks applicable to PMSCs.[4]

2. International Legal Order

Some authors have argued that "if there is a regulatory vacuum regarding PSCs, it exists under current international law".[5] There are indeed no norms at the international level specifically created to regulate PMSCs and their conduct. One can, however, find rules and principles of certain branches of international law that are applicable to the operations of these corporations and three types of States playing distinctive roles in their use. In general terms, it is true that intergovernmental initiatives for regulating the PMSC industry were met with some criticism.[6] International regulation is rather fragmented and incoherent and is not directly binding upon PMSCs. It fails to address the rather weak capacity of States to enforce international rules and standards and to take into account a variety of interests and needs of thirds parties, such as communities of PMSC personnel, local communities of areas where they operate, or victims of PMSC misconduct. Finally, it does not adequately deal with the distinction between the nature, role, and impact of mercenarism and those of PMSC activities.

---

[4] UN Human Rights Council, *Draft of a Possible Convention on Private Military and Security Companies (PMSCs) for Consideration and Action by the Human Rights Council*, Prepared by the Working Group on the use of mercenaries as a means of violating human rights and impeding the exercise of the right of peoples to self-determination, May 13, 2011, First Session, Geneva, May 23-27, 2011, UN Doc. A/HRC/WG.10/1/2, Preamble, p. 3.

[5] Sarah Percy, (2006) *Regulating the Private Security Industry*, Adelphi Series, Volume 46, Issue 384, Oxon: Routledge, p. 41.

[6] James Cockayne and Others, (2009) *Beyond Market Forces: Regulating the Global Security Industry*, New York: International Peace Institute, pp. 19-20.

The focus of the following Sections will be on not only general rules of public international law, but also relevant regulations from its two branches: international humanitarian and human rights law, as two "constituent elements of present-day international law".[7] In addition, international regulatory initiatives on business and human rights and specific legal instruments with a particular focus on PMSCs will be covered.

2.1 General rules of public international law

We can start by assessing the web of rules and principles of public international law regulating the conduct of States and applicable to the operation of PMSCs in conflict zones. It is no longer surprising to discover that nowadays not only States, but also PMSCs acting through their employees on behalf of governments and other non-State actors can resort to the use of armed force. Therefore, these companies are potentially capable of breaching the legal rule prohibiting the use of force laid down in the United Nations Charter. Article 2(4) of the Charter provides that in their international relations Member States of the United Nations shall not use force or threaten to use force against territorial integrity or political independence of other members or contrary to the purposes of the Organization. There are two exceptions to this prohibition contained in Article 51 of the Charter concerning the individual or collective right to self-defense and Chapter VII of the Charter regarding the primary responsibility of the UN Security Council for the maintenance of international peace and security.

There are also the Draft Articles on State Responsibility of the International Law Commission (hereinafter: ILC).[8] This document is in essence a codification of the law of State responsibility that is rightly considered to be "a regime, or a province, within the framework of public international law".[9] The Draft Articles are applicable

---

7 Christian Tomuschat, "Specificities of Human Rights Law and International Humanitarian Law Regarding State Responsibility", in Robert Kolb and Gloria Gaggioli (Eds.), (2013) *Research Handbook on Human Rights and Humanitarian Law*, Cheltenham and Northampton: Edward Elgar Publishing, p. 198.

8 References will be made to the Draft Articles on Responsibility of States for Internationally Wrongful Acts with Commentaries (hereinafter: Commentary to DARS as adopted in 2001), as found in Report of the International Law Commission on the Work of Its Fifty-Third Session, UN General Assembly Official Records, Fifty-Sixth Session, Supplement No. 10, UN Doc. A/56/10 (2001), reprinted in *Yearbook of the International Law Commission*, 2001, Volume II, Part 2, UN Doc. A/CN.4/SER.A/2001/Add.1 (Part 2); also reprinted in James Crawford (Ed.), (2002) *The International Law Commission's Articles on State Responsibility: Introduction, Text and Commentaries*, Cambridge: Cambridge University Press. The text of the Draft Articles on Responsibility of States as published in *Yearbook of the International Law Commission*, 2001, Volume II, Part 2, UN Doc. A/CN.4/SER.A/2001/Add.1 (Part 2) will be preferred.

9 Martin Scheinin, "Just Another Word? Jurisdiction in the Roadmaps of State Responsibility", in Malcolm Langford, Wouter Vandenhole, Martin Scheinin, and Willem van Genugten (Eds.), (2013) *Global Justice, State Duties: The Extraterritorial Scope of Economic, Social and Cultural Rights in International Law*, Cambridge: Cambridge University Press, p. 217.

to the acts and omissions of States and are of relevance for the functioning of PMSCs that are employed by States in the sense of States being responsible for their activities under certain circumstances. They contain general principles of State responsibility, such as those concerning legal consequences of its invocation, and more specific provisions relating to different types of attribution relevant for the State practice of employing PMSCs. While under international law States remain responsible for the official acts of their numerous agents and organs, they may also face responsibility if PMSCs and other private persons and entities, for instance, perform certain activities on behalf of these States or function under their control. This legal instrument reflecting rules and principles of customary international law relating to the issue of international responsibility will be applied to the functioning of PMSCs that operate on the basis of a contract with States for the purposes of identifying different modes of attribution and breaches of international law.

2.2 Specific legal instruments focused on PMSCs

In the first place, there is the 2010 Draft International Convention on PMSCs prepared by the UN Working Group on the Use of Mercenaries. In its Preamble, the Draft International Convention reaffirms principles and rules of international human rights and humanitarian law stressing that their violation may lead to the responsibility of not only PMSCs and intergovernmental organizations, but also States.[10] Further, it recalls the above-mentioned ILC Draft Articles on State Responsibility and acknowledges States' duty to prevent and to punish human rights violations.[11] The UN Draft Convention contains quite detailed provisions relating to the question of international responsibility for PMSC conduct, the requirement of adopting legislative measures and establishing oversight and monitoring mechanisms, and other matters which are of importance for the question of State responsibility. Currently, the Draft Convention is not widely supported by States.[12] It remains to be seen whether the Convention will prove to be a success in the future and whether States will be willing to sign and to ratify it. However, an international convention creating State obligations in the process of deploying PMSCs in their capacity of private businesses

---

[10] UN Human Rights Council, *Draft of a Possible Convention on Private Military and Security Companies (PMSCs) for Consideration and Action by the Human Rights Council*, Prepared by the Working Group on the use of mercenaries as a means of violating human rights and impeding the exercise of the right of peoples to self-determination, May 13, 2011, First Session, Geneva, May 23-27, 2011, UN Doc. A/HRC/WG.10/1/2, Preamble, p. 2.

[11] Ibidem, pp. 2-3.

[12] International Service for Human Rights, "Human Rights Council: Little Support for Proposed Convention Regulating Private Security Companies", September 23, 2010, <http://www.ishr.ch/news/human-rights-council-little-support-proposed-convention-regulating-private-security-companies>, accessed on July 10, 2014.

and addressing the issue of State control exercised over them might be regarded as a welcome development.[13]

In the spirit of the establishment of the UN Draft Convention for PMSCs, there have been calls from the international community for an international treaty specifically focused on transnational corporations and human rights.[14] Being supported by more than 80 States and a large number of civil society organizations,[15] this international regulatory instrument might be able to address some of the issues and responsibility gaps existing in the current international legal framework. However, it is not yet certain whether the proposal will be embraced by other States and, more importantly, whether a possible treaty will be ratified by them.

In 2006, the Swiss government and the International Committee of the Red Cross jointly launched the so-called "Swiss Initiative". It resulted in the adoption of the Montreux Document that can be signed by States. This significant instrument, that, according to its drafters, can be seen as a "milestone" or "bible",[16] is the product of the Swiss Initiative and the first major step made towards addressing the issues of the PMSC industry at the international level. It was endorsed by 17 States[17] on September 17, 2008 and constitutes an intergovernmental attempt to regulate various relationships between States and PMSCs that operate in armed conflicts. Addressing legal issues concerning PMSCs active on the battlefield, the Montreux Document recalls different types of international legal obligations of not only contracting, but also host and home States, and sets out what PMSCs, their personnel, and superiors are obliged to do. In addition, it specifies various good practices of three types of States – hiring States, host States, and home States – providing guidance with regard to their operations. The Montreux Document is, however, not a convention and is not legally binding. Moreover, it does not represent a final regulatory solution to a myriad of problems

---

13 Olivier De Schutter, "The Challenge of Imposing Human Rights Norms on Corporate Actors", in Olivier De Schutter (Ed.), (2006) *Transnational Corporations and Human Rights*, Oxford and Portland, Oregon: Hart Publishing, p. 38.

14 See Republic of Ecuador, "Statement on Behalf of a Group of Countries at the 24rd Session of the Human Rights Council", including among others Ecuador, Cuba, Nicaragua, Pakistan, and Sri Lanka, September 2013, <http://business-humanrights.org/media/documents/statement-unhrc-legally-binding.pdf>, accessed on May 21, 2014.

15 Surya Deva, "The Human Rights Obligations of Business: Reimagining the Treaty Business", an extended version of the paper presented at the Workshop on Human Rights and Transnational Corporations: Paving the Way for a Legally Binding Instrument, organized by the Permanent Missions of Ecuador and South Africa to the United Nations, Geneva, March 11-12, 2014, <http://business-humanrights.org/media/documents/reimagine_int_law_for_bhr.pdf>, accessed on May 21, 2014.

16 James Cockayne, (2008) "Regulating Private Military and Security Companies: The Content, Negotiation, Weaknesses and Promise of the Montreux Document", *Journal of Conflict & Security Law*, Volume 13, Issue 3, p. 401.

17 Afghanistan, Angola, Australia, Austria, Canada, China, France, Germany, Iraq, Poland, Sierra Leone, South Africa, Sweden, Switzerland, the UK, Ukraine, and the US.

associated with the use of PMSCs in armed conflict.[18] Taking international law as its basis, the instrument provides a restatement of the law applicable in the context of PMSC employment by States and gives necessary guidance concerning various legal and practical aspects of the operation of these corporations in the field. The Montreux Document is thus intended to remind States of an array of their obligations under international law and does not create new obligations for States or PMSCs.[19]

Currently, 50 States, the European Union, the OSCE, and NATO support the initiative, which still cannot be seen as fully comprehensive considering the limited range of participating countries, the will of which to engage in the process and to sign the Montreux Document is decisive for the effectiveness of the instrument.[20] However, it is a significant number of States assembled in a short period of time that acknowledge that States do have certain obligations in relation to the use of PMSCs and that control, oversight, and accountability mechanisms need improving.[21] Being quite critical about the Montreux Document, the UN Working Group on the Use of Mercenaries stressed that "while it is a good promotional document on existing international humanitarian law, the document has nevertheless failed to address the regulatory gap in the responsibility of States with respect to the conduct of private military and security companies and their employees".[22] Indeed, the Document has some weaknesses: it applies only to situations of armed conflict, does not refer to human rights conventions and specific human rights standards, and does not address all aspects of State responsibility, such as different notions of State control that can be exercised over PMSCs.[23] It does, however, serve as a useful guidance tool in identi-

---

18 Montreux Document, Foreword, p. 5.

19 Sorcha MacLeod, "The Role of International Regulatory Initiatives on Business and Human Rights for Holding Private Military and Security Contractors to Account", in Francesco Francioni and Natalino Ronzitti (Eds.), (2011) *War by Contract: Human Rights, Humanitarian Law, and Private Contractors*, Oxford: Oxford University Press, p. 351.

20 In addition to 17 States, as of July 18, 2014 these countries have signed the Montreux Document after September 17, 2008: Albania, Belgium, Bosnia and Herzegovina, Bulgaria, Chile, Costa Rica, Croatia, Cyprus, Czech Republic, Denmark, Ecuador, Finland, Former Yugoslav Republic of Macedonia, Georgia, Greece, Hungary, Iceland, Italy, Japan, Jordan, Kuwait, Liechtenstein, Lithuania, Luxembourg, The Netherlands, New Zealand, Norway, Portugal, Qatar, Slovenia, Spain, Uganda, and Uruguay. See Swiss Federal Department of Foreign Affairs, "Participating States of the Montreux Document", February 21, 2014, <http://www.eda.admin.ch/eda/en/home/topics/intla/humlaw/pse/parsta.html>, accessed on July 18, 2014.

21 Corinna Seiberth, (2013) *Private Military and Security Companies in International Law – A Challenge for Non-binding Norms: the Montreux Document and the International Code of Conduct for Private Security Service Providers*, Cambridge, Antwerp, Portland: Intersentia, pp. 234-244, 276.

22 UN Human Rights Council, *Promotion and Protection of All Human Rights, Civil, Political, Economic, Social and Cultural Rights, Including the Right to Development*, Report of the Working Group on the use of mercenaries as a means of violating human rights and impeding the exercise of the right of peoples to self-determination, January 21, 2009, UN Doc. A/HRC/10/14, p. 11, para. 44.

23 Corinna Seiberth, (2013) *Private Military and Security Companies in International Law – A Challenge for Non-binding Norms: the Montreux Document and the International Code of Conduct for Private Security Service Providers*, Cambridge, Antwerp, Portland: Intersentia, pp. 239-240.

fying existing State duties with regard to the use of PMSCs and providing guidelines on the activities of States.[24] This particular legal instrument assists in the implementation of hard law obligations of States and has a potential of progressively developing international law. Being currently a reaffirmation of existing international law, some recommendations laid down in the Montreux Document would potentially become part of customary international law or European regional custom if more States were to join the initiative and a more consistent, uniform, and extensive State practice could be achieved.[25] As a source of "generally respected standards", it can also potentially influence the adoption of other forms of regulation.[26]

2.3 International human rights law

International human rights law is a substantial branch of public international law aimed at international protection of human rights. International humanitarian law or the law of armed conflict is a distinct branch of public international law that cannot be equated with human rights law. The former applies in all situations, including times of war and peace, while the latter is applicable to the situations of armed conflict.[27] Besides different traditions and historical developments between international humanitarian and human rights law, one can discern another fundamental difference. The first branch of international law is established to regulate the conduct of States parties to a war or another armed conflict in order to grant a minimum level of protection to certain groups of people.[28] Human rights, on the other hand, aim to protect individuals from State interference, but their scope has been significantly extended to include not only negative obligations of States, but also positive rights and duties to protect from interference by non-State actors.[29]

The International Court of Justice (hereinafter: ICJ) made a distinction between two branches of international law, indicating that international human rights law forms *lex generalis* and international humanitarian law constitutes *lex specialis*.[30]

---

24 Ibidem, p. 11, para. 43.

25 Corinna Seiberth, (2013) *Private Military and Security Companies in International Law – A Challenge for Non-binding Norms: the Montreux Document and the International Code of Conduct for Private Security Service Providers*, Cambridge, Antwerp, Portland: Intersentia, pp. 244, 254-255.

26 James Cockayne, (2008) "Regulating Private Military and Security Companies: The Content, Negotiation, Weaknesses and Promise of the Montreux Document", *Journal of Conflict & Security Law*, Volume 13, Issue 3, pp. 402, 427.

27 David Turns, "The Law of Armed Conflict (International Humanitarian Law)", in Malcolm D. Evans (Ed.), (2010) *International Law*, Third Edition, Oxford: Oxford University Press, p. 816.

28 Christian Tomuschat, (2010) "Human Rights and International Humanitarian Law", *The European Journal of International Law*, Volume 21, Issue 1, p. 16.

29 Ibidem.

30 International Court of Justice, *Legality of the Threat or Use of Nuclear Weapons*, Advisory Opinion, July 8, 1996, ICJ Reports 1996, pp. 239-240, paras. 24-25; International Court of Justice, *Legal*

This *lex specialis* applies to all armed conflicts and to all parties in those conflicts.[31] It is now generally accepted that both branches of international law remain applicable during international and non-international armed conflicts.[32] In the *Construction of a Wall* Advisory Opinion given in 2004, the ICJ stated that the conventions on human rights do not cease to apply during the situations of armed conflict and further clarified:

> "As regards the relationship between international humanitarian law and human rights law, there are thus three possible situations: some rights may be exclusively matters of international humanitarian law; others may be exclusively matters of human rights law; yet others may be matters of both these branches of international law."[33]

This reasoning of the Court was confirmed in the *Case Concerning Armed Activities on the Territory of the Congo* from 2005, in which the ICJ used the same formula as in the *Construction of a Wall* Advisory Opinion and applied legal rules from both branches of international law.[34] International judicial bodies and courts have frequently referred to human rights instruments in the context of both international and non-international armed conflicts. Not only the ICJ, but also the UN Human Rights Committee, the European Court of Human Rights (hereinafter: ECtHR), and the Inter-American Court of Human Rights (hereinafter: IACtHR) did so in their jurisprudence.[35] There is also sufficient State practice suggesting that human rights are applicable during armed conflicts.[36]

The Montreux Document is a significant confirmation of the concurrent application of international humanitarian law and human rights in armed conflicts.[37]

Consequences of the Construction of a Wall in the Occupied Palestinian Territory, Advisory Opinion, July 9, 2004, ICJ Reports 2004, p. 178, para. 106.

31 David Turns, "The Law of Armed Conflict (International Humanitarian Law)", in Malcolm D. Evans (Ed.), (2010) *International Law*, Third Edition, Oxford: Oxford University Press, p. 817.

32 Hannah Tonkin, (2011) *State Control over Private Military and Security Companies in Armed Conflict*, Cambridge: Cambridge University Press, p. 143.

33 International Court of Justice, *Legal Consequences of the Construction of a Wall in the Occupied Palestinian Territory*, Advisory Opinion, July 9, 2004, ICJ Reports 2004, p. 178, para. 106.

34 International Court of Justice, *Case Concerning Armed Activities on the Territory of the Congo (Democratic Republic of the Congo v. Uganda)*, Judgment (Merits), December 19, 2005, ICJ Reports 2005, para. 216.

35 Christian Tomuschat, (2010) "Human Rights and International Humanitarian Law", *The European Journal of International Law*, Volume 21, Issue 1; Hans-Joachim Heintze, (2004) "On the Relationship Between Human Rights Law Protection and International Humanitarian Law", *International Review of the Red Cross*, Volume 86, Issue 856; Hannah Tonkin, (2011) *State Control over Private Military and Security Companies in Armed Conflict*, Cambridge: Cambridge University Press, p. 144.

36 Jean-Marie Henckaerts and Louise Doswald-Beck, (2005) *Customary International Humanitarian Law, Volume I: Rules*, ICRC, Cambridge: Cambridge University Press, pp. 303-305.

37 Hannah Tonkin, (2011) *State Control over Private Military and Security Companies in Armed Conflict*, Cambridge: Cambridge University Press, p. 145.

Compliance with two essential areas of international law and existing legal State obligations stemming from them is clearly promoted by the Montreux Document.[38] One of its provisions acknowledges the relevance of both branches of international law in armed conflicts and mentions the fact:

> "That certain well-established rules of international law apply to States in their relations with private military and security companies (PMSCs) and their operation during armed conflict, *in particular under international humanitarian law and human rights law*."[39] (emphasis added)

The Montreux Document provides a number of human rights obligations of hiring, host, and home States that are binding upon them in armed conflicts.[40] These duties are to be taken into consideration in addition to the obligations from the law of armed conflict, as the body of international human rights law can be said to guarantee certain additional procedural rights of high significance, such as the right to an individual remedy, and international procedures for individual complaint and redress.[41]

There is a spectrum of human rights norms applicable to the conduct of PMSCs and the three types of States associated with their use. PMSCs often find themselves in various conflict zones and other places around the globe and can breach numerous human rights instruments, which under certain circumstances might trigger international responsibility. In this regard, human rights treaties form an essential tool for assessing a variety of instances of PMSC misconduct and its qualification as violations of human rights law by States. The question of State responsibility might arise when these violations are to be attributed to those States or when States fail to comply with certain positive obligations they might have under this branch of international law. As PMSCs can, under certain circumstances, resort to the use of armed force and engage in the provision of coercive services, a number of human rights conventions are of particular importance. Human rights protected by these treaties can be violated by private contractors and their employees. States may also violate international law by failing to comply with the standards of human rights law. Only when there is a breach of an international obligation of a State, its responsibility can be incurred. With regard to PMSCs active in conflict areas, a number of human rights treaties are of significance. Those Conventions are: the International Covenant on Civil and Political Rights, (hereinafter: ICCPR), the European Convention on Human Rights (hereinafter: ECHR), the American Convention on Human Rights (hereinafter: ACHR), the

---

38 Montreux Document, Preface, p. 9, para. 2.

39 Ibidem, Preface, p. 9, para. 1.

40 Ibidem, Part One "Pertinent International Legal Obligations Relating to Private Military and Security Companies", pp. 11-13, paras. 4, 7, 10, 15.

41 Hannah Tonkin, (2011) *State Control over Private Military and Security Companies in Armed Conflict*, Cambridge: Cambridge University Press, p. 203.

African Charter on Human and Peoples' Rights (hereinafter: ACHPR), the UN Convention Against Torture (hereinafter: CAT), the UN Convention on the Rights of the Child (hereinafter: CRC), and the UN Convention on the Elimination of All Forms of Discrimination Against Women (hereinafter: CEDAW). In addition, it is also possible that PMSCs engage in breaches of other human rights instruments, such as the UN Convention on the Elimination of All Forms of Racial Discrimination, and their imputation to States and non-compliance with positive obligations would also lead to State responsibility. The central focus of the study is on the main regional human rights Conventions, the CAT, the CEDAW, and the CRC.

In the context of the use of PMSCs in conflict areas, international human rights law undoubtedly creates certain State obligations in relation to the misconduct of PMSCs, just as any other private person or entity, but the extent of applicability of this particular branch of international law appears to be limited due to a number of issues associated with it. In the first place, States must adhere to human rights conventions by signing and ratifying them in order to become bound by their provisions. It is true that some human rights and prohibitions constitute rules of customary international law and even *jus cogens* norms,[42] such as the prohibition of torture, genocide, slavery and the slave trade, aggression, racial discrimination, apartheid, and the right to self-determination. The protection of other human rights and freedoms is limited and depends on the willingness of States to become party to human rights treaties, which create legal obligations for States Parties, just as any other international treaty.[43] All of these conventions have not yet achieved a level of universal ratification and it might be too early to speak about the universal acceptance of all human rights standards. Additionally, States may also make reservations with regard to the scope of application of these conventions and some particular human rights and thus significantly reduce the reach of international human rights law.

Secondly, various international and regional conventions established in the field of human rights law contain clauses enabling States to derogate from their provisions in particular situations. Such clauses can be found in Article 4(1) ICCPR, Article 15(1) ECHR, and Article 27(1) ACHR. In order to take measures derogating from human rights obligations, States need to satisfy six requirements.[44] There must be a public emergency that threatens the life of the nation; the adopted measures must be required for addressing the urgent situation; they must not involve discrimination on the grounds of race, color, sex, religion, language, or social origin; these measures

---

42 Olivier De Schutter, (2010) *International Human Rights Law*, Cambridge: Cambridge University Press, p. 65.

43 Christian Tomuschat, (2008) *Human Rights: Between Idealism and Realism*, Second Edition, Oxford: Oxford University Press, pp. 40-41.

44 Olivier De Schutter, (2010) *International Human Rights Law*, Cambridge: Cambridge University Press, p. 514.

can only be taken if they are not inconsistent with other State obligations under international law; a State may not derogate from certain rights that are non-derogable; and this State must immediately inform other States Parties to the human rights conventions about its derogation.[45]

Finally, there is an issue of extraterritorial applicability of this part of international law making it less useful with regard to not only hiring, but also home States.[46] Originally, the application of human rights treaties is subject to one crucial limitation: the jurisdiction of States in various human rights regimes appears to be territorially limited. Jurisdiction can be defined as "the right to prescribe and enforce rules against others".[47] It is one of the important aspects of State sovereignty or even "a manifestation of state sovereignty"[48] and is essentially territorial.[49] The Permanent Court of International Justice (hereinafter: PCIJ) indicated in 1927:

> "Now the first and foremost restriction imposed by international law upon a State is that – failing the existence of a permissive rule to the contrary – it may not exercise its power in any form in the territory of another State. In this sense jurisdiction is certainly territorial; it cannot be exercised by a State outside its territory except by virtue of a permissive rule derived from international custom or from a convention."[50]

The notion of jurisdiction has been laid down in all three major human rights Conventions: in the European Convention on Human Rights in Article 1, in the American Convention on Human Rights in Article 1, and in the International Covenant on Civil and Political Rights in Article 2(1). It might be an issue if States hire PMSCs to operate outside their own territory, while it would not constitute a problem if hiring States are also host States where PMSCs carry out their activities. The ICJ understood the importance of the question of extraterritorial applicability of human rights conventions and looked into it.[51] The Court stated in its *Case Concerning Armed Activities*

---

45 Ibidem.

46 Hannah Tonkin, (2009) "Common Article 1: A Minimum Yardstick for Regulating Private Military and Security Companies", *Leiden Journal of International Law*, Volume 22, Issue 4, p. 780.

47 Vaughan Lowe, (2007) *International* Law, Oxford: Oxford University Press, p. 171.

48 Derek W. Bowett, (1982) "Jurisdiction: Changing Patterns of Authority over Activities and Resources", *British Yearbook of International Law*, Volume 53, Issue 1, p. 1.

49 Frederick A. Mann, (1984) "The Doctrine of International Jurisdiction Revisited After Twenty Years", *Collected Courses of the Hague Academy of International Law (Recueil des Cours)*, Volume 186, p. 20; Malcolm N. Shaw, (2008) *International Law*, Sixth Edition, Cambridge: Cambridge University Press, pp. 652-653; Cedric M.J. Ryngaert, (2008) *Jurisdiction in International Law*, Oxford: Oxford University Press, pp. 5-6.

50 Permanent Court of International Justice, *The Case of the S.S. "Lotus" (France v. Turkey)*, Judgment, September 7, 1927, PCIJ Reports, Series A, No. 10, pp. 18-19.

51 Michał Gondek, (2009) *The Reach of Human Rights in a Globalising World: Extraterritorial Application of Human Rights Treaties*, School of Human Rights Research Series, Volume 32, Antwerp and Portland: Intersentia, p. 213.

on the Territory of the Congo that international human rights instruments apply "in respect of acts done by a State in the exercise of its jurisdiction outside its own territory", especially in the occupied territories.[52] It implies that a form of territorial control exercised by a State might suffice to consider human rights treaties applicable. As pointed out by Monica Hakimi, this test considering State control over a particular territory represents quite a narrow approach and fails to take into account State obligations extending beyond a territory controlled by a State.[53]

From a moral perspective, it appears that human rights conventions must be applicable to a States' conduct taking place outside the States' own borders or having a human rights impact on the individuals living outside their territory.[54] States have a great potential to positively or negatively affect human rights.[55] They should not be able to breach human rights instruments abroad given that the same type of conduct is prohibited within those States themselves.[56] From a legal point of view, it would also be wrong,[57] but the matter seems to be more complicated. On the one hand, human rights treaties having an explicit jurisdiction clause may not apply extraterritorially and thus do not impose obligations upon States Parties for the possible misconduct of private persons and entities abroad.[58] On the other hand, human rights conventions lacking such a clause may seem to imply State duties in relation to the human rights abuses of individuals and entities committed outside their territory and jurisdiction.[59] In order to clarify this issue, the main human rights treaties should be considered in order to find out what the territorial limitations of their application might be, as the standards laid down therein may differ from the position of the ICJ. This must be done because the question of whether certain persons belong to the jurisdiction of a

---

52 International Court of Justice, *Case Concerning Armed Activities on the Territory of the Congo (Democratic Republic of the Congo v. Uganda)*, Judgment (Merits), December 19, 2005, ICJ Reports 2005, para. 216.

53 Monica Hakimi, (2010) "State Bystander Responsibility", *The European Journal of International Law*, Volume 21, Issue 2, pp. 377-378.

54 Michał Gondek, (2009) *The Reach of Human Rights in a Globalising World: Extraterritorial Application of Human Rights Treaties*, School of Human Rights Research Series, Volume 32, Antwerp and Portland: Intersentia, p. 3.

55 Malcolm Langford, Wouter Vandenhole, Martin Scheinin, and Willem van Genugten, "Introduction: An Emerging Field", in Malcolm Langford, Wouter Vandenhole, Martin Scheinin, and Willem van Genugten (Eds.), (2013) *Global Justice, State Duties: The Extraterritorial Scope of Economic, Social and Cultural Rights in International Law*, Cambridge: Cambridge University Press, p. 3.

56 Rick Lawson, "Life after Bankovic: On the Extraterritorial Application of the European Convention on Human Rights", in Fons Coomans and Menno T. Kamminga (Eds.), (2004) *Extraterritorial Application of Human Rights Treaties*, Antwerp and Oxford: Intersentia, p. 97.

57 Ibidem.

58 Nadia Bernaz, (2013) "Enhancing Corporate Accountability for Human Rights Violations: Is Extraterritoriality the Magic Potion?", *Journal of Business Ethics*, Volume 117, Issue 3, p. 504.

59 Ibidem.

State is not an issue of attributing instances of unlawful conduct to States, but an issue belonging to the domain of the primary rules of international law.[60]

Michał Gondek observes that there is a noticeable "cross-fertilization" between different human rights frameworks leading to many similarities in the interpretation of human rights treaties.[61] It is correct to point out that extraterritorial human rights duties are taken into account by the main human rights adjudicatory bodies when States exercise effective control over a particular territory or have physical power or control over individuals.[62] Fons Coomans and Menno Kamminga conclude that the supervisory bodies of the ICCPR, the ECHR, and the ACHR do not consider the notions of "jurisdiction" and "territory" to be the same, meaning that these human rights instruments do not apply exclusively to the conduct of States Parties within their own territory.[63] In this respect, there are two main categories of situations with regard to the applicability of these three treaties on which there is a general agreement between the UN Human Rights Committee, the European Court of Human Rights (hereinafter: ECtHR), and the Inter-American Commission on Human Rights (hereinafter: the Inter-American Commission or IACommHR).[64] In the first place, the human rights bodies seem to agree that when a State exercises power and authority or control over individuals, for instance, by detaining them on the territory of a third State, the human rights treaties apply to its conduct.[65] Secondly, if a State exercises effective control over the territory of a foreign State, for instance, during a military occupation, it is legally bound by its obligations stemming from the relevant treaties.[66] These positions of the main three human rights bodies and other human rights mechanisms will be explored in detail below.

---

60 Michał Gondek, (2009) *The Reach of Human Rights in a Globalising World: Extraterritorial Application of Human Rights Treaties*, School of Human Rights Research Series, Volume 32, Antwerp and Portland: Intersentia, p. 222.

61 Ibidem, p. 368.

62 Wouter Vandenhole, (2011) "Beyond Territoriality: The Maastricht Principles on Extra-Territorial Obligations in the Area of Economic, Social and Cultural Rights", *Netherlands Quarterly of Human Rights*, Volume 29, Issue 4, p. 430.

63 Fons Coomans and Menno T. Kamminga, "Comparative Introductory Comments on the Extraterritorial Application of Human Rights Treaties", in Fons Coomans and Menno T. Kamminga (Eds.), (2004) *Extraterritorial Application of Human Rights Treaties*, Antwerp and Oxford: Intersentia, p. 3.

64 Ibidem, pp. 3-4.

65 Ibidem, p. 4; Maarten den Heijer and Rick Lawson, "Extraterritorial Human Rights and the Concept of 'Jurisdiction'", in Malcolm Langford, Wouter Vandenhole, Martin Scheinin, and Willem van Genugten, (2014) *Global Justice, State Duties: The Extraterritorial Scope of Economic, Social and Cultural Rights in International Law*, Cambridge: Cambridge University Press, pp. 165, 172-182.

66 Fons Coomans and Menno T. Kamminga, "Comparative Introductory Comments on the Extraterritorial Application of Human Rights Treaties", in Fons Coomans and Menno T. Kamminga (Eds.), (2004) *Extraterritorial Application of Human Rights Treaties*, Antwerp and Oxford: Intersentia, p. 3; Maarten den Heijer and Rick Lawson, "Extraterritorial Human Rights and the Concept of 'Jurisdiction'", in Malcolm Langford, Wouter Vandenhole, Martin Scheinin, and Willem van Genugten, (2014) *Global Justice,*

The International Covenant on Civil and Political Rights

The ICCPR contains an application clause in Article 2(1) stating that:

> "Each State Party to the present Covenant undertakes to respect and to ensure to all individuals *within its territory* and *subject to its jurisdiction* the rights recognized in the present Covenant, without distinction of any kind, such as race, colour, sex, language, religion, political or other opinion, national or social origin, property, birth or other status." (emphasis added)

What this means is that if persons are not within the territory of a State and are not subject to its jurisdiction, that particular State cannot be held responsible for possible violations of the ICCPR.[67] In one of its early cases, the influential *Lopez Burgos* case, the UN Human Rights Committee looked into extraterritorial obligations of States under the ICCPR and noted:

> "The reference in article 1 of the Optional Protocol to 'individuals subject to its jurisdiction' does not affect the above conclusion because the reference in that article is not to the place where the violation occurred, but rather to *the relationship between the individual and the State in relation to a violation of any of the rights set forth in the Covenant, wherever they occurred*."[68] (emphasis added)

Despite the fact that the Committee referred to the Optional Protocol, its reasoning regarding the question of jurisdiction and the relationship between States and individuals is also applicable to Article 2(1) ICCPR.[69] The Committee continued by giving more guidance on the application of this provision observing that:

> "Article 2 (1) of the Covenant places an obligation upon a State party to respect and to ensure rights 'to all individuals within its territory and subject to its jurisdiction', but it does not imply that the State party concerned cannot be held accountable for violations of rights under the Covenant which its agents commit upon the territory of another

---

State Duties: The Extraterritorial Scope of Economic, Social and Cultural Rights in International Law, Cambridge: Cambridge University Press, pp. 165-172.

67 Dominic McGoldrick, "Extraterritorial Application of the International Covenant on Civil and Political Rights", in Fons Coomans and Menno T. Kamminga (Eds.), (2004) *Extraterritorial Application of Human Rights Treaties*, Antwerp and Oxford: Intersentia, p. 43.

68 UN Human Rights Committee, *Sergio Euben Lopez Burgos v. Uruguay*, July 29, 1981, Communication No. R.12/52, UN Doc. Supp. No. 40 (A/36/40) at 176 (1981), para. 12.2; see also UN Human Rights Committee, *Lilian Celiberti de Casariego v. Uruguay*, July 29, 1981, Communication No. R.13/56 (July 17, 1979), UN Doc. Supp. No. 40 (A/36/40) at 185 (1981), para. 10.2.

69 Michał Gondek, (2009) *The Reach of Human Rights in a Globalising World: Extraterritorial Application of Human Rights Treaties*, School of Human Rights Research Series, Volume 32, Antwerp and Portland: Intersentia, p. 137.

State, whether with the acquiescence of the Government of that State or in opposition to it."[70]

In this regard, the UN Human Rights Committee held that:

"... it would be unconscionable to so interpret the responsibility under article 2 of the Covenant as to permit a State party to perpetrate violations of the Covenant on the territory of another State, which violations it could not perpetrate on its own territory."[71]

In its Advisory Opinion on the *Construction of a Wall*, the ICJ observed that:

"... [W]hile the jurisdiction of States is primarily territorial, it may sometimes be exercised *outside the national territory*. Considering the object and purpose of the International Covenant on Civil and Political Rights, it would seem natural that, even when such is the case, States parties to the Covenant *should be bound to comply* with its provisions."[72] (emphasis added)

The ICJ referred to the cases of the Human Rights Committee, more specifically the *Lopez Burgos* case and the *Celiberti de Casariego* case, and considered the above statement to be consistent with the constant practice of the Committee. The Court took into account the *travaux préparatoires* of the ICCPR and considered them not to contain any limitations to the extraterritorial applicability of the Covenant.[73] It admitted that the ICCPR indeed applies in relation to acts performed by a State exercising its jurisdiction outside its national territory.[74]

In its General Comment 31 from 2004, the Committee took a clear position and provided, with respect to the requirement of Article 2(1) of the Covenant to respect and to ensure the rights of the Covenant, that:

"States Parties are required by article 2, paragraph 1, to respect and to ensure the Covenant rights to all persons who may be within their territory and to all persons subject to their jurisdiction. This means that a State party must respect and ensure the rights laid down in the Covenant to *anyone within the power or effective control* of that State Party, even if not situated within the territory of the State Party... the enjoyment of Covenant rights is not limited to citizens of States Parties but must also be available to

---

70 UN Human Rights Committee, *Sergio Euben Lopez Burgos v. Uruguay*, July 29, 1981, Communication No. R.12/52, UN Doc. Supp. No. 40 (A/36/40) at 176 (1981), para. 12.3; see also UN Human Rights Committee, *Lilian Celiberti de Casariego v. Uruguay*, July 29, 1981, Communication No. R.13/56 (July 17, 1979), UN Doc. Supp. No. 40 (A/36/40) at 185 (1981), para. 10.3.

71 Ibidem, para. 12.3; ibidem, para. 10.3.

72 International Court of Justice, *Legal Consequences of the Construction of a Wall in the Occupied Palestinian Territory*, Advisory Opinion, July 9, 2004, ICJ Reports 2004, p. 179, para. 109.

73 Ibidem.

74 Ibidem, p. 180, para. 111.

> all individuals, regardless of nationality or statelessness, such as asylum seekers, refugees, migrant workers and other persons, who may find themselves in the territory or subject to the jurisdiction of the State Party. This principle also applies to *those within the power or effective control* of the forces of a State Party acting outside its territory, regardless of the circumstances in which such power or effective control was obtained, such as forces constituting a national contingent of a State Party assigned to an international peace-keeping or peace-enforcement operation."[75] (emphasis added)

According to the Committee, States are obliged to respect and to ensure respect of the rights laid down in the ICCPR to all persons residing within their territory and subject to their jurisdiction. Essentially, it means that they have not only positive, but also negative human rights obligations.[76] In this regard, the concept of jurisdiction implies the exercise of power or effective control over persons and not over a territory.[77] Thus, when PMSCs are hired by States to run prisons, as was the case with a number of private contractors employed by the United States to operate at the Abu Ghraib detention facility in Iraq, and have detainees of the host State in their power or effective control, these hiring States will be obliged to comply with both negative and positive obligations of the ICCPR. It is, however, also arguable that the degree of full control over persons is not required to conclude that the ICCPR is to be applied extraterritorially. Following the reasoning of the Committee in the *Nunez v. Uruguay* case,[78] the *Ibrahim Gueye v. France* case,[79] and other "passport cases", it can be assumed that States' exercise of control over a certain aspect of an individual's life abroad is sufficient.[80] It might be a relationship between citizen and State, or former membership of a person in the official armed forces and State pension, as it was the case in two above-mentioned cases. It can also be argued that other short-term relationships between hiring States and citizens of host States may justify the territorial application of the ICCPR, such as the provision of security by States to a local population and further delegation of this task to PMSCs. This, in conjunction with the reasoning of the ICJ in the *Construction of a Wall* Advisory Opinion and the *Case Concerning Armed Activities on the Territory of the Congo*, indicates that the ICCPR actually creates legal obligations for hiring and home States with respect to the conduct of PMSCs when they operate overseas.

---

75 UN Human Rights Committee, *General Comment 31*, May 26, 2004, UN Doc. CCPR/C/21/Rev.1/Add.13, para. 10.

76 Michał Gondek, (2009) *The Reach of Human Rights in a Globalising World: Extraterritorial Application of Human Rights Treaties*, School of Human Rights Research Series, Volume 32, Antwerp and Portland: Intersentia, p. 233.

77 Ibidem.

78 UN Human Rights Committee, *Nunez v. Uruguay*, July 22, 1983, UN Doc. Supp. No. 40 (A/38/40).

79 UN Human Rights Committee, *Ibrahim Gueye v. France*, April 6, 1989, UN Doc. CCPR/C/35/D/196/1985.

80 Hannah Tonkin, (2011) *State Control over Private Military and Security Companies in Armed Conflict*, Cambridge: Cambridge University Press, pp. 205-206.

A different view of the UN Human Rights Committee regarding the exercise of effective control over a particular territory that can trigger extraterritorial application of the ICCPR has been provided in its views on the Israeli occupation of Palestinian territories. Concerning the situations of occupation, it is to be noted that not only the UN human rights bodies, but also the ECtHR and the ICJ consider occupying powers to be under an obligation to apply their existing human rights duties to all activities realized on the occupied territory.[81] In its concluding observations on the initial Israeli report, the UN Human Rights Committee expressed its deep concerns about the Israeli denial of responsibility to apply the Covenant in the Palestinian territories.[82] It took into consideration not only the duration of the presence of Israel in these territories, but also "its ambiguous attitude towards their future status" and "the exercise of effective jurisdiction by Israeli security forces therein".[83] The view of the Committee was that "under the circumstances, the Covenant must be held applicable to the occupied territories and those areas of southern Lebanon and West Bekaa where Israel exercises effective control".[84] In the concluding observation on the second report submitted by Israel, the Committee emphasized the view held in its initial report and concluded that:

> "… the provisions of the Covenant apply to the benefit of the population of the Occupied Territories, for all conduct by its authorities or agents in those territories that affect the enjoyment of rights enshrined in the Covenant and fall within the ambit of state responsibility of Israel under the principles of public international law."[85]

The Committee noted that Palestinians remained under control of Israeli security forces and could not enjoy the rights and freedoms laid down in the Covenant.[86] It also took the position that a State can be responsible for a breach of the ICCPR when the enjoyment of rights is affected by the conduct of State authorities or agents. This "issue" has been addressed by the Committee in its Comment No. 31 specifically dealing with power and effective control over individuals.[87]

---

81 Maarten den Heijer and Rick Lawson, "Extraterritorial Human Rights and the Concept of 'Jurisdiction'", in Malcolm Langford, Wouter Vandenhole, Martin Scheinin, and Willem van Genugten, (2014) *Global Justice, State Duties: The Extraterritorial Scope of Economic, Social and Cultural Rights in International Law*, Cambridge: Cambridge University Press, p. 168.

82 UN Human Rights Committee, *Concluding Observations of the Human Rights Committee: Israel*, August 18, 1998, UN Doc. CCPR/C/79/Add.93, para. 10.

83 Ibidem.

84 Ibidem.

85 UN Human Rights Committee, *Concluding Observations of the Human Rights Committee: Second Periodic Report of Israel*, August 5, 2003, CCPR/CO/78/ISR, para. 11.

86 UN Human Rights Committee, *Concluding Observations of the Human Rights Committee: Israel*, August 18, 1998, UN Doc. CCPR/C/79/Add.93, para. 13.

87 UN Human Rights Committee, *General Comment 31*, May 26, 2004, UN Doc. CCPR/C/21/Rev.1/Add.13, para. 10; Michał Gondek, (2009) *The Reach of Human Rights in a Globalising World:*

The European Convention on Human Rights

The next step is to look at the European Convention on Human Rights. Article 1 of the Convention provides that "the High Contracting Parties shall secure to everyone within their jurisdiction the rights and freedoms defined in Section I of this Convention". The ECtHR indicated in the *Banković* case that "the jurisdictional competence of a State is primarily territorial", but stressed that jurisdiction can be exercised extraterritorially.[88] This case was an important turning point in the case law of the Court and will be discussed below. For now, it might be useful to briefly set out its position and the view of the European Commission of Human Rights with regard to the extraterritorial applicability of the ECHR starting with the notion of jurisdiction depending on the State control over individuals.

In its early decisions, the Commission dealt with the situation of occupation of northern Cyprus by Turkey and elaborated on the notion "within their jurisdiction" laid down in Article 1 ECHR in its first decision from 1975. It concluded that:

> "... this term is not, as submitted by the respondent Government, equivalent to or limited to the national territory of the High Contracting Party concerned. It is clear from the language, in particular of the French text, and the object of this Article, and from the purpose of the Convention as a whole, that the High Contracting Parties are bound to secure the said rights and freedoms to *all persons under their actual authority and responsibility, whether that authority is exercised within their own territory or abroad*."[89] (emphasis added)

With regard to the actors whose conduct might constitute a violation of human rights law, the Court provided:

> "... that nationals of a State, including registered ships and aircrafts, are partly within its jurisdiction wherever they may be, and that authorised agents of a State, including diplomatic or consular agents and armed forces, not only remain under its jurisdiction when abroad but bring any other persons of property 'within the jurisdiction' of that State, to the extent that they exercise *authority over such persons or property*. Insofar as, by their acts or omissions, they affect such persons or property, the responsibility of the State is engaged."[90] (emphasis added)

Extraterritorial Application of Human Rights Treaties, School of Human Rights Research Series, Volume 32, Antwerp and Portland: Intersentia, p. 236.

88 European Court of Human Rights, *Banković and Others v. Belgium and Others*, Decision, December 12, 2001, Application No. 52207/99, Reports 2001-XII, paras. 59, 61, 63.

89 European Commission of Human Rights, *Cyprus v. Turkey*, Decision, May 26, 1975, Application Nos. 6780/74 and 6950/75, para. 8.

90 Ibidem.

In the *Loizidou v. Turkey* case, the ECtHR noted that the notion of jurisdiction used in Article 1 ECHR extends beyond the national territory of States Parties to the ECHR.[91] Acts and omissions of these States having effects outside their national borders can lead to State responsibility. In this regard, the focus has been shifted from control over persons to control over a certain territory. Referring to the Judgment on Preliminary Objections, it stated:

> "Of particular significance to the present case the Court held, in conformity with the relevant principles of international law governing State responsibility, that the responsibility of a Contracting Party could also arise when as a consequence of military action – whether lawful or unlawful – it exercises *effective control of an area outside its national territory*."[92] (emphasis added)

The Court found that Turkey had detailed control over the authorities of the Turkish Republic of Northern Cyprus (hereinafter: TRNC) and stationed a large number of its troops actively operating there.[93] Turkey was obliged to comply with its human rights obligations on the territory of northern Cyprus because its army exercised "effective overall control" over that part of the island.[94] Importantly, these human rights duties apply not only with regard to the acts of Turkish troops, officials, and the authorities of the TRNC, but also acts of the local administration and private actors who violate human rights of Greek Cypriots or Turkish Cypriots.[95]

After the *Loizidou v. Turkey* case, the *Banković* case was the most influential decision of the ECtHR, which set a very high standard for establishing jurisdiction and had significant impact on jurisprudence of other human rights bodies.[96] The Court acknowledged the primarily territorial notion of jurisdiction and stressed that only in exceptional cases acts of States performed abroad or producing effects outside their

---

91 European Court of Human Rights, *Loizidou v. Turkey*, Preliminary Objections, March 23, 1995, Application No. 15318/89, Series A, No. 310, para. 62; European Court of Human Rights, *Loizidou v. Turkey*, Judgment, December 18, 1996, Application No. 15318/89, Reports 1996-VI, para. 52.

92 European Court of Human Rights, *Loizidou v. Turkey*, Judgment, December 18, 1996, Application No. 15318/89, Reports 1996-VI, para. 52.

93 Ibidem, para. 56.

94 Ibidem.

95 European Court of Human Rights, *Cyprus v. Turkey*, Judgment, May 10, 2001, Application No. 25781/94, Reports 2001-IV, paras. 77, 81.

96 Maarten den Heijer and Rick Lawson, "Extraterritorial Human Rights and the Concept of 'Jurisdiction'", in Malcolm Langford, Wouter Vandenhole, Martin Scheinin, and Willem van Genugten, (2014) *Global Justice, State Duties: The Extraterritorial Scope of Economic, Social and Cultural Rights in International Law*, Cambridge: Cambridge University Press, pp. 176-177, 190; Michał Gondek, (2009) *The Reach of Human Rights in a Globalising World: Extraterritorial Application of Human Rights Treaties*, School of Human Rights Research Series, Volume 32, Antwerp and Portland: Intersentia, pp. 150-151; Alexander Orakhelashvili, (2003) "Restrictive Interpretation of Human Rights Treaties in the Recent Jurisprudence of the European Court of Human Rights", *The European Journal of International Law*, Volume 14, Issue 3, p. 538.

own territory can be regarded as the exercise of jurisdiction.[97] This is the case when a State exercises "all or some of the public powers normally to be exercised" by that State through the effective control over a certain territory and its inhabitants abroad during military occupation or through "the consent, invitation or acquiescence of the Government of that territory".[98]

In the recent decisions *Issa v. Turkey*,[99] *Ilaşcu v. Moldova and Russia*,[100] and *Öcalan v. Turkey*,[101] the ECtHR built on the reasoning provided in the *Banković* case and elaborated on the scope of extraterritorial State obligations. It found that the provisions of the ECHR may apply extraterritorially when a State exercises effective overall control over a certain territory outside its national territory or over individuals. In *Issa and Others v. Turkey*, concerning the murder of Iraqi individuals during a Turkish military operation, the Court observed:

> "The Court does not exclude the possibility that, as a consequence of this military action, the respondent State could be considered *to have exercised, temporarily, effective overall control* of a particular portion of the territory of northern Iraq. Accordingly, if there is a sufficient factual basis for holding that, at the relevant time, the victims were within that specific area, it would follow logically that they were within the jurisdiction of Turkey..."[102] (emphasis added)

Obviously, in these scenarios a State can exercise effective overall control only temporarily over a certain territory. It does not necessarily have to be the detailed control exercised through the State agents, armed forces, or subordinate local administration on the territory of a foreign State. Even the overall control of the area could be enough to hold a State responsible for violating the Convention's rights and freedoms.[103] Significantly, a State may be held accountable for breaching these provisions in relation to individuals who are to be found on the territory of another State, but are placed under the first State's authority and control through the State agents lawfully or unlawfully operating on the territory of the second State.[104] Time and again, the ECtHR believed that:

---

[97] European Court of Human Rights, *Banković and Others v. Belgium and Others*, Decision, December 12, 2001, Application No. 52207/99, Reports 2001-XII, para. 67.

[98] Ibidem, para. 71.

[99] European Court of Human Rights, *Issa and Others v. Turkey*, Judgment, November 16, 2004, Application No. 31821/96, ECHR 629.

[100] European Court of Human Rights, *Ilaşcu v. Moldova and Russia*, Judgment, July 8, 2004, Application No. 48787/99, Reports 2004-VII.

[101] European Court of Human Rights, *Öcalan v. Turkey*, Judgment, May 12, 2005, Application No. 46221/99, Reports 2005-IV.

[102] European Court of Human Rights, *Issa and Others v. Turkey*, Judgment, November 16, 2004, Application No. 31821/96, ECHR 629, para. 74.

[103] Ibidem, paras. 69-70.

[104] Ibidem, para. 71.

"Accountability in such situations stems from the fact that Article 1 of the Convention cannot be interpreted so as to allow a State party to perpetrate violations of the Convention on the territory of another State, which it could not perpetrate on its own territory."[105]

In the *Ilaşcu v. Moldova and Russia* case, the ECtHR further broadened the "effective overall control" notion by declaring that the extraterritorial applicability of the Convention is also possible in situations when one State supports the establishment of a separatist State within the territory of another State by military, political, and other means.[106] In the *Öcalan v. Turkey* case, the ECtHR set the facts of this case apart from the *Banković* case and made clear that if individuals are under effective State authority and are subject to its authority and control, they are essentially brought within its jurisdiction, as provided in Article 1 ECHR.[107]

The American Convention on Human Rights

Article 1 of the American Convention on Human Rights stipulates:

"The States Parties to this Convention undertake to respect the rights and freedoms recognized herein and to ensure to *all persons subject to their jurisdiction* the free and full exercise of those rights and freedoms, without any discrimination for reasons of race, color, sex, language, religion, political or other opinion, national or social origin, economic status, birth, or any other social condition." (emphasis added)

While the application of the Convention extends to all persons subject to the jurisdiction of States Parties to the ACHR, the American Declaration on the Rights and Duties of Man – another significant human rights instrument within the framework of the American Organization of States applicable to States that failed to ratify the Convention – does not contain a jurisdiction clause. In one case, the Inter-American Commission on Human Rights noted with regard to the extraterritorial applicability of the American Declaration that:

"While the extraterritorial application of the American Declaration has not been placed at issue by the parties, the Commission finds it pertinent to note that, under certain circumstances, the exercise of its jurisdiction over acts with an extraterritorial locus

---

105 Ibidem, para. 71.

106 European Court of Human Rights, *Ilaşcu v. Moldova and Russia*, Judgment, July 8, 2004, Application No. 48787/99, Reports 2004-VII, para. 312; see also Michał Gondek, (2009) *The Reach of Human Rights in a Globalising World: Extraterritorial Application of Human Rights Treaties*, School of Human Rights Research Series, Volume 32, Antwerp and Portland: Intersentia, p. 189.

107 European Court of Human Rights, *Öcalan v. Turkey*, Judgment, May 12, 2005, Application No. 46221/99, Reports 2005-IV, para. 93.

> will not only be consistent with but required by the norms which pertain... While this most commonly refers to persons within a state's territory, it may, under given circumstances, refer to conduct with *an extraterritorial locus* where the person concerned is present in the territory of one state, but subject to the control of another state – usually through the acts of the latter's agents abroad. In principle, the inquiry turns not on the presumed victim's nationality or presence within a particular geographic area, but on whether, under the specific circumstances, the State observed the rights of a person subject to *its authority and control*."[108] (emphasis added)

In addition, the Commission looked for inspiration in the jurisprudence of the European Court and Commission of Human Rights and concluded on another occasion in relation to the American Convention that:

> "[It] does not believe, however, that the term 'jurisdiction' in the sense of Article 1(1) is limited to or merely coextensive with national territory. Rather, the Commission is of the view that a state party to the American Convention may be responsible under certain circumstances for the acts and omissions of its agents which produce effects or are undertaken *outside that state's own territory*... the High contracting Parties are bound to secure the said rights and freedoms to all persons under *their actual authority* and *responsibility*, whether that authority is exercised within their own territory or abroad... This understanding of jurisdiction – and therefore responsibility for compliance with international obligations – as a notion linked to *authority* and *effective control*, and not merely to territorial boundaries, has been confirmed and elaborated on in other cases decided by the European Commission and Court."[109] (emphasis added)

In the *Alejandre v. Cuba* case from 1999, the Inter-American Commission on Human Rights mentioned the fact that the notion of jurisdiction under the American Convention also includes extraterritorial actions violating the Convention's rights and freedoms. American States were found to be under an obligation to respect rights of persons subject to their jurisdiction, which are usually individuals finding themselves within a territory of a State in question.[110] In certain cases, it can, however, refer to extraterritorial acts "when the person is present in the territory of a state but subject to the control of another state, generally through the actions of that state's agents abroad".[111] In this regard, one should focus not on the nationality of the alleged victims or their presence in certain geographical areas, but on the question of whether

---

[108] Inter-American Commission on Human Rights, *Coard v. United States*, Report No. 109/99, Case No. 10.951, September 29, 1999, para. 37.

[109] Inter-American Commission on Human Rights, *Saldano v. Argentina*, Petition, Report No. 38/99, March 11, 1999, paras. 17–19.

[110] Inter-American Commission on Human Rights, *Alejandre v. Cuba*, Report No. 86/99, Case No. 11.589, September 29, 1999, para. 23.

[111] Ibidem.

under given circumstances the State has observed the rights of persons subject to its authority and control.[112] This means that jurisdiction is established when States breach their negative obligations under the Convention or the Declaration and there is no necessity to establish effective control exercised by a given State over certain territory or alleged victims of the violations.[113]

The African Charter on Human and Peoples' Rights

The African Charter on Human and Peoples' Rights does not mention the concepts of jurisdiction or territory in its respective provision regarding the obligations of States Parties. Article 1 ACHPR stresses:

> "Every individual shall be entitled to the enjoyment of the rights and freedoms recognised and guaranteed in the present Charter without distinction of any kind such as race, ethnic group, colour, sex, language, religion, political or any other opinion, national and social origin, fortune, birth or any status."

While the lack of jurisdiction and territoriality clauses in the African Charter appears to be a significant argument supporting the idea that the scope of application of the treaty obligations is not limited in accordance with these two criteria, some might believe that States Parties are not required to realize human rights laid down in the Charter abroad.[114] Although it does not mean that the human rights treaty applies only in the territory and within jurisdiction of States Parties and may imply the extraterritorial applicability of the ACHPR, most cases concerning extraterritorial obligations of African States that had been brought before the African Commission on Human and Peoples' Rights (hereinafter: the African Commission) were dismissed due to the lack of jurisdiction.[115] In the more recent cases, the Commission took the position that international responsibility can be in order when States commit human rights violations abroad.[116] This changed attitude of the supervisory body indicates that State obligations arising from the ACHPR do indeed have an extraterritorial

---

112 Ibidem.

113 Hannah Tonkin, (2011) *State Control over Private Military and Security Companies in Armed Conflict*, Cambridge: Cambridge University Press, p. 208.

114 Takele Soboka Bulto, (2011) "Patching the 'Legal Black Hole': the Extraterritorial Reach of States' Human Rights Duties in the African Human Rights System", *South African Journal of Human Rights*, Volume 27, Issue 2, pp. 257-258.

115 Ibidem, pp. 258-260.

116 See African Commission on Human and Peoples' Rights, *D.R. Congo v. Burundi, Rwanda and Uganda*, Communication No. 227/99 (2003); African Commission on Human and Peoples' Rights, *Association pour la Sauvegarde de la Paix au Burundi v. Kenya, Uganda, Rwanda, Tanzania, Zaire (DRC), Zambia*, Communication No. 157/96 (2003).

dimension and African States Parties can be held responsible for violating human rights abroad.[117]

The UN Convention Against Torture

It deserves mentioning that, given the actual human rights protected by the CAT and the wording of the Convention, its extraterritorial application remains "relatively unchallenged".[118] The core obligation on States under this treaty is found in Article 2(1) CAT pronouncing that:

> "Each State Party shall take effective legislative, administrative, judicial or other measures to prevent acts of torture in any territory under its jurisdiction."

This particular provision refers to the obligation on every State Party to the Convention to take an array of measures "in any territory under its jurisdiction". Similar wording is also used in Articles 5, 11, 12, 13, and 16 of the CAT. The General Comment No. 2 of the Committee Against Torture touches upon the extraterritorial application of the Convention. It states:

> "Article 2, paragraph 1, requires that each State party shall take effective measures to prevent acts of torture not only in its sovereign territory but also 'in any territory under its jurisdiction.' The Committee has recognized that 'any territory' includes all areas where the State party exercises, directly or indirectly, in whole or in part, *de jure or de facto effective control*, in accordance with international law. The reference to 'any territory' in article 2, like that in articles 5, 11, 12, 13 and 16, refers to prohibited acts committed not only on board a ship or aircraft registered by a State party, but also during military occupation or peacekeeping operations and in such places as embassies, military bases, detention facilities, or other areas over which a State exercises *factual or effective control*. The Committee notes that this interpretation reinforces article 5, paragraph 1 (b), which requires that a State party must take measures to exercise jurisdiction 'when the alleged offender is a national of the State.' The Committee considers that the scope of 'territory' under article 2 must also include situations where a State party exercises, directly or indirectly, *de facto or de jure control over persons in detention*."[119] (emphasis added)

---

[117] Takele Soboka Bulto, (2011) "Patching the 'Legal Black Hole': the Extraterritorial Reach of States' Human Rights Duties in the African Human Rights System", *South African Journal of Human Rights*, Volume 27, Issue 2, pp. 278.

[118] Michał Gondek, (2009) *The Reach of Human Rights in a Globalising World: Extraterritorial Application of Human Rights Treaties*, School of Human Rights Research Series, Volume 32, Antwerp and Portland: Intersentia, p. 247.

[119] UN Committee Against Torture, *General Comment No. 2, Implementation of Article 2 by States Parties*, January 24, 2008, UN Doc. CAT/C/GC/2, para. 16.

According to the Committee, the obligation to prevent torture applies not merely in the territory of States Parties, but also extends to all other areas where they exercise in a direct or indirect manner effective control that can be both *de jure* or *de facto*. When a State exercises effective control over a territory of another State or persons abroad, it is required to meet not only negative, but also positive obligations under the CAT.[120] The prohibited acts can thus be committed by a State in the situations of military occupation and peace operations during which States exercise sufficient level of factual or effective control. In these scenarios, it is not uncommon for States to resort to the use of PMSCs, which may be implicated in torture and other related forms of abuse. Importantly, such places as embassies, military bases, and detention facilities where PMSCs can be active also fall under the term "any territory" under the jurisdiction of States, when the latter have a factual or effective control. The same is true for the situations when States directly or indirectly exercise *de jure* or *de facto* control over persons in detention: the use of PMSCs in prisons could serve as a perfect example of such a construction.

The UN Convention on the Rights of the Child

Concerning the CRC, Article 2(1) specifically provides:

> "States Parties shall respect and ensure the rights set forth in the present Convention to each child *within their jurisdiction* without discrimination of any kind, irrespective of the child's or his or her parent's or legal guardian's race, colour, sex, language, religion, political or other opinion, national, ethnic or social origin, property, disability, birth or other status." (emphasis added)

The Committee on the Rights of the Child has yet not extensively dealt with the issue of extraterritorial applicability of the CRC.[121] In its concluding observations on the initial report of Israel, however, it referred to the responsibility of Israel to implement the CRC in the occupied Palestinian territories.[122] The Committee pointed out that the discrimination of Palestinian children and inequalities in their enjoyment of their rights and freedoms still persisted in these territories and that there were allegations and complaints of torture, ill-treatment, and inhuman or degrading practices of these

---

120 Manfred Nowak, "Obligations of States to Prevent and Prohibit Torture in an Extraterritorial Perspective", in Mark Gibney and Sigrun Skogly (Eds.), (2010) *Universal Human Rights and Extraterritorial Obligations*, Philadelphia: University of Pennsylvania Press, p. 28.

121 Michał Gondek, (2009) *The Reach of Human Rights in a Globalising World: Extraterritorial Application of Human Rights Treaties*, School of Human Rights Research Series, Volume 32, Antwerp and Portland: Intersentia, p. 256.

122 UN Committee on the Rights of the Child, *Concluding Observations: Israel*, October 9, 2002, UN Doc. CRC/C/15/Add.195, para. 2.

children.[123] This is a clear indication of the applicability of the CRC in the occupied territories under effective State control.

The UN Convention on the Elimination of All Forms of Discrimination Against Women

The extraterritorial application of this Convention cannot easily be judged given that its provisions are more "policy-oriented" and that the Committee on the Elimination of Discrimination Against Women has not elaborated upon this issue.[124] While this particular treaty can be of relevance for determining an array of positive obligations of States in relation to PMSCs, together with the CRC it is of a less practical value than the above-mentioned human rights instruments. Therefore, the extraterritorial reach of the obligations laid down in this particular treaty will not be covered in detail.

Interim conclusion

In conclusion, it must be acknowledged that there are two sufficiently clear types of situations in which most human rights treaties have extraterritorial application. Those are the cases when a State party to a human rights convention exercises: (a) effective control over a territory of another State or (b) power and authority over individuals. Some authors reduce them to the exercise of "effective control" by State authorities over territories and persons abroad.[125] It might be argued that if a State has effective control over a certain foreign territory, it also has a capacity to control and to influence individuals residing in this territory. Indeed, Daniel Augenstein and David Kinley analyzed extraterritorial human rights obligations of States and observed that "what is decisive for extra-territorial human rights obligations is states asserting *de facto* power over the individual rights-holder".[126] This power over individuals might be an overarching concept that triggers extraterritorial State obligations under human rights law. In addition, there are scenarios in which there is no common agreement among human rights bodies regarding the issue of extraterritorial applicability of

---

123 Ibidem, paras. 24, 26, 36.

124 Michał Gondek, (2009) *The Reach of Human Rights in a Globalising World: Extraterritorial Application of Human Rights Treaties*, School of Human Rights Research Series, Volume 32, Antwerp and Portland: Intersentia, p. 257.

125 Manfred Nowak, "Obligations of States to Prevent and Prohibit Torture in an Extraterritorial Perspective", in Mark Gibney and Sigrun Skogly (Eds.), (2010) *Universal Human Rights and Extraterritorial Obligations*, Philadelphia: University of Pennsylvania Press, p. 14.

126 Daniel Augenstein and David Kinley, "When Human Rights 'Responsibilities' Become 'Duties': The Extra-territorial Obligations of States that Bind Corporations", in Surya Deva and David Bilchitz (Eds.), (2013) *Human Rights Obligations of Business: Beyond the Corporate Responsibility to Respect?*, Cambridge: Cambridge University Press, p. 291.

human rights instruments.[127] Instances of extraterritorial killings that are not preceded by arrest belong to this category. It is evident that if States exercise effective control over a certain territory due to a situation of a military occupation or have power and authority over individuals detained in prisons run by them abroad, human rights treaties will apply to them extraterritorially. It means that such States may face international responsibility if human rights violations are committed by PMSCs and their personnel operating on foreign soil.

2.4 International humanitarian law

As is made clear by some experts, claims that PMSCs operate in a legal vacuum seem to be based on the assumption that legal norms of international humanitarian law regulating the use of mercenaries are not applicable to the employees of these firms.[128] The law of armed conflict is "one of the most highly codified parts of international law"[129] and is said to include all rules of international law that restrict the means and methods of warfare and regulate the treatment of individuals, such as civilians and military personnel, in international and non-international armed conflicts. It is applicable in all armed conflicts and to all parties to these conflicts, but its application to PMSCs has a number of implications. In the first place, PMSCs can and on many occasions do breach rules and principles of international humanitarian law, which might lead to the international responsibility of States if the attribution of this unlawful conduct is possible. Secondly, this branch of international law also regulates the conduct of hostilities in international and non-international armed conflict and confers a certain legal status on the employees of PMSCs. Determining the type of an armed conflict and this actual position of PMSC personnel on the battlefield by examining international humanitarian law is crucial for understanding an array of rights, privileges, and immunities they might have. Additionally and more importantly, this legal status can also have consequences for the State responsibility that is central to the current study. Finally, the law of armed conflict contains positive obligations that must be observed by States and in case of non-compliance with these duties they might be held internationally responsible.

Many rules of the law of armed conflict of pertinent relevance for the State use of PMSCs are found in customary international law, the four Geneva Conventions

---

127 Fons Coomans and Menno T. Kamminga, "Comparative Introductory Comments on the Extraterritorial Application of Human Rights Treaties", in Fons Coomans and Menno T. Kamminga (Eds.), (2004) *Extraterritorial Application of Human Rights Treaties*, Antwerp and Oxford: Intersentia, p. 4.

128 Eugenio Cusumano, "Policy Prospects for Regulating Private Military and Security Companies", in Francesco Francioni and Natalino Ronzitti (Eds.), (2011) *War by Contract: Human Rights, Humanitarian Law, and Private Contractors*, Oxford: Oxford University Press, p. 31.

129 Malcolm N. Shaw, (2008) *International Law*, Sixth Edition, Cambridge: Cambridge University Press, pp. 1167-1168.

of 1949[130] that are regarded as forming part of customary international law[131] and their Additional Protocols, and the Hague Regulations of 1907. The law of armed conflict, however, does not specifically address the employment of PMSCs by States and there is no rule of international humanitarian law prohibiting this employment. PMSCs and their personnel often are in a position to breach the law of armed conflict by, for instance, engaging in indiscriminate shootings or mistreating prisoners of war that can result in a State being held responsible for such misconduct. For the purposes of the current study, the emphasis is put on the so-called Geneva law that is not concerned with the questions of the legality of war and the use of force, but seeks to grant a certain level of protection to the victims of armed conflict. In order to reduce and to eliminate where possible the horrors of armed struggles between States and to protect vulnerable persons, this theoretical line of international humanitarian law focuses on a number of categories of individuals: the sick and wounded in the armed forces in the field and at sea, prisoners of war, and civilians.[132] Scattered over various legal instruments, such as the four Geneva Conventions, Additional Protocols, and customary international law, the Geneva law is different from the Hague law, which seeks to regulate the means and methods of conducting war.[133] In this study, the Hague law will largely be left aside while relevant rules and principles of the Geneva law will be discussed and analyzed in detail.

2.5 International regulatory initiatives on business and human rights

An interesting set of norms regarding the corporate social responsibility of PMSCs is contained in the 2003 UN Norms on the Responsibilities of Trans-national Corporations and Other Business Enterprises with Regard to Human Rights (hereinafter: the Norms). They were drafted by the Sub-Commission on the Promotion and Protection of Human Rights, the main subsidiary body of the former UN Commission

---

130 The First Geneva Convention for the Amelioration of the Condition of the Wounded and Sick in Armed Forces in the Field of August 12, 1949, 75 UNTS 31 ("the First Geneva Convention"); the Second Geneva Convention for the Amelioration of the Condition of Wounded, Sick and Shipwrecked Members of Armed Forces at Sea of August 12, 1949, 75 UNTS 85 ("the Second Geneva Convention"); the Third Geneva Convention relative to the Treatment of Prisoners of War of August 12, 1949, 75 UNTS 135 ("the Third Geneva Convention"); the Fourth Geneva Convention relative to the Protection of Civilian Persons in Time of War of August 12, 1949, 75 UNTS 287 ("the Fourth Geneva Convention").

131 David Turns, "The Law of Armed Conflict (International Humanitarian Law)", in Malcolm D. Evans (Ed.), (2010) *International Law*, Third Edition, Oxford: Oxford University Press, p. 816.

132 Vincent Chetail, (2003) "The Contribution of the International Court of Justice to International Humanitarian Law", *International Review of the Red Cross*, Volume 85, Issue 850, p. 238.

133 Leslie C. Green, (2008) *The Contemporary Law of Armed Conflict*, Melland Schill Studies in International Law, Third Edition, Manchester: Manchester University Press, p. 39; Françoise Bugnion, (2001) "Droit de Genève et Droit de la Haye", *International Review of the Red Cross*, Volume 83, Issue 844, pp. 901-922.

on Human Rights, and focus on various obligations of transnational corporations and other business entities stemming from international and domestic human rights law. The Norms seek to address the gaps existing in the international legal framework and aim at having a binding effect on corporations.[134] They do not constitute a legally binding document, however, as they have not been approved by the UN Commission on Human Rights and are essentially a failed attempt to lay down human rights standards for transnational corporations and other businesses. As a restatement of existing international obligations of business and States, the Norms could serve as a guide in identifying responsibilities of businesses and States. In addition to the secondary responsibility of these businesses in relation to human rights standards, Article 1 of the Norms recognizes the primary responsibility of States "to promote, secure the fulfillment of, respect, ensure respect of and protect human rights recognized in international as well as national law, including ensuring that transnational corporations and other business enterprises respect human rights". Thus, the Norms also emphasize the State duties to ensure that PMSCs respect human rights while conducting their operations in armed conflicts and peace operations. Given that they have not been adopted at international level, the Norms will not be observed for the purposes of the present study. Instead, the focus will be shifted to the following regime of significance for States and PMSCs.

This important initiative established at the level of the United Nations is called the Guiding Principles on Business and Human Rights: Implementing the United Nations "Protect, Respect and Remedy" Framework.[135] After the proposal of the Sub-Commission was not adopted, the Commission requested the UN Secretary-General to appoint a Special Representative with a mandate to look into the issue of business and human rights.[136] Eventually, the "Protect, Respect and Remedy" framework developed by the Special Representative of the UN Secretary-General on the issue of human rights and transnational corporations and other business enterprises, John Ruggie, introduced significant principles relating to the regulation of corporate behavior. Essentially, the Guiding Principles as endorsed by the Human Rights Council in its Resolution 17/4 on June 16, 2011 form an important conceptual and policy framework having some great value for the international community regarding their focus on preventing and addressing human rights violations associated with the

---

134 Pini P. Miretski and Sascha-Dominik Bachmann, (2012) "The UN 'Norms on the Responsibility of Transnational Corporations and Other Business Enterprises with Regard to Human Rights': A Requiem", *Deakin Law Review*, Volume 17, Issue 1, p. 14.

135 Report of the Special Representative of the Secretary-General on the Issue of Human Rights and Transnational Corporations and Other Business Enterprises, John Ruggie, *Guiding Principles on Business and Human Rights: Implementing the United Nations 'Protect, Respect and Remedy' Framework*, UN Human Rights Council, Seventeenth Session, March 21, 2011, UN Doc. A/HRC/17/31.

136 John G. Ruggie, (2007) "Business and Human Rights: The Evolving International Agenda", *The American Journal of International Law*, Volume 101, Issue 4, p. 821.

business operations of transnational and other corporations, such as PMSCs actively exploring new employment opportunities in different countries.[137]

The Guiding Principles can be divided into three main categories: the State duty to protect human rights, the corporate responsibility to respect human rights, and the access to remedy. The first section of the Guiding Principles is entitled "the State duty to protect human rights". One of the founding principles of this section forming the standard of conduct elaborates on the notion suggesting that States must protect individuals against human rights violations committed on their territory and in their jurisdiction by third parties, including business enterprises. The essence of the principle can be found in the requirement to take appropriate steps "to prevent, investigate, punish and redress" such human rights abuses by adopting effective policies, legislation, regulations, and adjudication. It is specifically stated that States can be found to be in breach of their human rights obligations when actions or omissions of private actors, for example, PMSCs, constitute human rights abuses and can be attributed to States in accordance with the rules and principles of general public international law. Significantly, the Guiding Principles also refer to the due diligence obligations of States by providing that the State responsibility may be incurred when they fail to take appropriate measures to prevent, investigate, punish, and redress human rights violations committed by private actors. The operational principles under the first section address general State regulatory and policy functions, the State-business nexus, business respect for human rights in conflict-affected areas of the world, and the process of ensuring policy coherence. In this regard, one of the main operational standards requires States to take additional steps in order to protect against human rights abuses. Such violations might be committed by business entities that are owned or controlled by those States or that receive substantial support and services from State agencies, where appropriate, by requiring human rights due diligence. States should also be able to establish effective oversight, monitoring, and accountability mechanisms for the protection of human rights when they contract with or legislate for business enterprises that carry out services that may potentially lead to human rights abuses.

While the above-mentioned provision is relevant to the State practice of using PMSCs engaging in a broad scope of activities, another operational standard might be considered of greater importance for private contractors performing military and security services. This is the duty of States to ensure that businesses respect human rights in conflict-affected areas.[138] It means that in addition to the obligation of respecting rules and principles of international humanitarian law and international

---

[137] OHCHR, "New Guiding Principles on Business and Human Rights Endorsed by the UN Human Rights Council", Press Release, June 16, 2011, <http://www.business-humanrights.org/media/documents/ruggie/ruggie-guiding-principles-endorsed-16-jun-2011.pdf>, accessed on May 25, 2014.

[138] Report of the Special Representative of the Secretary-General on the Issue of Human Rights and Transnational Corporations and Other Business Enterprises, John Ruggie, *Guiding Principles on Business*

criminal law, States should oversee the functioning of businesses in these contexts and ensure that they do not violate norms of international human rights law. Perhaps, the essential role in this respect must be played by the home States where PMSCs and other corporations are established, given the fact that host States lacking effective control over those companies are often unable to ensure respect for human rights in conflict zones.[139] The second founding principle of the first section of the Guiding Principles incorporates the idea that States should clearly establish the expectation that all business entities operating on their territory or within their jurisdiction respect human rights norms in the course of their activities. The second and third sections of the Guiding Principles address such major issues as the corporate responsibility of businesses to respect human rights and the access to remedy. While the former is mainly concerned with the provision of clear operational instructions to corporations on how to identify relevant human rights standards, to comply with them and to demonstrate their commitment to respect those norms, the latter aims at ensuring the access to remedy for those affected by human rights abuses through judicial and non-judicial means.

These Guiding Principles are applicable to all States and corporate entities irrespective of their size, location, structure, or other parameters but do not create new international legal obligations or undermine and limit State obligations under international human rights law.[140] They seek to enhance standards and practices with regard to businesses, such as PMSCs, and human rights and form an important source of principles and standards of good practice for both States and business entities. Concerning the State practice of employing PMSCs in areas affected by conflicts, State obligations to protect human rights and to provide access to effective remedies are of crucial importance: they indicate an array of measures that must be taken by States for avoiding and addressing human rights violations that might be committed by PMSCs in conflict zones. These duties will be used in the following Chapters to delineate the scope of State responsibility for PMSC misconduct. Additionally, States are encouraged to adopt national action plans to implement the Guiding Principles.[141] These plans prepared by some governments might be a useful instrument in identifying the position of States with regard to an array of their duties in relation to PMSCs. They can potentially provide some "traces of evidence" leading to conclusions regarding the State practice of compliance or non-compliance with international law in situations when PMSCs are deployed in conflict zones.

---

and Human Rights: Implementing the United Nations 'Protect, Respect and Remedy' Framework, UN Human Rights Council, Seventeenth Session, March 21, 2011, UN Doc. A/HRC/17/31, pp. 10-11.

139 Ibidem.

140 Ibidem, p. 6.

141 OHCHR, "State National Action Plans", <http://www.ohchr.org/EN/Issues/Business/Pages/NationalActionPlans.aspx>, accessed on May 30, 2014.

3. Domestic Legal Order

Internal law of States is irrelevant when characterizing certain conduct of a State as an internationally wrongful act and cannot be invoked as a justification for non-compliance with State obligations under international law.[142] Domestic law of States involved in the use of PMSCs in conflicts forms, however, a source of regulations and policies that needs to be evaluated. These States are not only hiring and home States, but also host States. Their national laws can regulate the activities of PMSCs to a certain extent, establish particular licensing and certification regimes that might require companies to comply with the rules and principles of international law, address possible contractual and other relationships between States and PMSCs, and contain references to certain concepts that need to be assessed when examining the question of attribution and State responsibility. Such concepts that are found in the national laws and regulations include the notion of governmental authority, the content of which needs to be determined for the purpose of application of certain provisions of the Draft Articles.

Generally speaking, national legislation of States focusing on PMSCs has faced some criticism due to its deficiencies and gaps.[143] National laws and regulations are first of all criticized for being often outdated, failing to reflect the realities of the PMSC industry, such as the use of a variety of sub-contracts by the companies, and unable to provide operational standards for it.[144] They do not effectively regulate "off-shore activities" of private contractors and seem to serve interests of home States rather than host States and their local communities.[145] Regulatory instruments of host States cannot be enforced against foreign PMSCs that can appear to operate outside their reach.[146] Domestic regulation established by States is not sufficiently supported by administrative frameworks and bureaucratic incentives and can be said to fail to effectively address serious human rights violations and breaches of the law of armed conflict due to the lack of appropriate enforcement mechanisms.[147] Domestic criminal laws fail to prevent human rights violations, breaches of international humanitarian law, and other standards.[148] They cannot address systematic labor rights violations

---

142 Articles 3 and 32 DARS; see also Article 27 of the 1969 Vienna Convention on the Law of Treaties, UN Doc. A/CONF.39/27.

143 James Cockayne and Others, (2009) *Beyond Market Forces: Regulating the Global Security Industry*, New York: International Peace Institute, pp. 18-19; Kathleen M. Jennings, (2006) *Armed Services: Regulating the Private Military Industry*, New Security Programme, Fafo-report 532, p. 28, <http://www.fafo.no/pub/rapp/532/532.pdf>, accessed on May 8, 2011.

144 James Cockayne and Others, (2009) *Beyond Market Forces: Regulating the Global Security Industry*, New York: International Peace Institute, pp. 18-19.

145 Ibidem.

146 Ibidem, p. 19.

147 Ibidem, p. 18.

148 See, for instance, Cedric M.J. Ryngaert, (2008) "Litigating Abuses Committed by Private Military Companies", *The European Journal of International Law*, Volume 19, Issue 5, p. 1037.

committed by PMSCs and the role of some of these non-State actors in organized crime and trafficking.[149] Taking into consideration the Iraq War following the invasion by the US-led coalition in 2003, one might observe that national law of involved States appeared to be largely ineffective in ensuring that PMSCs acted in accordance with international humanitarian and human rights law.[150]

The activities of PMSCs and their personnel are regulated to a different extent in various countries depending on the extensiveness of the use of PMSCs, their perception in given societies, and constitutional particularities of States in question.[151] These companies are generally not prohibited by national laws and regulations.[152] Most European States have some sort of regulation for the operation of PMSCs on their territory.[153] It is frequently stated in the literature, however, that the only two home States that so far have extensively regulated the export of private military and security services are the United States and South Africa.

The relationship of the United States with PMSCs has been described by International Alert as "perhaps the most mature" one.[154] In one of the main hiring and home States of PMSCs, the US, at the federal level there is a lack of a clear-cut legal regime regulating PMSCs and the federal states have locally enacted legislative acts aimed at the provision of security services.[155] There is a rather complex regulatory framework consisting of a variety of statutes and regulations focusing on this specific industry. In the first place, there are certain possibilities to hold PMSC personnel criminally accountable for their misconduct. They can be subjected to court-martial on the basis of the Uniform Code of Military Justice that is also applicable to persons who serve

149 James Cockayne and Others, (2009) *Beyond Market Forces: Regulating the Global Security Industry*, New York: International Peace Institute, p. 18.

150 Robert McCorquodale and Penelope Simons, (2007) "Responsibility Beyond Borders: State Responsibility for Extraterritorial Violations by Corporations of International Human Rights Law", *Modern Law Review*, Volume 70, Issue 4, p. 610.

151 Ottavio Quirico, "A Comparative Overview of European and Extra-European National Regulation of Private Military and Security Services", in Christine Bakker and Mirko Sossai (Eds.), (2012) *Multilevel Regulation of Military and Security Contractors: The Interplay Between International, European and Domestic Norms*, Oxford and Portland, Oregon: Hart Publishing, p. 114.

152 Antenor Hallo de Wolf, (2012) *Reconciling Privatization with Human Rights*, School of Human Rights Research Series, Volume 49, Cambridge, Antwerp and Portland: Intersentia, p. 483.

153 Michael Cottier, (2006) "Elements for Contracting and Regulating Private Security and Military Companies", *International Review of the Red Cross*, Volume 88, Issue 863, p. 646.

154 Chaloka Beyani and Damian Lilly, "Regulating Private Military Companies: Options for the UK Government", International Alert, August 2001, p. 33, <http://psm.du.edu/media/documents/reports_and_stats/ngo_reports/intlalert_beyani_regulating_pmcs_2001.pdf>, accessed on October 19, 2013.

155 Ottavio Quirico, "A Comparative Overview of European and Extra-European National Regulation of Private Military and Security Services", in Christine Bakker and Mirko Sossai (Eds.), (2012) *Multilevel Regulation of Military and Security Contractors: The Interplay Between International, European and Domestic Norms*, Oxford and Portland, Oregon: Hart Publishing, p. 107.

with or accompany the armed forces.[156] Additionally, the 2000 US Military Extraterritorial Jurisdiction Act (hereinafter: MEJA) established jurisdiction of the Federal Court for prosecuting members of the armed forces and persons employed by or accompanying them abroad.[157] The criminal offenses committed by these individuals must take place outside the territory of the US and must be punishable by imprisonment for more than one year if committed in the US. The 2004 amendment of the Act extended its reach to include contractors employed by agencies other than the Department of Defense. The regime for issuing licenses in the US is based on two particular instruments: the International Traffic in Arms Regulation (hereinafter: ITAR) introduced in 1998 and the Arms Export Control Act of 1968 (hereinafter: AECA).[158] Under the ITAR, companies need to apply for a license before concluding contracts with foreign clients and every application undergoes an elaborate process of internal assessment by the US government.

In reality, a large number of other States also have regulatory frameworks applicable to PMSCs and this number is growing. In the UK, the Private Security Industry Act 2001 regulates PMSC activities at the national level.[159] The UK government decided not to establish further extensive laws on PMSCs and their activities but to opt for self-regulation as a way of addressing the issues of this particular sector and to adhere to international regulatory norms regulating the industry.[160] The UK

---

[156] Ottavio Quirico, "The Criminal Responsibility of Private Military and Security Company Personnel under International Humanitarian Law", in Francesco Francioni and Natalino Ronzitti (Eds.), (2011) *War by Contract: Human Rights, Humanitarian Law, and Private Contractors*, Oxford: Oxford University Press, p. 441; Micaela Frulli, "Immunity for Private Military Contractors: Legal Hurdles or Political Snags?", in Francesco Francioni and Natalino Ronzitti (Eds.), (2011) *War by Contract: Human Rights, Humanitarian Law, and Private Contractors*, Oxford: Oxford University Press, p. 458.

[157] Sarah Percy, (2006) *Regulating the Private Security Industry*, Adelphi Series, Volume 46, Issue 384, Oxon: Routledge, p. 28; Kristine Huskey and Scott Sullivan, "United States: Law and Policy Governing Private Military Contractors After 9/11", in Christine Bakker and Mirko Sossai (Eds.), (2012) *Multilevel Regulation of Military and Security Contractors: The Interplay Between International, European and Domestic Norms*, Oxford and Portland, Oregon: Hart Publishing, pp. 355-356.

[158] Kristine Huskey and Scott Sullivan, "United States: Law and Policy Governing Private Military Contractors After 9/11", in Christine Bakker and Mirko Sossai (Eds.), (2012) *Multilevel Regulation of Military and Security Contractors: The Interplay Between International, European and Domestic Norms*, Oxford and Portland, Oregon: Hart Publishing, p. 347; Sarah Percy, (2006) *Regulating the Private Security Industry*, Adelphi Series, Volume 46, Issue 384, Oxon: Routledge, p. 26.

[159] Antenor Hallo de Wolf, (2012) *Reconciling Privatization with Human Rights*, School of Human Rights Research Series, Volume 49, Cambridge, Antwerp and Portland: Intersentia, p. 483; Alexandra Bohm, Kerry Senior, and Adam White, "The United Kingdom", in Christine Bakker and Mirko Sossai (Eds.), (2012) *Multilevel Regulation of Military and Security Contractors: The Interplay Between International, European and Domestic Norms*, Oxford and Portland, Oregon: Hart Publishing, p. 314.

[160] Alexandra Bohm, Kerry Senior, and Adam White, "The United Kingdom", in Christine Bakker and Mirko Sossai (Eds.), (2012) *Multilevel Regulation of Military and Security Contractors: The Interplay Between International, European and Domestic Norms*, Oxford and Portland, Oregon: Hart Publishing, p. 316. See the Green Paper: UK Foreign and Commonwealth Office, *Private Military Companies: Options for Regulation 2001-02*, February 2002, HCP 577, London: The Stationary Office, para. 76,

government clearly understood that attempting to tightly regulate the activities of PMSCs comprising a strongly growing market or to prohibit them on the territory of the State itself would simply drive them to third countries.[161] And this is exactly what happened with some PMSCs after South Africa introduced strict regulatory measures. The South African Foreign and Military Assistance Act was introduced in 1998 and, in fact, banned all mercenary activity by creating a distinction between mercenary activity and foreign military assistance.[162] The National Conventional Arms Control Committee is a body that issues licenses to PMSCs, which can also be revoked. What various companies, such as Executive Outcomes, did was to relocate to other States or to even become part of the illegal market for private military and security services.[163] Executive Outcomes simply ceased its business activities and from it other companies emerged that continued with the provision of military and security services with the same personnel.[164]

4. SELF-REGULATION AND TRANSNATIONAL PRIVATE REGULATORY REGIMES

Besides the branches of international law and domestic legislation regulating the conduct of PMSCs described above, transnational private regulatory frameworks are applicable to these firms. In some of these self-regulatory initiatives, not only companies, but also States and other stakeholders take an active stance and develop several rules aimed at the vibrant PMSC industry. As indicated by some experts, these regimes might emanate from the strong desire of companies to improve their image and to safeguard contracts with potential clients and remain voluntary guidelines that are not suitable for implementation and enforcement.[165] Indeed, they cannot be perceived as the sole and final solution to various problems inherently connected to the

---

<http://www.official-documents.gov.uk/document/hc0102/hc05/0577/0577.pdf>, accessed on February 10, 2014.

161 See UK Parliament, Select Committee on Foreign Affairs, Ninth Report, Session 2001-2002, para. 102, <http://www.parliament.the-stationery-office.co.uk/pa/cm200102/cmselect/cmfaff/922/92209.htm>, accessed on October 19, 2013; Sarah Percy, (2006) *Regulating the Private Security Industry*, Adelphi Series, Volume 46, Issue 384, Oxon: Routledge, pp. 25, 38.

162 Deborah D. Avant, (2005) *The Market for Force: The Consequences of Privatizing Security*, Cambridge: Cambridge University Press, p. 161; Eugenio Cusumano, "Policy Prospects for Regulating Private Military and Security Companies", in Francesco Francioni and Natalino Ronzitti (Eds.), (2011) *War by Contract: Human Rights, Humanitarian Law, and Private Contractors*, Oxford: Oxford University Press, p. 28.

163 Eugenio Cusumano, "Policy Prospects for Regulating Private Military and Security Companies", in Francesco Francioni and Natalino Ronzitti (Eds.), (2011) *War by Contract: Human Rights, Humanitarian Law, and Private Contractors*, Oxford: Oxford University Press, p. 28.

164 Deborah D. Avant, (2005) *The Market for Force: The Consequences of Privatizing Security*, Cambridge: Cambridge University Press, p. 163.

165 Peter W. Singer, (2008) *Corporate Warriors: The Rise of the Privatized Military Industry*, Ithaca and London: Cornell University Press, p. 239.

PMSC sector. In general, transnational private regulation in the field of PMSCs has significant limitations and has attracted substantial criticism.[166] This type of regulation, especially trade associations, serves to a large extent the interests of the PMSC industry rather than the interests of third parties and does not represent an appropriate forum for addressing serious breaches of international humanitarian and human rights law. It is not widely supported by States, fails to safeguard the elements of transparency and accountability, and does not have credible monitoring, oversight, and enforcement instruments.[167]

Anton Katz, an independent expert and Chair-Rapporteur of the UN Working Group on the Use of Mercenaries, correctly points out that "these initiatives are not legally binding and cannot be considered as complete solutions for the problems concerning PMSCs".[168] Some of these regimes, however, have a certain promise of effectiveness and capacity for influencing the actual behavior of PMSCs given that they seek to establish not only coherent oversight, monitoring, and enforcement mechanisms, but also to improve legitimacy of the schemes and quality of the required standards. Seen in the broader context of law existing on different levels, these voluntary initiatives might not only serve as a reflection of State practice and views adopted by the multitude of actors on the international plane, but also indicate a certain direction in which rules concerning international responsibility and other related matters will develop in the years to come. Their enforcement tools – if operating transparently and providing information to the general public – that are employed against companies violating codes of conduct can be used to shed light on the actual unlawful conduct of PMSC personnel, their relationship with States, and the involvement of the latter in the commission of human rights violations. In addition, voluntary codes of conduct generally lacking strong enforcement structures can still have a certain legal effect

---

[166] James Cockayne and Others, (2009) *Beyond Market Forces: Regulating the Global Security Industry*, New York: International Peace Institute, p. 19.

[167] Nicola Jägers, Evgeni Moyakine, and Willem van Genugten, "Transnational Private Regulation, Private Security Companies and Human Rights: Promise of Effectiveness?", Case Study of the Center for Transboundary Legal Development (CTLD), Tilburg University, for the Hague Institute for the Internationalization of Law (HiiL), September 2012 (on file with the author). Details on the case study and broader research project "Transnational Private Regulation: Constitutional Foundations and Governance Design" are available at www.privateregulation.eu; see also Nicola Jägers, (2012) "Regulating the Private Security Industry: Connecting the Public and the Private Through Transnational Private Regulation", *Human Rights and International Legal Discourse*, Volume 6, Issue 1 and Evgeni Moyakine, (2015) "From National and International Frustrations to Transnational Triumph? Hybrid Transnational Private Regulatory Regimes in the Private Military and Security Companies' Industry and Their Effectiveness in Ensuring Compliance with Human Rights", *Pacific McGeorge Global Business & Development Law Journal*, Volume 28, Issue 1 (*forthcoming*).

[168] OHCHR, Working Group on the Use of Mercenaries as a Means of Violating Human Rights and Impeding the Exercise of the Right of Peoples to Self-Determination, "UN Expert Group Calls for Robust International Regulation of Private Military and Security Companies", November 4, 2013, <http://www.ohchr.org/EN/NewsEvents/Pages/DisplayNews.aspx?NewsID=13940&LangID=E>, accessed on November 5, 2013.

when the so-called process of "redeployment" takes place.[169] This means that essentially voluntary rules acquire a degree of being legally binding over time when their provisions are included in the contracts between States and PMSCs and are observed by various "gatekeepers", such as trade associations, investors, and States. In this regard, norms and standards forming the core of codes of conduct become requirements that not only PMSCs, but also States must comply with.

There are hybrid regimes bringing together a variety of stakeholders other than PMSCs. They constitute a useful platform for a constructive dialogue on the regulation of PMSC industry and compliance of its representatives with international law. In the first place, among the major multi-stakeholder initiatives is the International Code of Conduct for Private Security Service Providers (hereinafter: the ICoC), now regarded as the most comprehensive initiative setting standards for the industry.[170] It is a transnational private regulatory regime engaging three categories of stakeholders: governments, civil society organizations, and PMSCs. It was established as the second part of the Swiss Initiative, the first part being the Montreux Document. While the latter can only be signed by States, the ICoC is signed by corporations from the PMSC industry. Other participants may issue a political statement of support towards the initiative. Instead of mapping and clarifying State obligations with regard to PMSCs, the Code seeks to address duties of PMSCs to respect international humanitarian and human rights law and to act according to these rules and principles. As of November 2013, the ICoC has been signed by 708 companies, including Academi, G4S, and Triple Canopy. The signatory corporations commit to endorse the Montreux Document and the UN "Protect, Respect and Remedy" framework applicable to them.[171] They are expected to provide their security services responsibly by supporting the rule of law, respecting human rights, and protecting the clients' interests.[172] In this regard, PMSCs are also required to fulfill their humanitarian

169 Nicola Jägers, (2012) "Regulating the Private Security Industry: Connecting the Public and the Private Through Transnational Private Regulation", *Human Rights and International Legal Discourse*, Volume 6, Issue 1, pp. 82-83; see also Nicola Jägers, Evgeni Moyakine, and Willem van Genugten, "Transnational Private Regulation, Private Security Companies and Human Rights: Promise of Effectiveness?", Case Study of the Center for Transboundary Legal Development (CTLD), Tilburg University, for the Hague Institute for the Internationalization of Law (HiiL), September 2012 (on file with the author) and Evgeni Moyakine, (2015) "From National and International Frustrations to Transnational Triumph? Hybrid Transnational Private Regulatory Regimes in the Private Military and Security Companies' Industry and Their Effectiveness in Ensuring Compliance with Human Rights", *Pacific McGeorge Global Business & Development Law Journal*, Volume 28, Issue 1 (*forthcoming*).

170 Rebecca DeWinter-Schmitt, "Human Rights and Self-Regulation in the Apparel Industry", in Simon Chesterman and Angelina Fisher (Eds.), (2009) *Private Security, Public Order: The Outsourcing of Public Services and Its Limits*, Oxford: Oxford University Press, p. 148.

171 International Code of Conduct for Private Security Service Providers (ICoC), Preamble, para. 3, <http://www.icoc-psp.org/uploads/INTERNATIONAL_CODE_OF_CONDUCT_Final_without_Company_Names.pdf>, accessed on June 24, 2014.

172 Ibidem.

responsibilities towards those who might be affected by the conducted business activities.[173] The provisions of the ICoC not only focus on some general commitments of private contractors, such as making compliance with the Code an integral part of their contracts with personnel and sub-contractors, but also establish a set of more specific principles. These specific provisions concern the conduct of PMSC employees, for instance, regarding their use of force, apprehension and detention of individuals, and various forms of prohibition – torture, slavery and forced labor, human trafficking, and others. Other specific commitments focus on management and governance and cover such aspects as the incorporation of the Code into company policies, selection and vetting of personnel, management of weapons, and incident reporting.

In general, the ICoC can be perceived as an instrument establishing a number of principles which should be followed by signatory corporations, creating a basis for transforming these principles into standards and governance and oversight mechanisms.[174] This first significant international mechanism was launched in Geneva on September 19-20, 2013 and is called the International Code of Conduct Association (hereinafter: the ICoCA). An agreement on the charter of the ICoCA, the so-called Articles of Association, was reached among participants on February 19-22, 2013. It specifies that the Association is a multi-stakeholder initiative in the form of a non-profit Association that operates as an independent legal entity with legal capacity according to Swiss law.[175] The Association aims to promote, to govern, and to oversee implementation of the ICoC. It also promotes "the responsible provision of security services and respect for human rights and national and international law in accordance with the Code".[176] This independent governance and oversight mechanism monitors, reports, and assesses the participants' performance[177] and maintains a third party complaints process.[178] There are three particular stakeholder pillars within the Association: private security companies and private security service providers; governments; and civil society organizations.[179] In June 2014, the Association had among its members 162 companies,[180] six States,[181] and 13 civil society organizations.[182]

---

[173] Ibidem, para. 4.
[174] Ibidem, para. 5.
[175] Articles of Association, Article 1.1.
[176] Ibidem, Article 2.2.
[177] Ibidem, Article 12.
[178] Ibidem, Article 13.
[179] Ibidem, Article 3.1.
[180] ICoCA, Members of the ICoC Association: Private Security Service Providers, <http://www.icoca.ch/assets/icoca-member-companies_june-2014.pdf>, accessed on June 24, 2014.
[181] ICoCA, Members of the ICoC Association: States, <http://www.icoca.ch/assets/icoca-member-states_january-20142.pdf>, accessed on June 24, 2014.
[182] ICoCA, Members of the ICoC Association: Civil Society Organizations, <http://www.icoca.ch/assets/icoca-member-csos_january-20142.pdf>, accessed on June 24, 2014.

In 2011, the US Department of Defense awarded a contract to ASIS International – a global organization of security practitioners[183] – to develop standards for improving performance and accountability of private security service providers in line with the ICoC.[184] ASIS is an accredited standards developer of the American National Standards Institute (hereinafter: ANSI). It developed four sets of standards applicable to the industry: the ANSI/ASIS PSC.1-2012 Management System for Quality of Private Security Company Operations – Requirements with Guidance; the ANSI/ASIS PSC.2-2012 Conformity Assessment and Auditing Management Systems for Quality of Private Security Operations; the ANSI/ASIS PSC.3-2013 Maturity Model for the Phased Implementation of a Quality Assurance Management System for Private Security Providers; and the ANSI/ASIS PSC.4-2013 Quality Assurance and Security Management for Private Security Companies Operating at Sea – Guidance.[185] Two countries have adopted the ANSI/ASIS PSC.1-2012 standard:[186] the United States and the United Kingdom.[187] The question remains, however, as to whether it will also be used by other countries in their contracts with private contractors.

The Voluntary Principles on Security and Human Rights (hereinafter: the VPs) constitute the first transnational private regulatory initiative seeking to address human rights issues associated with the field of security provision. It came into existence by a joint effort of the US and UK governments in 2000, given that both governments were concerned with human rights violations resulting from the operations of various representatives of the extractive industry. The VPs do not have a corporate form: instead, a non-profit association was set up in The Hague, The Netherlands, to manage financial and administrative aspects of the regime. The initiative engages three types of stakeholders from three different pillars: the government pillar,[188] the

---

183 ASIS International, "Who We Are", <https://www.asisonline.org/About-ASIS/Who-We-Are/Pages/default.aspx>, accessed on June 24, 2014.

184 ASIS International, "ASIS Awarded Contract with US Department of Defense to Develop Standard to Improve Performance and Accountability of Private Security Service Providers", March 16, 2011, <https://www.asisonline.org/News/Press-Room/Press-Releases/2011/Pages/PSPProviderStandard.aspx>, accessed on June 24, 2014.

185 ASIS International, "ANSI/ASIS PSC Series: Private Security Service Provider Standards", Executive Summary, <https://www.asisonline.org/Standards-Guidelines/Standards/Documents/Executive%20SummaryANSIASIS%20PSC%20series.pdf>, accessed on July 18, 2014.

186 ASIS International, *Management System for Quality of Private Security Company Operations – Requirements with Guidance: ANSI/ASIS PSC.1-2012*, <http://www.acq.osd.mil/log/PS/p_vault/Item_1997-PSC_1_STD.PDF>, accessed on July 18, 2014.

187 US Office of the Deputy Assistant Seretary of Defense, "Private Security Companies (PSCs)", <http://www.acq.osd.mil/log/PS/psc.html>, accessed on July 18, 2014; UK Foreign and Commonwealth Office, "Private Security Companies", Written Ministerial Statement, December 17, 2012, <http://www.publications.parliament.uk/pa/cm201213/cmhansrd/cm121217/wmstext/121217m0001.htm>, accessed on June 24, 2014.

188 Nine States: Australia, Canada, Colombia, Ghana, The Netherlands, Norway, Switzerland, United Kingdom, and United States of America; See VPs, Fact Sheet – General, <http://www.voluntaryprinciples.org/wp-content/uploads/2014/05/VPs_Fact_Sheet_-_May_2014.pdf>, accessed on June 24, 2014.

NGO pillar,[189] and the corporate pillar.[190] While the companies from the extractive sector form the primary relevant industry, PMSCs belong to the secondary relevant industry indirectly addressed by the regime.[191] The Voluntary Principles are, in fact, the first hybrid initiative involving different types of stakeholders that indirectly tackles provision of security by PMSCs and is often considered a blueprint for other regimes, such as the ICoC. They have been created to specifically guide oil, gas, and mining companies in their interactions with public and private security providers and to make risk assessments during their operations in such a manner as to ensure respect for human rights. In making risk assessments, companies are required to take a number of factors into account: identification of security risks and risks posed by the potential for violence, assessment of human rights records and reputation of security providers, rule of law in operational context, making of conflict analysis, and examination of the risks of equipment transfers.[192] Most importantly for the PMSC industry, the guiding voluntary principles concerning the interactions of companies with private security service providers require them to conduct operations in a specific manner. They should, for example, maintain high levels of professional and technical proficiency, especially in the local use of force and firearms, and act in a lawful manner complying with not only international guidelines, such as the UN Principles on the Use of Force and Firearms by Law Enforcement Officials and the UN Code of Conduct for Law Enforcement Officials, but also emerging best practices of companies, NGOs, and governments.

Besides the hybrid regimes discussed above, there are trade associations largely serving interests of their members and promoting general standards of their operation. The International Stability Operations Association is an American trade association based in Washington, D.C. Established in April 2001 as the International Peace Operations Association, it specifically focuses on American PMSCs and positions itself as "a global partnership of private sector and nongovernmental organizations

---

189 10 NGOs: The Fund for Peace, Global Rights, Human Rights Watch, International Alert, LITE-Africa, Pact, Partners for Democratic Change International, Partnership Africa Canada, Pax, and Search for Common Ground. See VPs, Fact Sheet – General, <http://www.voluntaryprinciples.org/wp-content/uploads/2014/05/VPs_Fact_Sheet_-_May_2014.pdf>, accessed on June 24, 2014.

190 25 companies: Anglo American, AngloGold Ashanti, Barrick Gold Corporation, BG Group, BHP Billiton, BP, Chevron, ConocoPhillips, ExxonMobil, Freeport-McMoRan Copper & Gold, Hess Corporation, IAMGold, Marathon Oil Company, Newmont Mining Corporation, Norsk Hydro, Occidental Petroleum Corporation, PanAust, Premier Oil, Repsol, Rio Tinto, Shell, Statoil, Talisman Energy, Total, and Tullow Oil. See VPs, Fact Sheet – General, <http://www.voluntaryprinciples.org/wp-content/uploads/2014/05/VPs_Fact_Sheet_-_May_2014.pdf>, accessed on June 24, 2014.

191 Voluntary Principles on Security and Human Rights, Multi-Stakeholder Initiative Comparison Chart, <http://www.voluntaryprinciples.org/wp-content/uploads/2013/02/VPs_ICoC_EITI_Comparison_Chart.pdf>, accessed on February 9, 2014.

192 See Voluntary Principles on Security and Human Rights, <http://www.voluntaryprinciples.org/wp-content/uploads/2013/03/voluntary_principles_english.pdf>, accessed on February 9, 2014.

providing critical services in fragile environments worldwide".[193] As of June 2014, 49 PMSCs, such as Academi and DynCorp International, are part of the regime.[194] ISOA is a useful platform for bringing together member corporations with various industry stakeholders, such as States, NGOs, and other businesses.[195] Companies becoming members are required to sign the Code of Conduct that lists the standards of their operation. It specifies that the activities of PMSCs belonging to the association are guided by not only these ethical standards laid down in the Code, but also rules of international humanitarian and human rights law to be found in, *inter alia*, the 1949 Geneva Conventions, the 1948 Universal Declaration of Human Rights (hereinafter: UDHR), and other legal instruments. The Code of Conduct provides standards with regard to respect of human rights by its signatory companies, transparency, accountability, the use of force, and other important matters. They should, for instance, respect the dignity of all human beings and comply with all applicable rules of the law of armed conflict and human rights law, and perform their activities "with integrity, honesty and fairness".[196]

In the UK, PMSCs functioning overseas and meeting the required standards of membership can become part of the British Association of Private Security Companies (hereinafter: BAPSC). This is a private security sector regime that was created in February 2006 in the UK by the representatives of the PMSC industry. Seeking to represent and to promote the interests of PMSCs based in the UK and providing services overseas, the initiative has as its goal raising the standards of operation of these companies and ensuring their compliance with the law of armed conflict and human rights.[197] Currently, the membership of the trade association comprises 18 corporations, such as Aegis Defense Services and Black Pearl Maritime Security Management.[198] Self-regulation and cooperation with the UK government and international organizations are considered to be important elements of tackling various issues associated with the industry. The Charter of BAPSC provides:

> "The Association determines that it can only achieve its objectives through effective self-regulation and transparent engagement with UK Government departments and relevant International Organisations. The Association believes that it is only through

[193] ISOA, "About Us", <http://www.stability-operations.org/?page=About>, accessed on June 24, 2014.
[194] See ISOA, "Our Members", <http://www.stability-operations.org/?page=Members>, accessed on June 24, 2014.
[195] ISOA, "Benefits of Membership", <http://www.stability-operations.org/?page=Benefits>, accessed on June 24, 2014.
[196] ISOA, Code of Conduct, Under 1.1 and 2.1, <http://www.stability-operations.org/?page=Code>, accessed on June 24, 2014.
[197] BAPSC, <http://www.bapsc.org.uk/>, accessed on February 9, 2014.
[198] See BAPSC, <http://www.bapsc.org.uk/?membership=list-of-members>, accessed on February 9, 2014.

effective self-regulation that the Members will enhance their position and be able to achieve differentiation from non Members in the same industry sector."[199]

Companies belonging to the regime commit to complying with a number of principles laid down in the Charter. As part of these commitments, they need to ensure that their appropriate personnel are trained in a way corresponding with each particular mission and in accordance with applicable legislation of the appropriate country. These PMSCs should also refuse to conclude contracts for the provision of security services if such agreements might lead to conflicts with human rights law.

In 2009, the UK government organized a public consultation on the possible ways to promote high standards of operation of representatives of the PMSC industry.[200] In this regard, the BAPSC was mentioned as a potential partner. Two years later, however, the government decided to instead team up with a new special interest group called Security in Complex Environments Group (hereinafter: SCEG) that was created within ADS, the trade organization for the UK's aerospace, defense, security, and space industries.[201] Many UK-based companies providing security belong to this interest group that seeks to engage with the UK government on the issues concerning the PMSC industry.[202]

In sum, transnational private regulatory schemes constitute a significant body of standards requiring PMSCs to comply with international humanitarian and human rights law. As has been shown, there are notable drawbacks of this type of regulation specifically focusing on the obligations of companies. It does not address the responsibilities of States, but can be legally binding to a certain degree when the process of redeployment takes place and codes of conduct attain a status of "hard law" requirements for the operation of both PMSCs and States.

5. Hiring contracts

For a couple of decades, PMSCs have been providing their numerous services to States and other actors on the international plane on the basis of a contract as a very significant tool capable not only of laying down the tasks and obligations of those firms, but also of regulating their actual performance on the ground by setting out

---

199 BAPSC, <http://www.bapsc.org.uk/?keydocuments=charter>, accessed on February 9, 2014.

200 Stephanie Brown, (2010) "Bottom-up Lawmaking and the Regulation of Private Military and Security Companies", *Cuadernos de Derecho Transnacional*, Volume 2, Issue 1, p. 71.

201 SCEG, <https://www.adsgroup.org.uk/pages/24204653.asp>, accessed on July 18, 2014; UK Foreign and Commonwealth Office, "Promoting High Standards in the Private Military and Security Company Industry", June 21, 2011, <https://www.gov.uk/government/news/promoting-high-standards-in-the-private-military-and-security-company-industry--2>, accessed on October 10, 2014.

202 There are currently 80 members belonging to SCEG, including companies and UK government departments. See SCEG, "SCEG Members", <https://www.adsgroup.org.uk/pages/93099176.asp>, accessed on July 18, 2014.

obligations and rights of the parties to it.[203] For that matter, PMSCs are required to be registered on the territory of a State of their establishment, such as the United States and the United Kingdom, and to have a valid license for every contract in order to be able to work for a legitimate government.[204]

Contracts as instruments of military and security privatization are concluded between PMSCs and States opting to outsource some of their core governmental functions or other tasks to private firms.[205] They form the legal basis for the operation of private contractors, providing for the terms of their employment and conduct and regulating other matters.[206] In this respect, hiring contracts are to be perceived as legal instruments regulating the provision of services by PMSCs and their relationships with clients benefiting from these activities. A contract as a legal tool is in principle a mechanism that can be used by States to regulate the activities of the PMSCs they employ.[207] By specifying obligations of companies and their personnel, it can contain various terms that might also include references to the voluntary codes of conduct and hard law rules and principles applicable to the activities of PMSCs.[208] The drafting of a contract can thus be useful in bringing norms and standards of not only a voluntary nature, but also of international law, such as the law of armed conflict and human rights, into the realm of PMSCs in their capacity of private business entities.[209] While drafting the contracts, States can and should also comply with their numerous obligations under international law in relation to PMSCs by, for instance, setting up oversight and monitoring mechanisms and introducing requirements concerning

[203] Lindsey Cameron and Vincent Chetail, (2013) *Privatizing War: Private Military and Security Companies under Public International Law*, Cambridge: Cambridge University Press, p. 264.
[204] James T. Gathii, (2009-2010) *Commercializing War: Private Military and Security Companies, Mercenaries and International Law*, Legal Studies Research Paper Series, No. 49, Albany Law School, p. 31.
[205] Laura A. Dickinson, "Contract as a Tool for Regulating Private Military Companies", in Simon Chesterman and Chia Lehnardt (Eds.), (2007) *From Mercenaries to Market: The Rise and Regulation of Private Military Companies*, Oxford: Oxford University Press, p. 218.
[206] Michael Cottier, (2006) "Elements for Contracting and Regulating Private Security and Military Companies", *International Review of the Red Cross*, Volume 88, Issue 863, p. 640.
[207] Laura A. Dickinson, "Contract as a Tool for Regulating Private Military Companies", in Simon Chesterman and Chia Lehnardt (Eds.), (2007) *From Mercenaries to Market: The Rise and Regulation of Private Military Companies*, Oxford: Oxford University Press; Michael Cottier, (2006) "Elements for Contracting and Regulating Private Security and Military Companies", *International Review of the Red Cross*, Volume 88, Issue 863, p. 640; Cedric M.J. Ryngaert, *Private Military Contractors: Regulatory Options*, Policy Brief No. 8, May 2008, pp. 3-4. <https://ghum.kuleuven.be/ggs/publications/policy_briefs/pb05.pdf>, accessed on May 10, 2014.
[208] Michael Cottier, (2006) "Elements for Contracting and Regulating Private Security and Military Companies", *International Review of the Red Cross*, Volume 88, Issue 863, pp. 642-643.
[209] Laura A. Dickinson, "Contract as a Tool for Regulating Private Military Companies", in Simon Chesterman and Chia Lehnardt (Eds.), (2007) *From Mercenaries to Market: The Rise and Regulation of Private Military Companies*, Oxford: Oxford University Press, p. 218; Michael Cottier, (2006) "Elements for Contracting and Regulating Private Security and Military Companies", *International Review of the Red Cross*, Volume 88, Issue 863, pp. 642-643.

training of PMSC employees.[210] In addition, a contract can potentially address such issues as the responsibility of companies and their employees for instances of possible misconduct of PMSCs. If a contract concluded between States and PMSCs contains references to the standards of international humanitarian and human rights law that must be complied with, it can indirectly be enforced in national courts on the basis of negligence claims.[211]

States deploying their military personnel on the territory of another State or requiring the provision of military and security services on their own territory make contact with PMSCs and conclude contracts with these companies that are capable of rapidly coming in action. It is also the case that States providing troops for a peacekeeping mission quite often turn their attention to the PMSCs that in the capacity of civilian contractors can efficiently and effectively carry out various tasks. This use of PMSCs by States for the purposes of conducting peace operations can go hand in hand with the practice of international and regional organizations authorizing the deployment of these peace operations to directly hire PMSCs.[212] Also, in other contexts beyond peacekeeping missions, these organizations need to rely on private contractors. In addition, other entities, such as non-governmental organizations and corporations, negotiate with PMSCs and enter into contractual relationships with them. In all these instances, contracts remain the central tool of service provision and the main instrument of establishing a framework of their operation addressing rights and duties of private contractors and their employees.

The United States, the United Kingdom, and Canada are among countries that are the main customers of these corporations and conclude most contracts with them. The United States Department of Defense is apparently the main customer of PMSCs in the world, investing major financial resources in the use of private contractors,[213] though other governmental agencies of the United States, such as the State Department and the Agency for International Development, also frequently contract PMSCs to operate in conflict situations. Contracts awarded to PMSCs range from several thousands to millions of US dollars. For instance, it is often stated that Aegis Defence Services Ltd., a British private security firm, was hired by the United States in 2004 to manage the Reconstruction Operations Centers in Iraq and to provide security to the members of the Project & Contracting Office and was awarded the biggest single

---

[210] Ibidem; ibidem, pp. 643-646.

[211] Cedric M.J. Ryngaert, (2008) "Litigating Abuses Committed by Private Military Companies", *The European Journal of International Law*, Volume 19, Issue 5, pp. 1039-1040.

[212] Åse Gilje Østensen, (2011) *UN Use of Private Military and Security Companies: Practices and Policies*, SSR Paper 3, DCAF, pp. 12-14, 15-17.

[213] Eugenio Cusumano, "Policy Prospects for Regulating Private Military and Security Companies", in Francesco Francioni and Natalino Ronzitti (Eds.), (2011) *War by Contract: Human Rights, Humanitarian Law, and Private Contractors*, Oxford: Oxford University Press, p. 14.

security services contract, worth $293,000,000.[214] To illustrate the scope of contracting further, it is worth mentioning that in the period from 2000 to 2007 PMSC Blackwater had contracts with the US government worth over $1,059 billion.[215]

While the contract as a tool for hiring PMSCs can be an efficient instrument enabling hiring States to raise the required personnel at short notice, there are many issues associated with the practice of using it. It has been illustrated by some scholars that there are difficulties with identifying PMSCs operating on the ground and working for a large number of entities, as they are rarely hired directly by those clients and rarely offer their expertise directly to them. Most PMSCs are not directly employed, but are sub-contracted by other PMSCs to provide military and security services to the prime contractor employees: this employment by means of sub-contracts blurs the whole picture in the chain of employers and employees. For example, from 60 PMSCs working in Iraq, according to the Iraqi Coalition Provisional Authority only eight had direct contracts with it.[216] The remaining companies were sub-contracted. It must also be added that besides most contracts of States and other actors with PMSCs remaining vague and unspecified, they are also not freely available to the general public due to privacy implications.[217] This makes it impossible to thoroughly examine the terms of the contracts in most cases. By the same token, PMSCs and their parent companies tend to use all means to avoid publishing any details about contracts, salaries, and numbers of employees.[218] This makes it practically impossible to analyze contractual provisions that could be of importance for the assessment of the legal status of PMSCs and their employees, the nature of PMSC tasks, the extent of accountability of companies and their employees, and perhaps the scope of responsibility of their clients for the compliance of PMSCs with international and national law. It is interesting to know that the United States Department of Defense, Department of State, or other agencies do not have complete data on the costs of employing PMSCs, while even at contract level only a small amount of information is available to the employers.[219] In addition, contracts are strict legal tools that cannot easily be

---

214 Jan Stöber, "Contracting in the Fog of War… Private Security Providers in Iraq: A Principal-Agent Analysis", in Thomas Jäger and Gerhard Kümmel (Eds.), (2007) *Private Military and Security Companies: Chances, Problems, Pitfalls and Prospects*, Wiesbaden: VS Verlag für Sozialwissenschaften, p. 130.

215 David Isenberg, (2009) *Shadow Force: Private Security Contractors in Iraq*, Westport and London: Praeger Security International, p. 21.

216 Ibidem, p. 26.

217 Laura A. Dickinson, "Contract as a Tool for Regulating Private Military Companies", in Simon Chesterman and Chia Lehnardt (Eds.), (2007) *From Mercenaries to Market: The Rise and Regulation of Private Military Companies*, Oxford: Oxford University Press, p. 220.

218 David Isenberg, (2009) *Shadow Force: Private Security Contractors in Iraq*, Westport and London: Praeger Security International, p. 57.

219 Jan Stöber, "Contracting in the Fog of War… Private Security Providers in Iraq: A Principal-Agent Analysis", in Thomas Jäger and Gerhard Kümmel (Eds.), (2007) *Private Military and Security Companies: Chances, Problems, Pitfalls and Prospects*, Wiesbaden: VS Verlag für Sozialwissenschaften, p. 127.

adapted to changing operational circumstances and needs, which are often required by the environments in which PMSCs function.[220]

It is doubtful whether the practice of relying on individuals hired by a private company and having no other obligations than those laid down in a contract leads to a better performance of their duties and compliance with rules and principles of both national and international law. Entrusted with significant tasks having a military and security character, PMSC personnel can in theory quit their assigned jobs and avoid being prosecuted for desertion.[221] It is worth mentioning that the incidents of PMSCs breaching their contracts and leaving conflict zones have occurred in the past: some companies performing logistical services had suspended their activities and left Iraq because the local environment in which they operated became too dangerous.[222] Refusing to act in the interests of their clients, PMSCs might only face legal consequences for the breaches of contract and, in fact, a risk of not being hired again.[223] Moreover, the possibility exists that PMSCs get involved in local politics and even cooperate with local criminals participating in criminal networks. As pointed out by the United States Senate Armed Services Committee, some ties have been found between PMSCs working on behalf of the United States in Afghanistan – more specifically, Afghan security guards – Afghan warlords, and even Taliban insurgents.[224] Finally, the capacity of many PMSCs to exercise force and thus to violate international law may require those actors operating within the military chain of command of a contracting State.[225] In reality, employees hired by PMSCs and working on the basis of a contract are rarely prosecuted for their illegal activities. In many scenarios, they are most probably not able to establish a relationship based on trust and understanding with the local population and will have to face many difficulties affecting the outcome of a mission. In conflict zones, where lives of many people are in danger, it may seem contradictory to hire PMSCs that often overcharge, do not perform all

---

[220] Eugenio Cusumano, "Policy Prospects for Regulating Private Military and Security Companies", in Francesco Francioni and Natalino Ronzitti (Eds.), (2011) *War by Contract: Human Rights, Humanitarian Law, and Private Contractors*, Oxford: Oxford University Press, p. 18.

[221] Peter W. Singer, (2003) "Peacekeepers, Inc.", *Policy Review*, Volume 119, p. 66; US Department of Defense Instruction 1100.22, Policy and Procedures for Determining Workforce Mix, April 12, 2010, p. 50, <http://www.dtic.mil/whs/directives/corres/pdf/110022p.pdf>, accessed on May 5, 2011.

[222] Eugenio Cusumano, "Policy Prospects for Regulating Private Military and Security Companies", in Francesco Francioni and Natalino Ronzitti (Eds.), (2011) *War by Contract: Human Rights, Humanitarian Law, and Private Contractors*, Oxford: Oxford University Press, p. 18.

[223] Peter W. Singer, (2003) "Peacekeepers, Inc.", *Policy Review*, Volume 119, p. 65.

[224] US Senate, Committee on Armed Services, *Report of the Inquiry into the Role and Oversight of Private Security Contractors in Afghanistan*, September 28, 2010, <http://www.wbir.com/pdf/SASC.PSCReport.100710.pdf>, accessed on May 1, 2011.

[225] Eugenio Cusumano, "Policy Prospects for Regulating Private Military and Security Companies", in Francesco Francioni and Natalino Ronzitti (Eds.), (2011) *War by Contract: Human Rights, Humanitarian Law, and Private Contractors*, Oxford: Oxford University Press, p. 18.

duties, hide possible failures, and, in general, form part of the industry lacking any culture of peacekeeping.[226]

6. STANDARD OPERATING PROCEDURES, RULES OF ENGAGEMENT, AND STATUS OF FORCES AGREEMENTS

In addition to contracts concluded between PMSCs and States and other entities defining their mutual relationship, there are also standard operating procedures and rules of engagement that can be issued to the staff of PMSCs by the contracting governments. It is established practice that rules of engagement are created and distributed to armed forces deployed abroad, while PMSC functioning is merely governed by the terms of a contract existing between it and a government.[227] Rules of engagement can be issued by the hiring States to PMSCs as part of the contract or separately. For example, the contract between the PMSC Sandline and the government of Papua New Guinea included reference to the adherence to "the laws and rules of engagement relating to armed conflict".[228] In addition, when the use of PMSCs in the current conflict in Iraq became a reality, the US military units usually had to establish their own rules of engagement and rules on the use of force. Members of the PMSC personnel cooperating with them were required to merely read those rules issued by the combatant commander and to sign them. Changes to the rules were briefed to PMSC employees only in a very cursory manner before the actual mission.[229] Nowadays, rules of engagement are more carefully drafted and are usually observed by the actors involved. In the view of some legal scholars, rules of engagement may be regarded as essential State instructions to the operational activities of PMSCs.[230] Those rules

226 Peter W. Singer, (2003) "Peacekeepers, Inc.", *Policy Review*, Volume 119, p. 66.

227 Rebecca Weiner, "Private Military Contractors Come With Strings Attached", *Belfer Center Newsletter*, Winter 2005-2006, Belfer Center for Science and International Affairs, Harvard Kennedy School, <http://belfercenter.ksg.harvard.edu/publication/19402/private_military_contractors_come_with_strings_attached.html>, accessed on May 7, 2011; Phillip Carter, (2004) "Hired Guns: What to Do About Military Contractors Run Amok", <http://www.slate.com/id/2098571/>, accessed on May 8, 2011; Kathleen M. Jennings, (2006) *Armed Services: Regulating the Private Military Industry*, New Security Programme, Fafo-report 532, p. 20, <http://www.fafo.no/pub/rapp/532/532.pdf>, accessed on May 8, 2011; US Department of Defense Instruction 3020.41, *Contractor Personnel Authorized to Accompany the U.S. Armed Forces*, October 3, 2005, Section 6.3.5.3.4, <http://fas.org/irp/doddir/dod/i3020_41.pdf>, accessed on July 2, 2014.

228 Patrick Cullen, (2000) "Keeping the New Dog of War on a Tight Leash: Assessing Means of Accountability for Private Military Companies", *Conflict Trends*, Issue 1, p. 38, <http://www.accord.org.za/publications/conflict-trends/downloads.html?start=40>, accessed on May 10, 2011.

229 Jeffrey S. Thurnher, (2008) "Drowning in Blackwater: How Weak Accountability over Private Security Contractors Significantly Undermines Counterinsurgency Efforts", *Army Lawyer*, Volume 64, p. 72.

230 University Centre for International Humanitarian Law, *Expert Meeting on Private Military Contractors: Status and State Responsibility for Their Actions*, Report, Geneva, August 29-30, 2005, pp. 19-20, <http://www.geneva-academy.ch/docs/expert-meetings/2005/2rapport_compagnies_privees.pdf>, accessed on October 10, 2014.

may contain norms governing the operation of private contractors on the territory of another State and contain references to basic obligations under international humanitarian law and national law of the host States.[231] They also usually contain rules on the use of force seeking to limit abuses and violations of international law.[232]

Quite importantly for the PMSC industry, there are also status of forces agreements (hereinafter: SOFA) that might be applicable to private contractors.[233] SOFAs are multilateral or bilateral agreements concluded on a temporary or a permanent basis in order to regulate the military and civilian presence of one State on the territory of another State.[234] These agreements essentially set up a framework of operation for foreign forces of a sending State, when these forces are deployed on the territory of another country, for instance, under the auspices of an international organization, such as the United Nations, NATO, the European Union, or the African Union.[235] The military or police forces of a sending State can employ PMSCs and other companies for the provision of various tasks.[236] It is traditionally the case that sending States lack legislation regulating the activities of private contractors abroad, while host or receiving States are also not capable of coordinating their activities and exercising jurisdiction.[237] Additionally, PMSCs are different from members of armed forces enjoying various immunities as State organs.[238] Therefore, SOFA agreements are important for

---

[231] Emanuela-Chiara Gillard, (2006) "Business Goes to War: Private Military / Security Companies and International Humanitarian Law", *International Review of the Red Cross*, Volume 88, Issue 863, p. 548.

[232] Surabhi Ranganathan, (2010) "Between Complicity and Irrelevance? Industry Associations and the Challenge of Regulating Private Security Contractors", *Georgetown Journal of International Law*, Volume 41, Issue 2, pp. 322, 338.

[233] Giuseppe Melis and Alessio Persiani, "The Regulation of Private Military and Security Companies: Tax Aspects", in Francesco Francioni and Natalino Ronzitti (Eds.), (2011) *War by Contract: Human Rights, Humanitarian Law, and Private Contractors*, Oxford: Oxford University Press, pp. 599-600.

[234] Dieter Fleck (Ed.), (2001) *The Handbook of the Law of Visiting Forces*, Oxford: Oxford University Press; Michael N. Schmitt, (2005) "Humanitarian Law and Direct Participation in Hostilities by Private Contractors or Civilian Employees", *Chicago Journal of International Law*, Volume 5, Issue 2, p. 516; Micaela Frulli, "Immunity for Private Military Contractors: Legal Hurdles or Political Snags?", in Francesco Francioni and Natalino Ronzitti (Eds.), (2011) *War by Contract: Human Rights, Humanitarian Law, and Private Contractors*, Oxford: Oxford University Press, p. 451; Giuseppe Melis and Alessio Persiani, "The Regulation of Private Military and Security Companies: Tax Aspects", in Francesco Francioni and Natalino Ronzitti (Eds.), (2011) *War by Contract: Human Rights, Humanitarian Law, and Private Contractors*, Oxford: Oxford University Press, p. 599.

[235] University Centre for International Humanitarian Law, *Expert Meeting on Private Military Contractors: Status and State Responsibility for Their Actions*, Report, Geneva, August 29-30, 2005, p. 30, <http://www.geneva-academy.ch/docs/expert-meetings/2005/2rapport_compagnies_privees.pdf>, accessed on October 10, 2014.

[236] Dieter Fleck, "Status of Forces in Peace Operations", in Terry D. Gill and Dieter Fleck (Eds.), (2010) *The Handbook of the International Law of Military Operations*, Oxford: Oxford University Press, p. 145.

[237] Terry D. Gill and Dieter Fleck, "Private Contractors and Security Companies", in Terry D. Gill and Dieter Fleck (Eds.), (2010) *The Handbook of the International Law of Military Operations*, Oxford: Oxford University Press, p. 491.

[238] Ibidem, p. 493.

addressing the operation of PMSCs, for these agreements are based on the idea that the military and civilian personnel are present on the territory of a host State in the interests of both a sending State and a host State that usually do not have legislation effectively regulating these companies.[239] In fact, they establish the legal framework for the operation of armed forces and members of civilian personnel functioning in a foreign country, dealing with a large spectrum of issues, such as the status, rights, and obligations of deployed individuals. Significantly for the practice of employing PMSCs, these SOFA agreements address the exercise of criminal jurisdiction over the members of military and civilian personnel of a sending State, making references to monitoring, supervision, and regulation aspects of their operation.[240] In most scenarios, SOFAs contain provisions stipulating that the host governments are allowed to exercise primary jurisdiction over the soldiers and employees working for PMSCs of a sending State, meaning that the authorities have a right to exercise jurisdiction when members of foreign personnel violate their domestic laws.[241] However, members of the PMSC personnel can be excluded from local prosecution[242] on the basis of two main exceptions with regard to criminal cases that are negotiated in the SOFA agreements: usually, sending States have primary jurisdiction when the criminal offense in question is committed by their nationals in the performance of the official duty and when it is committed by a national of a sending State against another national.[243] For example, the United States provided immunity from local jurisdiction

239 Dieter Fleck, "Status of Forces in Peace Operations", in Terry D. Gill and Dieter Fleck (Eds.), (2010) *The Handbook of the International Law of Military Operations*, Oxford: Oxford University Press, p. 145.

240 Human Rights First, *Private Security Contractors at War: Ending the Culture of Impunity*, Report, 2008, New York, p. 24; Montreux Document, Part Two "Good Practices Relating to Private Military and Security Companies", pp. 20, 24, paras. 22, 23, 51, 52; James Cockayne, (2008) "Regulating Private Military and Security Companies: The Content, Negotiation, Weaknesses and Promise of the Montreux Document", *Journal of Conflict & Security Law*, Volume 13, p. 417; Micaela Frulli, "Immunity for Private Military Contractors: Legal Hurdles or Political Snags?", in Francesco Francioni and Natalino Ronzitti (Eds.), (2011) *War by Contract: Human Rights, Humanitarian Law, and Private Contractors*, Oxford: Oxford University Press, p. 452; Dieter Fleck, "Status of Forces in Peace Operations", in Terry D. Gill and Dieter Fleck (Eds.), (2010) *The Handbook of the International Law of Military Operations*, Oxford: Oxford University Press, p. 492.

241 Micaela Frulli, "Immunity for Private Military Contractors: Legal Hurdles or Political Snags?", in Francesco Francioni and Natalino Ronzitti (Eds.), (2011) *War by Contract: Human Rights, Humanitarian Law, and Private Contractors*, Oxford: Oxford University Press, p. 452; Andrea Atteritano, "Liability in Tort of Private Military and Security Companies: Jurisdictional Issues and Applicable Law", in Francesco Francioni and Natalino Ronzitti (Eds.), (2011) *War by Contract: Human Rights, Humanitarian Law, and Private Contractors*, Oxford: Oxford University Press, p. 476.

242 Shantanu Chakrabarti, (2009) *Privatisation of Security in the Post-Cold War Period: An Overview of its Nature and Implications*, Institute for Defence Studies and Analyses Monograph Series, No. 2, New Delhi: Magnum Custom Publishing, p. 24; Sarah Percy, (2009) "Private Security Companies and Civil Wars", *Civil Wars*, Volume 11, Issue 1, p. 60.

243 Micaela Frulli, "Immunity for Private Military Contractors: Legal Hurdles or Political Snags?", in Francesco Francioni and Natalino Ronzitti (Eds.), (2011) *War by Contract: Human Rights, Humanitarian Law, and Private Contractors*, Oxford: Oxford University Press, p. 452; Ieva Miluna, "Jurisdictional

to the personnel of its PMSCs by Order 17 of the Coalition Provisional Authority in Iraq, which was later replaced by a SOFA agreement.[244] A provision of the Order 17 stipulated that:

> "Contractors shall be immune from Iraqi legal process with respect to acts performed by them pursuant to the terms and conditions of a Contract or any sub-contracts thereto."[245]

The recent US-Iraqi SOFA stipulates that Iraq as a host State must have the primary jurisdiction over the personnel of the US private contractors functioning on its territory.[246] Before this agreement was made between the two States, the personnel of the US PMSC Blackwater involved in the shooting incident in Baghdad on September 16, 2007 were regarded as being unaccountable, enjoying immunity from the local prosecution by the Iraqi authorities.[247] Even nowadays, the value of the new SOFA

Competence and Applicable Law with Regard to Private Military and Security Companies", in Christine Bakker and Mirko Sossai (Eds.), (2012) *Multilevel Regulation of Military and Security Contractors: The Interplay Between International, European and Domestic Norms*, Oxford and Portland, Oregon: Hart Publishing, p. 532.

244 Agreement Between the United States of America and the Republic of Iraq On the Withdrawal of the United States Forces from Iraq and the Organization of Their Activities during Their Temporary Presence in Iraq, Signed on November 17, 2008, <http://www.state.gov/documents/organization/122074.pdf>, accessed on June 10, 2014; Eugenio Cusumano, "Policy Prospects for Regulating Private Military and Security Companies", in Francesco Francioni and Natalino Ronzitti (Eds.), (2011) *War by Contract: Human Rights, Humanitarian Law, and Private Contractors*, Oxford: Oxford University Press, p. 27.

245 See Section 4 "Contractors", Paragraph 3 of the Coalition Provisional Authority Order Number 17, *Status of the Coalition Provisional Authority, MNF – Iraq, Certain Missions and Personnel in Iraq*, June 27, 2004, p. 5, <http://www.iraqcoalition.org/regulations/20040627_CPAORD_17_Status_of_Coalition__Rev__with_Annex_A.pdf>, accessed on May 5, 2014.

246 Article 12(2) of the Agreement Between the United States of America and the Republic of Iraq On the Withdrawal of the United States Forces from Iraq and the Organization of Their Activities during Their Temporary Presence in Iraq, Signed on November 17, 2008, <http://www.state.gov/documents/organization/122074.pdf>, accessed on June 10, 2014; M. Cherif Bassiouni, (2010) "Legal Status of US Forces in Iraq from 2003-2008", *Chicago Journal of International Law*, Volume 11, Issue 1; Terry D. Gill and Dieter Fleck, "Private Contractors and Security Companies", in Terry D. Gill and Dieter Fleck (Eds.), (2010) *The Handbook of the International Law of Military Operations*, Oxford: Oxford University Press, p. 493; James Cockayne, (2008) "Regulating Private Military and Security Companies: The Content, Negotiation, Weaknesses and Promise of the Montreux Document", *Journal of Conflict & Security Law*, Volume 13, p. 417; Jenny S. Lam, (2009) "Accountability for Private Military Contractors Under the Alien Tort Statute", *California Law Review*, Volume 97, Issue 5, p. 1462.

247 Terry D. Gill and Dieter Fleck, "Private Contractors and Security Companies", in Terry D. Gill and Dieter Fleck (Eds.), (2010) *The Handbook of the International Law of Military Operations*, Oxford: Oxford University Press, p. 493; Michael Scheimer, (2009) "Separating Private Military Companies from Illegal Mercenaries in International Law: Proposing an International Convention for Legitimate Military and Security Support that Reflects Customary International Law", *American University International Law Review*, Volume 24, Issue 3, p. 622.

agreement is arguable considering related questions of accountability of PMSC employees under the US domestic legislation for the crimes committed in Iraq.[248]

Thus, while standard operating procedures, rules of engagement, and status of forces agreements are an important source of norms governing the operation of PMSCs in zones affected by conflicts, some of these rules and procedures might exempt the personnel of PMSCs from local criminal proceedings in case of international law violations. Such arrangements might create a sense of impunity of civilian contractors, especially when the contracting States are not willing to prosecute them for the alleged misconduct.

7. CONCLUSION

Having distinguished different categories of regulatory instruments applicable to PMSCs and States associated with their use, it is clear that there are still gaps in the current regulatory landscape of PMSCs and their activities. Domestic legislation of all three types of States to a certain extent involved in the functioning of PMSCs in conflict areas is not sufficient to extensively regulate their conduct and to hold individuals, companies, and States responsible for possible breaches of international law. Transnational private regulatory initiatives and hiring contracts are a reasonably useful tool in regulating the functioning of PMSCs on the ground, but as such do not present a solution to the problem of State responsibility. Similarly, the international legal framework lacks sufficient capacity to address the issues of control and accountability in relation to PMSCs active on the battlefield.[249] Various legal instruments available at the international level that are applied individually simply cannot effectively tackle these problems. What is clearly needed, is a more integrated approach.

It is to be noted that an intelligent combination of legal measures can offer the possibility to hold States responsible for breaches of international law committed by private contractors and their employees. The rules and principles of different branches of international law are the main tool for determining whether a violation of international law indeed took place. In this regard, the law of armed conflict and international human rights law can serve as a tool for determining a potential breach

---

[248] Michael Scheimer, (2009) "Separating Private Military Companies from Illegal Mercenaries in International Law: Proposing an International Convention for Legitimate Military and Security Support that Reflects Customary International Law", *American University International Law Review*, Volume 24, Issue 3, p. 645; Jon Cadieux, (2008) "Regulating the United States Private Army: Militarizing Security Contractors", *California Western International Law Journal*, Volume 39, Issue 1, pp. 206-207; Don Mayer, (2010) "Peaceful Warriors: Private Military Security Companies and the Quest for Stable Societies", *Journal of Business Ethics*, Volume 89, p. 391.

[249] Corinna Seiberth, (2013) *Private Military and Security Companies in International Law – A Challenge for Non-binding Norms: the Montreux Document and the International Code of Conduct for Private Security Service Providers*, Cambridge, Antwerp, Portland: Intersentia, pp. 122, 271.

of international law. Customary international law on international responsibility reflected by the Draft Articles on State Responsibility is instructive in providing a general framework of the doctrine and establishing the possibility of attaching potential misconduct of private contractors and their employees to States. A range of legal instruments established on the international plane and international humanitarian and human rights law give some clarity regarding a spectrum of positive obligations States have when contracting with PMSCs or having them registered or operating on national territory. For the purposes of the current study, the Montreux Document as an international initiative specifically focusing on PMSCs and State obligations in relation to the operations of PMSCs presents a valuable instrument. It is not legally binding but can and will be used to identify and to interpret the content of State obligations with regard to private contractors and the scope of their application. In addition, domestic legislation, processes within transnational private regulatory regimes, contracts, standard operating procedures, rules of engagement, and SOFA agreements can be useful in creating a full picture of the particular misconduct under analysis and its perpetrators, clarifying certain concepts, and to a certain extent establishing often blurred lines of responsibility.

CHAPTER IV
LEGAL STATUS OF PMSC EMPLOYEES

It is sometimes stated that PMCs/PSCs have no status under international law... From an international humanitarian law point of view this assertion is misleading.

Emanuela-Chiara Gillard[1]

1. INTRODUCTION

In Chapter II of this study, PMSCs themselves as corporate entities have been scrutinized. It is, however, important to also examine members of their personnel, whose legal status in armed conflicts appears to be quite ambiguous and controversial from the perspective of the law of armed conflict. The assessment of the legal status from the point of view of international law[2] is a crucial step to be taken in the course of analyzing possible relationships between States and PMSCs, which may influence the determination of State responsibility and its scope for the unlawful actions of PMSCs, and examination of a set of positive obligations of the States arising out of international humanitarian and human rights law. This Chapter will include analysis of the legal status of employees working for PMSCs and carrying out their tasks in the situations of armed conflict, focusing on the legal consequences of that classification under international humanitarian law and its relevance for the application of the doctrine of State responsibility. In this regard, it will consider the participation of PMSCs in both international and non-international armed conflicts.

2. LEGAL STATUS OF PMSC PERSONNEL AND STATE RESPONSIBILITY

The factual legal status of PMSC personnel is perhaps the area where the greatest concern lies regarding their rights, duties, and immunities on a mission. A number of fundamental questions can be posed with regard to the position of the employees working

---

1 Emanuela-Chiara Gillard, (2006) "Business Goes to War: Private Military / Security Companies and International Humanitarian Law", *International Review of the Red Cross*, Volume 88, Issue 863, p. 530.

2 Rebecca R. Vernon, (2003-2004) "Battlefield Contractors: Facing the Tough Issues", *Public Contract Law Journal*, Volume 33, Issue 2, p. 400.

for PMSCs in conflict areas. What is the legal status of military and security experts as members of the PMSC personnel under international humanitarian law? How does the nature of an armed conflict – international or non-international – affect this legal qualification? Can those individuals be attacked during the hostilities and do adversary powers have to treat them as prisoners of war if captured? Do they enjoy privileges and protection as civilians on the battlefield? There is no straightforward answer to all those questions and it becomes apparent that circumstances under which PMSCs operate require a careful examination from the perspective of the law of armed conflict.

While the assessment made below appears to resemble an issue of an entirely different nature and scope than the main research question concerned with the application of the doctrine of State responsibility, it may be surprising to discover that the legal status of PMSC employees does actually affect the relationships between States' personnel engaged in hostilities and can in practice give rise to international responsibility. Currently, PMSCs are increasingly hired by States and when they are employed in situations of armed conflict, the legal status of employees of these corporations needs to be understood.

In the first instance, the process of determining the legal status of the members of PMSC personnel is relevant for assessing the question of State responsibility, as it may be useful in establishing links between PMSC employees and hiring States and indicates the degree to which the former are involved in armed conflicts, the extent of their integration in State armed forces, and the level of control exercised by States. In this regard, the legal status of PMSC employees can be examined in order to assess whether a certain State is to be held responsible for the acts of these individuals constituting breaches of international law.

Secondly, perhaps the most important provision of the Geneva Conventions forming part of international humanitarian law is Common Article 1 to all four 1949 Geneva Conventions. The types of positive obligations that can be derived from this provision will be examined in Chapter VI, but at this stage it suffices to say that it contains a requirement directed to the States Parties to the Conventions to "undertake to respect and to ensure respect" of international humanitarian law "in all circumstances". Article 1 common to Geneva Conventions, containing a rule of customary international law, makes clear that States cannot disregard their obligations under the law of armed conflict by employing PMSCs and giving them the authority to carry out the most essential State functions.[3] In international armed conflicts, States hire private entities whose personnel might be considered combatants or might not be allowed to take an active part in hostilities, because they are civilians and do not enjoy the right of being lawful combatants who have a privilege of not being prosecuted for their military actions and

---

[3] Louise Doswald-Beck, "Private Military Companies under International Humanitarian Law", in Simon Chesterman and Chia Lehnardt (Eds.), (2007) *From Mercenaries to Market: The Rise and Regulation of Private Military Companies*, Oxford: Oxford University Press, p. 132.

the entitlement to prisoner-of-war status. It is clear that if PMSC employees can qualify as persons with strong connections to specific States, for instance, falling within the scope of definition of lawful combatants functioning under a high degree of State control, the States in question may be considered as being under a negative duty not to encourage them to violate rules and principles of international humanitarian law. In addition, States may also have a positive duty to ensure that no violations of the law of armed conflict are committed by their armed forces or groups and persons controlled by those States. Some notion of this positive duty closely linked to the due diligence idea can be found in the ICRC Commentary to Article 91 of the First Additional Protocol.[4] Article 91 states that a party to an armed conflict acting in violation of the Geneva Conventions or the First Additional Protocol shall be liable to pay compensation. The Commentary further stipulates that State responsibility for such breaches exists not only when members of the State armed forces commit them, but also when persons who do not belong to the State military apparatus do so. A State will be responsible for the unlawful acts of a private individual if "it has not taken such preventive or repressive measures as could reasonably be expected to have been taken in the circumstances".[5] In other words, it will be the case when the involved State did not act with due diligence to prevent such illegal activities or to take care of their repression. In situations when States hire PMSCs participating in armed conflicts and having a legal status closely linked to those States as entities exercising some degree of control, and when they fail to take into account these two types of duties – negative and positive – resting upon them, the question of their responsibility will become very prominent. Such reasoning covers the obligation on States to ensure that PMSCs are properly trained and operate on basis of clear and precise rules of engagement.[6]

Thirdly and equally importantly, the issue of State responsibility may also arise if PMSCs hired by a State and operating under its control fail to respect the main principle of distinction between combatants and civilians, as incorporated into the body of international humanitarian law. Depending on the legal status of PMSC employees, some of them might have the right to qualify as lawful combatants and to enjoy the prisoner-of-war protection, while others must be seen as civilians who also enjoy a respective form of protection. It is not merely a theoretical matter to assume that members of those PMSCs may target experts working for the PMSCs hired by an adversary State, considering them to be lawful combatants, while those individuals

---

4 Protocol Additional to the Geneva Conventions of August 12, 1949, and relating to the Protection of Victims of International Armed Conflicts, June 8, 1977, 1125 UNTS 3 ("First Additional Protocol").

5 Claude Pilloud and Others, (1987) *Commentary on the Additional Protocols of 8 June 1977 to the Geneva Conventions of 12 August 1949*, International Committee of the Red Cross, Geneva: Martinus Nijhoff Publishers, pp. 1057-1058, para. 3660.

6 Louise Doswald-Beck, "Private Military Companies under International Humanitarian Law", in Simon Chesterman and Chia Lehnardt (Eds.), (2007) *From Mercenaries to Market: The Rise and Regulation of Private Military Companies*, Oxford: Oxford University Press, p. 132.

are to be regarded and protected as civilians during an armed conflict. Some confusion may also arise and disturb the lawful operations of PMSCs when they treat representatives of other corporations working for enemy States as unlawful combatants or civilians and do not grant to them the right to be prisoners of war. In order to avoid this type of problem during the functioning of the above-mentioned firms and to ensure that they act in accordance with international humanitarian law, it is of the utmost importance to provide more clarity regarding the status of the members of PMSC personnel in armed conflicts. For States may be held responsible for this misconduct on the part of the hired PMSCs in accordance with public international law.

Finally, if PMSC personnel meet all the requirements for qualification as mercenaries, States that became party to some international legal instruments addressing the issue of mercenaries and mercenarism can bear international responsibility. This might be the case if States do not comply with the provisions of these specific treaties by, for instance, using these individuals or not preventing their activities, and thus violate international law.

3. Definition of PMSC employees

It might appear obvious that the classification and definition of PMSCs is not essential for the analysis of the legal status of persons employed by those companies under international humanitarian law. The services and activities performed by those individuals remain, however, of crucial importance. Military and security experts working for PMSCs are, in fact, employees of those businesses who perform their duties on the basis of an employment contract they have with the company concerned. The Montreux Document identifies them as "persons employed by, through direct hire or under a contract with, a PMSC, including its employees and managers".[7] The notion clearly includes the factual relations between PMSCs and their personnel – direct contracts or sub-contracts – characterizing those individuals as employees of those firms.

The typical representatives of this group of people are as diverse and different as the companies themselves.[8] They come from all possible countries around the world: the United States, the United Kingdom, South Africa, Israel, Ukraine, and many others. These men and women are experts in the provision of a broad range of services, from logistics management and dog training to actual military combat. In many cases, these specialists are former personnel of national military personnel having various ranks and are usually hired by PMSCs. The practice of employing ex-military personnel and security experts means that companies can drastically reduce the

[7] Montreux Document, Preface, p. 10, para. 9(b).

[8] Peter W. Singer, (2008) *Corporate Warriors: The Rise of the Privatized Military Industry*, Ithaca and London: Cornell University Press, p. 76.

investment costs in skills of their employees, who have already been trained and evaluated by a State.[9] It is also much easier for PMSCs to perform screening of their personnel, taking into account screening requirements of the armed forces and units. For these reasons, it is understandable that some businesses, such as MPRI, are well known for their policies of employing only ex-members of national armed forces.

4. Legal Status under International Humanitarian Law

4.1 International armed conflicts

Mr. José Luis Gómez del Prado, Chairperson-Rapporteur of the UN Working Group on the Use of Mercenaries, pointed out that individuals employed by PMSCs "cannot usually be considered as mercenaries" and that a great variety of their activities are "not covered by the Geneva Conventions or the International Convention against the recruitment, use, financing and training of mercenaries".[10] The analysis in the foregoing Chapter indicated that the first part of this statement appears to be true. However, whether the second part of the statement made by the Chairperson-Rapporteur is correct, is arguable. Although it is true that the phenomenon of privatization of war and security associated with the employment of PMSCs by States and even non-State entities is a recent phenomenon, the participation in armed conflicts of persons that do not officially belong to the official State forces is not unusual.[11] To a certain extent, mercenary activities taking place in armed conflicts are regulated by a number of international conventions and rules of international humanitarian law, while no reference made to PMSCs can be found in the law of armed conflict or customary international law. Therefore, the position of PMSC employees on the battlefield and their treatment under international law remain largely uncertain.[12] In this regard, one may acknowledge that some legal rules and principles of international humanitarian law are nevertheless applicable to this new type of private entity providing military and security services for many clients and performing outsourced activities that were previously carried out solely by regular armed forces.[13]

---

9 Ibidem, p. 77.

10 OHCHR, "It's High Time to Close the Legal Gap for Private Military and Security Contractors", <http://www.ohchr.org/EN/NewsEvents/Pages/DisplayNews.aspx?NewsID=10000&LangID=E>, accessed on May 3, 2011.

11 Louise Doswald-Beck, "Private Military Companies under International Humanitarian Law", in Simon Chesterman and Chia Lehnardt (Eds.), (2007) *From Mercenaries to Market: The Rise and Regulation of Private Military Companies*, Oxford: Oxford University Press, p. 115.

12 Rebecca R. Vernon, (2003-2004) "Battlefield Contractors: Facing the Tough Issues", *Public Contract Law Journal*, Volume 33, Issue 2, p. 401.

13 Mirko Sossai, "Status of Private Military and Security Company Personnel in the Law of International Armed Conflict", in Francesco Francioni and Natalino Ronzitti (Eds.), (2011) *War by Contract: Human Rights, Humanitarian Law, and Private Contractors*, Oxford: Oxford University Press, p. 197;

The law of armed conflict applies to the situations of both armed conflicts and peace operations, in which PMSCs perform their numerous tasks. It is worth mentioning that those scenarios must reach the threshold of an "armed conflict", the definition of which is not provided in the Geneva Conventions and their Additional Protocols. The ICRC Commentary suggests that the term "armed conflict" is a much more general expression for the notion of "war".[14] It means "any difference arising between two States and leading to the intervention of armed forces... even if one of the Parties denies the existence of a state of war".[15] Applying this concept, such operations as peacekeeping missions, humanitarian relief, and counterinsurgency support may fall below the threshold of an armed conflict.[16] In those instances, protections safeguarded by the law of armed conflict will not be applicable to the persons or groups of persons employed in these contexts. But if States hire PMSCs to perform their duties in an armed conflict, international humanitarian law will apply. In this regard, the law of armed conflict makes a distinction between an international armed conflict and a non-international armed conflict that can seem difficult to grasp.[17] According to the ICRC, the former essentially takes place when "there is resort to armed force between two or more States",[18] and the latter as "protracted armed confrontations occurring between governmental armed forces and the forces of one or more armed groups, or between such groups arising on the territory of a State".[19] These confrontations of a non-international armed conflict must have a certain minimum level of intensity and there must be a minimum of organization of the parties involved in the conflict. In general, legal rules of the law of armed conflict do not touch upon the question of legitimacy or lawfulness of PMSCs participating in both international and non-international armed conflicts or the State practice of hiring them. It largely focuses on the behavior of the personnel of PMSCs operating in armed conflicts, regulating the manner in which hostilities are conducted.[20]

Emanuela-Chiara Gillard, (2006) "Business Goes to War: Private Military / Security Companies and International Humanitarian Law", *International Review of the Red Cross*, Volume 88, Issue 863, p. 530.

14 Jean S. Pictet, (1952) *Commentary, I Geneva Convention for the Amelioration of the Condition of the Wounded and Sick in Armed Forces in the Field*, Geneva: International Committee of the Red Cross, p. 32.

15 Ibidem.

16 Michael J. Davidson, (2000) "Ruck Up: An Introduction to the Legal Issues Associated with Civilian Contractors on the Battlefield", *Public Contract Law Journal*, Volume 29, Issue 2, p. 248.

17 James G. Stewart, (2003) "Towards a Single Definition of Armed Conflict in International Humanitarian Law: A Critique of Internationalized Armed Conflict", *International Review of the Red Cross*, Volume 85, Issue 850, p. 313.

18 ICRC, "How is the Term 'Armed Conflict' Defined in International Humanitarian Law?", Opinion Paper, March 2008, p. 5, <http://www.icrc.org/eng/assets/files/other/opinion-paper-armed-conflict.pdf>, accessed on February 10, 2014.

19 Ibidem.

20 Emanuela-Chiara Gillard, (2006) "Business Goes to War: Private Military / Security Companies and International Humanitarian Law", *International Review of the Red Cross*, Volume 88, Issue 863, pp. 528-529.

Specific circumstances of the employment of PMSCs and their actual functioning on the basis of a contract must be analyzed in order to identify how the body of international humanitarian law applies in relation to the persons working for PMSCs. The law of armed conflict remains unclear and unspecified with regard to the phenomenon of PMSCs, meaning that the future State practice of using companies may be able to resolve a spectrum of unsolved issues, especially those relating to the actual legal status of PMSCs in armed conflicts and peace operations. For now, "due to the absence of effective national or international laws, the legal status of private security providers and other military contractors... is ambiguous and highly political".[21]

4.1.1 Combatants with prisoner-of-war status

One of the most significant legal issues of the involvement of PMSCs in armed conflicts relates to their legal status on the ground as combatants or non-combatants in international armed conflicts. These two types of legal status represent two sides of the same coin and form an essential distinction made in international humanitarian law.[22] Parties to a conflict are under an obligation to distinguish between combatants and non-combatants, which is an essential norm of customary international law[23] and even "the greatest triumph of International Law".[24] Importantly, individuals cannot be both combatants and non-combatants.[25] Depending on the status of a party involved in an armed conflict as a lawful combatant or a civilian, an individual belonging to that party faces a number of important legal consequences during his or her participation in the conflict.[26] Usually, combatants include members of the belligerent's armed forces and others who directly participate in hostilities. The notion of direct participation in hostilities has been clarified by the ICRC in its interpretive guidance. It is

---

21 Jan Stöber, "Contracting in the Fog of War... Private Security Providers in Iraq: A Principal-Agent Analysis", in Thomas Jäger and Gerhard Kümmel (Eds.), (2007) *Private Military and Security Companies: Chances, Problems, Pitfalls and Prospects*, Wiesbaden: VS Verlag für Sozialwissenschaften, p. 127.

22 Lester Nurick, (1945) "The Distinction Between Combatant and Noncombatant in the Law of War", *The American Journal of International Law*, Volume 39, p. 680; Rebecca R. Vernon, (2003-2004) "Battlefield Contractors: Facing the Tough Issues", *Public Contract Law Journal*, Volume 33, Issue 2, p. 401; Michael N. Schmitt, (2005) "Humanitarian Law and Direct Participation in Hostilities by Private Contractors or Civilian Employees", *Chicago Journal of International Law*, Volume 5, Issue 2, p. 522; Won Kidane, (2010) "The Status of Private Military Contractors under International Humanitarian Law", *Denver Journal of International Law and Policy*, Volume 38, Issue 3, p. 377.

23 Jean-Marie Henckaerts and Louise Doswald-Beck, (2005) *Customary International Humanitarian Law, Volume I: Rules*, ICRC, Cambridge: Cambridge University Press, p. 3.

24 James M. Spaight, (1911) *War Rights on Land*, London: Macmillan and Co., p. 37.

25 Leslie C. Green, (2008) *The Contemporary Law of Armed Conflict*, Third Edition, Manchester: Manchester University Press, p. 129.

26 Antonio Cassese, (2005) *International Law*, Second Edition, Oxford: Oxford University Press, p. 408.

applicable to situations of both international and non-international armed conflicts,[27] and in order for a certain act to be perceived as direct participation in hostilities, three cumulative criteria must be met.[28] (1) The act must be likely to adversely affect the military operations or military capacity of a party to an armed conflict or, alternatively, to inflict death, injury, or distruction on persons or objects protected against direct attack (threshold of harm). (2) There must be a direct causal link between the act and the harm likely to result either from that act, or from a coordinated military operation of which that act constitutes an integral part (direct causation). (3) The act must be specifically designed to directly cause the required threshold of harm in support of a party to the conflict and to the detriment of another (belligerent nexus).

In armed conflicts, civilians are protected by the law of armed conflict and cannot be directly attacked, as confirmed by the ICJ in the *Legality of the Threat or Use of Nuclear Weapons* Advisory Opinion.[29] Article 51(3) of the First Additional Protocol to the 1949 Geneva Conventions, which constitutes a norm of a customary international law and to which no reservation has been made,[30] explicitly states:

> "Civilians shall enjoy the protection afforded by this section, unless and for such time as they take a direct part in hostilities."

Thus, civilians are not allowed to get directly involved in armed conflicts, otherwise they will be labeled as unlawful combatants or unprivileged belligerents and will lose their immunity from possible attack.[31] If PMSC employees were to qualify as civilians, their direct involvement in armed conflicts would make them lose the protection

---

[27] ICRC, *Interpretive Guidance on the Notion of Direct Participation in Hostilities under International Humanitarian Law*, Prepared by Nils Melzer, May 2009, p. 41, <https://www.icrc.org/eng/assets/files/other/icrc-002-0990.pdf>, accessed on May 20, 2014.

[28] Ibidem, p. 46.

[29] Leslie C. Green, (2008) *The Contemporary Law of Armed Conflict*, Third Edition, Manchester: Manchester University Press, p. 256; International Court of Justice, *Legality of the Threat or Use of Nuclear Weapons*, Advisory Opinion, July 8, 1996, ICJ Reports 1996, para. 78.

[30] Jean-Marie Henckaerts and Louise Doswald-Beck, (2005) *Customary International Humanitarian Law, Volume I: Rules*, ICRC, Cambridge: Cambridge University Press, pp. 19-20.

[31] Michael N. Schmitt, (2005) "Humanitarian Law and Direct Participation in Hostilities by Private Contractors or Civilian Employees", *Chicago Journal of International Law*, Volume 5, Issue 2, p. 522; Louise Doswald-Beck, "Private Military Companies under International Humanitarian Law", in Simon Chesterman and Chia Lehnardt (Eds.), (2007) *From Mercenaries to Market: The Rise and Regulation of Private Military Companies*, Oxford: Oxford University Press, pp. 127-128; Mirko Sossai, "Status of Private Military and Security Company Personnel in the Law of International Armed Conflict", in Francesco Francioni and Natalino Ronzitti (Eds.), (2011) *War by Contract: Human Rights, Humanitarian Law, and Private Contractors*, Oxford: Oxford University Press, p. 204; Vesna Crnić-Grotić, "Non-State Actors and International Humanitarian Law: The Status of Private Military Companies", in Budislav Vukas and Trpimir M. Šošić (Eds.), (2010) *International Law: New Actors, New Concepts – Continuing Dilemmas; Liber Amicorum Božidar Bakotić*, Leiden [etc.]: Brill, p. 460; Antonio Cassese, (2005) *International Law*, Second Edition, Oxford: Oxford University Press, p. 409.

granted to civilians and they would be directly targeted by parties to these conflicts as long as they take a direct part in the hostilities.[32] It is, however, possible that they could be seen as combatants. In this respect, it must be observed that only members of PMSC staff hired by States and not other possible entities can be considered combatants because it is essentially only members of State armed forces that can qualify as such.[33] Recent studies show that almost 80% of all PMSC contracts are concluded with other actors than States.[34] Only this fact precludes a large number of individuals working for PMSCs from being classified as combatants. The Dutch Advisory Council on International Affairs appears to acknowledge that in most scenarios PMSCs engaging in both offensive and defensive operations will most probably not qualify as combatants having the right to prisoner-of-war status.[35] However, it is often far from easy to say whether PMSC employees constitute combatants or civilians under international humanitarian law, as "they seem to fall in a murky area between the two".[36]

In the law of armed conflict, the so-called primary status of individuals as combatants participating in armed conflicts defines their secondary status.[37] For instance, lawful combatants are allowed to take a direct part in hostilities and to acquire the prisoner-of-war status, which is not applicable to civilians.[38] This concept of prisoner of war belongs to the body of international humanitarian law applicable only in international armed conflicts having an inter-State character, as non-international armed conflicts are limited to the conflicts between governments and rebels or hostilities between rebels.[39] In effect, this means that combatants having the status of prisoners

---

32 Vesna Crnić-Grotić, "Non-State Actors and International Humanitarian Law: The Status of Private Military Companies", in Budislav Vukas and Trpimir M. Šošić (Eds.), (2010) *International Law: New Actors, New Concepts – Continuing Dilemmas; Liber Amicorum Božidar Bakotić*, Leiden [etc.]: Brill, p. 461.

33 Emanuela-Chiara Gillard, (2006) "Business Goes to War: Private Military / Security Companies and International Humanitarian Law", *International Review of the Red Cross*, Volume 88, Issue 863, p. 532.

34 Ibidem; Olivier De Schutter, "The Responsibility of States", in Simon Chesterman and Angelina Fisher (Eds.), (2009) *Private Security, Public Order: The Outsourcing of Public Services and Its Limits*, Oxford: Oxford University Press, p. 27.

35 Dutch Advisory Council on International Affairs, *Employing Private Military Companies: A Question of Responsibility* (Dutch: *De Inhuur van Private Militaire Bedrijven: Een Kwestie van Verantwoordelijkheid*), No. 59, December 2007, The Hague, p. 16, <http://www.aiv-advies.nl/ContentSuite/upload/aiv/doc/webversie_AIV_59eng(1).pdf>, accessed on May 9, 2011.

36 David Isenberg, (2009) *Shadow Force: Private Security Contractors in Iraq*, Westport and London: Praeger Security International, p. 77; Shannon Bosch, (2007) "Private Security Contractors and International Humanitarian Law – A Skirmish for Recognition in International Armed Conflicts", *African Security Review*, Volume 16, Issue 4, p. 36.

37 Knut Ipsen, "Combatants and Non-Combatants", in Dieter Fleck (Ed.), (2008) *The Handbook of International Humanitarian Law*, Oxford: Oxford University Press; Won Kidane, (2010) "The Status of Private Military Contractors under International Humanitarian Law", *Denver Journal of International Law and Policy*, Volume 38, Issue 3, p. 377.

38 Christopher Greenwood, (2006) *Essays on War in International Law*, London: Cameron May, p. 99.

39 Malcolm N. Shaw, (2008) *International Law*, Sixth Edition, Cambridge: Cambridge University Press, pp. 1172-1176; Louise Doswald-Beck, "Private Military Companies under International Humanitarian

of war cannot be prosecuted by capturing States for their participation in hostilities and may not be tried before their domestic courts for performing specific military actions during armed conflicts, such as killing and carrying firearms.[40] This secondary status of combatants emanates from their primary status and has far-reaching consequences for their functioning in a conflict.

On the one hand, the fundamental distinction between combatants, for instance, the members of State armed forces, and civilians seeks to limit civilian casualties by granting to some persons the status of civilians and protecting them as far as possible.[41] On the other hand, it reflects one of the basic ideas stemming from the law of armed conflict, which provides that the members of State armies and thus organs of States must benefit from the status of combatants and have certain immunities associated with it, such as a privilege of not being subjected to prosecution by a capturing State and not being tried for their combat activities.[42]

There are a number of reasons to distinguish between lawful combatants and civilians in armed conflicts and to establish the legal status of the personnel employed by PMSCs. First of all, it is important to know whether they are entitled to actively take part in armed conflicts and, thus, to lawfully participate in hostilities. This is necessary in order to ensure that hostilities are waged between disciplined armed troops of States respecting the laws and customs of armed conflict.[43] Secondly, there is a necessity to examine whether employees of PMSCs can be prosecuted for their participation in armed conflicts and face trial on the territory of States for their illegal activities. Thirdly, knowledge of the actual status of the PMSC personnel on the battlefield is required because parties to armed conflicts must be aware whether those individuals can be targeted and attacked by them and whether they must be granted the rights and privileges of the prisoners of war.[44] Thus, it follows from the foregoing that States can become responsible for the actions of PMSCs, if these are hired to take part in an international armed conflict and deny prisoner-of-war status to captured

---

Law", in Simon Chesterman and Chia Lehnardt (Eds.), (2007) *From Mercenaries to Market: The Rise and Regulation of Private Military Companies*, Oxford: Oxford University Press, p. 116.

[40] Louise Doswald-Beck, "Private Military Companies under International Humanitarian Law", in Simon Chesterman and Chia Lehnardt (Eds.), (2007) *From Mercenaries to Market: The Rise and Regulation of Private Military Companies*, Oxford: Oxford University Press, p. 116.

[41] Christopher Greenwood, (2006) *Essays on War in International Law*, London: Cameron May, p. 99; Won Kidane, (2010) "The Status of Private Military Contractors under International Humanitarian Law", *Denver Journal of International Law and Policy*, Volume 38, Issue 3, p. 380.

[42] Louise Doswald-Beck, "Private Military Companies under International Humanitarian Law", in Simon Chesterman and Chia Lehnardt (Eds.), (2007) *From Mercenaries to Market: The Rise and Regulation of Private Military Companies*, Oxford: Oxford University Press, p. 116.

[43] Won Kidane, (2010) "The Status of Private Military Contractors under International Humanitarian Law", *Denver Journal of International Law and Policy*, Volume 38, Issue 3, p. 380.

[44] Lindsey Cameron, (2006) "Private Military Companies: Their Status under International Humanitarian Law and Its Impact on Their Regulation", *International Review of the Red Cross*, Volume 88, Issue 863, p. 582.

employees of other private contractors, when it must be conferred on them under given circumstances. According to Article 12 of the 1949 Third Geneva Convention relative to the Treatment of the Prisoners of War,[45] prisoners of war remain in the hands of an enemy State and the detaining power concerned is also responsible for the treatment they receive. If PMSCs form part of the official State forces, act under instructions, direction, or control of a State, or exercise elements of its governmental authority, and guard and interrogate detained PMSC employees with a prisoner-of-war status interfering with their rights and privileges as prisoners of war, the State in question will be regarded as breaching one of the most essential principles of international humanitarian law. In this situation, that State might be held responsible for such a violation of international law. Finally, considerations made regarding the legal status of PMSC employees are instrumental in establishing whether these individuals belong to a State organ and whether their conduct can lead to international responsibility of a State in question. If members of PMSC personnel are essentially members of State armed forces and lawful combatants, their acts and omissions violating international law will necessarily incur State responsibility.

Many governments hiring PMSCs often assert that these firms need to be perceived as "civilian contractors". This inevitably entails that the members of their personnel should not be considered as combatants.[46] In order to clarify this position, it is essential to take a closer look at this issue and to establish the status of PMSC employees. The factual legal status of the PMSC staff seems to be dependent upon the nature of the relationships between private contractors and States and the nature of the military and security activities performed. As a result, the legal status needs to be determined on a case-by-case basis.[47] Considering once again the massive growth of the PMSC industry and the uncertainty regarding the position of their employees in armed conflicts, three essential issues containing a number of fundamental questions are to be addressed. First of all, it must be established whether the personnel of PMSCs can qualify as combatants who can participate in hostilities and be targeted by the adversary. Secondly, more clarity needs to be provided regarding their potential entitlement to prisoner-of-war status. And finally, it must be determined whether

---

45 University Centre for International Humanitarian Law, *Expert Meeting on Private Military Contractors: Status and State Responsibility for Their Actions*, Report, Geneva, August 29-30, 2005, p. 9, <http://www.geneva-academy.ch/docs/expert-meetings/2005/2rapport_compagnies_privees.pdf>, accessed on October 10, 2014.

46 Lindsey Cameron, (2006) "Private Military Companies: Their Status under International Humanitarian Law and Its Impact on Their Regulation", *International Review of the Red Cross*, Volume 88, Issue 863, p. 575.

47 Emanuela-Chiara Gillard, "Private Military/Security Companies: The Status of Their Staff and Their Obligations under International Humanitarian Law and The Responsibilities of States in Relation to Their Operation", in Andrew Alexandra, Deane-Peter Baker, and Marina Caparini (Eds.), (2008) *Private Military and Security Companies: Ethics, Policies and Civil-Military Relations*, London and New York: Routledge, p. 160.

they fall under the category of civilians and can nevertheless be targeted by a party to an armed conflict.

In the first place, a great deal of attention is to be devoted to the classification of PMSCs as combatants. If the actual legal status of PMSCs is that of lawful combatants, the personnel of those firms are allowed to carry out military activities, may be targeted by their enemy unlike protected civilian groups, and enjoy immunities of the prisoners of war upon their capture.[48] In order for the individuals employed by private contractors to be regarded as combatants, they must meet criteria for this qualification laid down in the provisions of the First Additional Protocol to the 1949 Geneva Conventions and the 1949 Third Geneva Convention relative to the Treatment of the Prisoners of War.[49] The reason for examining both international legal documents is that the First Additional Protocol has not been ratified by 27 States so far,[50] even though legal experts seem to agree that its most essential rules reflect customary international law or will be absorbed into the body of customary international law quickly.[51]

The first step is to investigate whether PMSCs can fall under the category of the members of State armed forces and thus automatically acquire the status of combatants. This basic principle containing reference to the concept of armed forces and the main distinction between combatants and non-combatants in armed conflicts is to be found in Article 3 of the 1899 Convention with Respect to the Laws and Customs of War on Land, which states that "the armed forces of the belligerent parties may consist of combatants and non-combatants".[52] In addition, the Article provides that "in case of capture by the enemy both have a right to be treated as prisoners of war".[53] It is evident that both combatants and non-combatants can be members of the State armed forces and be entitled to the status of prisoner of war. Negotiations over the First Additional Protocol to the 1949 Geneva Conventions changed the situation and provided a strict definition laid down in Article 43 of the First Additional Protocol, which is often used to assess the status of certain individual as regular armed troops:

---

48 Emanuela-Chiara Gillard, (2006) "Business Goes to War: Private Military / Security Companies and International Humanitarian Law", *International Review of the Red Cross*, Volume 88, Issue 863, p. 531.

49 University Centre for International Humanitarian Law, *Expert Meeting on Private Military Contractors: Status and State Responsibility for Their Actions*, Report, Geneva, August 29-30, 2005, p. 9, <http://www.geneva-academy.ch/docs/expert-meetings/2005/2rapport_compagnies_privees.pdf>, accessed on October 10, 2014.

50 Ratifications of the First Additional Protocol, <http://www.icrc.org/ihl.nsf/WebSign?ReadForm&id=470&ps=P>, accessed on May 9, 2011.

51 Christopher Greenwood, (2006) *Essays on War in International Law*, London: Cameron May, pp. 192-193; Mirko Sossai, "Status of Private Military and Security Company Personnel in the Law of International Armed Conflict", in Francesco Francioni and Natalino Ronzitti (Eds.), (2011) *War by Contract: Human Rights, Humanitarian Law, and Private Contractors*, Oxford: Oxford University Press, p. 198.

52 Article 3 of the Convention with Respect to the Laws and Customs of War on Land, July 29, 1899.

53 Ibidem.

> "1. The armed forces of a Party to a conflict consist of all organized armed forces, groups and units which are under a command responsible to that Party for the conduct of its subordinates, even if that Party is represented by a government or an authority not recognized by an adverse Party. Such armed forces shall be subject to an internal disciplinary system which, inter alia, shall enforce compliance with the rules of international law applicable in armed conflict.
>
> 2. Members of the armed forces of a Party to a conflict (other than medical personnel and chaplains covered by Article 33 of the Third Convention) are combatants, that is to say, they have the right to participate directly in hostilities.
>
> 3. Whenever a Party to a conflict incorporates a paramilitary or armed law enforcement agency into its armed forces it shall so notify the other Parties to the conflict."[54]

This provision indicates that only States are allowed to establish armed forces and to deploy them as their official agents in international relations.[55] It also clearly maintains no distinction between the combatants belonging to the regular armed forces and those of other armed groups.[56] Article 43(1) suggests that the armed troops of a State may include not only officially raised armed forces, but also groups and units falling under their command, being responsible to that State for their unlawful conduct. With the exclusion of the members of medical personnel and chaplains, individuals belonging to the armed forces of a State that is party to an armed conflict are, according to the second Paragraph of this provision, combatants having a right to take a direct part in hostilities and to be treated accordingly. If PMSCs fall within a definition of paramilitary entities or armed law enforcement agencies and States participating in armed conflicts opt to incorporate them into their military apparatus, these States are required to notify other parties to the conflict about this incorporation, as provided in Article 43(3).

Article 44(1) of the First Additional Protocol prescribes that any combatant, as identified on the basis of criteria of Article 43, who falls into the power of the adverse party to a conflict, will be granted a prisoner-of-war status. In addition, Article 44(3) further specifies that it is required to promote the protection of the civilian population from the effects of hostilities. It must be done by the combatants who are obliged to distinguish themselves from civilians with some sort of an external sign when they make preparatory arrangements for an attack or are in the process of conducting a

---

54 Article 43 of the First Additional Protocol.

55 Won Kidane, (2010) "The Status of Private Military Contractors under International Humanitarian Law", *Denver Journal of International Law and Policy*, Volume 38, Issue 3, pp. 378-389.

56 University Centre for International Humanitarian Law, *Expert Meeting on Private Military Contractors: Status and State Responsibility for Their Actions*, Report, Geneva, August 29-30, 2005, p. 9, <http://www.geneva-academy.ch/docs/expert-meetings/2005/2rapport_compagnies_privees.pdf>, accessed on October 10, 2014.

military operation.[57] If combatants cannot distinguish themselves from the civilian population in this way, they need to openly carry their arms in the preparation of a military operation and during the actual military engagement. As Article 44(4) stipulates, a failure to comply with the described criteria will result in the deprivation of the right to be a lawful combatant and a prisoner of war.

Analysis of Article 43 of the First Additional Protocol reveals that one of the requirements, which can hardly be satisfied by PMSCs, is the criterion of being "under a command responsible" to a party to a conflict for the conduct of subordinates.[58] This condition, in fact, requires that States employing PMSCs subject them to their criminal jurisdiction, as has been clarified during the Expert Meeting on Private Military Contractors in Geneva.[59] The experts agreed that in the most scenarios this requirement cannot be regarded as fulfilled because many States are not able to exercise effective criminal jurisdiction over the PMSC personnel hired by them. If Article 43 is narrowly interpreted, then the impossibility of successful application of this provision becomes even more apparent given that private contractors and their personnel should operate within the official military chain of command. The membership of armed forces is primarily regulated by the domestic legislation of a State providing for the formal incorporation of individuals or entities into the national military apparatus.[60] Considering the reluctance of States to formally incorporate PMSCs into the structures of their armed forces, this situation should be perceived as merely theoretical.[61] The foregoing means that in this case Article 43 would not be applicable and thus the personnel of PMSCs would escape qualification of being members of armed forces.[62]

While Article 43 of the First Additional Protocol uses a more flexible notion and does not require the existence of a military chain of command,[63] there is also

---

57 Louise Doswald-Beck, "Private Military Companies under International Humanitarian Law", in Simon Chesterman and Chia Lehnardt (Eds.), (2007) *From Mercenaries to Market: The Rise and Regulation of Private Military Companies*, Oxford: Oxford University Press, p. 120.

58 Ibidem.

59 University Centre for International Humanitarian Law, *Expert Meeting on Private Military Contractors: Status and State Responsibility for Their Actions*, Report, Geneva, August 29-30, 2005, pp. 9-10, <http://www.geneva-academy.ch/docs/expert-meetings/2005/2rapport_compagnies_privees.pdf>, accessed on October 10, 2014.

60 Mirko Sossai, "Status of Private Military and Security Company Personnel in the Law of International Armed Conflict", in Francesco Francioni and Natalino Ronzitti (Eds.), (2011) *War by Contract: Human Rights, Humanitarian Law, and Private Contractors*, Oxford: Oxford University Press, p. 200.

61 Michael N. Schmitt, (2005) "Humanitarian Law and Direct Participation in Hostilities by Private Contractors or Civilian Employees", *Chicago Journal of International Law*, Volume 5, Issue 2, pp. 511, 524; Montreux Document, "The Main Rules and Good Practices of the Montreux Document", p. 36.

62 Louise Doswald-Beck, "Private Military Companies under International Humanitarian Law", in Simon Chesterman and Chia Lehnardt (Eds.), (2007) *From Mercenaries to Market: The Rise and Regulation of Private Military Companies*, Oxford: Oxford University Press, p. 121.

63 Mirko Sossai, "Status of Private Military and Security Company Personnel in the Law of International Armed Conflict", in Francesco Francioni and Natalino Ronzitti (Eds.), (2011) *War by Contract: Human Rights, Humanitarian Law, and Private Contractors*, Oxford: Oxford University Press, p. 201.

another approach to consider the applicability of Article 43 and more specifically the fulfillment of the requirements of the command responsible to a party to a conflict and thus belonging to that party. It has been argued that using this functional approach one can describe PMSCs as falling within the scope of the provision if a State hires PMSCs in order to carry out essential military functions and to participate directly in the hostilities and if both have a certain qualified factual link that is also an essential element for establishing State responsibility.[64]

In order to identify the existence of a factual link between States and PMSCs, a number of significant elements must be present in the ties between them. It is obvious that if a PMSC is employed by a State that is party to an armed conflict in order to provide core military tasks, it may be regarded as functioning on behalf of that State. In this respect, the central question is whether the company is responsible to the State for its activities. Importantly, an entity, such as a PMSC, can be responsible to a hiring State participating in a conflict if it reports to the State in question and if the latter has some form of a supervision and oversight mechanism capable of examining the activities of its employees. This factual relationship based on the concept of responsibility towards States, which must be present in relations between States and PMSCs, can be demonstrated on the basis of a contract concluded between two entities for the performance of military services. The mere conclusion of a contract does not, however, indicate that the requirement is fulfilled.[65] In the first place, the contract must clearly define the precise tasks to be performed by the PMSC. In addition, it can be argued that PMSCs in their capacity of private corporations are responsible for the performance of their duties laid down in the contractual agreements with States. PMSCs acting in derogation of a contract breach the terms of that agreement and can in theory face legal consequences of their actions in form of civil liability. It remains, however, unsettled whether this type of contractual relationship between States and PMSCs is satisfactory for establishing a factual link: one must look further for other possibilities, such as criminal jurisdiction of States over the personnel of PMSCs.

According to Louise Doswald-Beck, it is arguable that some form of criminal jurisdiction of a State over PMSCs is required in order to establish a factual relationship between hiring States and private contractors.[66] One may not find it too shocking that

---

64 Louise Doswald-Beck, "Private Military Companies under International Humanitarian Law", in Simon Chesterman and Chia Lehnardt (Eds.), (2007) *From Mercenaries to Market: The Rise and Regulation of Private Military Companies*, Oxford: Oxford University Press, p. 121; Mirko Sossai, "Status of Private Military and Security Company Personnel in the Law of International Armed Conflict", in Francesco Francioni and Natalino Ronzitti (Eds.), (2011) *War by Contract: Human Rights, Humanitarian Law, and Private Contractors*, Oxford: Oxford University Press, p. 200.

65 Mirko Sossai, "Status of Private Military and Security Company Personnel in the Law of International Armed Conflict", in Francesco Francioni and Natalino Ronzitti (Eds.), (2011) *War by Contract: Human Rights, Humanitarian Law, and Private Contractors*, Oxford: Oxford University Press, p. 201.

66 Louise Doswald-Beck, "Private Military Companies under International Humanitarian Law", in Simon Chesterman and Chia Lehnardt (Eds.), (2007) *From Mercenaries to Market: The Rise and*

the experts working for PMSCs seldom have the nationality of the hiring States and can even be sub-contracted by other companies. These circumstances pose many difficulties to the exercise of the criminal jurisdiction by States. Nevertheless, from recent developments regarding PMSCs and their exploitation by States, the conclusion can be drawn that there is a general trend towards the exercise of criminal jurisdiction over the personnel of those businesses. For instance, the adoption by the United States of the Uniform Code of Military Justice in 2006 extending jurisdiction over individuals accompanying armed forces or serving with them in the field, and the authorization of US courts-martial to instigate criminal proceedings on the basis of the Code indicate that some States are willing to hold perpetrators of criminal offenses among PMSCs responsible for their illegal activities.[67] It has been not entirely correctly acknowledged by Louise Doswald-Beck that if a State can be said to have a civil or criminal jurisdiction over a PMSC, then such a firm can most probably qualify as an entity responsible to a State party to a conflict.[68] This mere fact does not necessarily lead to the conclusion that there are factual links between PMSCs and hiring States. Three examined conditions are probably required: a clear contract with precise tasks to be carried out by PMSCs, the existence of oversight and monitoring mechanisms established by the State, and the criminal jurisdiction over PMSCs. The fulfillment of these conditions demonstrates a strong factual link between States and private corporations fighting on their behalf leading to the conclusion that they are to be considered as the members of armed forces within the meaning of Article 43 of the First Additional Protocol. According to current State practice, it is doubtful as to whether PMSCs actually meet the three above-mentioned requirements necessary for establishment of a factual link between a State and a company and so can be defined as members of the armed forces.

The following set of requirements laid down in Article 4(A) of the Third Geneva Convention relative to the Treatment of the Prisoners of War of 1949[69] should form another core provision in the analysis of the legal status of the PMSC personnel as combatants having prisoner-of-war status applicable to international armed conflicts.[70] It reads as follows:

Regulation of Private Military Companies, Oxford: Oxford University Press, p. 121; Mirko Sossai, "Status of Private Military and Security Company Personnel in the Law of International Armed Conflict", in Francesco Francioni and Natalino Ronzitti (Eds.), (2011) *War by Contract: Human Rights, Humanitarian Law, and Private Contractors*, Oxford: Oxford University Press, p. 203.

67 Mirko Sossai, "Status of Private Military and Security Company Personnel in the Law of International Armed Conflict", in Francesco Francioni and Natalino Ronzitti (Eds.), (2011) *War by Contract: Human Rights, Humanitarian Law, and Private Contractors*, Oxford: Oxford University Press, p. 203.

68 Louise Doswald-Beck, "Private Military Companies under International Humanitarian Law", in Simon Chesterman and Chia Lehnardt (Eds.), (2007) *From Mercenaries to Market: The Rise and Regulation of Private Military Companies*, Oxford: Oxford University Press, p. 121.

69 The Third Geneva Convention.

70 University Centre for International Humanitarian Law, *Expert Meeting on Private Military Contractors: Status and State Responsibility for Their Actions*, Report, Geneva, August 29-30, 2005, pp. 9-10,

"Prisoners of war, in the sense of the present Convention, are persons belonging to one of the following categories, who have fallen into the power of the enemy:

(1) Members of the armed forces of a Party to the conflict, as well as members of militias or volunteer corps forming part of such armed forces.

(2) Members of other militias and members of other volunteer corps, including those of organized resistance movements, belonging to a Party to the conflict and operating in or outside their own territory, even if this territory is occupied, provided that such militias or volunteer corps, including such organized resistance movements, fulfil the following conditions:

(a) that of being commanded by a person responsible for his subordinates;
(b) that of having a fixed distinctive sign recognizable at a distance;
(c) that of carrying arms openly;
(d) that of conducting their operations in accordance with the laws and customs of war.

(3) Members of regular armed forces who profess allegiance to a government or an authority not recognized by the Detaining Power.

(4) Persons who accompany the armed forces without actually being members thereof, such as civilian members of military aircraft crews, war correspondents, supply contractors, members of labour units or of services responsible for the welfare of the armed forces, provided that they have received authorization, from the armed forces which they accompany, who shall provide them for that purpose with an identity card similar to the annexed model.

(5) Members of crews, including masters, pilots and apprentices, of the merchant marine and the crews of civil aircraft of the Parties to the conflict, who do not benefit by more favourable treatment under any other provisions of international law.

(6) Inhabitants of a non-occupied territory, who on the approach of the enemy spontaneously take up arms to resist the invading forces, without having had time to form themselves into regular armed units, provided they carry arms openly and respect the laws and customs of war."

The first category of combatants can be found in Article 4(A)(1) of the Third Geneva Convention referring to "members of the armed forces of a Party to the conflict as well as members of militias or volunteer corps forming part of such armed forces". In this Paragraph, the term "members of armed forces" replaces the term "army" used in Article 1 of the 1907 Hague Convention Respecting the Laws and Customs of War on Land (hereinafter: the 1907 Hague Convention IV).[71] It encompasses all military

<http://www.geneva-academy.ch/docs/expert-meetings/2005/2rapport_compagnies_privees.pdf>, accessed on October 10, 2014.

71 The Hague Convention Respecting the Laws and Customs of War on Land, October 18, 1907 ("the 1907 Hague Convention IV").

personnel, including the members of land, sea, and air forces.[72] As the practice of some States indicates, militias and volunteer corps can be quite distinct from regular armed forces, whilst it is the case that members of those groups do belong to the State armies and thus must be regarded as falling under the scope of the provision.[73]

As observed above, the employees of PMSCs can qualify as members of armed forces eligible for having the status of prisoners of war, but only if they are formally incorporated into the regular armed forces of a certain State and form part of the official military State structures. Only on this occasion, the staff of PMSCs can be regarded as members of the official armed forces of a specific State. A sole fact of private military and security providers performing their tasks on the basis of a contract is not enough to conclude that an entity has been incorporated into the State apparatus.[74] Examining the operations of PMSCs, it appears that these firms act in their capacity of private legal entities having a distinct civilian identity. They are usually seen as separate structures from the State apparatus that often provide their expertise under the control, direction, and instructions of States.

The possibility of formal incorporation of PMSCs into the official State armies seems to have merely theoretical connotations, as States remain reluctant to confer on the employees of private contractors the status of members of the armed forces. This is because, according to various experts, the main purpose of the States hiring PMSCs is to avoid numerous "employer's responsibilities that members of armies and their dependents enjoy".[75] In fact, PMSC employees do not qualify as members of State armed forces, as they are not employed by means of legitimate procedures for enlistment or conscription. In addition, States always have other options to recruit civilians by enlisting those individuals into the army and by employing them as reservists, whose status changes from a civilian into a member of armed forces during an armed conflict.[76] It is also true that PMSCs usually operate outside the

---

[72] Jean S. Pictet, (1960) *Commentary, III Geneva Convention Relative to the Treatment of Prisoners of War*, Geneva: International Committee of the Red Cross, p. 51.

[73] Ibidem, pp. 51-52.

[74] Shannon Bosch, (2007) "Private Security Contractors and International Humanitarian Law – A Skirmish for Recognition in International Armed Conflicts", *African Security Review*, Volume 16, Issue 4, p. 38.

[75] Louise Doswald-Beck, "Private Military Companies under International Humanitarian Law", in Simon Chesterman and Chia Lehnardt (Eds.), (2007) *From Mercenaries to Market: The Rise and Regulation of Private Military Companies*, Oxford: Oxford University Press, p. 118; Sarah Percy, (2006) *Regulating the Private Security Industry*, Adelphi Series, Volume 46, Issue 384, Oxon: Routledge, p. 48; Lindsey Cameron, (2006) "Private Military Companies: Their Status under International Humanitarian Law and Its Impact on Their Regulation", *International Review of the Red Cross*, Volume 88, Issue 863, p. 583; Shannon Bosch, (2008) "Private Security Contractors and State Responsibility: Are States Exempt from Responsibility for Violations of Humanitarian Law Perpetrated by Private Security Contractors?", *The Comparative and International Law Journal of Southern Africa*, Volume 41, Issue 3, p. 373.

[76] Michael N. Schmitt, (2005) "Humanitarian Law and Direct Participation in Hostilities by Private Contractors or Civilian Employees", *Chicago Journal of International Law*, Volume 5, Issue 2, p. 524;

main chain of military command and have their own supervisory chains and disciplinary systems.[77] In addition, PMSCs often manage to function unsupervised due to the lack of effective oversight and monitoring mechanisms established by the hiring States. Host and home States are often also unable to exercise oversight and monitoring over PMSCs. As proposed by some authors, it can be possible that PMSCs fall under the military command of States regarding their relationship with the military apparatus. In this regard, a contract can potentially serve as a tool for controlling PMSCs and bringing them under a command responsible to a certain State leading to the acceptance of the claim that those PMSCs operate as the regular armed forces, being State agents.[78] The possibility of such a scheme is, however, highly debatable. In conclusion, it would appear that in practical terms, the formal incorporation of PMSCs into the official State structures will most probably not take place. Therefore, the status of lawful combatants will not be conferred on those individuals working for PMSCs[79] and will stay irrelevant to their operation on the ground.[80] As the analysis shows, Article 4(A)(1) has possibly no practical application to PMSCs hired by States.[81]

In addition to the Paragraph assessed above, Article 4(A)(2) of the Third Geneva Convention deserves to be analyzed considering the fact that in the view of many experts it might have great significance for the qualification of PMSCs as *de facto* combatants and prisoners of war upon capture.[82] This provision referring to

---

Sarah Percy, (2006) *Regulating the Private Security Industry*, Adelphi Series, Volume 46, Issue 384, Oxon: Routledge, p. 48.

77 Montreux Document, "The Main Rules and Good Practices of the Montreux Document", p. 36; Michael N. Schmitt, (2005) "Humanitarian Law and Direct Participation in Hostilities by Private Contractors or Civilian Employees", *Chicago Journal of International Law*, Volume 5, Issue 2, p. 525.

78 Dutch Advisory Council on International Affairs, *Employing Private Military Companies: A Question of Responsibility* (Dutch: *De Inhuur van Private Militaire Bedrijven: Een Kwestie van Verantwoordelijkheid*), No. 59, December 2007, The Hague, p. 34, <http://www.aiv-advies.nl/ContentSuite/upload/aiv/doc/webversie_AIV_59eng(1).pdf>, accessed on May 9, 2011.

79 Lindsey Cameron, (2006) "Private Military Companies: Their Status under International Humanitarian Law and Its Impact on Their Regulation", *International Review of the Red Cross*, Volume 88, Issue 863, p. 583; Louise Doswald-Beck, "Private Military Companies under International Humanitarian Law", in Simon Chesterman and Chia Lehnardt (Eds.), (2007) *From Mercenaries to Market: The Rise and Regulation of Private Military Companies*, Oxford: Oxford University Press, pp. 116-118; Emanuela-Chiara Gillard, "Private Military/Security Companies: The Status of Their Staff and Their Obligations under International Humanitarian Law and The Responsibilities of States in Relation to Their Operation", in Andrew Alexandra, Deane-Peter Baker, and Marina Caparini (Eds.), (2008) *Private Military and Security Companies: Ethics, Policies and Civil-Military Relations*, London and New York: Routledge, p. 163.

80 Montreux Document, "The Main Rules and Good Practices of the Montreux Document", p. 36.

81 Louise Doswald-Beck, "Private Military Companies under International Humanitarian Law", in Simon Chesterman and Chia Lehnardt (Eds.), (2007) *From Mercenaries to Market: The Rise and Regulation of Private Military Companies*, Oxford: Oxford University Press, p. 118.

82 Lindsey Cameron, (2006) "Private Military Companies: Their Status under International Humanitarian Law and Its Impact on Their Regulation", *International Review of the Red Cross*, Volume 88, Issue 863, p. 584; Louise Doswald-Beck, "Private Military Companies under International Humanitarian

"members of other militias and members of other volunteer corps, including those of organized resistance movements" implies the idea of conferring prisoner-of-war status on groups taking part in hostilities, such as partisans during the Second World War, who need to be protected if captured by their adversaries.[83] Article 4(A)(2) is, however, not limited to such resistance movements and can in theory be applied in relation to PMSCs. It requires the fulfillment of four specific conditions that are also in Article 1 of the 1907 Hague Convention IV. First three requirements are that of being under the command of a person responsible for his or her subordinates, that of having a distinctive sign on their uniform, which can be recognized at a distance, and that of carrying weapons openly. The fourth condition forming the core element of the definition requires members of groups to conduct their operations in accordance with international humanitarian law in order to fall under the scope of Article 4(A)(2). In this respect, it is worth recalling that the provision containing four cumulative requirements should be considered on a case-by-case or company-by-company basis, making it possible to determine whether certain individuals or groups of people can be classified as this particular type of combatant.[84]

In the first place, Article 4(A)(2) specifies that a group should belong to a party to the conflict, while not being an element of the official armed forces.[85] Although a number of scholars seem to point out that PMSCs cannot be considered as entities "belonging to a party to the conflict",[86] the ICRC Commentary evidently provides otherwise. According to the Commentary, it is essential to establish a *de facto* relationship between the State and the group in question.[87] This *de facto* relationship can be shown on the basis of the motivations or intentions of an armed group and the

---

Law", in Simon Chesterman and Chia Lehnardt (Eds.), (2007) *From Mercenaries to Market: The Rise and Regulation of Private Military Companies*, Oxford: Oxford University Press, p. 118; Montreux Document, "The Main Rules and Good Practices of the Montreux Document", p. 36.

[83] Jean S. Pictet, (1960) *Commentary, III Geneva Convention Relative to the Treatment of Prisoners of War*, Geneva: International Committee of the Red Cross, p. 52.

[84] Malcolm N. Shaw, (2008) *International Law*, Sixth Edition, Cambridge: Cambridge University Press, p. 1173; Montreux Document, "The Main Rules and Good Practices of the Montreux Document", p. 36; Lindsey Cameron, (2006) "Private Military Companies: Their Status under International Humanitarian Law and Its Impact on Their Regulation", *International Review of the Red Cross*, Volume 88, Issue 863, pp. 584-585.

[85] Katherine Del Mar, (2010) "The Requirement of 'Belonging' under International Humanitarian Law", *The European Journal of International Law*, Volume 21, Issue 1, p. 111.

[86] University Centre for International Humanitarian Law, *Expert Meeting on Private Military Contractors: Status and State Responsibility for Their Actions*, Report, Geneva, August 29-30, 2005, p. 11, <http://www.geneva-academy.ch/docs/expert-meetings/2005/2rapport_compagnies_privees.pdf>, accessed on October 10, 2014.

[87] Jean S. Pictet, (1960) *Commentary, III Geneva Convention Relative to the Treatment of Prisoners of War*, Geneva: International Committee of the Red Cross, p. 57; Emanuela-Chiara Gillard, (2006) "Business Goes to War: Private Military / Security Companies and International Humanitarian Law", *International Review of the Red Cross*, Volume 88, Issue 863, p. 534.

reaction of the State in question.[88] The existence of this significant link is, in fact, sufficient if it can be indicated that, even merely by tacit agreement, the members of the group conduct their operations on behalf of one of the parties to the conflict. An official declaration of that party to the conflict may follow, but it is not required that it formally authorizes the group. Therefore it is safe to assume that the personnel of PMSCs can satisfy the requirement of belonging to a party to an armed conflict if there is this type of *de facto* relationship between hiring States and PMSCs functioning on the basis of a contract that evidences the existence of such a link.

With regard to the four criteria laid down in Article 4(A)(2), all of them can potentially be met by the members of PMSC personnel. First of all, a group must be commanded by a person responsible for his or her subordinates. While resistance movements during the Second World War usually functioned under the command of regular officers of the armed forces, it is not necessary for a group to follow this trend in order to fulfill the first requirement. Instead of being a military expert, the leader may also be a civilian.[89] One of the crucial roles of the head of the group is to provide discipline by commanding his or her subordinates, but also to accept responsibility for their actions.[90] PMSC units conducting their operations on the ground can usually be described as hierarchically structured groups with their own chain of command and have appointed commanders performing those important tasks. Therefore, it might safely be assumed that in the most cases this condition can be regarded as fulfilled.

Secondly, members of the PMSC personnel must have fixed, distinctive signs visible at a distance. According to the current practice of employing private contractors, it is usually the case that governments provide PMSCs with distinctive signs, which are visible to the members of the armed forces parties to the conflict and civilians helping to identify those individuals and their role on the battlefield. Analysis of the State practice of hiring PMSCs illustrates that private contractors employed by States often have a specific recognition attribute in the form of a card. For instance, the US Department of Defense is known for issuing cards with its logo and a certain text on them, making sure that the PMSC personnel operating on its behalf are recognized

---

88 Katherine Del Mar, (2010) "The Requirement of 'Belonging' under International Humanitarian Law", *The European Journal of International Law*, Volume 21, Issue 1, p. 111.

89 Jean S. Pictet, (1960) *Commentary, III Geneva Convention Relative to the Treatment of Prisoners of War*, Geneva: International Committee of the Red Cross, p. 59; Louise Doswald-Beck, "Private Military Companies under International Humanitarian Law", in Simon Chesterman and Chia Lehnardt (Eds.), (2007) *From Mercenaries to Market: The Rise and Regulation of Private Military Companies*, Oxford: Oxford University Press, p. 119; Emanuela-Chiara Gillard, (2006) "Business Goes to War: Private Military / Security Companies and International Humanitarian Law", *International Review of the Red Cross*, Volume 88, Issue 863, p. 535.

90 Louise Doswald-Beck, "Private Military Companies under International Humanitarian Law", in Simon Chesterman and Chia Lehnardt (Eds.), (2007) *From Mercenaries to Market: The Rise and Regulation of Private Military Companies*, Oxford: Oxford University Press, p. 119.

as US Department of Defense contractors.[91] While other States and their agencies remain less scrupulous in following similar policies, if a State distributes special signs to the employed PMSC members, it might be enough to conclude that there is a formal link between the personnel of those PMSCs and that particular State.[92] However, the situation can be more difficult and less clear if PMSCs are sub-contracted by other business entities.[93]

The third requirement of carrying arms openly is quite a straightforward functional element depending on a particular factual situation.[94] Members of PMSCs might be allowed to carry their weapons openly and often do so in practice, but it is also possible that they would not always do so.

The last condition of acting in accordance with the laws and customs of war is fairly self-explanatory. It requires some training in this particular field of international law and respect of it by the representatives of a group. In this, there is a general trend to be observed in the attitude of a large number of PMSCs heading towards general recognition of the significance of rules and principles of international humanitarian law in their operational environments and creating an impression of acting according to those concepts. Willingness to comply with the rules of international humanitarian law comes to the forefront of, for instance, the ICoC as a recent multi-stakeholder voluntary initiative. However, the actual practice of PMSCs operating in conflict zones does not unquestionably make clear that all firms comply with the laws and customs of war.[95]

Some experts share the opinion that in practice PMSCs as groups are not or rarely able to satisfy the four cumulative criteria of Article 4(A)(2) in order to be considered *de facto* combatants.[96] Lindsey Cameron also notes that from a historical point of view it might not be appropriate to use this provision for the categorization of PMSCs.[97] It was originally intended to grant to partisans active in the Second

---

91 Ibidem.

92 Katherine Del Mar, (2010) "The Requirement of 'Belonging' under International Humanitarian Law", *The European Journal of International Law*, Volume 21, Issue 1.

93 Louise Doswald-Beck, "Private Military Companies under International Humanitarian Law", in Simon Chesterman and Chia Lehnardt (Eds.), (2007) *From Mercenaries to Market: The Rise and Regulation of Private Military Companies*, Oxford: Oxford University Press, p. 119.

94 Ibidem, p. 120.

95 Ibidem.

96 University Centre for International Humanitarian Law, *Expert Meeting on Private Military Contractors: Status and State Responsibility for Their Actions*, Report, Geneva, August 29-30, 2005, p. 10, <http://www.geneva-academy.ch/docs/expert-meetings/2005/2rapport_compagnies_privees.pdf>, accessed on October 10, 2014; Louise Doswald-Beck, "Private Military Companies under International Humanitarian Law", in Simon Chesterman and Chia Lehnardt (Eds.), (2007) *From Mercenaries to Market: The Rise and Regulation of Private Military Companies*, Oxford: Oxford University Press, pp. 119-120; Michael N. Schmitt, (2005) "Humanitarian Law and Direct Participation in Hostilities by Private Contractors or Civilian Employees", *Chicago Journal of International Law*, Volume 5, Issue 2, p. 531.

97 Lindsey Cameron, (2006) "Private Military Companies: Their Status under International Humanitarian Law and Its Impact on Their Regulation", *International Review of the Red Cross*, Volume 88, Issue 863, p. 586.

World War prisoner-of-war status. Unlike PMSCs, these persons can be perceived as belonging to groups fighting for liberation of occupied territories or be seen as similar to military personnel of defeated armies. Additionally, application of Article 4(A)(2) can create much confusion, as PMSC employees are individuals that after being concluded to constitute *de facto* combatants can be regarded at the later stage as mercenaries having no prisoner-of-war status.[98] However, the possibility of PMSCs fulfilling the conditions laid down in Article 4(A)(2) should not be set aside without discussion. A small minority of the experts working for private contractors operating on behalf States may still qualify as members of other militias and members of other volunteer corps, including those of organized resistance movements, falling within the scope of the provision.[99] It is not likely that members of the PMSC personnel would be able to comply with all above-mentioned requirements to be classified as combatants, but on some occasions it is certainly possible.[100] Thus, in the light of the foregoing analysis it appears to be clear that even if employees of PMSCs have not been formally incorporated into regular armed forces, they might fall within the scope of application of the above-mentioned provision.[101] Consequently, if those individuals fall within the scope of application of Article 4(A)(2), the perception is that they are engaged in activities requiring the exercise of governmental authority, such as participating in military operations, which are attributable to the respective States.[102]

4.1.2 Non-combatants with prisoner-of-war status

In addition to the rules relating to the combatant status of individuals having the right to be prisoners of war upon capture assessed above, Article 4(A)(4) of the Third Geneva Convention is also one of the most relevant provisions with regard to PMSCs and their status under international humanitarian law, which has also been recognized by the Dutch Advisory Council on International Affairs.[103] In fact, this Article specifically concerns the entitlement of non-combatants to prisoner-of-war status.

---

98 Ibidem.

99 Emanuela-Chiara Gillard, (2006) "Business Goes to War: Private Military / Security Companies and International Humanitarian Law", *International Review of the Red Cross*, Volume 88, Issue 863, p. 536.

100 Lindsey Cameron, (2006) "Private Military Companies: Their Status under International Humanitarian Law and Its Impact on Their Regulation", *International Review of the Red Cross*, Volume 88, Issue 863, pp. 585-587.

101 Louise Doswald-Beck, "Private Military Companies under International Humanitarian Law", in Simon Chesterman and Chia Lehnardt (Eds.), (2007) *From Mercenaries to Market: The Rise and Regulation of Private Military Companies*, Oxford: Oxford University Press, p. 119.

102 University Centre for International Humanitarian Law, *Expert Meeting on Private Military Contractors: Status and State Responsibility for Their Actions*, Report, Geneva, August 29-30, 2005, p. 13, <http://www.geneva-academy.ch/docs/expert-meetings/2005/2rapport_compagnies_privees.pdf>, accessed on October 10, 2014.

103 Dutch Advisory Council on International Affairs, *Employing Private Military Companies: A Question of Responsibility* (Dutch: *De Inhuur van Private Militaire Bedrijven: Een Kwestie van*

These non-combatants are defined as "persons who accompany the armed forces without actually being members thereof, such as civilian members of military aircraft crews, war correspondents, supply contractors, members of labor units or of services responsible for the welfare of the armed forces".[104] The provision determines the legal position of persons, such as PMSC guards, who do not form part of the official armed forces, cannot participate in hostilities, and are prohibited from being combatants.[105] Nevertheless, they are entitled to the status of prisoners of war, while offering specialized expertise to the armed forces.[106] According to the current practice, States tend to emphasize that the employees of private contractors acting on their behalf are only to be considered as this type of individual.[107]

It is clear that PMSC employees can and often do accompany regular armed forces without being members of those troops and being classified as lawful combatants. For many centuries, State armies have employed persons who did not belong to the State troops and did not actually fire their weapons in the course of hostilities. However, they fulfill an important task in taking care of specific needs of military experts and their equipment. As an example, the internal policies of a governmental agency of one of the major employers of PMSCs can provide some evidence of the current State practice in that regard. The US Department of Defense Instruction 3020.41 of December 11, 2011 stipulates that certain civilians supporting American troops need to be identified as "contractors authorized to accompany the force", and are entitled

Verantwoordelijkheid), No. 59, December 2007, The Hague, p. 16, <http://www.aiv-advies.nl/ContentSuite/upload/aiv/doc/webversie_AIV_59eng(1).pdf>, accessed on May 9, 2011.

104 Montreux Document, "The Main Rules and Good Practices of the Montreux Document", p. 36; Lindsey Cameron, (2006) "Private Military Companies: Their Status under International Humanitarian Law and Its Impact on Their Regulation", *International Review of the Red Cross*, Volume 88, Issue 863, pp. 587-594; Louise Doswald-Beck, "Private Military Companies under International Humanitarian Law", in Simon Chesterman and Chia Lehnardt (Eds.), (2007) *From Mercenaries to Market: The Rise and Regulation of Private Military Companies*, Oxford: Oxford University Press, pp. 123-124; Emanuela-Chiara Gillard, "Private Military/Security Companies: The Status of Their Staff and Their Obligations under International Humanitarian Law and The Responsibilities of States in Relation to Their Operation", in Andrew Alexandra, Deane-Peter Baker, and Marina Caparini (Eds.), (2008) *Private Military and Security Companies: Ethics, Policies and Civil-Military Relations*, London and New York: Routledge, pp. 163-164.

105 Rebecca R. Vernon, (2003-2004) "Battlefield Contractors: Facing the Tough Issues", *Public Contract Law Journal*, Volume 33, Issue 2, pp. 406-407.

106 Emanuela-Chiara Gillard, (2006) "Business Goes to War: Private Military / Security Companies and International Humanitarian Law", *International Review of the Red Cross*, Volume 88, Issue 863, pp. 536-537; Shannon Bosch, (2007) "Private Security Contractors and International Humanitarian Law – A Skirmish for Recognition in International Armed Conflicts", *African Security Review*, Volume 16, Issue 4, p. 40.

107 Shannon Bosch, (2007) "Private Security Contractors and International Humanitarian Law – A Skirmish for Recognition in International Armed Conflicts", *African Security Review*, Volume 16, Issue 4, p. 38.

to the status of persons who accompany the armed forces.[108] In accordance with the provisions of the instruction, applicable international law requires that such individuals have a right to be assigned prisoner-of-war status when they are captured in an international armed conflict.[109] Similarly, the US Army Field Manual No. 3-100.21 and Army Regulation 715-9 refer to contractors as civilians who are authorized to accompany armed forces and acknowledge that they are to have the prisoner-of-war status upon capture.[110]

The notion of persons who accompany the State armed forces is associated with a number of important aspects. It is maintained that persons falling under this category must be defined as civilians and cannot be targeted by the adversary in its direct attacks during the hostilities.[111] Their activities are also not seen as a direct participation in the hostilities.[112] This basic principle is supported by Article 50(1) of the 1977 First Additional Protocol to the 1949 Geneva Conventions stating that persons falling into the scope of Article 4(A)(4) of the Third Geneva Convention are civilians excluded from classification as combatants. The status of individuals accompanying the armed forces can be granted to certain people only by the official armed forces by means of a formal authorization. For these reasons, private contractors providing military and security services must have a "real link" with the armed forces of a State or, in other words, there must be some nexus between PMSCs and the armed forces, meaning that they need to be employed in order to support the armed forces and not other possible branches of that State.[113]

---

108 US Department of Defense Instruction 3020.41, *Contractor Personnel Authorized to Accompany the U.S. Armed Forces*, December 20, 2011, Enclosure 2: Procedures, 1(a)(1), <http://www.dtic.mil/whs/directives/corres/pdf/302041p.pdf>, accessed on January 5, 2014.

109 Ibidem, Enclosure 2: Procedures, 1(a)(1).

110 US Department of the Army, *Contractors on the Battlefield*, Field Manual No. 3-100.21, January 3, 2003, paras. 1-21, 2-33, 4-49, <http://fas.org/irp/doddir/army/fm3-100-21.pdf>, accessed on June 20, 2014; US Department of the Army, Army Regulation 715-9, Procurement: Operational Contract Support Planning and Management, June 20, 2011, Chapter 4: In-Theater Contractor and Contract Management, pp. 10-11, para. 4-2(a), <http://armypubs.army.mil/epubs/pdf/r715_9.pdf>, accessed on June 21, 2014.

111 Giulio Bartolini, "Private Military and Security Contractors as 'Persons who Accompany the Armed Forces'", in Francesco Francioni and Natalino Ronzitti (Eds.), (2011) *War by Contract: Human Rights, Humanitarian Law, and Private Contractors*, Oxford: Oxford University Press, p. 220.

112 Ibidem; Louise Doswald-Beck, "Private Military Companies under International Humanitarian Law", in Simon Chesterman and Chia Lehnardt (Eds.), (2007) *From Mercenaries to Market: The Rise and Regulation of Private Military Companies*, Oxford: Oxford University Press, p. 129.

113 Montreux Document, "The Main Rules and Good Practices of the Montreux Document", p. 36; University Centre for International Humanitarian Law, *Expert Meeting on Private Military Contractors: Status and State Responsibility for Their Actions*, Report, Geneva, August 29-30, 2005, p. 14, <http://www.geneva-academy.ch/docs/expert-meetings/2005/2rapport_compagnies_privees.pdf>, accessed on October 10, 2014; Louise Doswald-Beck, "Private Military Companies under International Humanitarian Law", in Simon Chesterman and Chia Lehnardt (Eds.), (2007) *From Mercenaries to Market: The Rise and Regulation of Private Military Companies*, Oxford: Oxford University Press, p. 124.

Article 4(A)(4) provides a number of activities, which can be performed by the persons accompanying the armed forces. These individuals are often referred to as "supply contractors".[114] The list is, however, not exhaustive and is merely used to describe the nature of functions carried out by this particular type of personnel.[115] The activities concerned belong to the provision of general support and are usually considered as indirect participation in hostilities.[116] The importance of this notion cannot be underestimated, for taking a direct part in hostilities will automatically deprive this group of individuals of their status of civilians and prisoners of war upon capture with a risk of them being directly targeted by the adversary as unlawful combatants, as Article 51(3) of the First Additional Protocol and Article 13(3) of the Second Additional Protocol[117] stipulate.[118] Furthermore, such a direct participation in armed conflicts may qualify as "unprivileged belligerency" and lead to criminal prosecution of the experts working for private contractors under the domestic law of a State having jurisdiction.[119] In addition, members of PMSC personnel directly taking part in the hostilities may trigger the occurrence of many unaddressed issues

---

[114] University Centre for International Humanitarian Law, *Expert Meeting on Private Military Contractors: Status and State Responsibility for Their Actions*, Report, Geneva, August 29-30, 2005, p. 14, <http://www.geneva-academy.ch/docs/expert-meetings/2005/2rapport_compagnies_privees.pdf>, accessed on October 10, 2014.

[115] Giulio Bartolini, "Private Military and Security Contractors as 'Persons who Accompany the Armed Forces'", in Francesco Francioni and Natalino Ronzitti (Eds.), (2011) *War by Contract: Human Rights, Humanitarian Law, and Private Contractors*, Oxford: Oxford University Press, p. 224.

[116] Ibidem, p. 223; University Centre for International Humanitarian Law, *Expert Meeting on Private Military Contractors: Status and State Responsibility for Their Actions*, Report, Geneva, August 29-30, 2005, p. 15, <http://www.geneva-academy.ch/docs/expert-meetings/2005/2rapport_compagnies_privees.pdf>, accessed on October 10, 2014.

[117] Protocol Additional to the Geneva Conventions of August 12, 1949, and relating to the Protection of Victims of Non-International Armed Conflicts, June 8, 1977, 1125 UNTS 609 ("Second Additional Protocol").

[118] Terry D. Gill and Dieter Fleck, "Private Contractors and Security Companies", in Terry D. Gill and Dieter Fleck (Eds.), (2010) *The Handbook of the International Law of Military Operations*, Oxford: Oxford University Press, p. 490; University Centre for International Humanitarian Law, *Expert Meeting on Private Military Contractors: Status and State Responsibility for Their Actions*, Report, Geneva, August 29-30, 2005, pp. 12, 14, <http://www.geneva-academy.ch/docs/expert-meetings/2005/2rapport_compagnies_privees.pdf>, accessed on October 10, 2014; Louise Doswald-Beck, "Private Military Companies under International Humanitarian Law", in Simon Chesterman and Chia Lehnardt (Eds.), (2007) *From Mercenaries to Market: The Rise and Regulation of Private Military Companies*, Oxford: Oxford University Press, p. 124.

[119] Michael N. Schmitt, (2005) "Humanitarian Law and Direct Participation in Hostilities by Private Contractors or Civilian Employees", *Chicago Journal of International Law*, Volume 5, Issue 2, pp. 520-521; Knut Dörmann, (2003) "The Legal Situation of 'Unlawful/Unprivileged Combatants'", *International Review of the Red Cross*, Volume 85, Issue 849; Emanuela-Chiara Gillard, (2006) "Business Goes to War: Private Military / Security Companies and International Humanitarian Law", *International Review of the Red Cross*, Volume 88, Issue 863, p. 531.

with respect to their responsibility and accountability for violations of international and national law.[120]

Generally speaking, the body of international humanitarian law does not maintain a distinction between offensive and defensive operations.[121] Article 49(1) of the First Additional Protocol specifies that the term "attacks" used in the Protocol encompasses "acts of violence against the adversary, whether in offence or in defence". Taking a closer look at the State practice of using this type of civilian personnel, it can be maintained that traditionally no duties are entrusted to the persons falling under the category of tasks that can directly cause harm to one of the parties to the conflict.[122] Accordingly, the US Department of Defense Instruction 3020.41 of December 20, 2011 briefly refers to possible services provided by PMSCs. It states that contingency contractor personnel may perform the tasks of "providing communications support; transporting munitions and other supplies; performing maintenance functions for military equipment, providing private security services... and providing logistic services such as billeting and messing".[123] The previous Instruction 3020.41 of October 3, 2005 specifically mentioned that defense contractors accompanying the US armed forces are not allowed to perform services for uniquely military functions,[124] pointing out that members of their personnel may not guard military structures, such as "U.S. or coalition military supply routes, military facilities, military personnel, or military property".[125] One can imagine that PMSC members who frequently provide services involving, for example, the transport of military equipment, munitions, or troops and reparation of weapons and work in proximity to the armed forces may be considered as directly participating in hostilities, as indicated by some legal scholars.[126]

---

120 Terry D. Gill and Dieter Fleck, "Private Contractors and Security Companies", in Terry D. Gill and Dieter Fleck (Eds.), (2010) *The Handbook of the International Law of Military Operations*, Oxford: Oxford University Press, p. 490.

121 Emanuela-Chiara Gillard, (2006) "Business Goes to War: Private Military / Security Companies and International Humanitarian Law", *International Review of the Red Cross*, Volume 88, Issue 863, p. 540.

122 Giulio Bartolini, "Private Military and Security Contractors as 'Persons who Accompany the Armed Forces'", in Francesco Francioni and Natalino Ronzitti (Eds.), (2011) *War by Contract: Human Rights, Humanitarian Law, and Private Contractors*, Oxford: Oxford University Press, p. 222.

123 US Department of Defense Instruction 3020.41, *Contractor Personnel Authorized to Accompany the U.S. Armed Forces*, December 20, 2011, Enclosure 2: Procedures, 1(a)(2), <http://www.dtic.mil/whs/directives/corres/pdf/302041p.pdf>, accessed on January 5, 2014.

124 US Department of Defense Instruction 3020.41, *Contractor Personnel Authorized to Accompany the U.S. Armed Forces*, October 3, 2005, Section 6.3.5.2, <http://fas.org/irp/doddir/dod/i3020_41.pdf>, accessed on July 2, 2014.

125 Ibidem, Section 6.3.5.

126 Emanuela-Chiara Gillard, (2006) "Business Goes to War: Private Military / Security Companies and International Humanitarian Law", *International Review of the Red Cross*, Volume 88, Issue 863, p. 540; Giulio Bartolini, "Private Military and Security Contractors as 'Persons who Accompany the Armed Forces'", in Francesco Francioni and Natalino Ronzitti (Eds.), (2011) *War by Contract: Human Rights, Humanitarian Law, and Private Contractors*, Oxford: Oxford University Press, pp. 224-225.

For the qualification of individuals employed by PMSC contractors or other persons as individuals who accompany armed forces, it is required under Article 4(A)(4) that they be authorized by the armed forces and possess an identity card provided by the authorities.[127] Although the absence of such a card does not entirely exclude the possibility of being entitled to the prisoner-of-war status, national authorities usually require that the identity card be issued to private contractors. The US Department of Defense Instruction 3020.41 of December 20, 2011 specifies that contractor personnel may be used in a non-combat role to support contingency operations as long as they are designated by the armed forces as contractors authorized to accompany the force and have an identity card.[128] The ICRC Commentary provides that the identity cards need to virtually correspond to a military uniform or a partisan's arm-band.[129] More importantly, it makes clear that the existence of an identity card merely serves as a proof of the authorization of the armed forces granted to individuals. Hence, it does not constitute an absolute requirement for falling within the scope of the provision, as the capacity in which those persons serve remains the central determining condition.[130] In this regard, Giulio Bartolini maintains that one should consider "the functions performed and circumstances in which the individual was captured alongside members of the armed forces" as main elements for conferring the status of persons accompanying armed forces on PMSCs.[131]

Therefore, if members of the PMSC personnel are not merely employed by a State on the basis of a contract, but provide specific services directly to the State regular armed forces after being authorized, they will most probably fall within the scope of Article 4(A)(4) of the Third Geneva Convention.[132] As is generally accepted, if PMSCs accompanying armed forces participate directly in hostilities, then they will be lawfully targeted by the enemy and will not be able to benefit from prisoner-of-war status.[133] This point of view is, however, not shared by some States, such as

---

[127] Michael J. Davidson, (2000) "Ruck Up: An Introduction to the Legal Issues Associated with Civilian Contractors on the Battlefield", *Public Contract Law Journal*, Volume 29, Issue 2, p. 243.

[128] US Department of Defense Instruction 3020.41, *Contractor Personnel Authorized to Accompany the U.S. Armed Forces*, December 20, 2011, Enclosure 2: Procedures, 1(a)(1), <http://www.dtic.mil/whs/directives/corres/pdf/302041p.pdf>, accessed on January 5, 2014.

[129] Jean S. Pictet, (1960) *Commentary, III Geneva Convention Relative to the Treatment of Prisoners of War*, Geneva: International Committee of the Red Cross, p. 65.

[130] Giulio Bartolini, "Private Military and Security Contractors as 'Persons who Accompany the Armed Forces'", in Francesco Francioni and Natalino Ronzitti (Eds.), (2011) *War by Contract: Human Rights, Humanitarian Law, and Private Contractors*, Oxford: Oxford University Press, p. 229.

[131] Ibidem.

[132] Louise Doswald-Beck, "Private Military Companies under International Humanitarian Law", in Simon Chesterman and Chia Lehnardt (Eds.), (2007) *From Mercenaries to Market: The Rise and Regulation of Private Military Companies*, Oxford: Oxford University Press, p. 124.

[133] Emanuela-Chiara Gillard, (2006) "Business Goes to War: Private Military / Security Companies and International Humanitarian Law", *International Review of the Red Cross*, Volume 88, Issue 863, pp. 531, 538; Louise Doswald-Beck, "Private Military Companies under International Humanitarian Law", in

the US, emphasizing that civilians accompanying armed forces that take direct part in hostilities are still entitled to prisoner-of-war status upon their capture.[134] Finally, the question may arise as to whether private contractors must be physically present on the territory where the regular State forces operate, but considering the purpose of Article 4(A)(4) this seems to be unnecessary.[135]

4.1.3 Civilians

Finally, it is worth observing that if PMSC employees do not fall under the qualification of combatants or non-combatants entitled to the prisoner-of-war status in accordance with Article 43 of the First Additional Protocol and Article 4(A) of the Third Geneva Convention, it is possible for them to have the status of civilians, as defined in the Fourth Geneva Convention. This category of people is different from the group of persons accompanying the armed forces, as shown above. The Fourth Geneva Convention grants protection to civilians who do not constitute members of the armed forces and do not have a right to qualify as prisoners of war,[136] provided they do not actively engage in armed operations.[137] Indeed, if civilians take arms and participate in hostilities, they forfeit their special privileges and protections.[138] With regard to PMSCs, the Dutch Advisory Council on International Affairs shares the view that their personnel in most cases must be seen as civilians.[139] It should be

---

Simon Chesterman and Chia Lehnardt (Eds.), (2007) *From Mercenaries to Market: The Rise and Regulation of Private Military Companies*, Oxford: Oxford University Press, p. 124; Shannon Bosch, (2007) "Private Security Contractors and International Humanitarian Law – A Skirmish for Recognition in International Armed Conflicts", *African Security Review*, Volume 16, Issue 4, p. 40.

134 US Department of Defense Instruction 1100.22, Policy and Procedures for Determining Workforce Mix, April 12, 2010, Enclosure 5, para. 2(d)(5)(a), <http://www.dtic.mil/whs/directives/corres/pdf/110022p.pdf>, accessed on September 12, 2013; Emanuela-Chiara Gillard, (2006) "Business Goes to War: Private Military / Security Companies and International Humanitarian Law", *International Review of the Red Cross*, Volume 88, Issue 863, p. 538.

135 Louise Doswald-Beck, "Private Military Companies under International Humanitarian Law", in Simon Chesterman and Chia Lehnardt (Eds.), (2007) *From Mercenaries to Market: The Rise and Regulation of Private Military Companies*, Oxford: Oxford University Press, p. 124; University Centre for International Humanitarian Law, *Expert Meeting on Private Military Contractors: Status and State Responsibility for Their Actions*, Report, Geneva, August 29-30, 2005, p. 15, <http://www.geneva-academy.ch/docs/expert-meetings/2005/2rapport_compagnies_privees.pdf>, accessed on October 10, 2014.

136 Louise Doswald-Beck, "Private Military Companies under International Humanitarian Law", in Simon Chesterman and Chia Lehnardt (Eds.), (2007) *From Mercenaries to Market: The Rise and Regulation of Private Military Companies*, Oxford: Oxford University Press, p. 125.

137 Shannon Bosch, (2007) "Private Security Contractors and International Humanitarian Law – A Skirmish for Recognition in International Armed Conflicts", *African Security Review*, Volume 16, Issue 4, p. 43.

138 Rebecca R. Vernon, (2003-2004) "Battlefield Contractors: Facing the Tough Issues", *Public Contract Law Journal*, Volume 33, Issue 2, pp. 405-406.

139 Dutch Advisory Council on International Affairs, *Employing Private Military Companies: A Question of Responsibility* (Dutch: *De Inhuur van Private Militaire Bedrijven: Een Kwestie van*

emphasized that the military and security experts of PMSCs in their capacity of "protected persons" under the Fourth Geneva Convention are covered by the customary rules with respect to the prohibition of torture, inhumane and degrading treatment, hostage-taking, and the right to receive a fair trial. If PMSCs perform their military and security tasks on behalf of one State and qualify to be considered civilians, then employees of private contractors working for other States indiscriminately attacking them breach international humanitarian law. This can lead to the international responsibility of the States involved if those violations can be attributed to them in accordance with rules and principles of general public international law.

4.1.4 Mercenaries

Having addressed the legal status of employees working for PMSCs falling within the above-mentioned categories of individuals, some attention should be given to an issue often fiercely debated in the academic circles.[140] As some scholars point out, PMSC personnel are frequently referred to as or are confused with "mercenaries"[141] or "seat-of-the-pants" mercenary businesses;[142] others seem to agree that PMSCs might become "modern-day mercenaries" acting on behalf of States, which often lack supervision over the actions of contractors.[143] The ambition of PMSCs representing the industry of military and security services to become major players in the field of the provision of military and security tasks can be reflected by the words of the former president of the PMSC Blackwater Gary Jackson, who stated: "I would

---

Verantwoordelijkheid), No. 59, December 2007, The Hague, p. 16, <http://www.aiv-advies.nl/ContentSuite/upload/aiv/doc/webversie_AIV_59eng(1).pdf>, accessed on May 9, 2011.

[140] See, among other studies, the following publications addressing the definition and the status of mercenaries and their relevance for PMSCs: Katherine Fallah, (2006) "Corporate Actors: The Legal Status of Mercenaries in Armed Conflict", *International Review of the Red Cross*, Volume 88, Issue 863, pp. 599-600; Lindsey Cameron, (2006) "Private Military Companies: Their Status under International Humanitarian Law and Its Impact on Their Regulation", *International Review of the Red Cross*, Volume 88, Issue 863, p. 577-582; Elke Krahmann, (2012) "From 'Mercenaries' to 'Private Security Contractors': The (Re)Construction of Armed Security Providers in International Legal Discourses", *Millennium: Journal of International Studies*, Volume 40, Issue 2.

[141] Lindsey Cameron, (2006) "Private Military Companies: Their Status under International Humanitarian Law and Its Impact on Their Regulation", *International Review of the Red Cross*, Volume 88, Issue 863, p. 577.

[142] Nils Rosemann, (2005) "The Privatization of Human Rights Violations – Business' Impunity or Corporate Responsibility? The Case of Human Rights Abuses and Torture in Iraq", *Non-State Actors and International Law*, Volume 5, p. 82; Sarah Percy, "Morality and Regulation", in Simon Chesterman and Chia Lehnardt (Eds.), (2007) *From Mercenaries to Market: The Rise and Regulation of Private Military Companies*, Oxford: Oxford University Press, p. 11; Andrew Clapham, (2006) *Human Rights Obligations of Non-State Actors*, Oxford: Oxford University Press, pp. 299-300, 302.

[143] Deborah D. Avant, (2004) "The Privatization of Security and Change in the Control of Force", *International Studies Perspectives*, Volume 5, Issue 2, pp. 154-155.

like to have the largest, most professional private army in the world."[144] In general, rules and principles of international law do not prohibit any form of aid or assistance, such as participation in armed conflicts, provision of security services, training, and advice, by companies specialized in military and security expertise.[145] Possible problems may, however, arise if the personnel of PMSCs qualify as actual mercenaries.

For clarification of this classification issue, it needs to be examined whether the use of the term "mercenary" in relation to private contractors and their employees is actually justified. In order to analyze the legal problem and its scale, it is required to find the actual definition of the term "mercenary" frequently used in relation to PMSCs. A broad notion of "mercenary" is based on the concept of a soldier that is paid to serve in a foreign army or other military organization. In the traditional sense of the term used, mercenaries have largely been understood as persons or groups of persons performing military services in exchange for financial or other benefits.[146] In this sense, mercenaries are to be considered as individuals or groups of people who participate in military operations, while not performing their legal duty or being led by patriotic feelings, but participating merely for financial gain. Obviously, a wide range of soldiers can potentially fall under this description. Following the trend of active employment of PMSCs in areas of conflict, it is evident that in practice States might hire corporate entities in order to avoid the application of the definition of mercenary acting in violation of their treaty obligations, which might necessarily lead to their international responsibility. Therefore, careful analysis of international law must provide more clarity regarding the qualification of mercenary potentially applicable to some groups performing military services, such as PMSCs.

While being a mercenary is not prohibited by the 1949 Geneva Conventions and Additional Protocols, and the phenomenon itself is elaborately addressed by some international conventions, the establishment of the Special Rapporteur on Mercenaries by the UN Human Rights Commission in 1987 was an indication of the need to examine global trends in the development of the concept through the course of history and to crystallize rules of customary international law in that regard.[147] In

---

144 Barry Yeoman, "Need an Army? Just Pick Up the Phone", *The New York Times*, April 2, 2004, <http://www.nytimes.com/2004/04/02/opinion/02YEOM.html?ex=1081912176&ei=1&en=bce2ecd0c21b8bb4>, accessed on April 24, 2011.

145 Luisa Vierucci, "Private Military and Security Companies in Non-International Armed Conflicts: Ius ad Bellum and Ius in Bello Issues", in Francesco Francioni and Natalino Ronzitti (Eds.), (2011) *War by Contract: Human Rights, Humanitarian Law, and Private Contractors*, Oxford: Oxford University Press, p. 243; Antenor Hallo de Wolf, (2012) *Reconciling Privatization with Human Rights*, School of Human Rights Research Series, Volume 49, Cambridge, Antwerp and Portland: Intersentia, p. 480.

146 John Keegan, (1993) *A History of Warfare*, London: Hutchinson, p. 228; Michael Scheimer, (2009) "Separating Private Military Companies from Illegal Mercenaries in International Law: Proposing an International Convention for Legitimate Military and Security Support that Reflects Customary International Law", *American University International Law Review*, Volume 24, Issue 3, p. 615.

147 OHCHR, "Special Rapporteur of the Commission on Human Rights on the Use of Mercenaries as a Means of Impeding the Exercise of the Right of Peoples to Self-Determination", <http://www2.ohchr.

2005, the UN Human Rights Council instituted the UN Working Group on the Use of Mercenaries with the mandate of assessing the use of mercenaries and mercenary-related activities. This expert body examining and monitoring the functioning of PMSCs and the impact of their activities on human rights confirmed in April 2011 that mercenaries still pose a serious threat to human rights, as evidenced in the recent events in Cote d'Ivoire and Libya.[148] In years to come, experts may expect a change of perception towards the legitimacy of PMSC operations when and if a new convention regulating their activities will come into being. A new international legal instrument may also affect the question of State responsibility for the unlawful actions of PMSCs and their personnel, as Article 4 of the Draft of a Possible Convention on Private Military and Security Companies developed by the same UN Working Group already goes into some specifics of the matter.[149] At the moment, many discussions on the use of PMSCs focus on the moral implications and regulation of private contractors, while PMSCs themselves struggle to maintain a legitimate face of the industry.[150] Therefore, the fundamental question arises as to whether and to what extent PMSCs can fall under the scope of the definition of mercenary.[151]

The first attempt to address the issue of mercenarism was made by the international community in 1907.[152] In the 1907 Convention Respecting the Rights and Duties of Neutral Powers and Persons in Case of War on Land (hereinafter: the 1907 Hague Convention V),[153] it was decided to incorporate the existing customary international law into a treaty for the purpose of regulating mercenary activities. Article 4 of the 1907 Hague Convention V provided that neutral States must allow neither mercenary expeditions to be formed on their territory, nor mercenary recruiting to take place. But those States were not required to ban mercenary activities completely,

---

org/english/issues/mercenaries/specialrap.htm>, accessed on June 19, 2011.

148 OHCHR, "Mercenaries Still Pose a Serious Threat to Human Rights", April 8, 2011, <http://www.ohchr.org/EN/NewsEvents/Pages/DisplayNews.aspx?NewsID=10923&LangID=E>, accessed on June 19, 2011.

149 UN Human Rights Council, *Draft of a Possible Convention on Private Military and Security Companies (PMSCs) for Consideration and Action by the Human Rights Council*, Prepared by the Working Group on the use of mercenaries as a means of violating human rights and impeding the exercise of the right of peoples to self-determination, May 13, 2011, First Session, Geneva, May 23-27, 2011, UN Doc. A/HRC/WG.10/1/2.

150 Sarah Percy, "Morality and Regulation", in Simon Chesterman and Chia Lehnardt (Eds.), (2007) *From Mercenaries to Market: The Rise and Regulation of Private Military Companies*, Oxford: Oxford University Press.

151 Katherine Fallah, (2006) "Corporate Actors: The Legal Status of Mercenaries in Armed Conflict", *International Review of the Red Cross*, Volume 88, Issue 863, p. 602.

152 Todd S. Milliard, (2003) "Overcoming Post-Colonial Myopia. A Call to Recognize and Regulate PMCs", *Military Law Review*, Volume 176, Issue 1, p. 20.

153 The Hague Convention Respecting the Rights and Duties of Neutral Powers and Persons in Case of War on Land, October 18, 1907 ("the 1907 Hague Convention V").

as Article 6 made clear that the responsibility of a State in question is not engaged when persons cross their borders in order to offer mercenary services to belligerents.

More than 40 years later, the 1949 Geneva Conventions failed to include the notion of mercenary.[154] Almost three decades later, the situation changed in a sense after the establishment of the 1977 First Additional Protocol to the 1949 Geneva Conventions containing one of the most significant provisions regulating mercenary activities, Article 47. Aiming to regulate the phenomenon of mercenarism in the law of armed conflict, Article 47 is the core legal rule when it comes to the identification of mercenaries. Therein lie not merely a number of requirements for the qualification of individuals or group of individuals as mercenaries, but also legal consequences of such a classification.[155] It is considered to reflect customary international humanitarian law that applies in international armed conflicts, similar to the 1977 Additional Protocol itself.[156] Article 47 of the First Additional Protocol reads as follows:

> "1. A mercenary shall not have the right to be a combatant or a prisoner of war.
>
> 2. A mercenary is any person who:
>
> (a) is specially recruited locally or abroad in order to fight in an armed conflict;
> (b) does, in fact, take a direct part in the hostilities;
> (c) is motivated to take part in the hostilities essentially by the desire for private gain and, in fact, is promised, by or on behalf of a Party to the conflict, material compensation substantially in excess of that promised or paid to combatants of similar ranks and functions in the armed forces of that Party;
> (d) is neither a national of a Party to the conflict nor a resident of territory controlled by a Party to the conflict;
> (e) is not a member of the armed forces of a Party to the conflict; and
> (f) has not been sent by a State which is not a Party to the conflict on official duty as a member of its armed forces."

From reading Article 47 of the First Additional Protocol condemning mercenary activities,[157] it becomes increasingly obvious that mercenaries are deprived of a privilege

---

154 Katherine Fallah, (2006) "Corporate Actors: The Legal Status of Mercenaries in Armed Conflict", *International Review of the Red Cross*, Volume 88, Issue 863, p. 604.

155 Peter W. Singer, (2008) *Corporate Warriors: The Rise of the Privatized Military Industry*, Ithaca and London: Cornell University Press, p. 41; University Centre for International Humanitarian Law, *Expert Meeting on Private Military Contractors: Status and State Responsibility for Their Actions*, Report, Geneva, August 29-30, 2005, pp. 22-23, <http://www.geneva-academy.ch/docs/expert-meetings/2005/2rapport_compagnies_privees.pdf>, accessed on October 10, 2014.

156 Jean-Marie Henckaerts and Louise Doswald-Beck, (2005) *Customary International Humanitarian Law, Volume I: Rules*, ICRC, Cambridge: Cambridge University Press, p. 391.

157 Todd S. Milliard, (2003) "Overcoming Post-Colonial Myopia. A Call to Recognize and Regulate PMCs", *Military Law Review*, Volume 176, Issue 1, p. 41; Zoe Salzman, (2007-2008) "Private Military Contractors and a Taint of a Mercenary Reputation", *New York University Journal of International Law and Politics*, Volume 40, Issue 3.

to be lawful combatants in armed conflicts and thus to benefit from prisoner-of-war status when captured.[158] Most significantly, the definition of mercenary laid down in Article 47 narrowly defines a mercenary and includes six conditions, all of which need to be met by an individual in order to qualify as a mercenary.[159] A person must not only fall outside the framework of the official armed forces of a Party to an armed conflict and not be national of a party to an armed conflict or resident of the territory under its control, he or she must also be recruited locally or abroad in order to fight in an armed conflict and to directly take part in the hostilities motivated by private gain. In addition, an individual must not be sent by a State not party to a conflict on an official duty as a member of its military personnel. Hence, it follows from the narrow definition of mercenary including all the above-mentioned requirements of Article 47 that need to be cumulatively satisfied, that it is highly improbable that PMSCs will be able to meet all six requirements.[160]

In relation to the State practice of using PMSCs, according to some legal experts, two conditions are problematic: one with regard to the fact that mercenaries must specially be recruited to fight and that they must be neither a national of a Party to the conflict nor a resident of territory controlled by a Party to the conflict. These two requirements can be fulfilled – in conjunction with other conditions – if a contract or rules of engagement of PMSCs state that they are allowed to take part in hostilities and if members of the military and security personnel of PMSCs can indeed be seen as persons who are not nationals of a Party to a conflict or residents of a territory controlled by such a Party to a conflict. In this particular case, employees of PMSCs will qualify as mercenaries. While the consequences of falling within the scope of application of Article 47 of the First Additional Protocol seem to be quite troubling, it is not very probable that it will easily happen in practice, as all the requirements cannot usually be fulfilled by PMSCs.[161] It means that employees working for PMSCs

---

158 Adam Ebrahim, (2010) "Going to War with the Army You Can Afford: The United States, International Law and the Private Military Industry", *Boston University International Law Journal*, Volume 28, pp. 206-207.

159 James T. Gathii, (2009-2010) *Commercializing War: Private Military and Security Companies, Mercenaries and International Law*, Legal Studies Research Paper Series, No. 49, Albany Law School, p. 11.

160 Louise Doswald-Beck, "Private Military Companies under International Humanitarian Law", in Simon Chesterman and Chia Lehnardt (Eds.), (2007) *From Mercenaries to Market: The Rise and Regulation of Private Military Companies*, Oxford: Oxford University Press, p. 122; Luisa Vierucci, "Private Military and Security Companies in Non-International Armed Conflicts: Ius ad Bellum and Ius in Bello Issues", in Francesco Francioni and Natalino Ronzitti (Eds.), (2011) *War by Contract: Human Rights, Humanitarian Law, and Private Contractors*, Oxford: Oxford University Press, p. 243; Todd S. Milliard, (2003) "Overcoming Post-Colonial Myopia. A Call to Recognize and Regulate PMCs", *Military Law Review*, Volume 176, Issue 1, p. 19; Katherine Fallah, (2006) "Corporate Actors: The Legal Status of Mercenaries in Armed Conflict", *International Review of the Red Cross*, Volume 88, Issue 863, p. 610.

161 University Centre for International Humanitarian Law, *Expert Meeting on Private Military Contractors: Status and State Responsibility for Their Actions*, Report, Geneva, August 29-30, 2005, p. 23, <http://www.geneva-academy.ch/docs/expert-meetings/2005/2rapport_compagnies_privees.pdf>, accessed on

are normally not exempt from the right to be a combatant or a prisoner of war on the sole basis of Article 47 of the 1977 First Additional Protocol to the 1949 Geneva Conventions.[162] As assessed by some authors, employees of PMSCs will probably not be able to benefit from the status of combatants, while most of them must be perceived as having the status of civilians under international humanitarian law.[163]

In addition to the Geneva law, in the body of contemporary international law, the concept of mercenarism – often defined as the selling of military services[164] – is regulated by the 1989 United Nations International Convention against the Recruitment, Use, Financing and Training of Mercenaries[165] (hereinafter: the UN International Convention on Mercenaries) and the 1977 Organization of African Unity Convention for the Elimination of Mercenarism in Africa[166] (hereinafter: the OAU Convention on Mercenarism). Building upon Africa's experiences with hired-guns and their interventions in a number of decolonizing States, these Conventions attempting to regulate military activities came into being in order to eliminate the possibility of mercenary involvement in State affairs. In both international legal documents, States appear to be aware of the potential dangers of employing mercenaries. Reference is made to mercenary activities not merely as offenses of grave concern, the commission of which is to be sanctioned, but also as acts violating the most essential principles of international law: those of sovereign equality, political independence, territorial integrity of States, and self-determination of peoples.[167]

As has been established, mercenaries participate in armed conflicts and are highly likely to breach rules of international humanitarian law. Therefore, the contemporary

---

October 10, 2014; Corinna Seiberth, (2013) *Private Military and Security Companies in International Law – A Challenge for Non-binding Norms: the Montreux Document and the International Code of Conduct for Private Security Service Providers*, Cambridge, Antwerp, Portland: Intersentia, p. 65; Lindsey Cameron and Vincent Chetail, (2013) *Privatizing War: Private Military and Security Companies under Public International Law*, Cambridge: Cambridge University Press, p. 71.

162 Michael Scheimer, (2009) "Separating Private Military Companies from Illegal Mercenaries in International Law: Proposing an International Convention for Legitimate Military and Security Support that Reflects Customary International Law", *American University International Law Review*, Volume 24, Issue 3, pp. 624-629.

163 Lindsey Cameron, (2006) "Private Military Companies: Their Status under International Humanitarian Law and Its Impact on Their Regulation", *International Review of the Red Cross*, Volume 88, Issue 863, pp. 593-594.

164 Benedict Sheehy and Jackson N. Maogoto, (2008) "The Private Military Company – Unravelling the Theoretical, Legal & Regulatory Mosaic", *ILSA Journal of International & Comparative Law*, Volume 15, Issue 1, p. 152.

165 UN General Assembly, International Convention against the Recruitment, Use, Financing and Training of Mercenaries, December 4, 1989, UN Doc. A/RES/44/34 ("UN International Convention on Mercenaries").

166 Organization of African Unity (OAU), OAU Convention for the Elimination of Mercenarism in Africa, July 3, 1977, CM/817(XXIX), Annex II Rev. 1 ("OAU Convention on Mercenarism").

167 See Annex to the UN International Convention on Mercenaries; see also Preamble of the OAU Convention on Mercenarism.

law of armed conflict provides a definition of mercenary that can be found in the 1977 First Additional Protocol to the 1949 Geneva Conventions and has, in fact, largely served as the basis for the classifications adopted in the two above-mentioned Conventions and in several respects mirrors them both.[168] Both Conventions – the 1989 UN International Convention on Mercenaries and the 1977 OAU Convention on Mercenarism – provide in their Article 1 practically the same definition as that found in contemporary international humanitarian law. However, there are some notable differences with Article 47 of the 1977 First Additional Protocol. In Article 1(1), the UN International Convention on Mercenaries broadens the scope of the term "mercenary" by omitting the condition of direct participation in hostilities. In Article 1(2), it also uses a fairly broad definition and indicates that individuals hired to overthrow legitimate governments and to undermine the territorial integrity of States must also be considered as mercenaries, if they meet a set of additional requirements.[169] Furthermore, according to Article 1(1) of the OAU Convention on Mercenarism, a mercenary must only be promised by or on behalf of a Party to a conflict material compensation and this compensation does not have to be substantially in excess of what is promised or paid to combatants, as stated in Article 47 of the First Additional Protocol. The provision also lists in Paragraph 2 the aims of mercenaries to oppose with armed violence "a process of self-determination stability or the territorial integrity of another State".

It is often alleged that acceptance of the definition of mercenary and its relationship to the concept of PMSCs is far from crystallized, whilst some States may consider the legal rules assessed above as not reflecting customary international law or being in accordance with the current State practice.[170] In this regard, however, one can note that the reasons for using mercenaries are the same as those for hiring PMSCs – the employment of force in exchange for financial remuneration given the lack of appropriate military resources and technical skills on the part of PMSC clients who are interested in the expertise of private contractors.[171]

Due to the nature of the entities themselves, a number of legal scholars come to the conclusion that one of the main differences between PMSCs and mercenaries is the element of corporatization of military and security services dominant in the

---

[168] Todd S. Milliard, (2003) "Overcoming Post-Colonial Myopia. A Call to Recognize and Regulate PMCs", *Military Law Review*, Volume 176, Issue 1, pp. 52, 58.

[169] Katherine Fallah, (2006) "Corporate Actors: The Legal Status of Mercenaries in Armed Conflict", *International Review of the Red Cross*, Volume 88, Issue 863, pp. 608-609.

[170] Jean-Marie Henckaerts and Louise Doswald-Beck, (2005) *Customary International Humanitarian Law, Volume I: Rules*, ICRC, Cambridge: Cambridge University Press, p. 392.

[171] Luisa Vierucci, "Private Military and Security Companies in Non-International Armed Conflicts: Ius ad Bellum and Ius in Bello Issues", in Francesco Francioni and Natalino Ronzitti (Eds.), (2011) *War by Contract: Human Rights, Humanitarian Law, and Private Contractors*, Oxford: Oxford University Press, p. 244.

structure of the former.[172] This means that PMSCs have a solid company structure and operate as legitimate businesses established on the territory of a certain State according to its domestic legislation.[173] A sophisticated business model of PMSCs as private corporations and the manner in which they operate contribute to the creation of a specific image of those firms having arguably gained both implicit and explicit legitimacy.[174] It can also be said that mercenaries have almost exclusively been used and are used in situations violating the right to self-determination of peoples and sovereignty of States. PMSCs, however, in general never intervene on behalf of governments breaching those significant principles of international law.[175] Another striking difference between mercenaries and PMSCs frequently stressed by experts lies in the fact that the illegal activities of the latter are openly sanctioned by States that hire, license, or permit the functioning of PMSCs on their behalf in one way or another.[176] In addition, Peter W. Singer identified a number of distinguishing characteristics of PMSCs, due to their corporatization, setting them apart from common mercenaries.[177] PMSCs are thought to be organized in forms of businesses and to have a corporate structure with a clear executive element, such as boards of directors and share-holdings. In the process of carrying out a wide range of military and security tasks, PMSCs are motivated by generating business profit rather than individual profit, as is the case with mercenaries. PMSCs are firms that compete on the global market offering their services to States and non-State entities without denying their own existence, as mercenaries often do, and maintain or employ all possible means

---

[172] Fred Schreier and Marina Caparini, (2005) *Privatising Security: Law, Practice and Governance of Private Military and Security Companies*, Occasional Paper No. 6, DCAF, p. 7.

[173] Benedict Sheehy and Jackson N. Maogoto, (2008) "The Private Military Company – Unravelling the Theoretical, Legal & Regulatory Mosaic", *ILSA Journal of International & Comparative Law*, Volume 15, Issue 1, p. 149; Peter W. Singer, (2008) *Corporate Warriors: The Rise of the Privatized Military Industry*, New York: Cornell University Press, p. 40; Shantanu Chakrabarti, (2009) *Privatisation of Security in the Post Cold-War Period: An Overview of its Nature and Implications*, Institute for Defence Studies and Analyses Monograph Series, No. 2, New Delhi: Magnum Custom Publishing, p. 6; Fred Schreier and Marina Caparini, (2005) *Privatising Security: Law, Practice and Governance of Private Military and Security Companies*, Occasional Paper No. 6, DCAF, p. 7.

[174] Benedict Sheehy and Jackson N. Maogoto, (2008) "The Private Military Company – Unravelling the Theoretical, Legal & Regulatory Mosaic", *ILSA Journal of International & Comparative Law*, Volume 15, Issue 1, p. 149.

[175] Luisa Vierucci, "Private Military and Security Companies in Non-International Armed Conflicts: Ius ad Bellum and Ius in Bello Issues", in Francesco Francioni and Natalino Ronzitti (Eds.), (2011) *War by Contract: Human Rights, Humanitarian Law, and Private Contractors*, Oxford: Oxford University Press, p. 244.

[176] Chia Lehnardt, "Private Military Companies and State Responsibility", in Simon Chesterman and Chia Lehnardt (Eds.), (2007) *From Mercenaries to Market: The Rise and Regulation of Private Military Companies*, Oxford: Oxford University Press, p. 139.

[177] Peter W. Singer, (2008) *Corporate Warriors: The Rise of the Privatized Military Industry*, New York: Cornell University Press, pp. 44-47.

to build a reputation of being law-abiding entities.[178] These companies focus on the provision of a much broader spectrum of services than mercenaries to a wider array of different clients. They mainly function on the basis of a contract concluded with their clients and are usually connected to greater corporations or financial holdings. The features of PMSCs assessed above and their main differences in comparison to ordinary mercenaries make quite clear that these new actors often employed by States reach a whole new level of legitimacy on the national and international levels.[179]

One should bear in mind that it is not a violation of the 1949 Geneva Conventions or 1977 Additional Protocols to be a mercenary, while mercenarism as such does not lead to the international criminal responsibility of individuals according to international humanitarian law.[180] One of the most significant consequences of being a mercenary is the denial of lawful combatant status and the deprivation of the immunity to be regarded as a prisoner of war in international armed conflicts.[181] As the 1977 First Additional Protocol does not seek to criminalize mercenary activity, a Party to a conflict may decide to prosecute a mercenary under its domestic laws for his or her direct participation in hostilities as an unlawful combatant. Obviously, if there are national laws dealing with the crime of mercenarism this person will face trial for this specific crime. The main aim of the 1989 UN International Convention on Mercenaries and the 1977 OAU Convention on Mercenarism is, however, to criminalize possible mercenary activity and resorting to it, and to prohibit the use of mercenaries. Both mercenary-specific treaties clearly make mercenaries illegal and address different types of participation in mercenary activities, such as financing and training of mercenaries.[182] If a mercenary or a member of PMSC personnel falling under such a definition is captured by a State party to these Conventions, this person will face significant legal consequences. In addition to the deprivation of the status of lawful combatant and prisoner of war, a mercenary will also be tried before national courts

---

178 Luisa Vierucci, "Private Military and Security Companies in Non-International Armed Conflicts: Ius ad Bellum and Ius in Bello Issues", in Francesco Francioni and Natalino Ronzitti (Eds.), (2011) *War by Contract: Human Rights, Humanitarian Law, and Private Contractors*, Oxford: Oxford University Press, p. 244.

179 Peter W. Singer, (2008) *Corporate Warriors: The Rise of the Privatized Military Industry*, New York: Cornell University Press, pp. 44-47.

180 Lindsey Cameron, (2006) "Private Military Companies: Their Status under International Humanitarian Law and Its Impact on Their Regulation", *International Review of the Red Cross*, Volume 88, Issue 863, p. 577.

181 University Centre for International Humanitarian Law, *Expert Meeting on Private Military Contractors: Status and State Responsibility for Their Actions*, Report, Geneva, August 29-30, 2005, p. 27, <http://www.geneva-academy.ch/docs/expert-meetings/2005/2rapport_compagnies_privees.pdf>, accessed on October 10, 2014; Katherine Fallah, (2006) "Corporate Actors: The Legal Status of Mercenaries in Armed Conflict", *International Review of the Red Cross*, Volume 88, Issue 863, p. 606.

182 See Articles 2, 3, and 4 of the UN International Convention on Mercenaries and Article 1(2) of the OAU Convention on Mercenarism.

of those States for the crime of mercenarism and other crimes falling within their national jurisdiction, such as murder and carrying of arms.[183]

4.2 Non-international armed conflicts

The issue of the legal status of PMSC employees appears to be of major significance when PMSCs conduct military and security activities in international armed conflicts.[184] This is the case because the status of prisoner of war cannot be conferred on a person participating in non-international armed conflicts. The operation of PMSC guards in non-international armed conflicts, however, seems to pose similar legal questions as in international armed conflicts. In the view of legal scholars, most contractors are nowadays deployed in conflicts of a non-international character.[185] Thus, one should investigate what the actual status of the PMSC personnel operating in this type of conflict is, and how those individuals should be approached from the point of view of international humanitarian law. It is necessary to explore the matter, given the fact that PMSCs hired by States and deployed in an armed conflict of a non-international character, just as other parties to the conflict, must be aware of the legal position of employees of other private contractors and act accordingly. In the spirit of Common Article 3 of the Geneva Conventions and the Second Additional Protocol, those not taking an active part in the hostilities should be protected and treated humanely. The failure of PMSCs to comply with these provisions that can be attributed to the hiring States may trigger the responsibility of those States in accordance with the law of State responsibility.

4.2.1 Armed forces of the legitimate government

In Article 1 of the Second Additional Protocol concerning its material field of application, references are made to the armed conflicts between the armed forces of the State and the dissident armed forces or other organized armed groups. Thus, the

---

183 Louise Doswald-Beck, "Private Military Companies under International Humanitarian Law", in Simon Chesterman and Chia Lehnardt (Eds.), (2007) *From Mercenaries to Market: The Rise and Regulation of Private Military Companies*, Oxford: Oxford University Press, p. 123; University Centre for International Humanitarian Law, *Expert Meeting on Private Military Contractors: Status and State Responsibility for Their Actions*, Report, Geneva, August 29-30, 2005, p. 27, <http://www.geneva-academy.ch/docs/expert-meetings/2005/2rapport_compagnies_privees.pdf>, accessed on October 10, 2014.

184 Luisa Vierucci, "Private Military and Security Companies in Non-International Armed Conflicts: Ius ad Bellum and Ius in Bello Issues", in Francesco Francioni and Natalino Ronzitti (Eds.), (2011) *War by Contract: Human Rights, Humanitarian Law, and Private Contractors*, Oxford: Oxford University Press, p. 247.

185 Giulio Bartolini, "Private Military and Security Contractors as 'Persons who Accompany the Armed Forces'", in Francesco Francioni and Natalino Ronzitti (Eds.), (2011) *War by Contract: Human Rights, Humanitarian Law, and Private Contractors*, Oxford: Oxford University Press, p. 220.

first category of "fighters" in non-international armed conflicts is that of the persons belonging to the regular armed forces of the legitimate government.[186] As discussed, the question of defining PMSCs as members of regular armed forces is usually narrowed down to the assessment of the conditions laid down in Article 43 of the First Additional Protocol. The notion of the armed forces applicable to non-international armed conflicts appears to be similar to that of international armed conflicts.[187] As indicated above, the process of examining the operation of PMSCs as State organs using both the domestic law approach and the functional approach leads to the conclusion that members of private companies cannot easily be defined as members of regular armed forces and therefore cannot belong to the military apparatus of a legitimate government. If members of the PMSC personnel fall under the category of such persons, they can be directly attacked by the opposing party without being able to enjoy immunity from an attack.[188] This is a principle of customary international law applicable to non-international armed conflicts.

4.2.2 Armed opposition groups

Article 1 of the Second Additional Protocol provides a set of requirements for determining whether certain groups have the status of armed opposition groups.[189] They must be organized, act under responsible command, and exercise control over a part of a State territory that enables them to carry out sustained and concerted military operations and to implement the Second Additional Protocol. Rebel forces are highly unlikely to employ PMSC guards. In addition, violence they direct against the respective government is wrongful, as they lack the status and immunities of combatants.[190] Nevertheless, the assessment of conditions laid down in Article 1 needs to take place, taking into account the fact that PMSCs can potentially fall under the definition of organized armed groups when all the conditions listed in this provision are met.

First of all, a group must qualify as an organized group. It can be assumed that PMSCs fulfill this requirement given the high degree of their organization: within these corporations, hierarchical structures and clear divisions of roles and functions can be identified. This level of organization is sufficient to conclude that the condition is met. The PMSC units also operate under responsible command, meaning that

---

[186] Luisa Vierucci, "Private Military and Security Companies in Non-International Armed Conflicts: Ius ad Bellum and Ius in Bello Issues", in Francesco Francioni and Natalino Ronzitti (Eds.), (2011) *War by Contract: Human Rights, Humanitarian Law, and Private Contractors*, Oxford: Oxford University Press, p. 247.

[187] Ibidem, p. 249.

[188] Ibidem, p. 251.

[189] Emanuela-Chiara Gillard, (2006) "Business Goes to War: Private Military / Security Companies and International Humanitarian Law", *International Review of the Red Cross*, Volume 88, Issue 863, p. 546.

[190] Michael N. Schmitt, (2005) "Humanitarian Law and Direct Participation in Hostilities by Private Contractors or Civilian Employees", *Chicago Journal of International Law*, Volume 5, Issue 2, p. 522.

the second condition must be considered as fulfilled. In theory, private contractors are also capable of exercising such control over a part of the territory of a certain State, which enables them to conduct concerted and structured military operations and to implement the Second Additional Protocol. It seems clear that PMSCs may meet the above-mentioned requirements. It is, however, unlikely that in practice private military and security contractors will constitute organized armed groups given that they must be regarded as parties to non-international armed conflicts themselves and not merely act on behalf of such parties.[191] Indeed, legal scholars hold that there is no evidence indicating that PMSCs qualify as armed groups.[192] It has been shown that individuals working for a PMSC may contribute to the formation of an armed group when it has factual links to the parties opposing the legitimate government.[193] If PMSCs can be regarded as armed opposition groups, their personnel form a legitimate target for armed attacks. The Commentary to the Second Additional Protocol seems to leave no doubt on that point, stating that "those who belong to armed forces or armed groups may be attacked at any time".[194]

4.2.3 Civilians

Some experts maintain that PMSCs can have the status of civilians if they do not belong to the armed forces of a territorial State and if they accompany the armed forces of a State. The second scenario seems to be more problematic, as the notion of civilians accompanying armed forces stems from the field of international armed conflicts and it is unclear whether armed groups may be equated with armed forces or civilians actively taking part in the hostilities.[195] The fact remains, however, that the legal position of the most PMSC employees employed in non-international armed conflicts is that of civilians.[196] An important legal consequence of this qualification is that members of PMSC personnel are not allowed to be prisoners of war or enjoy

---

191 Emanuela-Chiara Gillard, (2006) "Business Goes to War: Private Military / Security Companies and International Humanitarian Law", *International Review of the Red Cross*, Volume 88, Issue 863, p. 546.

192 Luisa Vierucci, "Private Military and Security Companies in Non-International Armed Conflicts: Ius ad Bellum and Ius in Bello Issues", in Francesco Francioni and Natalino Ronzitti (Eds.), (2011) *War by Contract: Human Rights, Humanitarian Law, and Private Contractors*, Oxford: Oxford University Press, p. 252.

193 Ibidem.

194 Claude Pilloud and Others, (1987) *Commentary on the Additional Protocols of 8 June 1977 to the Geneva Conventions of 12 August 1949*, International Committee of the Red Cross, Geneva: Martinus Nijhoff Publishers, p. 1453, para. 4789.

195 Luisa Vierucci, "Private Military and Security Companies in Non-International Armed Conflicts: Ius ad Bellum and Ius in Bello Issues", in Francesco Francioni and Natalino Ronzitti (Eds.), (2011) *War by Contract: Human Rights, Humanitarian Law, and Private Contractors*, Oxford: Oxford University Press, p. 254.

196 Emanuela-Chiara Gillard, (2006) "Business Goes to War: Private Military / Security Companies and International Humanitarian Law", *International Review of the Red Cross*, Volume 88, Issue 863, p. 539.

immunities laid down in the Fourth Geneva Convention, but may benefit from the protection granted to them by Common Article 3 of the Geneva Conventions, the Second Additional Protocol, and the customary rules of international humanitarian law.[197] However, if captured by the enemy during participation in hostilities, PMSC members having the status of civilians will automatically lose their status and will possibly be tried under the domestic law of the State concerned. In addition, States on which behalf PMSCs operate and breach Article 3 common to the Geneva Conventions in relation to employees of adversary contractors will have to deal with the incurrence of State responsibility.

4.2.4 Mercenaries

As assessed above, Article 47 of the 1977 First Additonal Protocol is of relevance only in international armed conflicts and is not helpful in examining whether a person qualifies to be a mercenary given that it does not apply in situations of non-international armed conflicts.[198] The 1989 UN International Convention on Mercenaries and the 1977 OAU Convention on Mercenarism are, however, applicable in both international and non-international armed conflicts.[199] In the latter category of armed conflict, there is no right to be a prisoner of war and therefore mercenaries active in this type of conflict are not able to benefit from prisoner-of-war status.[200] Usually, PMSCs and the respective company personnel are not in the position to easily meet the criteria of the accepted definitions of mercenary, as negotiated and adopted in the final drafts of two above-mentioned Conventions.[201] Hence, one can state that it does not seem straightforward that the personnel of PMSCs will have the status of mercenaries while operating in the conflicts of a non-international character. Thus, a person who directly participates in the hostilities of a conflict of a non-international character and falls within the scope of the definition of mercenary is entitled to be protected by Common Article 3 of the Geneva Conventions, the Second Additional

---

[197] Ibidem, p. 541.

[198] Ibidem, p. 564.

[199] Lindsey Cameron, (2006) "Private Military Companies: Their Status under International Humanitarian Law and Its Impact on Their Regulation", *International Review of the Red Cross*, Volume 88, Issue 863, p. 578; Corinna Seiberth, (2013) *Private Military and Security Companies in International Law – A Challenge for Non-binding Norms: the Montreux Document and the International Code of Conduct for Private Security Service Providers*, Cambridge, Antwerp, Portland: Intersentia, p. 65; Lindsey Cameron and Vincent Chetail, (2013) *Privatizing War: Private Military and Security Companies under Public International Law*, Cambridge: Cambridge University Press, pp. 67-68.

[200] Jean-Marie Henckaerts and Louise Doswald-Beck, (2005) *Customary International Humanitarian Law, Volume I: Rules*, ICRC, Cambridge: Cambridge University Press, p. 395.

[201] Luisa Vierucci, "Private Military and Security Companies in Non-International Armed Conflicts: Ius ad Bellum and Ius in Bello Issues", in Francesco Francioni and Natalino Ronzitti (Eds.), (2011) *War by Contract: Human Rights, Humanitarian Law, and Private Contractors*, Oxford: Oxford University Press, p. 253.

Protocol, and the customary rules of international humanitarian law just as anyone else. He or she, however, may be prosecuted in accordance with domestic laws of the State, on which territory a mercenary operates, for fighting in a conflict and committing criminal offenses.

5. Conclusion

This Chapter sought to identify the legal status of PMSC employees. From the point of view of international humanitarian law, an attempt has been made to establish the actual legal status of the PMSC personnel deployed in conflict situations. The analysis is based on the basic distinction made in the law of armed conflict between international and non-international armed conflicts. In international armed conflicts, the importance of this assessment cannot be underestimated, for it can establish whether a person qualifies to be entitled to prisoner-of-war status or to the privileges of civilians. If a State hiring PMSCs fails to respect and to ensure respect for international humanitarian law and those employed companies violate the law of armed conflict protecting other PMSC employees, this may give rise to the international responsibility of the State concerned. Also, when States are party to the two mercenary-specific Conventions and PMSCs, especially those they hire and those that are active in their territory or under their jurisdiction, are in essense mercenaries, they might be held responsible for failing to comply with their obligations under these treaties.

Taking part in conflicts of an international character, private contractors are not likely to fall under the definition of members of the armed forces found in Article 43 of the First Additional Protocol to the 1949 Geneva Conventions. In a similar vein, it is not very useful to consider PMSC employees as "members of the armed forces of a Party to the conflict as well as members of militias or volunteer corps forming part of such armed forces" within the meaning of Article 4(A)(1) of the Third Geneva Convention. The qualification of PMSCs as "members of other militias and members of other volunteer corps, including those of organized resistance movements" according to Article 4(A)(2) of the Third Geneva Convention also does not appear to be unproblematic. The conclusion has been drawn that following the analysis made above it must be regarded as a possibility that a small number of PMSC employees can fall under the scope of the provision. It has also been stressed in the subsequent Sections that one of the most significant positions occupied by military and security experts working for PMSCs in conflict areas may be that of persons accompanying the armed forces, as mentioned in Article 4(A)(4) of the Third Geneva Convention. As noted, these individuals often working closely with State troops can fall under the scope of application of Article 4(A)(4), if they provide their services directly to the armed forces. The analysis has shown that employees of PMSCs may be classified as civilians protected by the Fourth Geneva Convention. Finally, it has been explored whether PMSC employees could constitute mercenaries, and established that in practice they cannot easily meet the set of requirements for being mercenaries under international humanitarian law.

In non-international armed conflicts, it is of crucial importance to observe the legal status of PMSCs present on the battlefield, as breaches of international humanitarian law applicable to this type of conflict by the members of private contractors in relation to other PMSCs can result in the responsibility of States. A case could be made for concluding that, during hostilities in non-international armed conflicts, it remains arguable whether PMSC employees meet the criteria for the definition of persons belonging to the armed forces of the legitimate government, an armed opposition group, or mercenaries. But they are highly likely to qualify as members of the civilian personnel covered by Common Article 3 of the Geneva Conventions, the Second Additional Protocol, and the customary rules of international humanitarian law. In this respect, the degree of their participation in hostilities during an armed conflict will determine whether they will keep their civilian status or will be considered unlawful combatants.

CHAPTER V
STATE RESPONSIBILITY UNDER THE DRAFT ARTICLES ON STATE RESPONSIBILITY

You cannot escape the responsibility of tomorrow by evading it today.

Abraham Lincoln[1]

1. INTRODUCTION

As was explored in the previous Chapters, nowadays the phenomenon of privatization of war and security forms the core subject of numerous debates in academic circles, while the whole industry of PMSCs shows clear signs of extensive growth and expansion.[2] Various States and even non-State entities, such as international organizations and corporations, more frequently rely on a wide array of military and security services provided by private companies, challenging international and transnational legal frameworks.[3] It has been established that these companies often seem to operate in practice and *de jure* in a "legal vacuum". Their employees potentially can and on many occasions do engage in the commission of international crimes and breach international humanitarian and human rights law without being prosecuted, while hiring States can avoid any form of responsibility by pointing out the absence of formal connections to the actual perpetrators of the crimes. Also, home and host States other than those employing PMSCs are not held responsible and remain at a considerable distance from what happens in the reality of deployment of private contractors by being physically removed from the actual situation on the ground or being unable or unwilling to get involved. Using this strategy of hiring private contractors in conflict-affected zones, States can escape international responsibility not only

1 US Department of State, Bureau of International Information Programs, Abraham Lincoln: A Legacy of Freedom, p. 61, <http://www.america.gov/media/pdf/books/lincoln.pdf>, accessed on March 5, 2011.

2 Jackson N. Maogoto and Benedict Sheehy, (2009) "Private Military Companies & International Law: Building New Ladders of Legal Accountability & Responsibility", *Cardozo Journal of Conflict Resolution*, Volume 11, pp. 99-102.

3 Laura A. Dickinson, "Contract as a Tool for Regulating Private Military Companies", in Simon Chesterman and Chia Lehnardt (Eds.), (2007) *From Mercenaries to Market: The Rise and Regulation of Private Military Companies*, Oxford: Oxford University Press, p. 217.

by successfully bypassing international humanitarian law, but also breaching human rights and fundamental freedoms without consequences.[4]

While there is still a spectrum of unsolved issues and uncertainties to be discovered regarding the responsibility of States for such acts and omissions, the body of contemporary international law is not irrelevant and can be used to incur international responsibility. As was pointed out above, domestic legal orders and transnational private regulatory frameworks are not able to coherently address instances of misconduct of PMSCs and their employees violating international law. Currently, the focus needs to shift to the responsibilities of States and the role of hiring, host, and home States as entities capable of guaranteeing respect for human rights and international humanitarian law. Although there is no extensive framework on the international responsibility of non-State actors, such as PMSCs and terrorists, customary rules and principles on the responsibility of States have been crystallized into a system that is clearer and more coherent due to the efforts of a number of International Law Commission's Special Rapporteurs.[5] Within the scope of the current study, it is most important to observe – from the perspective of international law – the possible legal consequences of the State practice of employing PMSCs and to examine the issue of international responsibility of hiring States for the unlawful conduct of private contractors violating international law. In addition, the responsibility of other States – host States and home States – that might be incurred for the unlawful conduct of PMSCs and their employees will be subjected to careful examination. These States can also be involved in the operations of PMSCs and must bear respective legal consequences. Hence, Chapter V will deal with the concept of State responsibility.

An array of considerations made in this Chapter will include discussion of the main historical developments of the doctrine of State responsibility and its legal framework. Furthermore, this part of the study will elaborate on the legal status of the Draft Articles on Responsibility of States for Internationally Wrongful Acts containing rules and principles with regard to the international responsibility of States, as laid down by the ILC. In addition, Chapter V will cover conditions for invoking the doctrine of State responsibility and possible legal consequences of its application. The most substantive part of Chapter V, addressing the main subject of the work, will concern the issue of attribution of internationally wrongful acts examined not merely from the general perspective of international law, but also from the specific point of view focusing on the State practice of employing PMSCs in conflict areas. It will deal with the provisions of the Draft Articles of relevance for different types

---

4 "US Uses Contractors to Bypass International Humanitarian Laws", *RT.com*, August 28, 2012, <http://rt.com/politics/us-contractors-international-blackwater-754/>, accessed on August 28, 2012; Human Rights Watch, *The Road to Abu Ghraib*, Report, June 9, 2004, <http://www.hrw.org/reports/2004/06/08/road-abu-ghraib>, accessed on August 28, 2012.

5 Robert p. Barnidge, Jr., (2008) *Non-State Actors and Terrorism: Applying the Law of State Responsibility and the Due Diligence Principle*, The Hague: T.M.C. Asser Press, p. 55.

of attribution, which are essential for attaching international responsibility to States. The main aim of this part of Chapter V will be to address possible circumstances and extent of attribution of unlawful conduct of PMSCs to States. Furthermore, it will also examine instances of misconduct committed by PMSCs and their personnel and establish whether they can be considered breaches of international law. Finally, the Chapter will conclude with the main findings and views on the possibility of holding States responsible for the violations of international law committed by PMSCs and their personnel from the angle of the DARS.

2. History, Framework, and Legal Status of the Draft Articles on Responsibility of States for Internationally Wrongful Acts

2.1 Historical developments

On November 21, 1947, the UN General Assembly adopted Resolution 174(II), setting up the International Law Commission and approving its Statute.[6] The Commission was entrusted with the task of "encouraging the progressive development of international law and its codification" in accordance with the Charter mandate.[7] After the first elections of November 3, 1948 and the first session of April 12, 1949, State responsibility was agreed upon as one of the topics to be discussed and codified without placing it on the list with high priority items.[8] At that particular moment, the importance of this specific subject could not be underestimated due to the fact that it is one of the major areas of international law which had already been chosen to be codified under the League of Nations during one of its conferences held in 1930.[9]

The process of drafting provisions on the issue of State responsibility took place under the chairmanship of a number of Special Rapporteurs. It commenced in 1956 under F.V. García Amador as Special Rapporteur from Cuba. When he left in 1961, Roberto Ago from Italy started to chair an inter-sessional subcommittee in 1962. In

---

6 Statute of the International Law Commission as adopted by the UN General Assembly in Resolution 174(II) of November 21, 1947, *Establishment of an International Law Commission*, UN Doc. A/RES/174(II) and as amended by Resolutions 485 (V) of December 12, 1950; 984 (X) of December 3, 1955; 985 (X) of December 3, 1955; and 36/39 of November 18, 1981.

7 James Crawford, (2002) *The International Law Commission's Articles on State Responsibility: Introduction, Text and Commentaries*, Cambridge: Cambridge University Press, p. 1.

8 International Law Commission, *Report of the International Law Commission on the Work of Its First Session, 12 April 1949*, Official Records of the General Assembly, Fourth Session, Supplement No. 10, (1949) Yearbook of the International Law Commission, Volume 1, UN Doc. A/CN.4/13 and Corr.1-3, p. 281; James Crawford, (2002) *The International Law Commission's Articles on State Responsibility: Introduction, Text and Commentaries*, Cambridge: Cambridge University Press, p. 1; International Law Commission, *State Responsibility*, <http://untreaty.un.org/ilc/summaries/9_6.htm>, accessed on March 1, 2011.

9 James Crawford, (2002) *The International Law Commission's Articles on State Responsibility: Introduction, Text and Commentaries*, Cambridge: Cambridge University Press, p. 1.

1963, Roberto Ago was appointed as Special Rapporteur of the International Law Commission and started producing a substantive number of reports on the concept of international responsibility. In 1979, a new Special Rapporteur was appointed: Willem Riphagen from The Netherlands. Replacing Riphagen, Gaetano Arangio-Ruiz from Italy was appointed to become Special Rapporteur in 1987. In 1996, the ILC adopted the complete set of Draft Articles on its first reading.[10] Finally, James Crawford from Australia served as Special Rapporteur from 1997 until 2001. On May 31, 2001, the International Law Commission adopted the Draft Articles on Responsibility of States for Internationally Wrongful Acts and the Commentary on the second reading during its fifty-third session.[11] At that session, two open-ended Working Groups were established to deal with remaining issues with regard to the rules on State responsibility and to examine Commentaries to the Draft Articles. Officially, the Draft Articles on Responsibility of States for Internationally Wrongful Acts consisting of 59 articles and Commentaries were finally adopted by the Commission on August 9, 2001. The ILC recommended the General Assembly to take note of the Articles and on December 12, 2001 the General Assembly referred to them in its Resolution 56/83 with the actual text of the DARS annexed to the Resolution.[12]

2.2 The Draft Articles on Responsibility of States for Internationally Wrongful Acts

The doctrine of State responsibility addressed in the Draft Articles prepared by the International Law Commission constitutes one of the fundamental principles of international law. Malcolm Shaw correctly points out that it arises out of "the nature of international legal system and the doctrines of State sovereignty and equality of States".[13] Ian Brownlie adds to the statement made by Shaw that the doctrine is, in fact, "a motor of foundation subject" having a "quasi-constitutional role" and firmly emphasizes that it "provides the foundation of the law of treaties and constitutes the

---

[10] International Law Commission, Draft Articles on State Responsibility with Commentaries Thereto Adopted by the International Law Commission on First Reading, as adopted in 1996, January 1997, <http://legal.un.org/ilc/texts/instruments/english/commentaries/9_6_1996.pdf>, accessed on May 8, 2014 (hereinafter: Commentary to DARS as adopted in 1996).

[11] Draft Articles on Responsibility of States for Internationally Wrongful Acts with Commentaries, in Report of the International Law Commission on the Work of Its Fifty-Third Session, UN General Assembly Official Records, Fifty-Sixth Session, Supplement No. 10, UN Doc. A/56/10 (2001); reprinted in *Yearbook of the International Law Commission*, 2001, Volume II, Part 2, UN Doc. A/CN.4/SER.A/2001/Add.1 (Part 2); also reprinted in James Crawford (Ed.), (2002) *The International Law Commission's Articles on State Responsibility: Introduction, Text and Commentaries*, Cambridge: Cambridge University Press; International Law Commission, *State Responsibility*, <http://untreaty.un.org/ilc/summaries/9_6.htm>, accessed on March 2, 2011.

[12] UN General Assembly, Resolution 56/83 of January 28, 2002, UN Doc. A/56/83.

[13] Malcolm N. Shaw, (2008) *International Law*, Sixth Edition, Cambridge: Cambridge University Press, p. 778.

most basic part of general international law".[14] This significant principle is based on the idea that a State is to be held responsible for an internationally wrongful act, which leads to a number of crucial legal consequences, such as an obligation on the responsible State to make reparation.

The rules on State responsibility concern "secondary rules" and not "primary rules" – rights and obligations – of international law. As reflected in the Draft Articles, the doctrine of State responsibility itself contains procedural and other important consequences, emerging as a result of a violation of specific substantive rules of international law by a State.[15] In the view of Roberto Ago:

> "... the rules relating to State responsibility are complementary to other substantive rules of international law – to those giving rise to legal obligations which States may be led to violate."[16]

Under Special Rapporteur Roberto Ago, the scope of the study was widened in order to include "all the forms of new legal relationship which may be established in international law by a State's wrongful act".[17] This decision of the ILC had one major consequence for the development of the doctrine of State responsibility and its codification. Namely, it created a distinction between the legal rules regarding the actual content of the substantive obligations of States under international law, the primary norms, and the law addressing possible consequences of the situations when those primary rules are breached by international legal persons, the secondary norms.[18] The Draft Articles constitute, in fact, a set of secondary rules aimed at governing potential violations of State obligations under international law and their possible legal consequences.

The structure of the ILC's Draft Articles and their most significant provisions are crucial for the examination of the question of international responsibility of States.

---

[14] Ian Brownlie, "State Responsibility and International Court of Justice", in Malgosia Fitzmaurice and Dan Sarooshi (Eds.), (2004) *Issues of State Responsibility Before International Judicial Institutions*, Oxford and Portland: Hart Publishing, p. 12.

[15] Malcolm N. Shaw, (2008) *International Law*, Sixth Edition, Cambridge: Cambridge University Press, pp. 778-779; Robert Rosenstock, (2002) "The ILC and State Responsibility", *The American Journal of International Law*, Volume 96, Issue 4, p. 792.

[16] International Law Commission, *Third Report on State Responsibility, by Mr. Roberto Ago, Special Rapporteur, the Internationally Wrongful Act of the State, Source of International Responsibility*, (1971) Yearbook of the International Law Commission, Volume II(1), UN Doc. A/CN.4/264 and Add.1-3, p. 219.

[17] Ibidem, pp. 210-211; Alain Pellet, "The ILC's Articles on State Responsibility for Internationally Wrongful Acts and Related Texts", in James Crawford, Alain Pellet, and Simon Olleson (Eds.), (2010) *The Law of International Responsibility*, Oxford: Oxford University Press, p. 76.

[18] Antonio Cassese, (2005) *International Law*, Second Edition, Oxford: Oxford University Press, p. 244; Gilbert Guillaume, "Overview of Part One of the Articles on State Responsibility", in James Crawford, Alain Pellet, and Simon Olleson (Eds.), (2010) *The Law of International Responsibility*, Oxford: Oxford University Press, p. 187.

When looking into the possibility of holding States responsible, one should be aware of a particular order of steps to be taken in order to attach a variety of instances of unlawful conduct to a State. In this respect, relevant parts of the Draft Articles relate to crucial concepts, such as internationally wrongful acts, international responsibility, its implementation, and legal consequences of internationally wrongful acts, and other issues that are required for determining State responsibility.

Part One entitled "the International Wrongful Act of a State" includes Chapters on "General Principles" (Chapter I), "Attribution of Conduct to a State" (Chapter II), "Breach of an International Obligation" (Chapter III), "Responsibility of a State in Connection with the Act of Another State" (Chapter IV), and "Circumstances Precluding Wrongfulness" (Chapter V). Article 1 of the Draft Articles, constituting, in fact, the core provision of the whole text, reads as follows:

> "Every internationally wrongful act of a State entails the international responsibility of that State."

International judicial bodies, such as the Permanent Court of International Justice and the International Court of Justice, have applied this principle on various occasions.[19] In the *Phosphates in Morocco* case, the PCIJ made a clear statement that when a State commits an internationally wrongful act against another State the international responsibility must be considered as being established "immediately as between the two States".[20] The Commentaries to the Draft Articles elaborating on the "basic principle underlying the articles as a whole"[21] accentuate that an internationally wrongful act of a State may consist of one or more actions, omissions, or combination of both. In this respect, the concept of an internationally wrongful conduct concerns all possible types of State conduct, including actions and omissions or failures to act.[22] In fact, States are to be perceived as entities operating on the international level and to

---

[19] See, for instance, Permanent Court of International Justice, *Phosphates in Morocco case (Italy v. France)*, Preliminary Objections, Judgment, June 14, 1938, PCIJ Series A/B, No. 74; International Court of Justice, *Gabčíkovo-Nagymaros Project (Hungary v. Slovakia)*, Judgment, September 25, 1997, ICJ Reports 1997; France-New Zealand Arbitration Tribunal, *Rainbow Warrior (New Zealand v. France)*, Decision, April 30, 1990, 20 RIAA 215; International Court of Justice, *Corfu Channel (United Kingdom v. Albania)*, Judgment (Merits), April 9, 1949, ICJ Reports 1949; International Court of Justice, *Case Concerning Military and Paramilitary Activities in and against Nicaragua* (Nicaragua v. United States of America), Judgment (Merits), June 27, 1986, ICJ Reports 1986.

[20] Permanent Court of International Justice, *Phosphates in Morocco case (Italy v. France)*, Preliminary Objections, Judgment, June 14, 1938, PCIJ Series A/B, No. 74, p. 28; Commentary to DARS as adopted in 2001, p. 32, Article 1, para. 2.

[21] Commentary to DARS as adopted in 2001, p. 32, Article 1, para. 1.

[22] See Article 2 DARS; Commentary to DARS as adopted in 2001, Article 1, para. 8; Commentary to DARS as adopted in 1996, Article 1, para. 14; James Crawford, (2013) *State Responsibility: The General Part*, Cambridge Studies in International and Comparative Law, No. 100, Cambridge: Cambridge University Press, p. 49.

be regarded as primary subjects of international law. As such, they are in a position to breach and often do breach numerous obligations arising out of the body of contemporary international law and resting upon them. Those obligations can be formulated, for instance, in human rights treaties, violations of which might result in application of the doctrine of State responsibility, as form of responsibility to be found on the international plane,[23] and lead to respective legal consequences, which will be discussed below. Thus, the content of international obligations of a specific State is decisive for determining what is to be seen as its breach of international law.[24] In this regard, one could state that there is no division in types of liability, such as those to be found in various national legal systems that are familiar with the concepts of crime, delict, tort, and contract.[25] In 1990, the Arbitral Tribunal indicated in the *Rainbow Warrior* arbitration case that:

> "... the general principles of International Law concerning State responsibility are equally applicable in the case of breach of treaty obligation, since in the international law field there is no distinction between contractual and tortuous responsibility, so that any violation of a State of any obligation, of whatever origin gives rise to State responsibility and consequently, to the duty of reparation."[26]

Article 2 DARS, which will be examined in more detail in this Chapter, addresses elements of an internationally wrongful act of a State. An internationally wrongful act is considered to have been committed if the conduct of a particular State is attributable to the State in question under international law and constitutes a breach of an international obligation of that State. The issue of attribution, as the core concept under discussion in the current study, is more thoroughly addressed in Chapter II of the Draft Articles and will be elaborated on below.

What can further be of relevance for the practice of using PMSCs by States and the assessment of international responsibility for their misconduct, are the provisions of the last two Chapters of Part One of the Draft Articles: Chapter IV and Chapter V. They relate respectively to responsibility of States in connection with the conduct of other States and circumstances precluding wrongfulness. Considering the scope of the current work, the provisions of those two Chapters do not require extensive discussion, but are to be taken into account for the complete overview of relevant legal rules concerning international responsibility for the illegal activities of PMSCs and will briefly be covered.

---

23 James Crawford, "The System of International Responsibility", in James Crawford, Alain Pellet, and Simon Olleson (Eds.), (2010) *The Law of International Responsibility*, Oxford: Oxford University Press, p. 17.

24 Ibidem, p. 20.

25 Ibidem, p. 21.

26 France-New Zealand Arbitration Tribunal, *Rainbow Warrior (New Zealand v. France)*, Decision, April 30, 1990, 20 RIAA 215, para. 75.

2.3 Legal status of the Draft Articles

The ILC's Draft Articles do not constitute a treaty, being an agreement under international law concluded between sovereign States. In fact, the Draft Articles still have the form of articles drafted and adopted by the International Law Commission and referred to in the Resolution 56/83 of the UN General Assembly. In this respect, it is worth emphasizing that in many instances – such as those relating to attribution, continuing wrongful acts, and reparation – they are regarded by various commentators and courts as a codification of rules and principles of customary international law on State responsibility.[27] In its case law, the ICJ, for example, looked into the substance of attaching genocidal acts to a State and indicated:

> "This is the state of customary international law, as reflected in the ILC Articles on State Responsibility."[28]

The ILC's Draft Articles are often considered as an international legal instrument progressively developing international law and constituting an authoritative text that has frequently been cited and used as a source on issues regarding State responsibility by international courts, tribunals, and other international judicial bodies.[29] Human rights bodies more often rely on the Draft Articles to apply the rules on admissibility under human rights conventions.[30] It might be assumed that the Draft Articles and Commentaries will also be used in the same vein by various international judicial bodies in the years to come.

---

[27] James Crawford and Simon Olleson, (2005) "The Continuing Debate on a UN Convention on State Responsibility", *International and Comparative Law Quarterly*, Volume 54, Issue 4, p. 968; Kaj Hobér, "State Responsibility and Attribution", in Peter Muchlinski, Federico Ortino, and Christoph Schreuer (Eds.), (2008) *The Oxford Handbook of International Investment Law*, Oxford: Oxford University Press, p. 553; Martin Scheinin, "Just Another Word? Jurisdiction in the Roadmaps of State Responsibility", in Malcolm Langford, Wouter Vandenhole, Martin Scheinin, and Willem van Genugten (Eds.), (2013) *Global Justice, State Duties: The Extraterritorial Scope of Economic, Social and Cultural Rights in International Law*, Cambridge: Cambridge University Press, p. 215.

[28] International Court of Justice, *Case Concerning the Application of the Convention on the Prevention and Punishment of the Crime of Genocide (Bosnia and Herzegovina v. Serbia and Montenegro)*, Judgment (Merits), February 26, 2007, ICJ Reports 2007, para. 401.

[29] James Crawford, (2002) "The ILC's Articles on Responsibility of States for Internationally Wrongful Acts: A Retrospect", *The American Journal of International Law*, Volume 96, Issue 4; Antonio Cassese, (2005) *International Law*, Second Edition, Oxford: Oxford University Press, p. 244; James Crawford and Simon Olleson, (2005) "The Continuing Debate on a UN Convention on State Responsibility", *International and Comparative Law Quarterly*, Volume 54, Issue 4, pp. 966-967.

[30] Martin Scheinin, "Just Another Word? Jurisdiction in the Roadmaps of State Responsibility", in Malcolm Langford, Wouter Vandenhole, Martin Scheinin, and Willem van Genugten (Eds.), (2013) *Global Justice, State Duties: The Extraterritorial Scope of Economic, Social and Cultural Rights in International Law*, Cambridge: Cambridge University Press, p. 215.

While many voices in debates were raised supporting the idea of establishing a convention on the basis of the Draft Articles, the decision of the ILC to recommend the General Assembly to merely make note of the Draft Articles and not to set in motion a law-making process on the diplomatic level is perhaps a very rational one.[31] It is a right step into the direction of making the Draft Articles an influential text, which should be observed by States and other legal persons on the international plane. As David Caron rightly asserts, the success of codification efforts is dependent upon three major factors. In the first place, there needs to be a substantial degree of consensus among States in order to codify a certain set of rules and principles of international law. In addition, the subject that is being codified must be sufficiently stable in order to prevent the codification from quickly becoming outdated and in need of adjustment. Finally, long-term utility can be achieved and is sustainable if there are mechanisms in the society making the adjustment and adaptation of the codification possible.[32] Carefully examining and applying these three factors, the author points out that there was no general consensus among members of the ILC on some core aspects of the whole State responsibility doctrine. This fact became obvious in the very beginning of the Commission's functioning when responsibility for injuries to foreigners was under discussion. Moreover, it might have been clear that the complicated notion of State responsibility – with all its various nuances and implications – must be regarded as capable of some kind of evolution, development, and changes that could occur in the near or distant future. For instance, the concept of governmental authority mentioned in the Draft Articles is left undefined, given the possibility of interpreting it in various manners depending on the legal background of potential interpreters and in accordance with political, cultural, and other developments of modern society. Therefore, it is obviously not an insightful proposition to codify rules on State responsibility and lay them down as a treaty: an authoritative statement of the law providing clarity and guidance in application of the concept of State responsibility that is able to evolve and to change with the years appears to be a more suitable and desirable option.[33]

Without doubt, the Draft Articles form one of the unique initiatives undertaken by the international community. David Caron carefully states "the paradox that they could have more influence as an ILC text than as a multilateral treaty"[34] suggesting that "the ILC's work on state responsibility will best serve the needs of the

31 James Crawford and Simon Olleson, (2005) "The Continuing Debate on a UN Convention on State Responsibility", *International and Comparative Law Quarterly*, Volume 54, Issue 4, pp. 959, 971.

32 David D. Caron, (2002) "The ILC Articles on State Responsibility: The Paradoxical Relationship Between Form and Authority", *The American Journal of International Law*, Volume 96, pp. 860-861.

33 Ibidem, p. 861; James Crawford, (2002) "The ILC's Articles on Responsibility of States for Internationally Wrongful Acts: A Retrospect", *The American Journal of International Law*, Volume 96, Issue 4.

34 David D. Caron, (2002) "The ILC Articles on State Responsibility: The Paradoxical Relationship Between Form and Authority", *The American Journal of International Law*, Volume 96, p. 857.

international community only if it is weighed, interpreted, and applied with much care".[35] Although many commentators from the academic world approved the Draft Articles and endorsed them, the fact remained that a large number of States were reluctant to recognize the text as an influential and binding legal document and expressed hostile views.[36] The last Special Rapporteur, James Crawford, highlighted in his reports many problems associated with the Draft Articles and did not go as far as to send the draft to a diplomatic conference with an aim to conclude a treaty. The ILC expressed its view on the Draft Articles and made a recommendation to the United Nations General Assembly to take note of the Articles on Responsibility of States for Internationally Wrongful Acts. The General Assembly took note of the Articles solely recommending them to the attention of States. Therefore, these rules and principles carefully drafted by the ILC became, instead of a treaty binding upon its signatories merely an instrument of soft law that forms an essential part of the process of progressive development of international law and one of its important features.[37] But this legal instrument still contains rules and principles of customary international law to be complied with by all States.

3. Conditions for Engaging State Responsibility

The question of State responsibility can appear to be rather simple on the first sight. As a principle, it might imply that if a State engages in a violation of its international obligation, it will face international responsibility for the breach in question. Or as Tim Hillier put it: "a corollary of binding legal obligations is legal responsibility for a breach of those obligations".[38] As noted by Nicola Jägers, this issue is, however, quite complicated and therefore requires close examination.[39] The first highly significant rule supported by the practice[40] is to be found in Article 1 of the ILC's Draft Articles. It clearly provides that every internationally wrongful act of a certain State implies the international responsibility of that particular State for that specific conduct. The concept of an internationally wrongful act committed by a State in its capacity as a primary subject of international law is "at the heart of the state responsibility

35 Ibidem, p. 873.

36 Gilbert Guillaume, "Overview of Part One of the Articles on State Responsibility", in James Crawford, Alain Pellet, and Simon Olleson (Eds.), (2010) *The Law of International Responsibility*, Oxford: Oxford University Press, p. 188.

37 Ibidem.

38 Tim Hillier, (1999) *Principles of Public International Law*, Second Edition, London and Sydney: Cavendish Publishing Limited, p. 161.

39 Nicola Jägers, (2002) *Corporate Human Rights Obligations. In Search of Accountability*, School of Human Rights Research Series, Volume 17, Antwerp: Intersentia, p. 138.

40 Malcolm N. Shaw, (2008) *International Law*, Sixth Edition, Cambridge: Cambridge University Press, p. 781.

regime"[41] and its two constituent elements have been laid down in Article 2 DARS stipulating the following:

> "There is an internationally wrongful act of a State when conduct consisting of an action or mission:
>
> (a) is attributable to the State under international law; and
> (b) constitutes a breach of an international obligation of the State."

The two above-mentioned conditions – conduct that is, firstly, attributable and, secondly, forms a breach of an international obligation of a State – need to be met in order to establish whether a State has committed an internationally wrongful act potentially leading to the international responsibility of that State. In the first place, one must be able to conclude that the behavior in question can be attributed to a State: this determination process is covered by Articles 4-11 DARS in Chapter II of Part One. Secondly, that conduct in question must be regarded as a breach of an international obligation of a State, as emphasized in Articles 12-15 DARS in Chapter III of Part One. While the first step of establishing an internationally wrongful act relating to the concept of attribution is often identified as a subjective element, the second element of the breach of an obligation is frequently referred to as an objective element.[42] In other words, there must be an attributable conduct that qualifies as wrongful in order for that conduct to be considered an internationally wrongful act.[43]

To some, the order of the two elements can appear to be illogical: one might assume that in the first place a violation of international law must be identified and that the second step needs to be the attribution of that unlawful conduct to the State in question. However, the ILC clearly decided to opt for the particular sequence that is to be found in Article 2 DARS. In fact, the reasoning of the Commission was perfectly in line with the case law of international judicial institutions, such as the International Court of Justice. The World Court stated in its famous *Tehran Hostages* case that it is necessary to assess both elements in that specific order:

> "First, it must determine how far, legally, the acts in question may be regarded as *imputable* to the Iranian State. Secondly, it must determine *the compatibility* or *incompatibility* with the obligations of Iran under treaties in force or any other rules of international law that may be applicable."[44] (emphasis added)

---

41 James Crawford, (2013) *State Responsibility: The General Part*, Cambridge Studies in International and Comparative Law, No. 100, Cambridge: Cambridge University Press, p. 215.

42 Brigitte Stern, "The Elements of An Internationally Wrongful Act", in James Crawford, Alain Pellet, and Simon Olleson (Eds.), (2010) *The Law of International Responsibility*, Oxford: Oxford University Press, p. 202.

43 Ibidem, p. 201.

44 International Court of Justice, *United States Diplomatic and Consular Staff in Tehran (United States of America v. Iran)*, Judgment, May 24, 1980, ICJ Reports 1980, p. 29.

It seems apparent that the issue of attribution must be addressed in order to examine whether certain conduct qualifies to be imputed to a State and that the next step needs to be that of considering whether that conduct was in accordance with rules and principles of international law that are expected to be complied with by States. However, some commentators, such as Pierre-Marie Dupuy, clearly indicate that the order of two elements, as laid down in Article 2 DARS, must be changed, for it is not the correct approach.[45] In the first instance, an examination must focus on the question of whether a breach of international law took place and then on the assessment whether it is attributable to the State in question. Interestingly enough, the International Court of Justice followed the reversed order of the two essential elements in one of its relatively recent cases, the *Bosnian Genocide* case. As Brigitte Stern emphasizes, it is understandable considering the fact that this approach was justified by the necessity of documenting and recording for history all atrocities that were committed in Bosnia.[46]

The crucial element of an internationally wrongful act, attribution, will be investigated in more detail below. We should first concentrate on the breach of an international obligation. Importantly, what is to be perceived as such is specified in Chapter III DARS. In this respect, Article 12 DARS provides in general terms the following:

> "There is a breach of an international obligation by a State when an act of that State is not in conformity with what is required of it by that obligation, regardless of its origin or character."

The Commentaries to the Draft Articles further make clear that international obligations can be established by a rule of customary international law, by a treaty, or by a general principle of international law that exists within the international legal order. In general, it is not significant what the origin of those obligations is, for there could be numerous sources of international obligations associated with various processes of establishing legal obligations in international law.[47] However, an act of a State cannot constitute a breach of an international obligation if the State is not bound by that obligation at the time of the occurrence of the act in question, as emphasized in Article 13 DARS.

In situations of conflict, more specifically armed conflicts and peace operations, both the law of armed conflict and human rights law apply. In order to establish a violation of an international obligation, a State must be bound by these parts of international law. As pointed out above, international humanitarian law, *lex specialis*, is

---

[45] Brigitte Stern, "The Elements of An Internationally Wrongful Act", in James Crawford, Alain Pellet, and Simon Olleson (Eds.), (2010) *The Law of International Responsibility*, Oxford: Oxford University Press, p. 202.

[46] Ibidem.

[47] Commentary to DARS as adopted in 2001, p. 55, Article 12, para. 3.

applicable in all situations of armed conflict, while the body of human rights law, *lex generalis*, in theory applies in both conflicts and times of peace. During armed conflicts, however, the scope of application of international human rights law might be significantly limited depending on the circumstances of a particular case, as observed in Chapter III of the present study. According to the DARS, its provisions are not applicable "where and to the extent that the conditions for the existence of an internationally wrongful act or the content or implementation of the international responsibility of a State are governed by special rules of international law". The issue of State responsibility may be tackled by the ICJ on the basis of the DARS. While its jurisprudence with regard to the law of armed conflict is rather scarce, the World Court has dealt with the responsibilities of States by using the framework of the law on State responsibility, as laid down in the ILC's Draft Articles.[48] However, the question can be raised as to whether the law of State responsibility can be applied in relation to international human rights law. Although Andrew Clapham argues with respect to the ECHR that the framework of public international law on international responsibility is not sufficient for the European human rights law system,[49] other scholars, including Nicola Jägers, take an opposing view.[50] The practice of human rights bodies does indeed seem to indicate that they refer to the law of State responsibility as codified by the ILC and therefore one needs to acknowledge that rules and principles of customary international law concerning State responsibility also have their relevance for the body of human rights law.

When the conduct of a State can qualify as an internationally wrongful act in accordance with international law and not the internal law of that State, as pointed out in Article 3 DARS, the international responsibility of that State will be invoked.

---

48 Vincent Chetail, (2003) "The Contribution of the International Court of Justice to International Humanitarian Law", *International Review of the Red Cross*, Volume 85, Issue 850, p. 267.

49 Andrew Clapham, "The 'Drittwirkung' of the Convention", in Ronald St. J. Macdonald, Franz Matscher and Herbert Petzold (Eds.), (1993) *The European System for the Protection of Human Rights*, Dordrecht, Boston, London: Martinus Nijhoff Publishers, p. 170; Danwood Mzikenge Chirwa, (2004) "The Doctrine of State Responsibility as a Potential Means of Holding Private Actors Accountable for Human Rights", *Melbourne Journal of International Law*, Volume 5, Issue 1, p. 9; see also Andrew Clapham, (2006) *Human Rights Obligations of Non-State Actors*, Oxford: Oxford University Press, pp. 317-318; Celina Romany, "State Responsibility Goes Private: A Feminist Critique of the Public/Private Distinction in International Human Rights Law", in Rebecca J. Cook (Ed.), (1994) *Human Rights of Women: National and International Perspectives*, Philadelphia: University of Pennsylvania Press, p. 96.

50 Nicola Jägers, (2002) *Corporate Human Rights Obligations. In Search of Accountability*, School of Human Rights Research Series, Volume 17, Antwerp: Intersentia, p. 146; Robert McCorquodale and Penelope Simons, (2007) "Responsibility Beyond Borders: State Responsibility for Extraterritorial Violations by Corporations of International Human Rights Law", *Modern Law Review*, Volume 70, Issue 4, pp. 601-602; Rick Lawson, "Out of Control. State Responsibility and Human Rights: Will the ILC's Definition of the 'Act of State' Meet the Challenges of the 21st Century?", in Monique Castermans-Holleman, Fried van Hoof, and Jacqueline Smith (Eds.), (1998) *The Role of the Nation-State in the 21st Century: Human Rights, International Organisations and Foreign Policy. Essays in Honour of Peter Baehr*, The Hague: Kluwer Law International, p. 115.

In addition, some States other than those actually engaging in the commission of internationally wrongful acts might bear international responsibility if their acts can be connected with the acts of States performing internationally wrongful acts, as stated in Chapter IV of Part One of the Draft Articles. Following the line of reasoning of Article 16, a State aiding or assisting another State in committing an internationally wrongful act is to be held responsible if it does so with knowledge of the circumstances of the internationally wrongful act and the act would be considered internationally wrongful if committed by the aiding or assisting State itself. While the issue of direction and control that can be exercised by a State over the commission of internationally wrongful acts will be discussed in the relevant parts of the work addressing the question of attribution, it must be added that in accordance with Article 17 of the Draft Articles, a State that directs and controls another State in its commission of an internationally wrongful act is internationally responsible for this act if the same two conditions, as laid down in previously discussed Article 16, are met. In the first place, the directing and controlling State must act with knowledge of the circumstances of the internationally wrongful act in question and, secondly, the act must be regarded internationally wrongful if committed by this State itself. Also, Article 18 stresses that a State that coerces another State to commit an act faces international responsibility for this particular act if the act would, but for the coercion, be seen as an internationally wrongful act of the coerced State and the coercing State operates in this instance with knowledge of the circumstances of the act. Under these circumstances, States other than those committing internationally wrongful acts themselves will be held internationally responsible, while Chapter IV of the Draft Articles is to be applied without prejudice to the State responsibility of States actually committing international wrongful acts or other States, as Article 19 of the Draft Articles suggests.

4. Implementation of State Responsibility, Circumstances Precluding Wrongfulness, and Legal Consequences

While international responsibility is supposed to be engaged directly after the commission of an internationally wrongful act as attributable misconduct in form of a breach of State obligations under international law, in practice it needs to be invoked by someone.[51] Part Three of the Draft Articles specifically deals with the issue of invocation of State responsibility. This part of the DARS addresses the question of what States can do about a violation of an international obligation by another State or States and does not cover the issue of State responsibility that may be at stake at

[51] James Crawford and Simon Olleson, "The Nature and Forms of International Responsibility", in Malcolm D. Evans (Ed.), (2010) *International Law*, Third Edition, Oxford: Oxford University Press, p. 467.

the level of non-State actors, such as individuals, companies, or international organizations.[52] According to Article 42 of the Draft Articles, one of the core concepts in relation to the State responsibility doctrine is that of an injured State. Article 42 DARS manifestly determines:

> "A State is entitled as an injured State to invoke the responsibility of another State if the obligation breached is owed to:
>
> (a) that State individually; or
> (b) a group of States including that State, or the international community as a whole, and the breach of the obligation:
>   (i) specially affects that State; or
>   (ii) is of such a character as radically to change the position of all the other States to which the obligation is owed with respect to the further performance of the obligation."

The above-cited provision suggests that the primary entitlement to implement State responsibility belongs to an injured State whose individual rights are affected by the conduct of another State.[53] It states that an injured State has a right to engage international responsibility against a State that committed an internationally wrongful act, if the obligation breached is owed to the injured State. It is also possible if the breached obligation is owed to a group of States, including the injured State, or the international community as a whole, if that breach specifically affects the injured State or radically changes the position of other States to which the obligation is owed regarding the further performance of that obligation. It is important to stress that the responsibility of a specific State arises independently of the invocation of the State responsibility by another State.[54] In this respect, the issue of both damage and the necessary causal link between a breach and that damage is not central to the process of establishing State responsibility, as may become clear on reading Article 1 of the Draft Articles.[55] At the same time, States other than the injured State can also be

---

[52] James Crawford, (2013) *State Responsibility: The General Part*, Cambridge Studies in International and Comparative Law, No. 100, Cambridge: Cambridge University Press, p. 541; Ineke Boerefijn, (2009) "Establishing State Responsibility for Breaching Human Rights Treaty Obligations: Avenues Under UN Human Rights Treaties", *Netherlands International Law Review*, Volume 56, Issue 2, p. 180.

[53] Edith Brown Weiss, (2002) "Invoking State Responsibility in the Twenty-First Century", *The American Journal of International Law*, Volume 96, Issue 4, pp. 800-802.

[54] James Crawford, "The System of International Responsibility", in James Crawford, Alain Pellet, and Simon Olleson (Eds.), (2010) *The Law of International Responsibility*, Oxford: Oxford University Press, p. 23.

[55] Brigitte Stern, "The Obligation to Make Reparation", in James Crawford, Alain Pellet, and Simon Olleson (Eds.), (2010) *The Law of International Responsibility*, Oxford: Oxford University Press, pp. 569-570; Brigitte Stern, "The Elements of An Internationally Wrongful Act", in James Crawford, Alain Pellet, and Simon Olleson (Eds.), (2010) *The Law of International Responsibility*, Oxford: Oxford University Press, p. 194.

regarded as legal entities capable of invoking international responsibility under certain circumstances. In this regard, Article 48(1) DARS provides:

> "Any State other than an injured State is entitled to invoke the responsibility of another State in accordance with paragraph 2 if:
>
> (a) the obligation breached is owed to a group of States including that State, and is established for the protection of a collective interest of the group; or
> (b) the obligation breached is owed to the international community as a whole."

The above-cited Paragraph emphasizes that States other than the injured State are also allowed to invoke international responsibility of the State breaching international law. In this respect, the breached obligation must be owed to the group of States including the State invoking responsibility and be meant to protect the interests of the whole group. Also, States that do not belong to groups with shared interests can be entitled to the invocation of State responsibility if the breached obligation is owed to the whole international community. This provision gives effect to the decision of the ICJ in the *Barcelona Traction* case.[56] These particular obligations owed to the international community are obligations of *erga omnes* character and are rules of *jus cogens*, such as the prohibition of torture, genocide, and slavery.[57] In addition to the engagement of State responsibility by States, in accordance with Article 33(2) DARS parties other than States are also allowed to invoke international responsibility to be addressed elsewhere.[58] The idea that States can raise claims even when they are not injured is not at variance with current trends of international judicial bodies, such as the European Court of Human Rights.[59]

In essence, the issue of State responsibility can be raised at various levels. At the international level, the International Court of Justice is open to the States Parties to the ICJ Statute, while this treaty clearly maintains that only States may act as parties to the cases before the Court.[60] The World Court can decide any case concerning

---

[56] Commentary to DARS as adopted in 2001, p. 127, Article 48, para. 8; Edith Brown Weiss, (2002) "Invoking State Responsibility in the Twenty-First Century", *The American Journal of International Law*, Volume 96, Issue 4, p. 804; James Crawford, (2013) *State Responsibility: The General Part*, Cambridge Studies in International and Comparative Law, No. 100, Cambridge: Cambridge University Press, p. 549.

[57] Olivier De Schutter, (2010) *International Human Rights Law*, Cambridge: Cambridge University Press, p. 65.

[58] James Crawford and Simon Olleson, "The Nature and Forms of International Responsibility", in Malcolm D. Evans (Ed.), (2010) *International Law*, Third Edition, Oxford: Oxford University Press, p. 467.

[59] Edith Brown Weiss, (2002) "Invoking State Responsibility in the Twenty-First Century", *The American Journal of International Law*, Volume 96, Issue 4, pp. 805-806.

[60] See Articles 34(1) and 35(1) of the Statute of the International Court of Justice; see also Edith Brown Weiss, (2002) "Invoking State Responsibility in the Twenty-First Century", *The American Journal of International Law*, Volume 96, Issue 4, p. 814.

international law and violations of international legal rules by States.[61] In the first place, legal proceedings against States breaching the law of armed conflict can be initiated before the ICJ. It is capable of addressing contentious cases between States regarding their international responsibility for violations of international humanitarian law, while individuals cannot directly start a case.[62] Additionally, the ICJ can also decide cases regarding breaches of international human rights law, similarly for other branches of international law.

Besides the ICJ, international and regional human rights bodies constitute valuable forums for both States and – more importantly – non-State actors to raise the matter of State responsibility.[63] In this vein, these actors, such as individuals who became victims of human rights breaches committed by private contractors, might seek to hold States accountable for their misconduct. This goal can be achieved by using individual complaints mechanisms. Additionally, several human rights instruments have avenues for States to engage international responsibility of other States by means of inter-State applications. Under most human rights regimes, these inter-State complaints procedures are conditional upon the acceptance of the competence of treaty monitoring bodies. Article 2 of the First Optional Protocol to the ICCPR[64] creates the possibility for individuals to claim a violation of human rights by a State Party before the UN Human Rights Committee if they have exhausted domestic remedies. The ICCPR itself contains Article 41, providing that a State Party recognizing the competence of the Committee can refer a claim of a human rights violation committed by another State Party if the former has brought the matter to the attention of the latter and the matter has not been addressed. According to Articles 34 and 35(1), the ECHR recognizes the right of persons, groups of persons, and non-governmental organizations to lodge individual applications with the ECtHR if they claim to be the victim of a violation of the human rights set out in the Convention by a State Party to it and if all domestic remedies have been exhausted. In addition, Article 33 stipulates that States may also start a case at the ECtHR if another State Party to the ECHR violates a provision of the Convention. In the context of the ACHR, Article 44 of the Convention provides that individuals, groups of individuals, and NGOs are allowed to lodge petitions with the Commission when a violation of the ACHR by a State Party takes place. Also, any State Party to the Convention may allege that

---

61 See Article 36(2) of the Statute of the International Court of Justice.

62 Emanuela-Chiara Gillard, "International Humanitarian Law and Extraterritorial State Conduct", in Fons Coomans and Menno T. Kamminga (Eds.), (2004) *Extraterritorial Application of Human Rights Treaties*, Antwerp and Oxford: Intersentia, p. 37.

63 Ineke Boerefijn, (2009) "Establishing State Responsibility for Breaching Human Rights Treaty Obligations: Avenues Under UN Human Rights Treaties", *Netherlands International Law Review*, Volume 56, Issue 2, p. 180.

64 Optional Protocol to the International Covenant on Civil and Political Rights, adopted and opened for signature, ratification, and accession by the UN General Assembly Resolution 2200A (XXI) of December 16, 1966 and entered into force on March 23, 1976.

another State has been engaged in a human rights violation, if the first State declares the Commission competent to receive and to examine such claims, as is mentioned in Article 45(1). According to Articles 55-58 of the ACHPR, individuals and NGOs can communicate their complaints regarding human rights breaches to the Commission when all domestic measures have been used. In Article 49, the ACHPR also makes it possible for States to refer a matter of human rights violations by other States to the Commission if it has not been dealt with by approaching the violating State itself. Within the legal framework of the CAT, in accordance with Article 22, there is an individual complaints mechanism that can be approached by persons of States Parties who have made a declaration under Article 22(1). The Committee will consider communications received from individuals when the same issue is not under consideration by other international investigations or settlements and complainants have exhausted all domestic remedies. Article 21 provides that a State recognizing the competence of the Committee may submit communications to this body if the case of torture or other related treatment has been brought to the attention of the State in question first and has not been settled to the State's satisfaction.

Finally, domestic courts of various States can be used by non-State actors, corporations, individuals, and NGOs to complain about breaches of international law with the purpose of establishing responsibility of States. Many cases concerning the question of State responsibility and reparation claims for human rights violations are basically decided by the national judges at the domestic level.[65] The Commentary to the ILC's Draft Articles specifically provides that the law of State responsibility is applicable to all international obligations of States, also those owed to individuals or groups.[66] In this respect, domestic courts apply national laws and – implicitly or explicitly – the rules of customary international law on State responsibility laid down in the Draft Articles which are supposed to be incorporated into the domestic legal order in various jurisdictions.[67]

Chapter V of Part One of the Draft Articles deals with six circumstances precluding wrongfulness that might provide "shield against an otherwise well-founded claim for the breach of an international obligation".[68] Those six circumstances are listed in Articles 20-25 DARS and include: consent, self-defense, countermeasures, *force majeure*, distress, and necessity. As the Commentaries to the Draft Articles suggest, these provisions are of general application and can apply to any internationally wrongful act.[69] In accordance with international practice, these traditional

---

[65] Stephan Wittich, (2013) "Domestic Courts and the Content and Implementation of State Responsibility", *Leiden Journal of International Law*, Volume 26, Issue 3, pp. 645-646.

[66] Commentary to DARS as adopted in 2001, p. 32, General Commentary, para. 5.

[67] Stephan Wittich, (2013) "Domestic Courts and the Content and Implementation of State Responsibility", *Leiden Journal of International Law*, Volume 26, Issue 3, p. 649.

[68] Commentary to DARS as adopted in 2001, p. 71, para. 1.

[69] Ibidem, p. 71, para. 2.

circumstances precluding wrongfulness apply to two types of international State obligations – positive and negative – a distinction that will more prominently be addressed below.[70] When a State committing an internationally wrongful act successfully invokes one or more circumstances precluding wrongfulness, the breached obligation is not annulled or terminated. The above-listed provisions rather contain excuses or justification grounds for non-performance of an international obligation, as indicated by the ICJ.[71] When factors justifying the application of circumstances precluding wrongfulness cease to exist, performance of respective international obligations must be resumed.

The engagement of international responsibility involves a number of legal consequences. According to Christian Tomuschat:

> "[T]he main legal consequence of a breach of a rule of international law is that secondary rights of reparation arise for the victim State in consonance with the ARS [the Draft Articles]."[72]

The actual legal consequences of State responsibility are listed in Part Two of the ILC's Draft Articles. In this regard, Article 28 DARS provides:

> "The international responsibility of a State which is entailed by an internationally wrongful act in accordance with the provisions of Part One involves legal consequences as set out in this Part."

In accordance with Part Two of the Draft Articles, the main legal consequences of the commission of an internationally wrongful act include the obligations to cease the wrongful conduct, as stipulated in Article 30 DARS, and to make reparations for the injury induced by that internationally wrongful act, as set out in Article 31 DARS.

Following the reasoning of Article 30 DARS, the State committing an internationally wrongful act is obliged not only to cease that act, but also to offer appropriate assurances and guarantees that it will not occur again, if specific circumstances of the case require.

As stated in Article 31 DARS, it is significant that another core legal consequence of the commission of internationally wrongful acts by States relates to the obligation

---

[70] Riccardo Pisillo-Mazzeschi, (1992) "The Due Diligence Rule and the Nature of the International Responsibility of States", *German Yearbook of International Law*, Volume 35, p. 45.

[71] Commentary to DARS as adopted in 2001, p. 71, para. 2; International Court of Justice, *Gabčíkovo-Nagymaros Project (Hungary v. Slovakia)*, Judgment, September 25, 1997, ICJ Reports 1997, p. 39, para. 48.

[72] Christian Tomuschat, "Specificities of Human Rights Law and International Humanitarian Law Regarding State Responsibility", in Robert Kolb and Gloria Gaggioli (Eds.), (2013) *Research Handbook on Human Rights and Humanitarian Law*, Cheltenham and Northampton: Edward Elgar Publishing, p. 199.

on the responsible State to make full reparation for injury caused by its unlawful conduct. In this regard, injury may consist of any type of damage, including material and moral damages, occurring as a result of an internationally wrongful act. In fact, if it is established that an internationally wrongful act has been committed, a new legal relationship arises and the responsible State will have to make reparations.[73] This notion has been affirmed by the Permanent Court of International Justice in its *Factory at Chorzów* case. The Court stated that:

> "It is a principle of international law that the breach of an engagement involves an obligation to make reparation in an adequate form. Reparation therefore is the indispensable complement of a failure to apply a convention and there is no necessity for this to be stated in the convention itself."[74]

Article 34 DARS lists three forms of reparation that can be made by the responsible State: restitution, compensation, and satisfaction. According to Article 35 DARS, a State responsible for the commission of an internationally wrongful act is regarded as being under an obligation to make restitution, which means that it has to re-establish the situation as it existed before the commission of that internationally wrongful act. Another form of reparation mentioned in Article 36 DARS is compensation. A State responsible for an internationally wrongful act is obliged to compensate for damage caused by it to the extent that that damage is not made good by restitution. In this case, the notion of damage covers both material and moral damage, as regulated in Article 31(2) DARS, while compensation covers any financially assessable damage, such as loss of profits to the extent that it has been established. Finally, Article 37 DARS provides that the responsible State is under an obligation to give satisfaction for the injury caused by its internationally wrongful act insofar as it is not made good by restitution or compensation. In this regard, the concept of satisfaction can take forms of an acknowledgment of the breach, an expression of regret, a formal apology, or another appropriate modality. It must be borne in mind, however, that satisfaction may not be out of proportion to the injury caused by an internationally wrongful act and may not be used to humiliate the responsible State.

Where an internationally wrongful act concerns a violation of a peremptory norm of general international law by a State, that violation leads to a number of other essential legal consequences for both the responsible State and other States. In accordance with Article 40(1) DARS, responsibility of a State for a serious breach of an obligation arising out of peremptory norms of general international law can be engaged. As stated in Paragraph 2 of the same provision, the breach is considered serious if

---

[73] Brigitte Stern, "The Obligation to Make Reparation", in James Crawford, Alain Pellet, and Simon Olleson (Eds.), (2010) *The Law of International Responsibility*, Oxford: Oxford University Press, p. 563.
[74] Permanent Court of International Justice, *Factory at Chorzów (Germany v. Poland)*, Jurisdiction, Judgment, July 26, 1927, PCIJ Reports, Series A, No. 9, p. 21.

it concerns a gross or systematic failure of the responsible State to comply with its obligations under international law. If the responsibility of a State for such a type of serious violation is established, States will have to cooperate in order to bring to an end that breach by any lawful means according to Article 41(1) DARS. Moreover, States are obliged to refrain from recognizing situations created by this type of serious breach as lawful and from rendering aid or assistance in maintaining the situation, as regulated in Article 41(2) DARS.

When PMSCs hired by particular States to carry out military and security tasks on their behalf engage in violating rules and principles of international law and if the commission of those internationally wrongful acts can be attributed to the hiring States, it will lead to the above-mentioned legal consequences not only for the responsible States, but possibly also for other States. For instance, if private contractors systematically commit torture prohibited under international law, which constitutes one of its important peremptory norms that may not be breached, and their illegal activities are attributed to a specific State employing them, States other than the responsible State will be under an obligation to combine their efforts in order to end those violations by any lawful means. They will also be obliged not to recognize those activities as lawful and not to cooperate with the responsible State in further commission of torture. In addition, the responsible State that used the PMSC in question will have to cease the commission of those internationally wrongful acts, to guarantee that they will not occur in the future, and to make reparations, which may take form of restitution, compensation, and satisfaction. The legal consequences of international responsibility of hiring States for the misconduct of PMSCs violating international law are emphasized in the Montreux Document specifically restating international obligations of States. The Document contains the following provision:

> "Contracting States have an obligation to provide reparations for violations of international humanitarian law and human rights law caused by wrongful conduct of the personnel of PMSCs when such conduct is attributable to the Contracting States in accordance with the customary international law of State responsibility."[75]

5. State Responsibility and Institutional Responsibility

While dealing with the question of State responsibility in the context of conflicts – armed conflicts and peace operations – it is important to observe that the misconduct of PMSCs violating international law can also lead to the responsibility of international organizations. This issue is specifically pertinent when peace operations are authorized by international organizations, such as the UN, or regional organizations, such as NATO. The ICJ made clear that international organizations are to be seen as

---

[75] Montreux Document, Part One "Pertinent International Legal Obligations Relating to Private Military and Security Companies", p. 12, para. 8.

subjects of international law, which are bound by a variety of obligations arising from general rules of international law, their constitutions, and international agreements concluded between these organizations and other parties.[76] While it has been determined by the World Court that the UN is an international person with an international legal personality,[77] it can be assumed that NATO can also qualify as such.[78] Article 2(a) of the 2011 ILC's Draft Articles on the Responsibility of International Organizations (hereinafter: DARIO) mentions that those are only international organizations possessing international legal personality that are capable of bearing international responsibility and the DARIO apply to international organizations having this international legal personality.[79] Similarly to the DARS, the DARIO provide that every internationally wrongful act of an international organization entails the international responsibility of the organization in question.[80]

As previously indicated, international and regional organizations rely extensively on the provision of military and security services by private contractors and might be held responsible for their illegal activities. According to the ICJ, the UN and other international organizations act through their agents and are responsible for any damage arising from the conduct of these agents acting in their official capacity.[81] The DARIO stress that the acts of organs and agents of international organizations performing their official functions are indeed to be considered acts of those organizations.[82] When the UN and other international and regional organizations directly employ PMSCs in peace operations and those business entities operate as organs and agents of the international organizations, the international organizations in question will bear responsibility for their violations of international law. If PMSCs constitute parts of the official organs of the organization, their conduct automatically becomes the conduct of this organization. When PMSCs act as agents of international organizations they must be charged by an organ of these organizations with "carrying out, or helping to carry out, one of its functions" or, put plainly, this organ must be acting

---

[76] International Court of Justice, *Interpretation of the Agreement of 25 March 1951 Between the WHO and Egypt*, Advisory Opinion, December 20, 1980, ICJ Reports 1980, para. 37.

[77] International Court of Justice, *Reparation for Injuries Suffered in the Service of the United Nations*, Advisory Opinion, April 11, 1949, ICJ Reports 1949, p. 179.

[78] Nigel D. White, "Institutional Responsibility for Private Military and Security Companies", in Francesco Francioni and Natalino Ronzitti (Eds.), (2011) *War by Contract: Human Rights, Humanitarian Law, and Private Contractors*, Oxford: Oxford University Press, p. 385.

[79] Article 2 DARIO and Commentaries; see Draft Articles on the Responsibility of International Organizations, With Commentaries, as Adopted by the International Law Commission at its Sixty-Third Session, (2011) Yearbook of the International Law Commission, Volume II(2), UN Doc. A/66/10, pp. 6, 8-9.

[80] Article 3 DARIO.

[81] International Court of Justice, *Difference Relating to Immunity from Legal Process of A Special Rapporteur of the Commission on Human Rights*, Advisory Opinion, April 29, 1999, ICJ Reports 1999, para. 66.

[82] Article 6(1) DARIO.

through the agents.[83] Thus, in the situations when PMSCs are employed by the organs of international organizations authorizing peace operations, such as the UN Security Council, and are used to carry out the mandate of these missions or are considered experts on mission, their misconduct will engage international responsibility of the respective organizations.[84]

While international organizations might directly be responsible for the conduct of hired PMSCs, the use of these companies in peace operations may further complicate the matters of responsibility, as the Member States acting under the authority of these organizations contribute personnel, including civilian contractors, to these missions. In this respect, both troop-contributing States and international organizations authorizing peace operations can bear international responsibility depending on who essentially has authority, command, and control over PMSCs.[85] Thus, the level of control carried out by the organization and States influences the determination to whom to attribute acts of private contractors. The attribution can be effective in relation to the international organization, the State, or both, as acknowledged by the Dutch Court of Appeal of The Hague and Supreme Court in the recent *Nuhanović* case.[86] Given that the current analysis focuses on the question of State responsibility, the issue of attribution will be extensively discussed in relation to States while a few comments will also be made on the level of control with regard to international organizations authorizing peace operations.

6. Attribution of Unlawful Conduct to States

The most controversial and perhaps the most interesting aspect of State responsibility concerns the issue of attribution – or imputability and imputation – of internationally wrongful acts to States.[87] The definition of attribution proposed by Luigi Condorelli and Claus Kress underlines its most essential aspects that clearly characterize the notion. The authors indicate that:

---

83 International Court of Justice, *Reparation for Injuries Suffered in the Service of the United Nations*, Advisory Opinion, April 11, 1949, p. 177.

84 Chia Lehnardt, "Peacekeeping", in Simon Chesterman and Angelina Fisher (Eds.), (2009) *Private Security, Public Order: The Outsourcing of Public Services and Its Limits*, Oxford: Oxford University Press, pp. 211-212.

85 Nigel D. White, "Institutional Responsibility for Private Military and Security Companies", in Francesco Francioni and Natalino Ronzitti (Eds.), (2011) *War by Contract: Human Rights, Humanitarian Law, and Private Contractors*, Oxford: Oxford University Press, p. 387.

86 Court of Appeal, *Hasan Nuhanović against the State of the Netherlands*, The Hague, July 5, 2011, Case No. 200.020.174/1, para. 5.9; Supreme Court of the Netherlands, *The State of the Netherlands v. Hasan Nuhanović*, First Chamber, The Hague, September 6, 2013, Case No. 12/03324, para. 3.11.2.

87 Antonio Cassese, (2005) *International Law*, Second Edition, Oxford: Oxford University Press, pp. 245-246.

> "'Attribution' (or 'imputation') is the term used to denote the legal operation having as its function to establish whether given conduct of a physical person, whether consisting of a positive action or an omission, is to be characterized, from the point of view of international law, as an 'act of the State'."[88]

In other words, it is a concept used to refer to a set of rules and principles containing standards and criteria that have been crystallized as customary international law and laid down in the Draft Articles on State Responsibility by the International Law Commission. These provisions are employed in the process of determining whether specific unlawful conduct is to be regarded as the conduct of a State itself.

As identified above, even if a breach of international law has been established, but cannot be attributed to a certain State, its international responsibility will not be invoked. On the other hand, if the unlawful conduct of private contractors working for PMSCs can be attributed to States in accordance with the rules and principles of international law, those States must bear international responsibility for their commission and its respective legal consequences.

The State responsibility under international law appears to be dependent upon the link between the State and persons or entities actually committing unlawful acts or omissions.[89] Only those activities that can be attached to States as sovereign entities functioning in the international arena in an objective manner can be attributed to them. The links between States and unlawful acts or omissions can be of legal, functional, and factual nature, or can be derived from the actions of State organs.[90] In this regard, international responsibility can be incurred not only by hiring States, but also host States and home States. In theory, all three types of States can have different links to PMSCs, whose misconduct is attributable to them. In practice, however, hiring States will have the most extensive relationship with private contractors performing their services for these particular States and will be closely connected to the actions of these companies and their employees.[91] Only with regard to contracting States does the Montreux Document confirm this idea and indicate that these States are responsible for violations of international law, such as the law of armed conflict

---

88 Luigi Condorelli and Claus Kress, "The Rules of Attribution: General Considerations", in James Crawford, Alain Pellet, and Simon Olleson (Eds.), (2010) *The Law of International Responsibility*, Oxford: Oxford University Press, p. 221.

89 Malcolm N. Shaw, (2008) *International Law*, Sixth Edition, Cambridge: Cambridge University Press, p. 786; Pierre-Marie Dupuy, (1992) *Droit International Public*, Paris: Dalloz-Sirey, p. 339.

90 Olivier De Frouville, "Attribution of Conduct to the State: Private Individuals", in James Crawford, Alain Pellet, and Simon Olleson (Eds.), (2010) *The Law of International Responsibility*, Oxford: Oxford University Press, p. 261.

91 Nicki Boldt, (2004) "Outsourcing War – Private Military Companies and International Humanitarian Law", *German Yearbook of International Law*, Volume 47, p. 538.

and human rights law, committed by PMSCs and their staff when these violations can be attributed to them in accordance with customary international law.[92]

The complicated notion of attribution of unlawful acts committed by the employees of PMSCs will now be further explored. The main aim of this examination is to establish the circumstances and extent of application of the doctrine of State responsibility for the illegal actions in which PMSCs and their employees might get involved considering the scope of their activities. As correctly observed by the Dutch Advisory Council on International Affairs in its advisory report *Employing Private Military Companies: A Question of Responsibility* from 2007, there are four essential types of attribution with regard to PMSCs used as four distinct alternatives to attach their wrongful conduct to States. These are Articles 4, 5, 8, and 11 DARS. In this regard, the attribution of internationally wrongful acts takes place in accordance with the provisions of the Draft Articles if: first, a company or person in question acts as a State organ (Article 4 DARS); secondly, a PMSC or person that not only is empowered to exercise elements of governmental authority (Article 5 DARS), but also exceeds its authority and contravenes instruction (Article 7 DARS); thirdly, a company or person acts under the instructions, direction, or control of a State (Article 8 DARS); and, finally, a State acknowledges or adopts the unlawful acts of a firm or person as its own (Article 11 DARS).[93] The restatement of existing international obligations of States in the Montreux Document also contains references to these modes of attribution without Article 11 DARS. In particular, a State appears to be responsible for the misconduct of PMSCs and their personnel if these corporations and their employees are:

> "a) incorporated by the State into their regular armed forces in accordance with its domestic legislation;
> b) members of organized armed forces, groups or units under a command responsible to the State;
> c) empowered to exercise elements of governmental authority if they are acting in that capacity (i.e. are formally authorized by law or regulation to carry out functions normally conducted by organs of the State); or
> d) in fact acting on the instructions of the State (i.e. the State has specifically instructed the private actor's conduct) or under its direction or control (i.e. actual exercise of effective control by the State over a private actor's conduct)."[94]

---

[92] Montreux Document, Part One "Pertinent International Legal Obligations Relating to Private Military and Security Companies", p. 12, para. 7.

[93] Dutch Advisory Council on International Affairs, *Employing Private Military Companies: A Question of Responsibility* (Dutch: *De Inhuur van Private Militaire Bedrijven: Een Kwestie van Verantwoordelijkheid*), No. 59, December 2007, The Hague, p. 21, <http://www.aiv-advies.nl/ContentSuite/upload/aiv/doc/webversie_AIV_59eng(1).pdf>, accessed on May 9, 2011.

[94] Montreux Document, Part One "Pertinent International Legal Obligations Relating to Private Military and Security Companies", p. 12, para. 7.

Additionally, it is also possible to attribute the conduct of PMSCs and their personnel to States under Articles 6, 9, and 10 DARS. Article 6 DARS deals with situations when a State organ is placed at the disposal of a State by another State and indicates that the conduct of such an organ will be considered an act of the State at whose disposal the organ has been placed if it exercises elements of governmental authority of this State. Article 9 DARS specifically states that the conduct of a person or a group of persons is regarded as an act of a State if this person or group of persons exercise elements of governmental authority in the absence or default of the official authorities and in circumstances requiring this exercise of elements of governmental authority. Article 10 DARS concerns the conduct of an insurrectional or other movement. According to this provision, if an insurrectional movement becomes the new government of a State, its conduct is considered an act of that State under international law. Also, the conduct of an insurrectional or other movement succeeding in establishing a new State in the territory of an existing State or a territory under its administration is to be perceived as an act of the new State. As also mentioned by Lindsey Cameron and Vincent Chetail, in addition to Article 11 DARS these three types of attribution could be employed in relation to PMSCs but are less relevant in practice, such as Articles 6 and 10 DARS, or there are no complications with regard to their substance and application, such as Article 9 DARS.[95] Following the reasoning of the Dutch Advisory Council on International Affairs, and deviating from the position of two above-mentioned scholars, Article 11 DARS will be elaborated upon below given that it might be relevant for the State practice of using PMSCs to a higher extent than Articles 6 and 10 DARS and requires some explanation in contrast to Article 9 DARS.

The core issue of attribution will be touched upon through the assessment of the relevant parts of the Draft Articles on State Responsibility for Internationally Wrongful Acts, namely Chapter II of Part One of the Draft Articles. This Chapter focuses particularly on specific circumstances under which the actual attribution of unlawful acts and omissions occurs, meaning that the illegal conduct is regarded as the conduct of a State.[96] In addition, case law of international judicial bodies, such as the International Court of Justice and the International Criminal Tribunal for the former Yugoslavia (hereinafter: ICTY), will be analyzed. Both case law and the Draft Articles form sources of valuable insight into the circumstances and the extent of the attribution of unlawful acts committed by PMSCs to States and need to be examined in a coherent and systematic manner.

[95] Lindsey Cameron and Vincent Chetail, (2013) *Privatizing War: Private Military and Security Companies under Public International Law*, Cambridge: Cambridge University Press, p. 223.
[96] Commentary to DARS as adopted in 2001, p. 38, para. 1.

6.1 Article 4 DARS

6.1.1 General application of the provision

States are "judicial abstractions"[97] or abstract legal entities[98] acting with the assistance of one or more persons or entities that can qualify as State agents. They "can only act by and through their agents and representatives".[99] Although States cannot be responsible for all unlawful acts that might be committed by their nationals, under international law they certainly have responsibility over the commission of internationally wrongful acts by their authorized officials and representatives. As Anzilotti puts it, "the act of the State is nothing but the activity of individuals that the law imputes to the State".[100]

One of the main provisions laid down in the Draft Articles focusing on the issue of attribution, is without doubt Article 4 DARS, which is the first mode of attribution normally used as a general point of departure.[101] This provision is concerned with the unlawful conduct of *de jure* State organs, which is attributable to States in question, and it reads as follows:

> "1. The conduct of any State organ shall be considered an act of that State under international law, whether the organ exercises legislative, executive, judicial or any other functions, whatever position it holds in the organization of the State, and whatever its character as an organ of the central Government or of a territorial unit of the State.
>
> 2. An organ includes any person or entity which has that status in accordance with the internal law of the State."

For the purpose of Article 4 DARS, the Commentaries define a State organ as "all the individual or collective entities, which make up the organization of the State and act on its behalf".[102] In this regard, it bears emphasizing that illegal acts or omissions – which are not purely private acts[103] – constituting internationally wrongful acts

---

97 James Crawford and Simon Olleson, "The Nature and Forms of International Responsibility", in Malcolm D. Evans (Ed.), (2010) *International Law*, Third Edition, Oxford: Oxford University Press, p. 452.

98 Malcolm N. Shaw, (2008) *International Law*, Sixth Edition, Cambridge: Cambridge University Press, p. 786.

99 Permanent Court of International Justice, *Case of Certain Questions Relating to Settlers of German Origin in the Territory Ceded by Germany to Poland*, Advisory Opinion, September 10, 1923, PCIJ Series B 1925, No. 6, p. 22.

100 Dionisio Anzilotti, (1929) *Cours de Droit International*, Paris: Editions Panthéon-Assas, p. 269.

101 Commentary to DARS as adopted in 2001, p. 40, Article 4, para. 2.

102 Ibidem, p. 40, Article 4, para. 1.

103 James Crawford and Simon Olleson, "The Nature and Forms of International Responsibility", in Malcolm D. Evans (Ed.), (2010) *International Law*, Third Edition, Oxford: Oxford University Press, p. 453.

that are committed by State organs will directly be imputed to the States in question, irrespective of the character of those organs, the position they hold, or function they perform. These persons or individual and collective entities forming State organs may belong to the judicial, executive, or legislative branches within a State or function in other possible spheres of activity. In the case *Difference Relating to Immunity from Legal Process of a Special Rapporteur of the Commission on Human Rights,* the International Court of Justice stressed that "according to a well-established rule of international law, the conduct of any organ of a State must be regarded as an act of that State".[104] This rule of international law belongs to the body of customary international law, which is widely acknowledged and recognized by States,[105] and closely relates to the significant "question of fact" of what is to be defined as a State organ.[106]

In addition, as laid down in Article 4(2) DARS, the internal law of States must provide that the actual legal status of persons and individuals is that of State organ. The provision seems to suggest that the domestic law of a specific State needs to be taken as a starting point in the process of identifying a certain person or entity as a State organ. It is, however, not the most significant matter under consideration and does not constitute the decisive criterion.[107] According to the Commentaries to the Draft Articles, it is not enough to refer to the sole internal law of a State in order to determine the status of its State organs.[108] As expressed in the Commentaries, the function and status of persons and entities within the State apparatus may not be decided upon merely on the basis of law, as practice also often frequently determines those two significant aspects. It is not beyond the limits of reasonable expectation that the domestic legislation of a State may avoid any classification of a certain body as a State organ and prevent that State from being held responsible for the unlawful conduct of the body.

There could be circumstances when specific persons or entities do not qualify to be State organs because the domestic law of a State does not confer on them the legal status of official organs of the State. However, those individuals or bodies can perform a wide array of functions on behalf of States, while being practically similar

---

104 International Court of Justice, *Difference Relating to Immunity from Legal Process of a Special Rapporteur of the Commission on Human Rights*, Advisory Opinion, April 29, 1999, ICJ Reports 1999, p. 87, para. 62; André Nollkaemper, (2003) "Concurrence Between Individual Responsibility and State Responsibility in International Law", *International & Comparative Law Quarterly*, Volume 52, Issue 3, pp. 616-617.

105 Malcolm N. Shaw, (2008) *International Law*, Sixth Edition, Cambridge: Cambridge University Press, p. 787.

106 Ian Brownlie, (1983) *System of the Law of Nations: State Responsibility*, Oxford: Oxford University Press, p. 136.

107 Djamchid Momtaz, "Attribution of Conduct to the State: State Organs and Entities Empowered to Exercise Elements of Governmental Authority", in James Crawford, Alain Pellet, and Simon Olleson (Eds.), (2010) *The Law of International Responsibility*, Oxford: Oxford University Press, p. 243.

108 Commentary to DARS as adopted in 2001, p. 42, Article 4, para. 11.

to State organs. This similarity is reflected in the nature of tasks carried out by them and the specific close relationship they might have with the States in question. Such persons or entities are frequently referred to as *de facto* organs due to their "complete dependence" on a State and the fact that they actually constitute its instruments through which the State acts, as pointed out by the International Court of Justice in the *Bosnian Genocide* case.[109] According to the Court:

> "... persons, groups of persons or entities may, for purposes of international responsibility, be equated with State organs even if that status does not follow from internal law, provided that in fact the persons, groups or entities act in *'complete dependence'* on the State, of which they are ultimately *merely the instrument*. In such a case, it is appropriate to look beyond legal status alone, in order to grasp the reality of the relationship between the person taking action, and the State to which he is so closely attached as to appear to be nothing more than *its agent*: any other solution would allow States to escape their international responsibility by choosing to act through persons or entities whose supposed independence would be purely fictitious." (emphasis added)

One needs to carefully assess all above-mentioned issues in order to find out under what circumstances and to what degree this particular type of attribution laid down in Article 4 DARS may find its application in relation to the illegal conduct of PMSCs and their employees, which will be dealt with in the following sub-Chapter.

6.1.2 Application of Article 4 DARS with regard to PMSCs

All State organs without exception can commit internationally wrongful acts.[110] The extent of application of Article 4 DARS with regard to PMSCs is such as to imply that PMSCs need to function as State organs or to be equated with them in order to trigger this particular type of attribution laid down in the provision, which will eventually lead to the acceptance of international responsibility of the State in question. In fact, it needs to be established whether PMSCs act in accordance with the internal law of a State as its official organs in order to make possible the imputation of their acts to the State under Article 4 DARS. Due to the fact that States constitute legal

---

109 International Court of Justice, *Case Concerning the Application of the Convention on the Prevention and Punishment of the Crime of Genocide (Bosnia and Herzegovina v. Serbia and Montenegro)*, Judgment (Merits), February 26, 2007, ICJ Reports 2007, para. 391; Chia Lehnardt, "Private Military Companies and State Responsibility", in Simon Chesterman and Chia Lehnardt (Eds.), (2007) *From Mercenaries to Market: The Rise and Regulation of Private Military Companies*, Oxford: Oxford University Press, p. 143.

110 Commentary to DARS as adopted in 2001, p. 40, Article 4, paras. 5-6; Djamchid Momtaz, "Attribution of Conduct to the State: State Organs and Entities Empowered to Exercise Elements of Governmental Authority", in James Crawford, Alain Pellet, and Simon Olleson (Eds.), (2010) *The Law of International Responsibility*, Oxford: Oxford University Press, p. 237.

entities acting through their agents and representatives,[111] PMSCs must have the same status as those officials, the acts of which must fall under the definition of the acts of States themselves.[112] The conduct of such State organs is considered an act of the State itself according to international law. This well-established rule of customary international law requires a determination of the actual position of PMSCs within the State apparatus and the State structures, which must take into consideration the relationship between the State in question and those legally established businesses.

In practice, PMSCs are often hired to perform numerous tasks associated with the executive power within the State structures – such as of police and regular armed forces – and carry out security and military services, such as participation in armed conflicts. Within the framework of PMSC involvement in conflict-affected areas around the globe, the most likely example of incorporation of PMSCs into the domain of State apparatus is that of their potential fusion with the regular armed forces. The most practical type of this incorporation can take place by internal laws of hiring States that contract with PMSCs for the provision of military and security services. It is, however, not impossible that also home and host States could engage in the similar practice. These two categories of States are most likely to also be contracting States making use of PMSC expertise.

According to the explanations provided in the Commentaries to the Draft Articles, the term "person or entity" used in Article 4 DARS to describe the notion of a State organ refers in a broad sense to "any natural or legal person, including an individual office holder, a department, commission or other body exercising public authority".[113] The regular armed forces of States exercising public authority do not merely participate in wars, but are also often engaged in many other activities, such as frontier policing and the maintenance of public order and security.[114] The military, in fact, form an essential part of the State apparatus and constitute the State organ, the acts of which can be attributed to States, taking into account the fact that the armed forces are created with an aim to maintain independence, authority, and safety of respective States.[115] Hence, the acts of regular armed forces in their capacity of State agents[116]

---

[111] Permanent Court of International Justice, *Case of Certain Questions Relating to Settlers of German Origin in the Territory Ceded by Germany to Poland*, Advisory Opinion, September 10, 1923, PCIJ Series B 1925, No. 6, p. 22.

[112] Chia Lehnardt, "Private Military Companies and State Responsibility", in Simon Chesterman and Chia Lehnardt (Eds.), (2007) *From Mercenaries to Market: The Rise and Regulation of Private Military Companies*, Oxford: Oxford University Press, p. 143.

[113] Commentary to DARS as adopted in 2001, p. 42, Article 4, para. 12.

[114] Ian Brownlie, (1983) *System of the Law of Nations: State Responsibility*, Oxford: Oxford University Press, p. 139.

[115] Lassa Oppenheim, (1920) *International Law: A Treatise*, Volume I. – Peace, Third Edition Edited by Ronald F. Roxburgh, Clark, New Jersey: The Lawbook Exchange Ltd, p. 607.

[116] Ian Brownlie, (1983) *System of the Law of Nations: State Responsibility*, Oxford: Oxford University Press, p. 140.

need to be equated with the acts of States, for they form part of official State organs. If members of the armed forces of a State engage in the commission of internationally wrongful acts, those activities will directly be attributed to the State in accordance with the principle laid down in Article 4 DARS.

PMSCs as such in their capacity of private corporations that clearly have a distinctive legal personality different from that of a State cannot be considered to be State organs.[117] Therefore, any form of unlawful conduct of PMSCs as private firms or their employees cannot be imputed to States under Article 4 DARS. However, if PMSCs cannot be regarded State organs, the question remains as to whether PMSCs can qualify as constituting part of a State organ and as such belonging to the military structures of a State.

As already pointed out, the 1977 First Additional Protocol to the 1949 Geneva Conventions defines armed forces of a State that is party to a conflict. In order to form part of the armed forces, a PMSC needs to be regarded as an organized armed force, group, or unit placed under a command responsible to a State party to a conflict for the conduct of its employees and subjected to an internal disciplinary system. In practice, a PMSC acts in its capacity of private legal entities separately from the State, but often under its control, directions, and instructions. Formal incorporation of PMSCs into official State armies seems to be a matter of solely theoretical value, for States remain reluctant to go so far as to confer on private contractors the status of the personnel of armed forces. It is not highly likely that PMSCs will be incorporated into official State structures by means of diffusion with the regular armed forces. However, it should still be regarded as a possibility and a potential form of cooperation between those private businesses and the military apparatus.

Some authors observe that PMSCs may be regarded as parts of official structures of State armed forces, referring to one of the famous incidents of involvement of the British PMSC Sandline, established in 1997 by Lieutenant-Colonel Tim Spicer, in the domestic affairs of Papua New Guinea. The government of that country was faced with the refusal of Australia to provide military training and equipment in the fight against a rebel group on the island of Bougainville.[118] Subsequently, Papua New Guinea turned to the private sector and concluded a contract with the above-mentioned company for the provision of military assistance in order to end the rebellion.[119] In this particular case, private contractors employed by Sandline were

---

[117] Oliver R. Jones, (2009) "Implausible Deniability: State Responsibility for the Actions of Private Military Firms", *Connecticut Journal of International Law*, Volume 24, Issue 2, p. 263.

[118] Peter W. Singer, (2008) *Corporate Warriors: The Rise of the Privatized Military Industry*, Ithaca and London: Cornell University Press, p. 177.

[119] See the contract concluded between the government of Papua New Guinea and Sandline obtained through the Project on Papua New Guinea at Australian National University in Peter W. Singer, (2008) *Corporate Warriors: The Rise of the Privatized Military Industry*, Ithaca and London: Cornell University Press, Appendix 2, pp. 263-270.

appointed as "special constables" of Papua New Guinea and had a number of rights and responsibilities identical to those of the regular State armed forces.[120] This was done in order to make sure that Sandline employees would not qualify as mercenaries.[121] The agreement concluded between the government and Sandline specifies with regard to the responsibilities of the State the following:

> "All officers and personnel of Sandline assigned to this contract shall be enrolled as Special Constables, but hold military ranks commensurate with those they hold within the Sandline command structure and shall be entitled to give orders to junior ranks as may be necessary for the execution of their duties and responsibilities."[122]

Sandline and its personnel were required to deliver a number of crucial services that are usually provided by the armed forces, including training, intelligence gathering and planning, directing, and participating in offensive operations aimed at neutralizing the Bougainville Revolutionary Army, and repossessing the Panguna mine.[123] Being non-citizens of Papua New Guinea, they were legally allowed to carry weapons, to arrest individuals, and to act in self-defense.[124] Furthermore, in the agreement Papua New Guinea granted to the commanders of Sandline a broad range of far-reaching competences that normally belong to the domain of State military. The contract provided that:

> "The State recognises that Sandline's commanders will have such powers as are required to efficiently and effectively undertaken their given roles, including but not limited to the powers to engage and fight hostile forces, repel attacks therefrom, arrest any persons suspected of undertaking or conspiring to undertake a harmful act, secure Sovereign assets and territory, defend the general population from any threat, and pro-actively protect their own and State Forces from any form of aggression or threat."[125]

Taking into account the fact that the internal law of a State does not function as the ultimate tool to classify a person or an entity as a State organ,[126] and that there is no difference for the application of the concept of attribution between the acts of State

---

[120] Damian Sturzaker and Craig Cawood, (1999) "The Sandline Affair: Illegality and International Law", *Australian International Law Journal*, p. 215; Oliver R. Jones, (2009) "Implausible Deniability: State Responsibility for the Actions of Private Military Firms", *Connecticut Journal of International Law*, Volume 24, Issue 2, p. 263.

[121] Peter W. Singer, (2008) *Corporate Warriors: The Rise of the Privatized Military Industry*, Ithaca and London: Cornell University Press, pp. 193-194.

[122] Ibidem, Appendix 2, p. 267.

[123] Ibidem, Appendix 2, pp. 263-264, 266.

[124] Ibidem, p. 194.

[125] Ibidem, Appendix 2, p. 267.

[126] Commentary to DARS as adopted in 2001, p. 42, Article 4, para. 11.

organs acting in their official capacity as superiors or subordinates,[127] one might conclude that the international responsibility of Papua New Guinea could have been engaged in this particular scenario.[128] In this regard, the terms of employment of Sandline by means of concluding a contract between the firm and the government, the temporal nature of the operation of private contractors, and other issues would have been less relevant if it could clearly have been indicated that Sandline employees were, in fact, operating as a part of the regular armed forces of Papua New Guinea and thus as its State organ.

Dealing with the situation involving the conduct of State organs, there is no relevance in whether soldiers of a certain country or PMSCs working as members of the armed forces perform military or security tasks: violations of international law committed by individuals belonging to the regular armed forces will successfully be attributed to the respective State in accordance with Article 4 DARS and will necessarily engage State responsibility.[129] Therefore, if a particular illegal conduct violating international law has been proved and sufficient evidence can be provided that private contractors that have committed those acts are employed by a PMSC forming part of or belonging to the armed forces of a specific State, then their conduct will be regarded as an act of a State organ. As a consequence, it will lead to the attribution of internationally wrongful acts to the respective State, which will be held responsible and will have to face the respective legal consequences. Taking into consideration the customary international law laid down in Article 3 of the Hague Convention IV and Article 91 of the 1977 First Additional Protocol to the 1949 Geneva Conventions, States parties to a conflict are liable to pay compensation for violating the law of armed conflict. They are responsible for all unlawful acts committed by those individuals that belong to their official armed forces. In this regard, no defense could be invoked that the offenders did not act in their actual capacity as soldiers or that they contravened instructions. As acknowledged above, it goes without saying that PMSCs as such do not normally constitute members of armed forces. In addition, the current practice of using these companies suggests that they are not incorporated into the official army structures, but operate merely on a contractual basis providing their services for different States or non-State actors.[130]

---

127 Ibidem, p. 41, Article 4, para. 7.

128 Oliver R. Jones, (2009) "Implausible Deniability: State Responsibility for the Actions of Private Military Firms", *Connecticut Journal of International Law*, Volume 24, Issue 2, p. 263.

129 Carsten Hoppe, (2008) "Passing the Buck: State Responsibility for Private Military Companies", *The European Journal of International Law*, Volume 19, Issue 5, pp. 990-991; Frits Kalshoven, (1991) "State Responsibility for Warlike Acts of the Armed Forces", *International and Comparative Law Quarterly*, Volume 40, Issue 4, p. 827.

130 Chia Lehnardt, "Private Military Companies and State Responsibility", in Simon Chesterman and Chia Lehnardt (Eds.), (2007) *From Mercenaries to Market: The Rise and Regulation of Private Military Companies*, Oxford: Oxford University Press, p. 143.

Even if PMSCs are not to be regarded as official State organs or parts of the regular State armed forces, it must be observed that under certain circumstances they can qualify as *de facto* State organs. As assessed in the famous cases of the International Court of Justice *Nicaragua*[131] and *Bosnian Genocide*,[132] the type of relationship between PMSCs and States can in some instances be one of complete dependence on one side and control on the other. In this regard, PMSCs just as other private entities can be considered *de facto* State organs and their unlawful conduct will necessarily be attributed to the States in question.[133] This doctrine of *de facto* State organs associated with the control tests developed by the ICJ and the ICTY is to be considered in conjunction with the assessment of the degree of control and will be addressed in the following Sections.

Concluding, one can state that Article 4 DARS should evidentially not be seen as a core provision in relation to the concept of attribution of internationally wrongful acts that might be committed by PMSC employees.[134] The theoretical possibility still exists that under certain circumstances PMSCs can fall under the qualification of the parts of official armed forces, but in reality it is usually not the case. It can be argued that perhaps the most significant provisions for attributing illegal actions of PMSCs to States are Article 5 DARS and Article 8 DARS.[135]

6.2 Article 5 DARS

6.2.1 General application of the provision

Another type of attribution that seems to be of relevance for attributing the unlawful conduct of PMSCs to States is laid down in Article 5 DARS, which provides:

> "The conduct of a person or entity which is not an organ of the State under article 4 but which is empowered by the law of that State to exercise *elements of the governmental authority* shall be considered an act of the State under international law, provided the person or entity is acting in that capacity in the particular instance." (emphasis added)

---

[131] International Court of Justice, *Case Concerning Military and Paramilitary Activities in and against Nicaragua (Nicaragua v. United States of America)*, Judgment (Merits), June 27, 1986, ICJ Reports 1986, paras. 109, 115.

[132] International Court of Justice, *Case Concerning the Application of the Convention on the Prevention and Punishment of the Crime of Genocide (Bosnia and Herzegovina v. Serbia and Montenegro)*, Judgment (Merits), February 26, 2007, ICJ Reports 2007, p. 391.

[133] Marko Milanović, (2009) "State Responsibility for Acts of Non-state Actors: A Comment on Griebel and Plücken", *Leiden Journal of International Law*, Volume 22, Issue 2, p. 309.

[134] Oliver R. Jones, (2009) "Implausible Deniability: State Responsibility for the Actions of Private Military Firms", *Connecticut Journal of International Law*, Volume 24, Issue 2, p. 264.

[135] Chia Lehnardt, "Private Military Companies and State Responsibility", in Simon Chesterman and Chia Lehnardt (Eds.), (2007) *From Mercenaries to Market: The Rise and Regulation of Private Military Companies*, Oxford: Oxford University Press, p. 143.

From this provision, two distinctive conditions can be derived. They need to be fulfilled in order to make the attribution on the basis of Article 5 DARS possible: in the first place, there must be an entity that exercises elements of governmental authority; secondly, that entity must be authorized to perform a governmental function by the law of a State.

Article 5 DARS appears to be specifically designed to deal with the conduct of persons and entities that do not constitute organs of a State, but are entitled by the domestic law to have public functions and to exercise governmental authority. It clearly deals with situations of parastatal entities and privatized corporations fulfilling certain governmental functions of a public or regulatory nature.[136] States tend to frequently employ private entities that are authorized to perform specific governmental tasks.[137] The term "entities" may include a broad spectrum of bodies that cannot qualify as State organs, but merely have certain authority in accordance with domestic laws of States. The Commentaries to the Draft Articles profoundly state that these bodies may be:

> "... *public corporations, semi-public entities, public agencies* of various kinds and even, in special cases, *private companies*, provided that in each case the entity is empowered by the law of the State to exercise functions of a public character normally exercised by State organs, and the conduct of the entity relates to the exercise of the governmental authority concerned."[138] (emphasis added)

As a few examples, one can mention central banks having their own distinct legal status within a State[139] or private and State-owned airlines that are authorized by States to deal with the issues of immigration control and quarantine, as indicated in the Commentaries to the Draft Articles.[140] Interestingly enough, the ILC's Commentaries explicitly provide another example of private security firms that might be employed by a State in order to have various public functions and to exercise elements of governmental authority, for instance, by acting as prison guards. In this regard, PMSCs may exercise core public powers, such as those of detention and

---

136 Commentary to DARS as adopted in 2001, p. 42, Article 5, para. 1; Malcolm N. Shaw, (2008) International Law, Sixth Edition, Cambridge: Cambridge University Press, p. 787.

137 Djamchid Momtaz, "Attribution of Conduct to the State: State Organs and Entities Empowered to Exercise Elements of Governmental Authority", in James Crawford, Alain Pellet, and Simon Olleson (Eds.), (2010) *The Law of International Responsibility*, Oxford: Oxford University Press, p. 244.

138 Commentary to DARS as adopted in 2001, p. 43, Article 5, para. 2.

139 Djamchid Momtaz, "Attribution of Conduct to the State: State Organs and Entities Empowered to Exercise Elements of Governmental Authority", in James Crawford, Alain Pellet, and Simon Olleson (Eds.), (2010) *The Law of International Responsibility*, Oxford: Oxford University Press, p. 244; Chia Lehnardt, "Private Military Companies and State Responsibility", in Simon Chesterman and Chia Lehnardt (Eds.), (2007) *From Mercenaries to Market: The Rise and Regulation of Private Military Companies*, Oxford: Oxford University Press, pp. 143-144.

140 Commentary to DARS as adopted in 2001, p. 43, Article 5, para. 2.

discipline, in accordance with domestic prison regulations and pursuant to a judicial sentence.[141]

However, the classification of an entity as public or as private, the degree of participation of a State in its capital or in the ownership of its assets do not constitute decisive criteria for the application of Article 5 DARS. The most important element in the considerations relating to the application of this provision is, according to the Commentaries, the characteristic feature of an entity as exercising elements of governmental authority: thus, the nature of the activity itself, and its connection to the State in question.[142] It is remarkable that Article 5 DARS does not define the notion of elements of governmental authority. The Commentaries clearly create an impression that the provision does not attempt to specify the precise scope of the notion for the purpose of attribution of the unlawful conduct of an entity to a certain State.[143] In this respect, the ILC decided that the term "governmental" is dependent upon "the particular society, its history and traditions".[144] It is also stressed that of particular significance are not merely the content of the exercised powers, but also the manner in which they have been conferred on an entity, the purposes for which they are exercised, and the extent of accountability of an entity to government for their exercise. One should understand that there is no general criterion to identify entities exercising elements of public power[145] and there is no exhaustive list of governmental functions or inherent State functions, as has been pointed out at the Meeting of Experts organized by the University Centre for International Humanitarian Law in Geneva.[146] However, it can be argued that, according to the case law of international judicial bodies, governmental functions may most certainly include the following:[147]

---

141 Ibidem.

142 Ibidem, p. 43, Article 5, para. 3; International Law Commission, *Third Report on State Responsibility, by Mr. Roberto Ago, Special Rapporteur, the Internationally Wrongful Act of the State, Source of International Responsibility*, (1971) Yearbook of the International Law Commission, Volume II(1), UN Doc. A/CN.4/264 and Add.1-3, para. 191; Chia Lehnardt, "Private Military Companies and State Responsibility", in Simon Chesterman and Chia Lehnardt (Eds.), (2007) *From Mercenaries to Market: The Rise and Regulation of Private Military Companies*, Oxford: Oxford University Press, p. 144.

143 Commentary to DARS as adopted in 2001, p. 43, Article 5, para. 6.

144 Ibidem.

145 Djamchid Momtaz, "Attribution of Conduct to the State: State Organs and Entities Empowered to Exercise Elements of Governmental Authority", in James Crawford, Alain Pellet, and Simon Olleson (Eds.), (2010) *The Law of International Responsibility*, Oxford: Oxford University Press, p. 244.

146 University Centre for International Humanitarian Law, *Expert Meeting on Private Military Contractors: Status and State Responsibility for Their Actions*, Report, Geneva, August 29-30, 2005, p. 16, <http://www.geneva-academy.ch/docs/expert-meetings/2005/2rapport_compagnies_privees.pdf>, accessed on October 10, 2014.

147 Oliver R. Jones, (2009) "Implausible Deniability: State Responsibility for the Actions of Private Military Firms", *Connecticut Journal of International Law*, Volume 24, Issue 2, p. 265.

the authority to maintain law and order,[148] to pursue governmental policy,[149] to confiscate property,[150] and to deal with immigration and customs.[151]

Finally, it must be added that Article 7 DARS might prove to be of relevance with regard to the practice of relying on PMSCs that constitute State organs in accordance with Article 4 DARS or persons and entities falling under Article 5 DARS. It maintains that, if a State organ, a person, or an entity exercising elements of governmental authority exceed their authority or contravene instructions, their conduct will be considered an act of the State in question. Dealing with the important issue of unauthorized or *ultra vires* State acts, the Commentaries to the ILC's Draft Articles state that such activities are to be attributed to the State granting to the persons or entities in question governmental authority.[152] This rule laid down in Article 7 DARS that has emerged due to the necessity for having more certainty and security in international relations[153] should also be taken into consideration in addition to Article 5 DARS when assessing the use of PMSCs performing a spectrum of State functions.

6.2.2 Application of Article 5 DARS with regard to PMSCs

It appears from the foregoing that the commission of illegal acts by official State organs will necessarily be attributed to the State to which those organs belong. Some authors argue that it is unacceptable that States can be held responsible for the illegal conduct of their forces officially operating on their behalf and protecting their interests, while at the same time they seem to be able to escape responsibility when they opt to employ PMSCs, which, in fact, can be performing similar functions.[154] Therefore, it is worth mentioning that the commission of internationally wrongful acts by

---

148 Iran – United States Claims Tribunal, *Yeager v. Government of the Islamic Republic of Iran*, (1987) 17 Iran – United States Claims Tribunal Reports, p. 102.

149 International Centre for Settlement of Investment Disputes, *Emilio Agustín Maffezini v. The Kingdom of Spain*, Decision on Jurisdiction (English translation), January 25, 2000, ICSID Case No. ARB/97/7, pp. 79-88; Iran – United States Claims Tribunal, *Hyatt International Corporation v. Government of the Islamic Republic of Iran*, (1985) 9 Iran – United States Claims Tribunal Reports, p. 91; Iran – United States Claims Tribunal, *Oil Field of Texas, Inc. v. Government of the Islamic Republic of Iran*, (1982) 1 Iran – United States Claims Tribunal Reports, p. 351.

150 Iran – United States Claims Tribunal, *Daley v. Government of the Islamic Republic of Iran*, (1988) 18 Iran – United States Claims Tribunal Reports, p. 237; Iran – United States Claims Tribunal, *Computer Sciences Corporation v. Government of the Islamic Republic of Iran*, (1986) 10 Iran – United States Claims Tribunal Reports, pp. 270-271; Iran – United States Claims Tribunal, *William L. Pereira Associates, Iran v. Government of the Islamic Republic of Iran*, (1984) 5 Iran – United States Claims Tribunal Reports, p. 206.

151 Iran – United States Claims Tribunal, *Yeager v. Government of the Islamic Republic of Iran*, (1987) 17 Iran – United States Claims Tribunal Reports, p. 110.

152 Commentary to DARS as adopted in 2001, p. 45, Article 7, para. 1.

153 Ibidem, p. 45, Article 7, para. 3.

154 Hin-Yan Liu, (2010) "Leashing the Corporate Dogs of War: The Legal Implications of the Modern Private Military Company", *Journal of Conflict & Security Law*, Volume 15, Issue 1, p. 164.

PMSCs falling within the scope of definition of a State organ or its part will, therefore, trigger the engagement of international responsibility of the respective State under Article 4 DARS. Although this specific type of attribution is unlikely to find its application with regard to PMSCs hired by States, another mode of attribution might prove to be of a greater practical relevance. Violations of international humanitarian and human rights law committed by private contractors can be considered to be attributable to a State if PMSCs are allowed to exercise elements of governmental authority, as stipulated in Article 5 DARS. In this respect, it is necessary to assess under what conditions and circumstances and to what extent PMSCs can be regarded as being authorized to carry out governmental tasks, which will make the attribution of the basis of Article 5 DARS possible.

As mentioned above, the most important aspect of this form of attribution lays in the nature of the activity performed by PMSCs and not the actual character – private or public – of the entity involved.[155] Even if a certain entity is to be considered a private firm, but is entitled to carry out governmental tasks having the authority of a governmental body, this circumstance must trigger the attribution of its unlawful conduct to the State concerned. The reasoning behind this rule is derived from the fact that the phenomenon of emerging corporations in the military and security spheres that exercise public State authority should not enable States to escape responsibility by hiring them and outsourcing the core State functions.[156] States should not be able to disregard their obligations under international law by using private corporations performing essential State activities. This idea was reflected in the reasoning of the UN Human Rights Committee that rightly declared:

> "... the contracting out to the private commercial sector of core State activities which involve the use of force and the detention of persons does not absolve a State party of its obligations under the Covenant, notably under articles 7 and 10."[157]

Therefore, in the first instance, it needs to be examined whether PMSCs can qualify as legally established businesses exercising elements of governmental authority or, as the Committee has put it, "core State activities".

In the *Emilio Agustín Maffezini v. The Kingdom of Spain* arbitration case (hereinafter: *Maffezini v. Spain* case), the Tribunal provided two criteria for determining

---

[155] Jackson N. Maogoto and Benedict Sheehy, (2009) "Private Military Companies & International Law: Building New Ladders of Legal Accountability & Responsibility", *Cardozo Journal of Conflict Resolution*, Volume 11, p. 110.

[156] Chia Lehnardt, "Private Military Companies and State Responsibility", in Simon Chesterman and Chia Lehnardt (Eds.), (2007) *From Mercenaries to Market: The Rise and Regulation of Private Military Companies*, Oxford: Oxford University Press, p. 144.

[157] UN Human Rights Committee, *Mr. Carlos Cabal and Mr. Marco Pasini Bertran v. Australia*, September 19, 2003, Communication No. 1020/2001, UN Doc. CCPR/C/78/D/1020/2001 (2003), para. 7.2.

whether a certain corporation exercises elements of governmental authority.[158] First of all, one should examine a structural relationship between the corporation and the State in question.[159] This particular analysis needs to be based on the assessment of the degree of control exercised by the State over the corporation or the level of ownership. According to the Commentaries to the Draft Articles, the scope of Article 5 DARS is clearly limited to entities that are specifically authorized to exercise governmental authority by internal law, meaning that the situations when they act under the direction or control of a State are excluded from the scope of its application, as this issue is addressed in Article 8 DARS.[160] However, the theory of State control can come into play when deciding on the degree of State involvement in the operation of a specific body. These two criteria – the degree of State control and the level of ownership – are without doubt of great significance, as they can indicate the possibility of attribution of the unlawful conduct of those entities to State, but it needs to be borne in mind that they do not constitute a set of decisive factors.[161] Secondly, it must be identified whether a corporation exercises functions that are to be seen as essentially State functions.[162] As specified by the ILC, "the Commission has come to the conclusion that the most appropriate solution is to refer to the real common feature which these entities have: namely that they are empowered, if only exceptionally and to a limited extent, to exercise specified functions which are akin to those normally exercised by organs of the State".[163] In other words, there must be a link between the activities of an entity exercising prerogatives of public authority and the State in

---

158 International Centre for Settlement of Investment Disputes, *Emilio Agustín Maffezini v. The Kingdom of Spain*, Decision on Jurisdiction (English translation), January 25, 2000, ICSID Case No. ARB/97/7, pp. 31-33, <http://icsid.worldbank.org/ICSID/FrontServlet?requestType=CasesRH&actionVal=showDoc&docId=DC565_En&caseId=C163>, accessed on March 1, 2011; Abby C. Smutny, "State Responsibility and Attribution: When is a State Responsible for the Acts of State Enterprises? Emilio Agustín Maffezini v. The Kingdom of Spain", in Todd Weiler (Ed.), (2005) *International Investment Law and Arbitration*, London: Cameron May.

159 International Centre for Settlement of Investment Disputes, *Emilio Agustín Maffezini v. The Kingdom of Spain*, Decision on Jurisdiction (English translation), January 25, 2000, ICSID Case No. ARB/97/7, pp. 31, 33, <http://icsid.worldbank.org/ICSID/FrontServlet?requestType=CasesRH&actionVal=showDoc&docId=DC565_En&caseId=C163>, accessed on March 1, 2011.

160 Commentary to DARS as adopted in 2001, p. 43, Article 5, para. 7.

161 Djamchid Momtaz, "Attribution of Conduct to the State: State Organs and Entities Empowered to Exercise Elements of Governmental Authority", in James Crawford, Alain Pellet, and Simon Olleson (Eds.), (2010) *The Law of International Responsibility*, Oxford: Oxford University Press, pp. 244-245.

162 International Centre for Settlement of Investment Disputes, *Emilio Agustín Maffezini v. The Kingdom of Spain*, Decision on Jurisdiction (English translation), January 25, 2000, ICSID Case No. ARB/97/7, pp. 31-33, <http://icsid.worldbank.org/ICSID/FrontServlet?requestType=CasesRH&actionVal=showDoc&docId=DC565_En&caseId=C163>, accessed on March 1, 2011.

163 International Law Commission, *Report of the International Law Commission on the Work of Its Twenty-Sixth Session, 6 May – 26 July 1974*, (1974) Yearbook of the International Law Commission, Volume II(1), UN Doc. A/9610/Rev.1, p. 282, para. 18.

question.[164] In this particular case, the Tribunal came to the conclusion that the firm was, in fact, created as a tool to pursue governmental policy and was for 88% owned by the State. Moreover, it was entitled to exercise typically governmental functions that are normally not supposed to be carried out by non-State entities and individuals. Therefore, the conclusion had to be drawn that the corporation did not have a solely commercial nature as a business entity, but was in practice an instrument of the State, through which it acted and exercised its sovereign powers.

The mere fact that PMSCs perform some core State functions does not necessarily lead to the conclusion that those companies exercise elements of governmental authority and that this matter automatically falls under the scope of Article 5 DARS. According to the conditions laid down in Article 5 DARS, the internal law of a State in question needs to empower PMSCs to exercise certain public functions that normally belong to the domain of official State organs. However, this criterion of the availability of domestic legislation conferring powers of public authority on certain entities cannot be regarded as the most decisive factor in the examination of this particular type of attribution.[165] What is of relevance is the establishment of a nexus between an entity and a State granting to it specific governmental powers. In the most cases, this nexus is to be found in the municipal law of the State conferring certain State functions on the entity. If the internal law of a State grants to that entity a right to exercise elements of governmental authority, then its acts must be attributable to that State, which is a principle acknowledged in international law.[166] This particular provision of the DARS is in practice of crucial importance for the attribution mainly to the hiring States that employ PMSCs.[167] It is, however, also possible that home and host States enact internal laws empowering PMSCs to perform elements of governmental authority, especially when these States also contract for the services of PMSCs.

For the application of this particular provision, the national law of a certain State needs to explicitly stipulate that a specific governmental authority is allowed to delegate its powers to PMSCs and when this authority hires PMSCs, the condition of having the internal law laid down in Article 5 DARS will be met.[168] This appears to

---

164 Djamchid Momtaz, "Attribution of Conduct to the State: State Organs and Entities Empowered to Exercise Elements of Governmental Authority", in James Crawford, Alain Pellet, and Simon Olleson (Eds.), (2010) *The Law of International Responsibility*, Oxford: Oxford University Press, p. 245.

165 Ibidem, p. 246.

166 Chia Lehnardt, "Private Military Companies and State Responsibility", in Simon Chesterman and Chia Lehnardt (Eds.), (2007) *From Mercenaries to Market: The Rise and Regulation of Private Military Companies*, Oxford: Oxford University Press, p. 144.

167 Charlotte Beaucillon, Julian Fernandez, and Hélène Raspail, "State Responsibility for Conduct of Private Military and Security Companies Violating Ius ad Bellum", in Francesco Francioni and Natalino Ronzitti (Eds.), (2011) *War by Contract: Human Rights, Humanitarian Law, and Private Contractors*, Oxford: Oxford University Press, p. 405.

168 University Centre for International Humanitarian Law, *Expert Meeting on Private Military Contractors: Status and State Responsibility for Their Actions*, Report, Geneva, August 29-30, 2005, p. 18,

be a strict requirement of establishing relationships between States and private actors operating under the authorization of States' legislative or executive organs.[169] The requirement is based on the idea that a State performing its sovereign functions and issuing laws and regulations can be held liable on the international plane for their creation.[170] Clearly, the main type of empowerment by a State will be "a delegation or authorization by or under the law of the State".[171] A logical assumption is that a broad rather than a narrow interpretation of the term "internal law" can and should be applied.[172] Establishing a criterion requiring solely the internal law to grant elements of governmental authority to certain entities and individuals will provide a potential possibility for States to avoid any responsibility for their actions performed by third persons and entities. If such a requirement is set and its narrow interpretation is used, States will have a leeway in the process of adjusting the extent of international responsibility by outsourcing their most essential State functions to individuals and entities and omitting to regulate the matter at the level of domestic legislation.[173] Therefore, when the concept of internal law in its broad sense is applied, the empowerment may also be realized by means of a legal instrument other than law, for instance, a contract concluded between a State and an entity.[174] Taking into account the fact that PMSCs are often licensed and certified by States to perform their military and security tasks and that hiring States conclude contracts with them specifying terms of their operation, one can assume that those instruments may be regarded as internal law conferring the exercise of governmental authority on certain States, if that notion is applied in its broad sense.[175]

It is worth mentioning that it is not entirely improbable that PMSCs can be entrusted with certain tasks similar to State functions normally carried out by State

---

<http://www.geneva-academy.ch/docs/expert-meetings/2005/2rapport_compagnies_privees.pdf>, accessed on October 10, 2014.

169 Charlotte Beaucillon, Julian Fernandez, and Hélène Raspail, "State Responsibility for Conduct of Private Military and Security Companies Violating Ius ad Bellum", in Francesco Francioni and Natalino Ronzitti (Eds.), (2011) *War by Contract: Human Rights, Humanitarian Law, and Private Contractors*, Oxford: Oxford University Press, p. 404.

170 Ibidem.

171 International Law Commission, *First Report on State Responsibility by Mr. James Crawford, Special Rapporteur*, (1998) Yearbook of the International Law Commission, Volume II(2), UN Doc. A/CN.4/490 and Add 1-7, p. 3.

172 Oliver R. Jones, (2009) "Implausible Deniability: State Responsibility for the Actions of Private Military Firms", *Connecticut Journal of International Law*, Volume 24, Issue 2, p. 268.

173 Djamchid Momtaz, "Attribution of Conduct to the State: State Organs and Entities Empowered to Exercise Elements of Governmental Authority", in James Crawford, Alain Pellet, and Simon Olleson (Eds.), (2010) *The Law of International Responsibility*, Oxford: Oxford University Press, p. 246.

174 Marina Spinedi, (2005) "Private Contractors: Responsabilité Internationale des Entreprises ou Attribution à l'Etat de la Conduite des Personnes Privées?", *International Law FORUM du Droit International*, Volume 7, p. 277.

175 Oliver R. Jones, (2009) "Implausible Deniability: State Responsibility for the Actions of Private Military Firms", *Connecticut Journal of International Law*, Volume 24, Issue 2, pp. 268-269.

organs and therefore fall under the scope of application of Article 5 DARS. One problematic matter raised in the context of application of the principle laid down in this provision is associated with the classification of PMSCs as non-State entities exercising elements of governmental authority. PMSCs specialize in the provision of a large number of services that cannot easily be said to involve the exercise of State authority due to the uncertainty regarding their range and nature.[176] It is hard to determine what constitutes the exercise of governmental authority considering the fact that there is no international consensus on this issue.[177] Although the International Law Commission provides some guidelines regarding the content of delegated powers, there is still much incertitude. The qualification of a certain authority as governmental is dependent on the "particular society, its history and traditions".[178] In this respect, not only the content of the delegated powers, but also the way of delegation, the purposes of their exercise, and the extent of accountability of PMSCs to a government need to be taken into consideration.[179] It appears that the application of Article 5 DARS highly depends on particular circumstances of a specific case, which need to be closely examined.

In order to find some illustrative examples of PMSCs in their capacity of non-State entities empowered to exercise elements of governmental authority, reference should be made to the Commentaries to the ILC's Draft Articles stipulating that those entities may include public corporations, semi-public entities, various public agencies, and private companies, such as PMSCs.[180] In practice, States do not own PMSCs, but can definitely be said to exercise some degree of control over them. Normally, a contract serves as a tool to have control over the operations carried out by private firms that perform military and security services for the hiring States. It is no secret that some States expressly limit the types of tasks to be performed by PMSCs for which they can be contracted. This is usually done in order to avoid this particular mode of attribution concerned with the performance of the functions of public character.[181]

---

[176] Abby C. Smutny, "State Responsibility and Attribution: When is a State Responsible for the Acts of State Enterprises? Emilio Agustín Maffezini v. The Kingdom of Spain", in Todd Weiler (Ed.), (2005) *International Investment Law and Arbitration*, London: Cameron May, p. 39; Shannon Bosch, (2008) "Private Security Contractors and State Responsibility: Are States Exempt from Responsibility for Violations of Humanitarian Law Perpetrated by Private Security Contractors?", *The Comparative and International Law Journal of Southern Africa*, Volume 41, Issue 3, pp. 374-375; Chia Lehnardt, "Private Military Companies and State Responsibility", in Simon Chesterman and Chia Lehnardt (Eds.), (2007) *From Mercenaries to Market: The Rise and Regulation of Private Military Companies*, Oxford: Oxford University Press, p. 144.

[177] Chia Lehnardt, "Private Military Companies and State Responsibility", in Simon Chesterman and Chia Lehnardt (Eds.), (2007) *From Mercenaries to Market: The Rise and Regulation of Private Military Companies*, Oxford: Oxford University Press, p. 144.

[178] Commentary to DARS as adopted in 2001, p. 43, Article 5, para. 6.

[179] Ibidem.

[180] Ibidem, p. 43, Article 5, para. 2.

[181] Shannon Bosch, (2008) "Private Security Contractors and State Responsibility: Are States Exempt from Responsibility for Violations of Humanitarian Law Perpetrated by Private Security Contractors?",

The US Department of Defense is, for instance, known for requiring in its instructions that "inherently governmental" functions cannot be outsourced to PMSCs.[182]

Nevertheless, it should be acknowledged that a large number of tasks carried out by PMSCs may be regarded by their nature as falling within the scope of the exercise of governmental authority. Some of those services are, in fact, essentially State functions that have been performed by States in the course of history and even nowadays are usually seen as the core of governmental powers. State authorities often hire PMSCs to carry out the tasks of traditional armed forces or other State organs, while denying them access to possible privileges and benefits of being a member of the military or belonging to other governmental structures.[183] The participation of, for instance, PMSC Blackwater in the offensive and defensive operations on the Iraqi soil may also be regarded as one of the main State prerogatives to use force and to carry out military operations.[184] It means that if Blackwater as a corporate entity was authorized by the US to perform tasks, which can be seen as governmental, the unlawful actions of its employees at Nisour Square will necessarily be attributed to the US. In this respect, two elements must be established: it is to be evaluated, in the first place, whether Blackwater had been exercising elements of governmental authority and, secondly, whether the company and its personnel were empowered to do so by US law.

The actual content of the notion of governmental authority appears to be dependent upon "the particular society, its history and traditions". There is no coherent list of what constitutes elements of governmental authority that can provide useful guidance in this regard.[185] During the Expert Meeting on PMSCs that took place in Geneva on August 29-30, 2005, one of the experts voiced his opinion that among intrinsic State functions one can find activities related to policing, prisons, and judicial administration. The ILC's Commentaries specifically refer to private security firms that might be contracted by States to provide prison guards and thus exercise

---

The Comparative and International Law Journal of Southern Africa, Volume 41, Issue 3, p. 374; Titia van der Maas, (2010) "Private Military Companies in het Amerikaanse en Britse defensiebeleid", *Vrede en Veiligheid*, Volume 39, Issue 4.

182 US Department of Defense Instruction 1100.22, Policy and Procedures for Determining Workforce Mix, April 12, 2010, p. 2, para. 4(c), <http://www.dtic.mil/whs/directives/corres/pdf/110022p.pdf>, accessed on September 12, 2013.

183 Shannon Bosch, (2007) "Private Security Contractors in International Humanitarian Law: A Skirmish for Recognition in International Armed Conflicts", *African Security Review*, Volume 16, Issue 4; Shannon Bosch, (2008) "Private Security Contractors and State Responsibility: Are States Exempt from Responsibility for Violations of Humanitarian Law Perpetrated by Private Security Contractors?", *The Comparative and International Law Journal of Southern Africa*, Volume 41, Issue 3, p. 377.

184 Oliver R. Jones, (2009) "Implausible Deniability: State Responsibility for the Actions of Private Military Firms", *Connecticut Journal of International Law*, Volume 24, Issue 2, p. 268.

185 University Centre for International Humanitarian Law, *Expert Meeting on Private Military Contractors: Status and State Responsibility for Their Actions*, Report, Geneva, August 29-30, 2005, p. 16, <http://www.geneva-academy.ch/docs/expert-meetings/2005/2rapport_compagnies_privees.pdf>, accessed on October 10, 2014.

governmental functions, such as "powers of detention and discipline pursuant to a judicial sentence or to prison regulations".[186] Other examples addressed in the Commentaries are less relevant to the State practice of employing PMSCs, such as the current instance of Blackwater taking care of security matters with regard to State officials and assets. As previously examined, the case law of international judicial bodies indicates that there are certain activities belonging to the realm of governmental functions: the authority to maintain law and order, to pursue governmental policy, to confiscate property, and to deal with immigration and customs. A number of experts in Geneva agreed that tasks associated with the conduct of hostilities require governmental authority to be exercised and that military functions in principle entail elements of governmental authority, as laid down in Article 5 DARS.[187]

Using the test of the *Maffezini v. Spain* case, one needs to identify two criteria necessary for establishing whether a certain non-State entity exercises elements of governmental authority. First, the structural relationship between the company and the US must be analyzed. This examination is based on the degree of State ownership of the firm or the extent of State control over it. Unlike the factual situation in the above-mentioned case, Blackwater is not a company owned by the US, but is a privately run business. It has not been created as a policy instrument of the US government. The level of control also cannot be said to be high due to the obvious lack of oversight and monitoring at the time of the incident. Therefore, the conclusion cannot be drawn that there is a clear structural relationship between Blackwater and the US.

Secondly, it is to be assessed whether the functions of Blackwater are typically or essentially State functions. The arising question is then: what are inherently governmental functions according to the US practice? In the US, there are laws, regulations, and policies addressing this pertinent problem.[188] An inherently governmental function is an activity that pursuant to law and policy must be carried out by federal government employees and cannot be outsourced because it is "intimately related to the public interest".[189] Within US federal law and policy, there are two main definitions which can be relevant for the purpose of determining the core of inherently governmental functions. The first one can be found in the Federal Activities Inventory

---

186 Commentary to DARS as adopted in 2001, p. 43, Article 5, para. 2.

187 University Centre for International Humanitarian Law, *Expert Meeting on Private Military Contractors: Status and State Responsibility for Their Actions*, Report, Geneva, August 29-30, 2005, p. 17, <http://www.geneva-academy.ch/docs/expert-meetings/2005/2rapport_compagnies_privees.pdf>, accessed on October 10, 2014.

188 US Federal Acquisition Regulation, 48 Code of Federal Regulations (C.F.R.) 7.503; US Department of Defense Instruction 1100.22, Policy and Procedures for Determining Workforce Mix, April 12, 2010; Defense Federal Acquisition Regulation Supplement.

189 John R. Luckey, Valerie Bailey Grasso, and Kate M. Manuel, "Inherently Governmental Functions and Department of Defense Operations: Background, Issues, and Options for Congress", CRS Report for Congress, February 1, 2010, p. 1, <http://www.fas.org/sgp/crs/natsec/R40641.pdf>, accessed on September 19, 2013.

Reform Act of 1998 (hereinafter: FAIR Act).[190] According to this act, an inherently governmental function is "a function so intimately related to the public interest as to require performance by Federal Government employees". The FAIR Act contains a list with examples of such functions.[191] These activities are in the first place "binding the United States to take, or not to take, action by contract, policy, regulation, authorization, order or otherwise". Secondly, they are "determining, protecting, and advancing U.S. economic, political, territorial, property, or other interests by military or diplomatic action, civil or criminal judicial proceedings, contract management, or otherwise". In addition, such functions are those "significantly affecting the life, liberty, or property interests of private persons". They can also be associated with "commissioning, appointing, directing or controlling officers or employees of the United States". Finally, to this list belongs the process of "exerting ultimate control over the acquisition, use, or disposition of the real or personal, tangible or intangible, property of the United States, including the collection, control or disbursement of appropriated and other federal funds". From this spectrum of governmental tasks are excluded such functions as collecting information for or providing advice, recommendations, and other ideas to Federal Government officials, and any function that is to be perceived as ministerial and internal in nature, such as building security, housekeeping, and warehouse operations.[192]

The second policy-oriented definition from the Office of Management and Budget Circular A-76 (hereinafter: OMB Circular) is based on the idea that an inherently governmental activity is "an activity that is so intimately related to the public interest as to mandate performance by government personnel".[193] The circular applies to executive departments, such as the Department of Defense or the Department of State,[194] and the examples used in it are the same as those of the FAIR Act.[195] In addition to this Act and the OMB Circular, the third important administrative law source of the definition of inherently governmental functions is the Federal Acquisition Regulation (hereinafter: FAR).[196] The definition laid down in the FAR is basically the same as the one used in the OMB Circular.[197] It stresses that contracts shall not be used to carry out inherently governmental functions[198] and provides a non-inclusive list with examples of such tasks.[199] As inherently governmental, one should consider, *inter alia*, the

---

190 Federal Activities Reform Act (1998), 31 United States Code (U.S.C.) 501, para. 5(2)(A).

191 Ibidem, paras. 5(2)(B)(i)-(v).

192 Ibidem, para. 5(2)(C).

193 Office of Management and Budget Circular A-76, Attachment A, para. (B)(1)(a), <http://www.whitehouse.gov/omb/circulars_a076_a76_incl_tech_correction>, accessed on September 10, 2013.

194 Ibidem, Attachment A, para. 5(a).

195 Ibidem, Attachment B, paras. (B)(1)(a)(1)-(4).

196 Federal Acquisition Regulation, 48 C.F.R., Subparts 2, 7.3 and 7.5.

197 Ibidem, para. 2.101.

198 Ibidem, para. 7.503(a).

199 Ibidem, para. 7.503(c).

direct conduct of criminal investigations and the control of prosecutions and performance of adjudicatory tasks other than arbitration or other types of alternative dispute resolution. Importantly for the State practice of using PMSCs in armed conflicts and peace operations, other crucial inherently governmental functions are "the command of military forces, especially the leadership of military personnel who are members of the combat, combat support, or combat service support role" and "the direction and control of intelligence and counter-intelligence operations".[200]

Also, the US Department of Defense instruction is useful in determining the scope of functions that could be considered inherently governmental.[201] It can be extracted from the instruction that combat operations are considered to be the domain of the government. The instruction provides that "when armed fighting or use of force is deemed necessary for national defense, the Department of Defense may authorize deliberate destruction and/or disruptive action against the armed forces or other military objectives of another sovereign government or against other armed actors on behalf of the United States".[202] Furthermore, it is stated that "the U.S. Government has exclusive responsibility for discretionary decisions concerning the appropriate, measured use of combat power, including the offensive use of destructive or deadly force on behalf of the United States".[203] The instruction further provides that "combat operations authorized by the U.S. Government entail the exercise of sovereign Government authority and involve substantial discretion".[204] In this regard, another instruction addresses the situations of PMSCs used by the government by stressing that "contractors performing private security functions are not authorized to perform inherently governmental functions. In this regard, they are limited to a defensive response to hostile acts or demonstrated hostile intent".[205]

More importantly, the US DoD instruction indicates under what circumstances provision of security by PMSCs to protect resources and operations in hostile and volatile environments can be perceived as an inherently governmental activity.[206] This is the case when PMSCs operate in hostile areas as part of "larger, totally integrated and cohesive" armed forces and perform missions in direct support of combat.[207] Provision

---

200 Ibidem, paras. 7.503(c)(3), 7.503(c)(8).

201 US Department of Defense Instruction 1100.22, Policy and Procedures for Determining Workforce Mix, April 12, 2010, Enclosure 4, para. 1(c), <http://www.dtic.mil/whs/directives/corres/pdf/110022p.pdf>, accessed on September 12, 2013.

202 Ibidem.

203 Ibidem, Enclosure 4, para. 1(c)(1)(a).

204 Ibidem, Enclosure 4, para. 1(c)(1)(b).

205 US Department of Defense Instruction 3020.50, Private Security Contractors (PSCs) Operating in Contingency Operations, July 22, 2009, Glossary, Part II. Definition, p. 12, <http://www.acq.osd.mil/log/PS/SPOT/DoD_policy/DODI_3020.50_PSCs.pdf>, accessed on September 12, 2013.

206 US Department of Defense Instruction 1100.22, Policy and Procedures for Determining Workforce Mix, April 12, 2010, Enclosure 4, para. 1(d), <http://www.dtic.mil/whs/directives/corres/pdf/110022p.pdf>, accessed on September 12, 2013.

207 Ibidem, Enclosure 4, para. 1(d)(1)(a).

of security is also inherently governmental if it is done "in environments where there is such a high likelihood of hostile fire, bombings, or biological or chemical attacks by groups using sophisticated weapons and devices that, in the judgment of the military commander, the situation could evolve into combat".[208] The same is true for "security actions that entail assisting, reinforcing, or rescuing PSCs or military units who become engaged in hostilities are IG [inherently governmental] because they involve taking deliberate, offensive action against a hostile force on behalf of the United States".[209] Additionally, providing security is inherently governmental "if, in the commander's judgment, an offensive response to hostile acts or demonstrated hostile intentions would be required to operate in, or move resources through, a hostile area of operation. Decisions to offensively respond to hostile acts or demonstrated hostile intentions (e.g., assault or preemptively attack) entail substantial discretion and are IG."[210] Finally, security tasks are inherently governmental "if, in the commander's judgment, decisions on the appropriate course of action would require substantial discretion, the outcome of which could significantly affect U.S. objectives with regard to the life, liberty, or property of private persons, a military mission, or international relations. Such actions typically require high-risk, on-the-spot judgments on the appropriate level of force, acceptable level of collateral damage, and whether the target is friend or foe in situations pivotal to U.S. interests."[211] PMSCs are, in fact, allowed to provide security services as long as they are not considered inherently governmental in accordance with the above-mentioned regulations.[212]

The instructions of the US DoD apply when private contractors are hired by the US DoD to carry out military and security activities. The Blackwater employees engaged in the incident at Nisour Square were sent to Iraq on the basis of a contract concluded between Blackwater and the US Department of State. It can be argued that in this scenario one should observe the FAIR Act, the Federal Acquisition Regulation, and the OMB Circular, despite the fact that the US DoD instruction might be in line with those but is specifically aimed at the US DoD. The examples of inherently governmental functions listed in the FAR implementing the FAIR Act are not to be seen as an exhaustive list of activities exclusively belonging to the State domain. The actual engagement in hostilities by Blackwater employees on behalf of the US would probably fall within the scope of tasks qualifying as the exercise of governmental authority given that it can be perceived as "the command of military forces, especially the leadership of military personnel who are members of the combat, combat support, or combat service support role".[213] The activities of Blackwater's employees

208 Ibidem, Enclosure 4, para. 1(d)(1)(b).
209 Ibidem, Enclosure 4, para. 1(d)(1)(c).
210 Ibidem, Enclosure 4, para. 1(d)(1)(d).
211 Ibidem, Enclosure 4, para. 1(d)(1)(e).
212 Ibidem, Enclosure 4, para. 1(d)(1)(f).
213 Federal Acquisition Regulation, 48 C.F.R., para. 7.503(c)(3).

that were performed in practice can hardly meet such qualification. The Raven 23 team involved in the Nisour Square massacre was part of Blackwater providing security services to US government employees active in Iraq. In this particular instance, the Raven 23 guards were giving back-up fire support and clearing the evacuation route for the other Blackwater team protecting US officials. Similarly, this type of action does not fit under the categories of examples provided in the FAIR Act and the OMB Circular and should most probably not be considered as involving "determining, protecting, and advancing U.S. economic, political, territorial, property, or other interests"[214] by military actions, which comes closer to what the Blackwater guards had been occupied with in Iraq.

However, there is a claim to be made in favor of the application of one of the provisions of the FAIR Act and the OMB Circular stressing that inherently governmental functions are those that "significantly affect the life, liberty, or property of private persons".[215] Although various types of security service provision performed by Blackwater for the US Department of State and, in particular, the activities of the Raven 23 team do not directly affect the life, liberty, or property of private individuals, it is not beyond the realm of possibility that their operations can have an indirect impact on not only the life of Iraqi population, but also liberty and property of Iraqi citizens. What happened at Nisour Square is, in fact, a perfect example of how security activities of PMSCs can indirectly affect the life of the local population of the countries where they operate, and have negative consequences for liberty and property of those private persons. Many Iraqi civilians were killed and injured while being stuck in the traffic and vehicles of many of them appeared to have been damaged. From the above examination, it appears that in principle private security services provided by the Raven 23 team would not qualify as the exercise of elements of governmental authority by means of a private corporation, if a strict interpretation of the provisions of the FAIR Act and the OMB Circular is applied. The general doctrine seems to underline the idea that the provision of security as such does not meet the criteria of being defined as the exercise of governmental authority.[216]

The closer examination used in combination with a less strict approach towards the interpretation of the FAIR Act and the OMB Circular, however, reveals that these functions indirectly affect important aspects of life of Iraqi population. The process of providing security to individuals in conflict zones, such as was the case in the Blackwater incident, belongs to a grey area of uncertainty suitable for interpretation.[217] In

---

214 Federal Activities Reform Act (1998), 31 U.S.C. 501, para. 5(2)(B)(ii); Office of Management and Budget Circular A-76, Attachment A, para. (B)(1)(a)(2), <http://www.whitehouse.gov/omb/circulars_a076_a76_incl_tech_correction>, accessed on September 10, 2013.

215 Ibidem, para. 5(2)(B)(iii); ibidem, Attachment A, para. (B)(1)(a)(3).

216 Lindsey Cameron and Vincent Chetail, (2013) *Privatizing War: Private Military and Security Companies under Public International Law*, Cambridge: Cambridge University Press, p. 202.

217 Ibidem, p. 203.

this respect, it can be argued that the activities and functions of Blackwater carried out in the areas of combat operations and having the potential of escalating into the use of force and direct involvement in armed hostilities can be labeled as the exercise of inherently governmental functions. This statement would support the idea that Blackwater indeed exercised elements of governmental authority, as laid down in Article 5 DARS. The company had been doing it on a consistent basis, had been empowered to perform these acts by the law of the United States, especially if one considers the contract and certification and licensing schemes as a legal tool authorizing PMSCs, and was acting in such a capacity through its Raven 23 team at the time that the events of the Nisour Square massacre took place. If it is indeed the case, the attribution ex Article 5 DARS would be possible and the United States as the hiring State would be held responsible if it can also be established that the conduct of Blackwater fell under qualification of a breach of an international obligation of the State in question.

Another example of PMSCs exercising tasks closely related to governmental authority involves two companies, CACI and Titan, employed by the United States to perform their tasks at the Abu Ghraib detention facility in Iraq and involved in the abuse of prisoners. These two PMSCs contracted by the US DoD at the Abu Ghraib prison were involved with at least two main types of activities: translation and interrogation.[218] CACI was hired to provide personnel for the interrogation of detainees, while Titan employees were occupied with the tasks of translation and interpretation. In the analysis of these activities, the laws and regulations of the US can provide a perspective on the actual State practice of this State and give an answer to the question of whether these functions imply the exercise of governmental authority. The FAIR Act, the FAR, the OMB Circular, and the US DoD Instruction 1100.22 contain references to the concept of inherently governmental functions and list certain examples of such tasks.

Taking into account the fact that Titan was hired by the government to provide translation services, they would not constitute the exercise of elements of governmental authority and the activities of this corporation would be deemed to be commercial. Interrogation is, however, an entirely different matter that is clearly associated with the array of functions performed by the government.[219] It can fall under the definition of tasks "significantly affecting the life, liberty, or property interests of private persons", as provided in the FAIR Act and the OMB Circular.[220] Under the FAR, the interrogation of detainees belongs to "the direction and control of intelligence and

---

[218] Katja Nieminen, (2004) "The Rules of Attribution and the Private Military Contractors at Abu Ghraib: Private Acts or Public Wrongs?", *Finnish Yearbook of International Law*, Volume 15, p. 299.

[219] Paul R. Verkuil, (2007) *Outsourcing Sovereignty: Why Privatization of Government Functions Threatens Democracy and What We Can Do About It*, Cambridge: Cambridge University Press, p. 130.

[220] Federal Activities Reform Act (1998), 31 U.S.C. 501, para. 5(2)(B)(iii); Office of Management and Budget Circular A-76, Attachment B, para. (B)(1)(a)(3), <http://www.whitehouse.gov/omb/circulars_a076_a76_incl_tech_correction>, accessed on September 10, 2013.

counter-intelligence operations".[221] According to Army Regulation 715-9 implementing the US DoD Instructions 1100.22 and 3020.41, the direction and control of these particular operations include interrogations and constitute inherently governmental functions.[222] Army Regulation 715-9 further stresses that private contractors may be employed to make drafts of interrogation plans and to engage in interrogations if there is sufficient monitoring and supervision exercised by government officials.[223] This does not appear to be the case in the incident involving CACI and Titan personnel that were not properly supervised.[224]

Interrogations carried out by employees of private contractors at the Abu Ghraib detention facility can be considered inherent activities of the State, given that individuals interrogating prisoners or prisoners-of-war must have a sufficient level of military training and education in the matters of international humanitarian and human rights law.[225] The US regulation suggests that these inherently governmental activities must be performed by government personnel and cannot be outsourced to contractors.[226] As some experts believe, it can be concluded that PMSC employees at the Abu Ghraib prison engaging in interrogations of prisoners in practice exercise elements of the essential governmental authority on the basis of a contract with the government that in combination with certification and licensing regimes might be seen as a legal instrument authorizing them.[227] The US, in fact, authorized this process of outsourcing of governmental functions to two PMSCs through its domestic laws and

---

221 Federal Acquisition Regulation, 48 C.F.R., para. 7.503(c)(8).

222 US Department of the Army, Army Regulation 715-9, Procurement: Operational Contract Support Planning and Management, June 20, 2011, Appendix B: Inherently Governmental Function Guidance, p. 22, para. B-2(d), <http://armypubs.army.mil/epubs/pdf/r715_9.pdf>, accessed on June 20, 2014.

223 Ibidem.

224 Major General Antonio N. Taguba, Report of the Article 15-6 Investigation of the 800th Military Police Brigade, March 2004, p. 26, para. 30, <http://www.dod.mil/pubs/foi/operation_and_plans/Detainee/taguba/TAGUBA_REPORT_CERTIFICATIONS.pdf>, accessed on October 15, 2013.

225 Antenor Hallo de Wolf, (2012) *Reconciling Privatization with Human Rights*, School of Human Rights Research Series, Volume 49, Cambridge, Antwerp and Portland: Intersentia, p. 508.

226 Federal Activities Reform Act (1998), 31 U.S.C. 501, para. 5(2)(A); Federal Acquisition Regulation, 48 C.F.R., para. 7.503(a); Office of Management and Budget Circular A-76, Attachment B, para. (B)(1)(a), <http://www.whitehouse.gov/omb/circulars_a076_a76_incl_tech_correction>, accessed on September 10, 2013; US Department of Defense Instruction 1100.22, April 12, 2010, p. 2, para. 4(d), <http://www.dtic.mil/whs/directives/corres/pdf/110022p.pdf>, accessed on September 12, 2013; US Department of the Army, Army Regulation 715-9, Procurement: Operational Contract Support Planning and Management, June 20, 2011, Appendix B: Inherently Governmental Function Guidance, p. 21, para. B-1(a), <http://armypubs.army.mil/epubs/pdf/r715_9.pdf>, accessed on June 20, 2014.

227 Chia Lehnardt, "Private Military Companies and State Responsibility", in Simon Chesterman and Chia Lehnardt (Eds.), (2007) *From Mercenaries to Market: The Rise and Regulation of Private Military Companies*, Oxford: Oxford University Press, p. 147; Antenor Hallo de Wolf, (2006) "Modern Condottieri in Iraq: Privatizing War from the Perspective of International Human Rights Law", *Indiana Journal of Global Legal Studies*, Volume 13, Issue 2, pp. 344-345.

policies and endorsed it by means of the contract.[228] The circumstances as identified above would make the attribution ex Article 5 DARS possible, as also pointed out by Antenor Hallo de Wolf.[229]

Concluding this sub-Chapter, it should be observed that this type of attribution addressing entities having public powers is also applicable in relation to persons or entities that do not belong to State organs, but exercise State functions and commit *ultra vires* acts by exceeding their authority or contravening instructions, as stipulated in Article 7 DARS. This means that if a PMSC is employed to carry out governmental functions and exceeds its delegated powers or contravenes instructions, its unlawful conduct will be regarded a State act and attributed to the State in question. Moreover, Article 9 DARS provides that the attribution of unlawful acts to States may occur if PMSCs exercise elements of governmental authority in the absence or default of the official State authorities. This provision is applicable only in a very limited number of scenarios, for instance, in armed conflicts and during foreign occupation, when there are no State authorities or they are not active.[230] In this context, a PMSC can perform governmental activities, such as police functions, the exercise of which is justified by the circumstances of a particular case, although these companies have not been authorized by the government to do so. Thus, possible unlawful actions of such a PMSC exercising elements of governmental authority in the absence or default of the official State authorities and in circumstances requiring the exercise of such functions will be considered an act of a State and attributed to it in accordance with rules and principles of international law.

6.3 Article 8 DARS

6.3.1 General application of the provision

The next provision under examination is Article 8 DARS, requiring significant attention and scrutiny. It becomes relevant when persons or entities do not constitute official State organs, as provided in Article 4 DARS, and do not exercise elements of governmental authority on the basis of internal law, which is regulated in Article 5 DARS. In principle, Article 8 DARS concentrates on the circumstances when a certain factual relationship exists between the wrongful acts committed by persons or groups of persons and the State.[231] To be more precise, it focuses on the conduct realized under instructions of a State, directed, or controlled by it and provides the following:

---

228 Antenor Hallo de Wolf, (2006) "Modern Condottieri in Iraq: Privatizing War from the Perspective of International Human Rights Law", *Indiana Journal of Global Legal Studies*, Volume 13, Issue 2, p. 344.

229 Antenor Hallo de Wolf, (2012) *Reconciling Privatization with Human Rights*, School of Human Rights Research Series, Volume 49, Cambridge, Antwerp and Portland: Intersentia, p. 508.

230 Commentary to DARS as adopted in 2001, p. 49, Article 9, para. 1.

231 Ibidem, p. 47, Article 8, para. 1.

> "The conduct of a person or group of persons shall be considered an act of a State under international law if the person or group of persons is in fact acting on *the instructions* of, or under *the direction* or *control* of, that State in carrying out the conduct." (emphasis added)

As stated in Article 8 DARS, if certain persons or entities do not belong to the official State organs, but act under instructions, direction, or control of State authorities, their illegal activities will be attributed to the State in question and trigger the application of the doctrine of State responsibility. This is a provision subjected to fierce academic debate[232] requiring an in-depth assessment.

Article 8 DARS uses term "persons or group of persons" reflecting the idea that the unlawful conduct might be committed not merely by legal persons, such as corporations, but also by a group lacking separate legal personality and operating collectively on a *de facto* basis.[233] The provision addresses three distinct circumstances: in the first place, it deals with the situations when persons, groups of persons, or entities act under instructions provided by States; in the second place, it focuses on the situations when those persons or groups of persons act under the direction of States; and finally, it aims at addressing situations when individuals or entities operate under State control. For the application of Article 8 DARS, it suffices when one of the three situations – instructions, direction, or control by States – has been established, given that these are not cumulative criteria.[234] The first two scenarios concern the instructions issued by a State and its direction, and must be concluded to be rather clear and concrete. The issue of instructions or directions given by a State does not pose many difficulties, for it involves the process of giving explicit orders and instructions to a person or an entity to realize a certain conduct.[235] Problems may arise, however, when dealing with the third type of situation specifically introduced in Article 8 DARS in form of a distinct legal construct connecting the misconduct of private actors to States:[236] the concept of control to be exercised by a State.[237]

In case of private individuals or entities acting under instructions provided by a State or its direction, State organs recruit or instigate those private persons or entities

---

232 Malcolm N. Shaw, (2008) *International Law*, Sixth Edition, Cambridge: Cambridge University Press, p. 787; Oliver R. Jones, (2009) "Implausible Deniability: State Responsibility for the Actions of Private Military Firms", *Connecticut Journal of International Law*, Volume 24, Issue 2, p. 269.

233 Commentary to DARS as adopted in 2001, p. 49, Article 8, para. 9.

234 Ibidem, p. 48, Article 8, para. 7.

235 Antonio Cassese, (2007) "The Nicaragua and Tadić Tests Revisited in Light of the ICJ Judgment on Genocide in Bosnia", *The European Journal of International Law*, Volume 18, Issue 4, p. 663.

236 Charlotte Beaucillon, Julian Fernandez, and Hélène Raspail, "State Responsibility for Conduct of Private Military and Security Companies Violating Ius ad Bellum", in Francesco Francioni and Natalino Ronzitti (Eds.), (2011) *War by Contract: Human Rights, Humanitarian Law, and Private Contractors*, Oxford: Oxford University Press, p. 406.

237 Commentary to DARS as adopted in 2001, p. 47, Article 8, paras. 2-3; Malcolm N. Shaw, (2008) International Law, Sixth Edition, Cambridge: Cambridge University Press, p. 790.

that act as "auxiliaries" outside official State structures and authorize violations of international law. The attribution based on this type of situation is widely accepted in international law. The relevant circumstances can relatively easily be identified and determined if there is enough evidence indicating specific instructions issued by the State in question.[238] Instructions imply that a State clearly specifies its intentions for the commission of certain acts breaching international law and in essence authorizes these illegal activities.[239] The concept of direction is strongly connected to the idea of subordination of persons or groups of persons to a State in the process of realizing a certain unlawful conduct and the requirement of a State, in fact, directing or leading the commission of these illegal activities step-by-step.[240] What a State needs to do, is to issue specific instructions directing the commission of unlawful acts violating international law. On the other hand, the unlawful conduct of private persons and entities controlled by a State is only attributable to that State if the control has been exercised with respect to the specific operation of which the illegal conduct was an integral part.[241] The determination of State involvement in the commission of unlawful acts by persons and entities controlled by a State appears, in fact, to be more problematic if one makes an attempt to establish a role of a certain State in carrying out those activities by means of controlling them.[242] In this regard, the ILC refers in the Commentaries to the Draft Articles to the principle of effectiveness in international law. This principle implies that the illegal conduct of private persons and entities, such as PMSCs, can be attributed to States if the above-mentioned circumstances arise indicating "a specific factual relationship" between those non-State actors and States. In this respect, "a real link" must be established between private persons or entities committing violations of international law and the State apparatus.[243]

The extent of control that needs to be exercised by a State over private persons, groups of persons, and entities, such as PMSCs and their contractors, is one of the core issues of the concept of attribution laid down in Article 8 DARS, but is also subject to debate.[244] Scholars agree that the notion of control incorporated in this

---

238 Commentary to DARS as adopted in 2001, p. 47, Article 8, para. 2.

239 International Court of Justice, *United States Diplomatic and Consular Staff in Tehran (United States of America v. Iran)*, Judgment, May 24, 1980, ICJ Reports 1980, para. 59.

240 Lindsey Cameron and Vincent Chetail, (2013) *Privatizing War: Private Military and Security Companies under Public International Law*, Cambridge: Cambridge University Press, p. 209.

241 Commentary to DARS as adopted in 2001, p. 47, Article 8, para. 3.

242 Chia Lehnardt, "Private Military Companies and State Responsibility", in Simon Chesterman and Chia Lehnardt (Eds.), (2007) *From Mercenaries to Market: The Rise and Regulation of Private Military Companies*, Oxford: Oxford University Press, pp. 148-149.

243 Commentary to DARS as adopted in 2001, p. 47, Article 8, para. 1.

244 Chia Lehnardt, "Private Military Companies and State Responsibility", in Simon Chesterman and Chia Lehnardt (Eds.), (2007) *From Mercenaries to Market: The Rise and Regulation of Private Military Companies*, Oxford: Oxford University Press, p. 149.

provision has an autonomous meaning and can be interpreted in different manners.[245] For the examination of the degree of control exercised by a State over certain persons acting unlawfully, recent case law of international judicial bodies – the International Court of Justice and the International Criminal Tribunal for the former Yugoslavia – has provided that there are basically two main types of control theories that can be used to determine the level of State control.[246]

The first test was introduced by the International Court of Justice in the *Nicaragua* case[247] from 1986: the "effective control" test. In this particular case, the ICJ had to determine whether the unlawful conduct of the contras violating international humanitarian law could be attributed to the United States in order to hold it responsible for those attrocities. The ICJ decided that the United States did not exercise the sufficient degree of control, which needed to be effective, in all the areas of activities performed by the contra rebels, in order to justify the attribution of their acts to the State.[248]

The ICTY also looked into the issue of attribution on the basis of Article 8 DARS in its *Tadić* case from 1999. The Tribunal had to decide whether the armed forces of the Bosnian-Serb Republic, the Republika Srpska, could be perceived as operating under the control of the authorities of the Federal Republic of Yugoslavia. While the Trial Chamber applied the same strict test having the same high threshold as the ICJ, the Appeals Chamber decided to introduce a new more flexible test called the "overall control" test with a lower threshold than the "effective control" theory established in the *Nicaragua* judgment.[249]

In a more recent case, the ICJ had to elaborate on the legal dispute between Bosnia and Herzegovina against Serbia and Montenegro in the so-called *Bosnian Genocide* case[250] from 2007, in which it had to establish whether the acts of the perpetrators of the genocide could be attributed to the State of Serbia and Montenegro. In its judgment, the ICJ reaffirmed the "effective control" test established in the *Nicaragua* case, which required the effective control that must be exercised by the State over the persons and groups of persons committing genocide. The particular type of attribution laid down in Article 8 DARS in combination with the three cases mentioned above requires careful consideration and elaboration for gathering necessary

---

245 Lindsey Cameron and Vincent Chetail, (2013) *Privatizing War: Private Military and Security Companies under Public International Law*, Cambridge: Cambridge University Press, p. 212.

246 Malcolm N. Shaw, (2008) *International Law*, Sixth Edition, Cambridge: Cambridge University Press, p. 790.

247 International Court of Justice, *Case Concerning Military and Paramilitary Activities in and against Nicaragua (Nicaragua v. United States of America)*, Judgment (Merits), June 27, 1986, ICJ Reports 1986.

248 Ibidem, pp. 62-64.

249 International Criminal Tribunal for the former Yugoslavia, *Prosecutor v. Tadić*, Appeals Chamber, Judgment, July 15, 1999, ICTY-IT-94-1-A.

250 International Court of Justice, *Case Concerning the Application of the Convention on the Prevention and Punishment of the Crime of Genocide (Bosnia and Herzegovina v. Serbia and Montenegro)*, Judgment (Merits), February 26, 2007, ICJ Reports 2007.

insights with regard to the particular concept of attribution, as incorporated in Article 8 DARS, and its relevance for the State practice of hiring PMSCs and internationally wrongful acts committed by the latter.

6.3.2 Application of Article 8 DARS with regard to PMSCs

Thus, it is possible that PMSCs do not qualify as entities forming part of the regular armed forces of a particular State and do not exercise elements of governmental authority. In this case, their illegal acts violating rules and principles of international law, more specifically international humanitarian and human rights law, are to be imputed to the State in question for the purpose of State responsibility, if the unlawful conduct was carried out under three distinctive circumstances used alternatively: (1) according to instructions of a State, (2) under its direction, or (3) control.[251] The actions of PMSCs taking place under one of these three conditions suggest that there is a real factual link between the State apparatus and firms under examination.[252] It should be understood to mean that any facts, which suggest that States have issued instructions towards or have given specific directions to PMSCs committing internationally wrongful acts, can engage the application of the type of attribution, as laid down in Article 8 DARS. It is generally clear that when States instruct PMSCs to commit violations of human rights or to breach the law of armed conflict, these States giving instructions will be responsible for the misconduct in question.[253] These instructions can be laid down into the contracts concluded with PMSCs, issued to private contractors on the ground, or both.[254] In case of direction, a State is required to issue specific instructions directing the commission of unlawful acts constituting violations of international law. As observed by some scholars, it might, however, also be argued that the State is responsible when it directs the operation of a PMSC and not the commission of particular types of misconduct that take place during that

---

[251] Article 8 DARS; Commentary to DARS as adopted in 2001, p. 47, Article 8, para. 1.

[252] Shannon Bosch, (2008) "Private Security Contractors and State Responsibility: Are States Exempt from Responsibility for Violations of Humanitarian Law Perpetrated by Private Security Contractors?", *The Comparative and International Law Journal of Southern Africa*, Volume 41, Issue 3, p. 376.

[253] Chia Lehnardt, "Private Military Companies and State Responsibility", in Simon Chesterman and Chia Lehnardt (Eds.), (2007) *From Mercenaries to Market: The Rise and Regulation of Private Military Companies*, Oxford: Oxford University Press, p. 148; Emanuela-Chiara Gillard, (2006) "Business Goes to War: Private Military / Security Companies and International Humanitarian Law", *International Review of the Red Cross*, Volume 88, Issue 863, p. 555; Charlotte Beaucillon, Julian Fernandez, and Hélène Raspail, "State Responsibility for Conduct of Private Military and Security Companies Violating Ius ad Bellum", in Francesco Francioni and Natalino Ronzitti (Eds.), (2011) *War by Contract: Human Rights, Humanitarian Law, and Private Contractors*, Oxford: Oxford University Press, p. 406; Hannah Tonkin, (2011) *State Control over Private Military and Security Companies in Armed Conflict*, Cambridge: Cambridge University Press, pp. 114-117.

[254] James Crawford, (2013) *State Responsibility: The General Part*, Cambridge Studies in International and Comparative Law, No. 100, Cambridge: Cambridge University Press, p. 145.

mission.[255] What is clear, is that both instructions given by a State and direction by a State will necessarily lead to the international responsibility of that State.

A note should be taken of a fact that all three cases – instructions, direction, and control – must have a connection to the misconduct constituting a violation of international law. The provision mentions that persons or groups of persons carry out a certain conduct and the Commentaries further provide that:

> "[I]t is made clear that the instructions, direction or control must relate to the conduct which is said to have amounted to an internationally wrongful act."[256]

It is important to observe that States may instruct, direct, or control lawful operations of private actors, but those persons or groups act beyond the limits of States' authorization and violate international law. As argued by Emanuela-Chiara Gillard, the attribution under Article 8 DARS is in order when a State essentially directs a PMSC to breach international law and not when a company is employed to carry out a lawful activity and its employees commit a violation of international humanitarian or human rights law during that lawful mission.[257]

When a State gives lawful instructions to a PMSC, the company might unlawfully act while carrying them out. In this case, the unlawful conduct will not be attributed to the State in question under Article 8 DARS.[258] The ILC supports this idea by stressing that:

> "In general a State, in giving lawful instructions to persons who are not its organs, does not assume the risk that the instructions will be carried out in an internationally unlawful way."[259]

In order to determine whether a particular State is responsible for the misconduct of PMSCs contravening instructions or directions of that State and breaching its international obligations, one should examine facts of that specific case. The Commentaries stress in this regard that:

> "Such cases can be resolved by asking whether the unlawful or unauthorized conduct was really incidental to the mission or clearly went beyond it."[260]

---

[255] Lindsey Cameron and Vincent Chetail, (2013) *Privatizing War: Private Military and Security Companies under Public International Law*, Cambridge: Cambridge University Press, p. 209.

[256] Commentary to DARS as adopted in 2001, p. 48, Article 8, para. 7.

[257] Emanuela-Chiara Gillard, (2006) "Business Goes to War: Private Military / Security Companies and International Humanitarian Law", *International Review of the Red Cross*, Volume 88, Issue 863, p. 555.

[258] Lindsey Cameron and Vincent Chetail, (2013) *Privatizing War: Private Military and Security Companies under Public International Law*, Cambridge: Cambridge University Press, p. 207.

[259] Commentary to DARS as adopted in 2001, p. 48, Article 8, para. 8.

[260] Ibidem, p. 48, Article 8, para. 8.

Thus, when the misconduct is incidental to the mission performed by a PMSC and takes place to achieve the goals of this mission, it can be attributed to the State who has instructed or directed this company to lawfully carry out the mission in its entirety. Similarly, if that particular conduct is not incidental to the mission or goes beyond the scope of State instructions and directions, the attribution will not be triggered.

If a PMSC authorized by a State, however, does not follow its lawful instructions and violates international law, these actions and omissions will be attributed to the State in question if it exercises effective control. In the Commentary, this position is reflected as follows:

> "[W]here persons or groups have committed acts under the effective control of a State, the condition for attribution will still be met even if particular instructions may have been ignored. The conduct will have been committed under the control of the State and it will be attributable to the State in accordance with article 8."[261]

It can be concluded that instructions to commit certain crimes and to violate international law need to be specifically given by States to PMSCs. It is also required that private contractors are specifically directed by States to engage in the commission of unlawful acts. Although in practice States would not go so far as to issue this type of instruction or to direct the unlawful conduct, it is not entirely improbable. On the other hand, the control exercised by States over PMSCs is of greater practical importance given the fact that many companies and their employees can fall under the category of entities and individuals operating under these particular circumstances and being controlled by the State machinery with its different faces. When States control PMSCs, the former will be responsible for the conduct of the latter acting as State agents if there is a sufficient degree of control.[262] Although it is more likely that in practice hiring States issue instructions or employ PMSCs operating under their direction or control, it is possible that also home and host States fall under the scope of Article 8 DARS.[263]

6.3.2.1 The "effective control" test established in the ICJ *Nicaragua* case

In the *Nicaragua* case, the International Court of Justice identified possible relationships between the contras that committed several violations of international humanitarian law and the Government of the United States. The World Court made

---

261 Ibidem, pp. 48-49, Article 8, para. 8.

262 Andrew Clapham, (2006) *Human Rights Obligations of Non-State Actors*, Oxford: Oxford University Press, p. 302.

263 Charlotte Beaucillon, Julian Fernandez, and Hélène Raspail, "State Responsibility for Conduct of Private Military and Security Companies Violating Ius ad Bellum", in Francesco Francioni and Natalino Ronzitti (Eds.), (2011) *War by Contract: Human Rights, Humanitarian Law, and Private Contractors*, Oxford: Oxford University Press, p. 407.

its considerations on the issue in terms of the degree of dependence of the contras on the United States on the one hand and the extent of the control exercised by the State on the other hand. These factors are of major significance for answering the question of whether the contras could be equated with organs of the United States or whether they must be regarded as persons acting on behalf of the Government.[264] For the ICJ judges, it was clear that the United States provided some form of aid, assistance, and support to the contras. In this regard, the cessation of that provision could be considered as "the only element of control".[265] In its further considerations, the Court stressed the following idea:

> "... there is no clear evidence of the United States having actually exercised such a degree of control in all fields as to justify treating the contras as acting on its behalf... For this conduct to give rise to legal responsibility of the United States, it would in principle have to be proved that the State had *effective control* of the military or paramilitary operations *in the course of which* the alleged violations were committed."[266] (emphasis added)

The "effective control" test is without doubt a strict one with a particularly high threshold. Only in rare circumstances can non-State actors that are not State agents and do not belong to the State apparatus be regarded as acting on behalf of States. Judge Ago stressed in his separate opinion that if private persons, groups of persons, or entities, such as PMSCs and their employees, are specifically charged by a State that gave clear instructions to commit a certain act or to carry out a specific task inconsistent with international law, they can be considered as acting in the name and on behalf of that particular State.[267] Only in these cases, one can see the misconduct of these persons and groups as the acts of a State in question. When the acts of non-State actors cannot be held to be acts of a State, the "effective control" test developed by the ICJ makes it possible to attribute the conduct to a State when it is established that a State has effective control with respect to each operation in which private individuals or groups commit the alleged violations. In other words, States need to exercise effective control over all activities of PMSCs and, in fact, authorize them.[268] These conditions being met will lead to the attribution of illegal acts of private contractors

---

264 International Court of Justice, *Case Concerning Military and Paramilitary Activities in and against Nicaragua (Nicaragua v. United States of America)*, Judgment (Merits), June 27, 1986, ICJ Reports 1986, para. 109.

265 Ibidem.

266 Ibidem, paras. 109, 115.

267 Separate Opinion of Judge M. Ago in International Court of Justice, *Case Concerning Military and Paramilitary Activities in and against Nicaragua (Nicaragua v. United States of America)*, Judgment (Merits), June 27, 1986, ICJ Reports 1986, paras. 16-17.

268 Brigitte Stern, "The Elements of An Internationally Wrongful Act", in James Crawford, Alain Pellet, and Simon Olleson (Eds.), (2010) *The Law of International Responsibility*, Oxford: Oxford University Press, p. 206.

and their personnel violating international humanitarian and human rights law to the respective States.[269]

In order to meet the criterion of being effective, the control must not be merely of a general character. When dealing with the question of attributing acts of the Unilaterally Controlled Latino Assets, the ICJ uses certain wording that could be helpful in determining whether the control is in practice effective. The Court thought that the imputability of attacks carried out by the contras was established and, without referring to the concept of effective control, indicated the following:

> "Although it is not proved that any United States military personnel took a direct part in the operations, agents of the United States *participated in the planning, direction, support and execution of the operations*. The execution was the task rather of the 'UCLAs' [Unilaterally Controlled Latino Assets], while United States nationals participated in the planning, direction and support."[270] (emphasis added)

The military operations were executed by the Unilaterally Controlled Latino Assets, but the United States agents took part in planning, directing, supporting, and executing them. Thus – as argued by Lindsey Cameron and Vincent Chetail referring to the *Nicaragua* case[271] – if a State provides this degree of "practical support" to a certain operation carried out by non-State actors in which the misconduct takes place or for the actual execution of this misconduct, it might be concluded to exercise effective control.[272]

With regard to the PMSCs hired by the United States, as one of the States relying heavily on the provision of military and security services by those firms, it should be noted that the relationship between private military and security companies and the US military could be summed up as one of "informal co-ordination". According to the experts, this often involves frequent meetings organized for the purpose of sharing information, coordinating, and resolving conflict situations in operations.[273] It appears that the US commanders probably do not have such a high degree of con-

---

[269] Chia Lehnardt, "Private Military Companies and State Responsibility", in Simon Chesterman and Chia Lehnardt (Eds.), (2007) *From Mercenaries to Market: The Rise and Regulation of Private Military Companies*, Oxford: Oxford University Press, p. 149.

[270] International Court of Justice, *Case Concerning Military and Paramilitary Activities in and against Nicaragua (Nicaragua v. United States of America)*, Judgment (Merits), June 27, 1986, ICJ Reports 1986, para. 86.

[271] Ibidem, para. 75.

[272] Lindsey Cameron and Vincent Chetail, (2013) *Privatizing War: Private Military and Security Companies under Public International Law*, Cambridge: Cambridge University Press, p. 213.

[273] Amanda Tarzwell, (2009) "In Search of Accountability: Attributing the Conduct of Private Security Contractors to the United States Under the Doctrine of State Responsibility", *Oregon Review of International Law*, Volume 11, Issue 1, p. 189; Chia Lehnardt, "Private Military Companies and State Responsibility", in Simon Chesterman and Chia Lehnardt (Eds.), (2007) *From Mercenaries to Market: The Rise and Regulation of Private Military Companies*, Oxford: Oxford University Press, p. 150.

trol over private military companies that might fall under the definition of the term "effective", as specified by the ICJ. In addition, according to the reports, some private security companies, such as those hired to guard the Abu Ghraib prison in Iraq, can even have governmental personnel under supervision of companies' employees.[274] In these situations, it would be extremely difficult to find sufficient evidence of effective control that could have been exercised by States over PMSCs in order to prove their acting on behalf of those particular governments. Moreover, the mere fact that a license has been granted by certain governments to PMSCs or that States conclude contracts with those companies cannot indicate that the high degree of control is exercised by hiring States, host States, or home States having licensing regimes, such as the United States, Israel, and South Africa. This is the case because the adequate oversight and monitoring mechanisms essential for the proper control over PMSC conduct are simply lacking or are not functional.[275]

6.3.2.2 The "overall control" test established in the ICTY *Tadić* case

Considering a very high threshold of the "effective control" test introduced in the ICJ *Nicaragua* case, the discussion must focus on the test with a lower threshold introduced in the *Tadić* case of the International Criminal Tribunal for the former Yugoslavia. This case can be highly important for the application of Article 8 DARS with regard to the PMSCs employed by States and can make the attribution on the basis of this provision easier.[276]

The ICTY judges did not apply the strict test introduced in the *Nicaragua* case. Instead, they opted for a new concept of the State control. According to the view of the ICTY, the "effective control" test must be regarded as unconvincing and unpersuasive. The reason for that can be found in the fact that international law renders any State responsible for acts breaching its rules and principles carried out by official State organs or organized persons or entities under the State control irrespective of whether a State gave specific instructions to those actors.[277] In addition, it has been stipulated that the "effective control" test should be declared as incompatible with international judicial and State practice, for they indicate that a lower degree of control than that required by the test used in the *Nicaragua* case is sufficient to attribute violations of international law to States. Judges believed that the "effective control" test could

---

[274] Major General George R. Fay, Investigating Officer, Article 15-6 Investigation of the Abu Ghraib Prison and 205th Military Intelligence Brigade, August 2004, pp. 51-52, <http://news.findlaw.com/hdocs/docs/dod/fay82504rpt.pdf>, accessed on October 15, 2013.

[275] Chia Lehnardt, "Private Military Companies and State Responsibility", in Simon Chesterman and Chia Lehnardt (Eds.), (2007) *From Mercenaries to Market: The Rise and Regulation of Private Military Companies*, Oxford: Oxford University Press, p. 150.

[276] International Criminal Tribunal for the former Yugoslavia, *Prosecutor v. Tadić*, Appeals Chamber, Judgment, July 15, 1999, ICTY-IT-94-1-A.

[277] Ibidem, para. 123.

not be considered the most suitable concept that ought to be applied in cases dealing with members of military or paramilitary groups,[278] but used a more flexible approach in order to establish the so-called "overall control" test. In fact, this test constituted a response of the judges to the increasing importance of non-State actors, such as PMSCs, in international law.[279] Given the purpose of the law of State responsibility based on the "realistic concept of accountability",[280] the rationale behind the rules and principles of international law on the issue of attribution of illegal activities of private persons and entities to States and the concept of attribution according to the "overall control" test is in the opinion of the Appeals Chamber the following:

> "The principles of international law concerning the attribution to States of acts performed by private individuals are not based on rigid and uniform criteria. These principles are reflected in Article 8 of the Draft on State Responsibility adopted on first reading by the United Nations International Law Commission and, even more clearly, in the text of the same provisions as provisionally adopted in 1998 by the ILC Drafting Committee... The rationale behind this rule is *to prevent States from escaping international responsibility* by having private individuals carry out tasks that may not or should not be performed by State officials, or by claiming that individuals actually participating in governmental authority are not classified as State organs under national legislation and therefore do not engage State responsibility. In other words, States are not allowed on the one hand to act *de facto* through individuals and on the other to disassociate themselves from such conduct when these individuals breach international law."[281] (emphasis added)

In situations when no single individual is to be regarded as a perpetrator of certain unlawful acts, but the private acts violating international law emanate from groups of persons or entities that are well-organized and hierarchically structured, as it is usually the case with contras or PMSCs, the flexible "overall control" test constitutes a suitable attribution theory and will be sufficient to impute those breaches to States in question exercising that specific degree of control.[282] The Appeals Chamber held that:

---

278 Ibidem, paras. 124-136.

279 Chia Lehnardt, "Private Military Companies and State Responsibility", in Simon Chesterman and Chia Lehnardt (Eds.), (2007) *From Mercenaries to Market: The Rise and Regulation of Private Military Companies*, Oxford: Oxford University Press, p. 151; Antonio Cassese, (2007) "The Nicaragua and Tadić Tests Revisited in Light of the ICJ Judgment on Genocide in Bosnia", *The European Journal of International Law*, Volume 18, Issue 4, pp. 665-667.

280 International Criminal Tribunal for the former Yugoslavia, *Prosecutor v. Tadić*, Appeals Chamber, Judgment, July 15, 1999, ICTY-IT-94-1-A, para. 121; Tal Becker, (2008) *Terrorism and the State: Rethinking the Rules of State Responsibility*, Oxford and Portland: Hart Publishing, p. 69.

281 International Criminal Tribunal for the former Yugoslavia, *Prosecutor v. Tadić*, Appeals Chamber, Judgment, July 15, 1999, ICTY-IT-94-1-A, para. 117.

282 Chia Lehnardt, "Private Military Companies and State Responsibility", in Simon Chesterman and Chia Lehnardt (Eds.), (2007) *From Mercenaries to Market: The Rise and Regulation of Private Military Companies*, Oxford: Oxford University Press, p. 150.

> "... control by a State over subordinate armed forces or militias or paramilitary groups may be of *an overall character* (and must comprise more than the mere provision of financial assistance or military equipment or training). This requirement, however, does not go so far as to include the issuing of specific orders by the State, or its direction of each individual operation. Under international law it is by no means necessary that the controlling authorities should plan all the operations of the units dependent on them, choose their targets, or give specific instructions concerning the conduct of military operations and any alleged violations of international humanitarian law. The control required by international law may be deemed to exist when a State (or, in the context of an armed conflict, the Party to the conflict) has a role in *organizing, coordinating or planning* the military actions of the military group, in addition to financing, training and equipping or providing operational support to that group. Acts performed by the group or members thereof may be regarded as acts of de facto State organs regardless of any specific instruction by the controlling State concerning the commission of each of those acts."[283] (emphasis added)

According to the "overall control" test applied in relation to the State use of PMSCs, the degree of control exercised by States over PMSCs can easily meet the set of criteria introduced in the *Tadić* case.[284] It is true that States do not always provide specific instructions for the functioning of PMSCs in every particular instance or exercise effective control and issue directions for every single PMSC operation,[285] such as issuing clear orders to assassinate individuals and to commit other violations of international law. In this regard, the control exercised by the hiring or other States over PMSCs can qualify under the definition of the "overall control" if it satisfies the criteria developed by the ICTY.

It is realistic for the conduct of PMSCs to be successfully attributed to States on the basis of the "overall control" test. More specifically, hiring States usually play a role in organizing, coordinating, planning, and financing the operations of PMSCs they hire. If one analyzes the State practice of using PMSCs for carrying out military and security tasks, it becomes obvious that hiring States not only pay for the services provided by PMSCs and finance their operations in accordance with the concluded contracts, but in practice can also take a modest part in the organization, coordination, and planning of their operations on the ground. PMSCs often cooperate with the regular armed forces and operate together with State troops in order to complete common objectives. If in case of this particular degree of State involvement in the functioning of PMSCs, the requirements for the "overall control" test are met, that

---

[283] International Criminal Tribunal for the former Yugoslavia, *Prosecutor v. Tadić*, Appeals Chamber, Judgment, July 15, 1999, ICTY-IT-94-1-A, para. 137.

[284] Oliver R. Jones, (2009) "Implausible Deniability: State Responsibility for the Actions of Private Military Firms", *Connecticut Journal of International Law*, Volume 24, Issue 2, p. 270.

[285] Chia Lehnardt, "Private Military Companies and State Responsibility", in Simon Chesterman and Chia Lehnardt (Eds.), (2007) *From Mercenaries to Market: The Rise and Regulation of Private Military Companies*, Oxford: Oxford University Press, p. 151.

control exercised by a State does not have to qualify as effective. State organs do not provide specific instructions for every activity of private contractors and there are no effective oversight and monitoring mechanisms available for examining their legal and illegal activities.[286] It is apparent that according to the revisited theory of attribution established in the *Tadić* case, under certain circumstances hiring States often can be regarded as having overall control over PMSCs and thus responsible for their unlawful activities. On the other hand, home and host States do not play such an extensive role in the organization, coordination, planning, and financing of PMSC activities and the degree of control exercised by them cannot easily meet the *Tadić* standard.

6.3.2.3 Reaffirmation of the "effective control" test in the ICJ *Bosnian Genocide* case

As follows from the assessment made above, while the application of the restrictive "effective control" test would most probably not lead to the attribution of internationally wrongful acts committed by PMSCs to States, one can argue that those acts can be regarded as imputable to States if there is sufficient evidence supporting the idea that they exercise overall control over PMSCs. However, it needs to be stressed that the view of the Appeals Chamber of the ICTY regarding its formulated "overall control" test was not used in the ICJ's *Case Concerning Armed Activities on the Territory of the Congo* from 2005. In its judgment, the Court referred to the "effective control" test to establish the responsibility of Uganda.[287] Importantly, the "overall control" test has specifically been rejected in the more recent judgment of the ICJ in the *Bosnian Genocide* case.[288]

In this case, after having established that in Bosnia and Herzegovina only the massacre in Srebrenica constituted genocide, the Court turned its attention to the determination of the status of perpetrators. Concluding that genocide was not perpetrated by persons or entities having the status of *de jure* organs of the Federal Republic of Yugoslavia nor that it was committed by persons or entities that could be equated with State organs considering their "complete dependence" on the State, the Court proceeded with the discussion of the issue of perpetrators acting in accordance with instructions or under control of the State.[289] It declared that:

---

286 Laura A. Dickinson, "Contract as a Tool for Regulating Private Military Companies", in Simon Chesterman and Chia Lehnardt (Eds.), (2007) *From Mercenaries to Market: The Rise and Regulation of Private Military Companies*, Oxford: Oxford University Press, pp. 223-224.

287 International Court of Justice, *Case Concerning Armed Activities on the Territory of the Congo (Democratic Republic of the Congo v. Uganda)*, Judgment (Merits), December 19, 2005, ICJ Reports 2005, para. 160.

288 International Court of Justice, *Case Concerning the Application of the Convention on the Prevention and Punishment of the Crime of Genocide (Bosnia and Herzegovina v. Serbia and Montenegro)*, Judgment (Merits), February 26, 2007, ICJ Reports 2007.

289 Ibidem, paras. 385-397.

"... the acts of genocide at Srebrenica cannot be attributed to the Respondent as having been committed by its organs or by persons or entities wholly dependent upon it, and thus do not on this basis entail the Respondent's international responsibility... the Court must now determine whether the massacres at Srebrenica were committed by persons who, though *not having the status of organs of the Respondent*, nevertheless acted on *its instructions* or under *its direction* or *control*..."[290] (emphasis added)

The Court made clear that Article 8 DARS constitutes a rule of customary international law[291] and stated that this provision needs to be understood in the light of the *Nicaragua* case from the year 1986 requiring the "effective control" test.[292] The Court noted that according to the strict test introduced in the *Nicaragua* case it must be proved that private persons or entities have acted following the instructions of States or operating under their effective control. The ICJ observed:

"It must however be shown that this 'effective control' was exercised, or that the State's instructions were given, *in respect of each operation* in which the alleged violations occurred, not generally in respect of overall actions taken by the persons or groups of persons having committed the violations."[293] (emphasis added)

Time and again, it has thus been confirmed that the control States have must be effective with regard to every operation in which the alleged violations of international law have occurred, and it is not sufficient if States exercise merely the overall degree of control over the actions of private actors.

By considering the judgment of the ICTY in the *Tadić* case, the ICJ came to the conclusion that the view held in that case could not be followed in its considerations in the *Bosnian Genocide* case. Indications provided by the Court revealed that the arguments in favor of the "overall control" test introduced by the Tribunal were unpersuasive and it refused to apply that test.[294] First of all, it found that the ICTY has criminal jurisdiction extending only to persons and did not rule on the questions of State responsibility, which was at stake in its case. Moreover, the ICJ held that the positions adopted by the Tribunal on matters of general international law are not the same as presented before it considering its criminal jurisdiction.[295] As the ICJ recalled, the ICTY employed the "overall control" theory in order to establish whether it had to deal with an international armed conflict. It stated further that the test might be perfectly suitable and appropriate for this type of determination, but

[290] Ibidem, paras. 395-396.
[291] Ibidem, para. 398.
[292] Ibidem, para. 399.
[293] Ibidem, para. 400.
[294] Ibidem, para. 404.
[295] Ibidem, para. 403.

should not be used in the case presented before the ICJ, focusing on the issue of State responsibility.[296]

Secondly and importantly, the ICJ made clear that the "overall control" test cannot be adopted in resolving two issues that are different in nature, stating that:

> "It should first be observed that logic does not require the same test to be adopted in resolving the two issues, which are very different in nature: the degree and nature of a State's involvement in an armed conflict on another State's territory which is required *for the conflict to be characterized as international*, can very well, and without logical inconsistency, differ from the degree and nature of involvement required to give rise to *that State's responsibility* for a specific act committed in the course of the conflict."[297] (emphasis added)

In addition to this argument, the Court declared that the theory of "overall control" appears to broaden the scope of the doctrine of State responsibility, going beyond the standards laid down by the ILC in Article 8 DARS. The ICJ pointed out:

> "It must next be noted that the 'overall control' test has the major drawback of broadening the scope of State responsibility well beyond the fundamental principle governing the law of international responsibility: a State is responsible only for its own conduct, that is to say the conduct of persons acting, on whatever basis, on its behalf... This is so where an organ of the State gave the instructions or provided the direction pursuant to which the perpetrators of the wrongful act acted or where it exercised effective control over the action during which the wrong was committed. In this regard the 'overall control' test is *unsuitable*, for it stretches too far, almost to breaking point, the connection which must exist between the conduct of a State's organs and its international responsibility."[298] (emphasis added)

In its judgment in the *Bosnian Genocide* case, the ICJ made a clear step back to the 1980s when it adopted the "effective control" test, deciding not to apply the "overall control" theory. This position of the Court can have significant consequences for the PMSCs operating under control of States and violating international law. According to the findings made in the *Nicaragua* case, the effective control of a State over PMSCs and their employees needs to be proved in order for their illegal acts to be attributed to that State and to give rise to State responsibility.[299] The views formulated by the Court make clear that the evidence must be present to indicate that PMSCs

---

296 Ibidem, para. 404.

297 Ibidem, para. 405.

298 Ibidem, para. 406; Priya Pillai, (2007) "The Genocide Convention Case: Confusion or Clarity?", *Australian International Law Journal*, Volume 14, p. 251.

299 International Court of Justice, *Case Concerning Military and Paramilitary Activities in and against Nicaragua (Nicaragua v. United States of America)*, Judgment (Merits), June 27, 1986, ICJ Reports 1986, para. 115.

have been acting under effective control of a State or have been following its detailed instructions. Hence, it must be established that this "effective control" was exercised or instructions were given by a State in relation to "each operation in which the alleged violations occurred, not generally in respect of the overall actions taken by the persons or groups of persons having committed the violations".[300]

As the above analysis illustrates, this strict test of effective control undoubtedly creates a much higher threshold for the attribution of unlawful acts of PMSCs that might commit internationally wrongful acts. There must be sufficient evidence of specific instructions and effective control exercised by a State over PMSCs that cannot easily be gathered and examined. Thus, if States need to issue specific instructions for every single operation carried out by private contractors who engage in the commission of internationally wrongful acts, which is normally never done in practice, then it might be assumed that the control exercised by those States will most probably not reach the required level of effective control. It is usually the case that PMSCs do not operate under strict control and monitoring of hiring, home, and host States, meaning that those States will be able to escape State responsibility when PMSCs engage in various violations of international law, such as breaches of international humanitarian and human rights law.

6.3.2.4 Other types of control

In addition to the ICJ, various other international and national judicial bodies elaborated on the concept of control and, more specifically, effective control. This notion and other types of tests were used in the process of establishing institutional and State responsibility for human rights violations and determining the scope of extraterritorial applicability of human rights treaties.

In the case law of the European Court of Human Rights, an "effective overall control test" over a certain territory has been established to determine whether conduct falls within the State jurisdiction and thus can entail international responsibility of the States in question.[301] The Court took into account the number of troops present in the foreign territory,[302] but did not consider what the threshold is to maintain sufficient numbers of armed forces being present in a certain territory. Therefore,

[300] International Court of Justice, *Case Concerning the Application of the Convention on the Prevention and Punishment of the Crime of Genocide (Bosnia and Herzegovina v. Serbia and Montenegro)*, Judgment (Merits), February 26, 2007, ICJ Reports 2007, para. 400.

[301] See European Court of Human Rights, *Loizidou v. Turkey*, Judgment, December 18, 1996, Application No. 15318/89, Reports 1996-VI, para. 56; European Court of Human Rights, *Cyprus v. Turkey*, Judgment, May 10, 2001, Application No. 25781/94, Reports 2001-IV, para. 77.

[302] European Court of Human Rights, *Loizidou v. Turkey*, Judgment, December 18, 1996, Application No. 15318/89, Reports 1996-VI, para. 56.

this test is rather unclear and imprecise.[303] It seems to be a combination of the ICJ's "effective control" test and the ICTY's "overall control" test, but the Court did not provide any reasons for combining the two and did not explain the actual meaning of the test. It remains to be seen how this particular notion of "effective overall control" will be developed further. In its subsequent decision in the *Banković* case, the ECtHR adopted the notion of "effective control" as opposed to the "effective overall control" test discussed above.[304] The Court seemed to imply that this level of control can be exercised by a State when it occupies a territory of a foreign State or when it acts on the basis of the consent, invitation, or acquiescence of another State.[305]

Furthermore, the ECtHR stated in the case of *Behrami* and *Saramati*[306] that one should opt for the "ultimate authority and control" test in order to attribute the acts of the Kosovo Force led by NATO to the United Nations instead of France as a contributing State.[307] In the view of the Court, it had to be established whether the UN Security Council had ultimate authority and control over the security mission and whether the operational control was merely delegated to NATO. In another case of the ECtHR, *Al-Jedda*,[308] it determined whether the United Kingdom could bear responsibility for the detention of the applicant by the United Kingdom forces in Iraq by means of attribution to the respective State and not to the UN that had not dispatched the coalition forces to Iraq. In this regard, the Court concluded that the UN had "neither effective control nor ultimate authority and control" over the conduct of the troops and that this conduct was not attributable to the UN.[309] In essence, it means that when a certain organization authorizes a peace operation and exercises ultimate authority and control over it, it should be held responsible for the misconduct of the personnel of PMSCs breaching international law.[310] In this respect, States providing

---

303 Michał Gondek, (2009) *The Reach of Human Rights in a Globalising World: Extraterritorial Application of Human Rights Treaties*, School of Human Rights Research Series, Volume 32, Antwerp and Portland: Intersentia, pp. 155, 168-169.

304 European Court of Human Rights, *Banković and Others v. Belgium and Others*, Decision, December 12, 2001, Application No. 52207/99, Reports 2001-XII, para. 71.

305 Michał Gondek, (2009) *The Reach of Human Rights in a Globalising World: Extraterritorial Application of Human Rights Treaties*, School of Human Rights Research Series, Volume 32, Antwerp and Portland: Intersentia, p. 173.

306 European Court of Human Rights, *Behrami and Behrami v. France and Saramati v. France, Germany and Norway*, Grand Chamber, Decision as to the Admissibility of Application No. 71412/01 and Application No. 78166/01, May 2, 2007.

307 Ibidem, para. 133.

308 European Court of Human Rights, *Al-Jedda v. the United Kingdom*, Grand Chamber, Judgment, July 7, 2011, Application No. 27021/08.

309 Ibidem, para. 84.

310 Nigel D. White, "Institutional Responsibility for Private Military and Security Companies", in Francesco Francioni and Natalino Ronzitti (Eds.), (2011) *War by Contract: Human Rights, Humanitarian Law, and Private Contractors*, Oxford: Oxford University Press, pp. 388-389.

these companies for the completion of a mission and having no control over it can escape international responsibility.

The UN Human Rights Committee also mentioned the concept of "effective control" when deciding upon the reach of human rights obligations of States Parties to the ICCPR. States can have this degree of control over persons[311] or a territory, such as in the territories of one State occupied by another.[312] This test, having the same wording, is used to determine whether human rights treaties apply extraterritorially and whether States acting beyond their own borders have to meet their human rights obligations under those treaties.

The national courts of The Netherlands and Belgium have further provided their insights on the issue of effective control exercised by States in peace operations. In the Belgian *Mukeshimana-Ngulinzira* case,[313] a decision to withdraw Belgian forces from the Kigali camp and to abandon the *de facto* refugee camp was enough to stress that the troop-contributing State was exercising effective control. In the *Nuhanović* case, the Dutch Supreme Court upheld the Hague Court of Appeal judgments[314] regarding the actions of Dutchbat in Srebrenica.[315] The Court of Appeal maintained that the standard for attributing the misconduct in this case to the UN or the State is the test of effective control and not the "command and control" criterion.[316] In this regard, not only the fact that the conduct in question constituted "the execution of a specific instruction, issued by the UN or the State", but also in the absence of such instructions "the power to prevent the conduct concerned" were seen as important considerations to establish whether the degree of effective control was exercised by The Netherlands.[317] Both the Court of Appeal and the Supreme Court appear to be on the same line of reasoning regarding the possibility of dual attribution to both the State and the UN as an international organization.[318] Leaving the question of whether the UN exercised effective control aside, the Court of Appeal concluded that

---

311 UN Human Rights Committee, *General Comment 31*, May 26, 2004, UN Doc. CCPR/C/21/Rev.1/Add.13, para. 10.

312 UN Human Rights Committee, *Concluding Observations of the Human Rights Committee: Israel*, August 18, 1998, UN Doc. CCPR/C/79/Add.93, para. 10.

313 Court of First Instance, *Mukeshimana-Ngulinzira and Others v. Belgian State and Others*, Brussels, ILDC 1604 (BE 2010), para. 38.

314 Court of Appeal, *Hasan Nuhanović against the State of the Netherlands*, The Hague, July 5, 2011, Case No. 200.020.174/1.

315 Supreme Court of the Netherlands, *The State of the Netherlands v. Hasan Nuhanović*, First Chamber, The Hague, September 6, 2013, Case No. 12/03324.

316 Court of Appeal, *Hasan Nuhanović against the State of the Netherlands*, The Hague, July 5, 2011, Case No. 200.020.174/1, para. 5.8.

317 Ibidem, para. 5.9.

318 Ibidem; Supreme Court of the Netherlands, *The State of the Netherlands v. Hasan Nuhanović*, First Chamber, The Hague, September 6, 2013, Case No. 12/03324, para. 3.11.2.

The Netherlands indeed possessed effective control over the conduct of Dutchbat,[319] which was a position shared by the Supreme Court.[320]

6.3.2.5 Which test to apply?

From reading the ICJ judgment in the *Bosnian Genocide* case, one can draw the conclusion that the "effective control" test is the leading theory according to the World Court and thus needs to be used in order to establish whether the illegal conduct of PMSCs is to be imputed to States under Article 8 DARS. Or as James Crawford put it: "this determination effectively ends the debate as to the correct standard of control to be applied under Article 8".[321] The Court obviously provided that, on the basis of a number of grounds, this was the most suitable test to analyze the attributability of genocide committed by the Bosnian Serb armed forces to the Federal Republic of Yugoslavia within the framework of the above-mentioned provision. After deciding that, it is, however, still arguable whether the "effective control" theory is to be seen as a more persuasive test to be followed with regard to the PMSCs operating under the control of States.[322] Article 8 DARS itself does not mention a level of control that must be in the hands of a State. The Netherlands was very much in favor of this approach and indicated that it creates the possibility of using both control tests and facilitates further progressive development of the law of State responsibility.[323] There is much to be said about the choice of a degree of control for the purposes of attribution.

Sharing the view of the ICTY in the *Tadić* case, Antonio Cassese elaborated upon both arguments set out above by the ICJ refusing to apply the "overall control" test and instead choosing for the application of the "effective control" test in its *Bosnian Genocide* judgment. Cassese believes that these arguments are not convincing and should be rejected.[324] In the first instance, the Court failed to demonstrate that Article 8 DARS, which in its view must be understood in the light of the *Nicaragua*

319 Court of Appeal, *Hasan Nuhanović against the State of the Netherlands*, The Hague, July 5, 2011, Case No. 200.020.174/1, para. 5.20.

320 Supreme Court of the Netherlands, *The State of the Netherlands v. Hasan Nuhanović*, First Chamber, The Hague, September 6, 2013, Case No. 12/03324, para. 3.13.

321 James Crawford, (2013) *State Responsibility: The General Part*, Cambridge Studies in International and Comparative Law, No. 100, Cambridge: Cambridge University Press, p. 156.

322 J. Craig Barker and Sandesh Sivakumaran, (2007) "Application of the Convention on the Prevention and Punishment of the Crime of Genocide (Bosnia and Herzegovina v. Serbia and Montenegro)", *International and Comparative Law Quarterly*, Volume 56, Issue 3, pp. 702-703; Antonio Cassese, (2005) *International Law*, Second Edition, Oxford: Oxford University Press, p. 249.

323 International Law Commission, *Comments and Observations Received from Governments*, (2001) Yearbook of the International Law Commission, Volume II(1), UN Doc. A/CN.4/515 and Add.1-3, p. 49.

324 Antonio Cassese, (2007) "The Nicaragua and Tadić Tests Revisited in Light of the ICJ Judgment on Genocide in Bosnia", *The European Journal of International Law*, Volume 18, Issue 4, p. 651; Jernej L. Černič, (2007) "Catching the Conscience of Judges", *Australian International Law Journal*, Volume 14, pp. 257-258.

judgment, reflects customary international law.[325] Extensively analyzing State practice and international judicial practice, the ICTY in the *Tadić* case drew the conclusion that the Nicaragua test is at variance with those two types of practice and is to be regarded as unpersuasive.[326] The ICJ, however, based this main assumption merely on its own authority in the *Nicaragua* case and the authority of the ILC without being able to indicate that the "effective control" test is well-grounded in State practice and *opinio juris*, as two necessary requirements for establishing rules of customary international law,[327] and international judicial practice. This was necessary in order to conclude that the concept of State responsibility is indeed overly broadened if the "overall control" test is used. Moreover, if no rules of customary international law could have been identified, the ICJ would have needed to determine whether the theory of effective control is to be found in the general principles on State responsibility or general principles of international law. In addition, the Court would also have needed to prove that the "overall control" test for identifying the actual nature of an armed conflict used by the ICTY and the "effective control" test required in the opinion of the ICJ for establishing State responsibility are two different theories applicable to distinct issues and are to be applied in different contexts.[328] Therefore, it may be concluded that the particularly high threshold of the "effective control" test appears to be unjustified and has no basis in State practice or case law.

In its *Tadić* case, the Tribunal carefully examined State practice and judicial practice and came to the conclusion that there are, in fact, two distinct situations in which two different control theories are applicable. In principle, the ICTY clearly managed to illustrate that circumstances may require the application of one or another test depending on the context of a specific situation. It held that:

> "The requirement of international law for the attribution to States of acts performed by private individuals is that the State exercises control over the individuals. *The degree of control may*, however, *vary* according to the factual circumstances of each case. The Appeals Chamber fails to see why in each and every circumstance international law should require a high threshold for the test of control. Rather, various situations may be distinguished."[329] (emphasis added)

---

[325] Antonio Cassese, (2007) "The Nicaragua and Tadić Tests Revisited in Light of the ICJ Judgment on Genocide in Bosnia", *The European Journal of International Law*, Volume 18, Issue 4, pp. 651, 653-655.
[326] International Criminal Tribunal for the former Yugoslavia, *Prosecutor v. Tadić*, Appeals Chamber, Judgment, July 15, 1999, ICTY-IT-94-1-A, paras. 124-145; Valentina Azarov, (2009) *Internationalisation and Attribution: The Tadić Analysis in the Genocide in Bosnia Case*, University of Westminster School of Law Research Paper No. 09-11, p. 20, <http://ssrn.com/abstract=1470800>, accessed on March 10, 2011.
[327] International Court of Justice, *North Sea Continental Shelf Case (Federal Republic of Germany v. Netherlands)*, Judgment, February 20, 1969, ICJ Reports 1969.
[328] Antonio Cassese, (2007) "The Nicaragua and Tadić Tests Revisited in Light of the ICJ Judgment on Genocide in Bosnia", *The European Journal of International Law*, Volume 18, Issue 4, p. 651.
[329] International Criminal Tribunal for the former Yugoslavia, *Prosecutor v. Tadić*, Appeals Chamber, Judgment, July 15, 1999, ICTY-IT-94-1-A, para. 117.

In the first instance, the ICTY dealt with the cases when private individuals or unorganized groups of persons are employed by a State in order to participate in the commission of a certain illegal act on the territory of another State.[330] As examples, it refers to such illegal activities as kidnapping of a State official, murdering a high-ranking State official, carrying out acts of sabotage, and others. According to the Tribunal, in these situations it must be necessary to indicate that a State gave explicit instructions to private persons regarding the commission of the crimes in question in order to determine whether they acted as *de facto* State organs. Furthermore, it is required to show that the State at issue has publicly made statements concerning the retroactive approval of the unlawful actions of those individuals. A generic authority of a State over those persons or groups will not suffice to conclude that the international responsibility of that State may be engaged, but specific instructions must be given in order to make the attribution possible. In a similar vein, the State responsibility will be incurred if private individuals or groups of individuals are entrusted with the task of performing lawful actions on behalf of a State, but carrying out their tasks breach certain international obligations of that State. In this type of situation, according to the rules on State responsibility for the illegal activities of State officials acting *ultra vires*, the State in question will be held responsible on account of its specific instructions towards private individuals functioning on its behalf.[331]

The second type of situation concerns cases of individuals forming organized and hierarchically structured groups, such as military units and armed bands of rebels.[332] An organized and hierarchically structured group of individuals is to be perceived as cardinally different from single persons acting on their own: these groups often have a structure, a chain of command, a set of rules and procedures they comply with, and certain symbols of authority. Members of organized groups usually act in conformity with the standards prevailing in those groups and are subjected to the authority of a group's leader. Taking into account these considerations, the ICTY pointed out that for the successful attribution of illegal activities of such groups, it is sufficient that they have acted under the overall control of a State. Therefore, if a certain structured group carries out its activities under the control of a State having an overall character, the international responsibility of that State may be engaged independently of whether or not particular operations of those individuals constituting violations of international law were explicitly imposed, requested, directed, or instructed by the State.[333] According to the ICTY:

---

330 Ibidem, para. 118.
331 Ibidem, para. 119.
332 Ibidem, para. 120.
333 Ibidem, para. 121.

> "... the rationale behind this legal regulation is that otherwise, States might easily shelter behind, or use as a pretext, their internal legal system or the lack of any specific instructions in order to disclaim international responsibility."[334]

It is instructive to note that the overall control exercised by a certain State over organized and hierarchically structured groups, such as militias, paramilitary units, and subordinate armed forces, can be realized by means of equipping and financing those groups, coordinating, and helping in the general planning of their military and other activities. In this regard, it should be noted that the scope of State's control having an overall character must go beyond the mere provision of financial assistance, military equipment, or training.[335] One must solely be able to establish that besides financing, training, equipping, and providing other operational support, the State in question plays a role in organizing, coordinating, or planning the military operations. As stated in the ICTY *Celebici* case, the requirements of "overall control" are met even when armed forces act on behalf of a controlling State and are able to make autonomous choices of means and tactics, but participate in a common strategy together with the controlling State.[336] In this respect, it is not necessary to indicate that that State plans all operations of a group or issues specific instructions and orders regarding the illegal conduct violating rules and principles of international law.

Apparently, there has been some disagreement among commentators regarding the approach taken by the ICJ in the *Bosnian Genocide* case and the proper application of two control theories.[337] Although some authors argue that the "overall control" test lacks support by case law and State practice,[338] it is safe to assume that both the "effective control" and the "overall control" tests are to be regarded as supported by customary international law – emanating from State practice and *opinio juris* – and international judicial practice.[339] But it is apparent that they both have a different purpose and scope of application. The Vice-President of the ICJ, Awn Shawkat Al-Khasawneh, acknowledged in his dissenting opinion in the *Bosnian Genocide* case that the ICJ chose the wrong path to apply the "effective control" theory to the situation

---

334 Ibidem, para. 123.

335 Ibidem, para. 137.

336 International Criminal Tribunal for the former Yugoslavia, *Prosecutor v. Delalic*, Appeals Chamber, Judgment, February 20, 2001, ICTY-IT-96-21-A, para. 47.

337 Ruth Wedgwood, "Slobodan Milosevic's Last Waltz", *The New York Times*, March 12, 2007, <http://www.nytimes.com/2007/03/12/opinion/12wedgwood.html?pagewanted=1&_r=1>, accessed on March 1, 2011; Antonio Cassese, "A Judicial Massacre: The International Court Has Set an Unrealistically High Standard of Proof for Finding Serbia Complicit in Genocide", *Guardian Unlimited*, February 27, 2007, <http://www.guardian.co.uk/commentisfree/2007/feb/27/thejudicialmassacreofsrebr>, accessed on March 1, 2011.

338 Oliver R. Jones, (2009) "Implausible Deniability: State Responsibility for the Actions of Private Military Firms", *Connecticut Journal of International Law*, Volume 24, Issue 2, p. 272.

339 Antonio Cassese, (2007) "The Nicaragua and Tadić Tests Revisited in Light of the ICJ Judgment on Genocide in Bosnia", *The European Journal of International Law*, Volume 18, Issue 4, p. 657.

different from the *Nicaragua* case.[340] He observed that in the case regarding Bosnian genocide there was "a unity of goals, unity of ethnicity and a common ideology" that could be established between the State and the actual perpetrators of genocide.[341] He pointed out that the rejection of the "overall control" test established in the *Tadić* case does not address the problem of a variety of activities that may be employed in armed conflicts requiring different approaches to the issue of attribution. The Vice-President indicated that "the test of control is a variable one",[342] an idea that seems to be supported by State practice and judicial precedent. Following his line of reasoning, the exclusive application of the "effective control" test disregarding varying degrees of control in specific legal contexts cannot be the correct approach. If one considers that when a shared objective between private persons and entities and States is the commission of international crimes and they have a common criminal purpose, then it might be unnecessary to apply the "effective control" test with its particularly high threshold, for "it gives States the opportunity to carry out criminal policies through non-State actors or surrogates without incurring direct responsibility therefore".[343] Therefore, to clarify the matter it must be determined which of two control theories is to be regarded as the most suitable test that is to be applied with regard to the State practice of employing PMSCs.

Essentially, the exact standard of attributing the conduct of persons or entities acting under the control of a State in general remains unclear and there is still no clarity regarding the application of one particular norm.[344] According to some authors, the established judicial practice and State practice seem to indicate that, depending on the context and circumstances, in which private persons or groups of persons operate under the control of a State, one should have a choice between the application of the "effective control" test or the "overall control" test, being different in their attitude towards degree of control.[345] Having examined the main features of the phenomenon of privatization of war and security and the State practice of hiring PMSCs, it needs to be said that those firms and their employees can be regarded as organized and hierarchically structured groups in terms of significance of this notion

---

340 Dissenting Opinion of the Vice-President of the International Court of Justice A.S. Al-Khasawneh in International Court of Justice, *Case Concerning the Application of the Convention on the Prevention and Punishment of the Crime of Genocide (Bosnia and Herzegovina v. Serbia and Montenegro)*, Judgment (Merits), February 26, 2007, ICJ Reports 2007, para. 36.

341 Ibidem.

342 Ibidem, para. 37.

343 Ibidem, para. 39.

344 Jackson N. Maogoto and Benedict Sheehy, (2009) "Private Military Companies & International Law: Building New Ladders of Legal Accountability & Responsibility", *Cardozo Journal of Conflict Resolution*, Volume 11, p. 112.

345 Tal Becker, (2008) *Terrorism and the State: Rethinking the Rules of State Responsibility*, Oxford and Portland: Hart Publishing, p. 70; Amanda Tarzwell, (2009) "In Search of Accountability: Attributing the Conduct of Private Security Contractors to the United States Under the Doctrine of State Responsibility", *Oregon Review of International Law*, Volume 11, Issue 1, p. 198.

to the "overall control" theory envisaged by the ICTY.[346] Therefore, one may argue that States are not required to exercise effective control over their activities, as it is the case with single individuals, and to instruct companies to commit violations of international law, which is highly unlikely to happen in practice. For the purpose of application of the "overall control" theory, one can certainly state that PMSCs can be equated with militias and paramilitary units operating on behalf of States and carrying out a broad range of military and security tasks, and thus fall within the scope of application of the "overall control" theory.

Assessing the criteria put forward by the ICTY, one should acknowledge that the requirement of having a structure inside PMSCs is normally met. Private contractors do not function as single individuals receiving orders from the States hiring them or any other States. On the contrary, they are organized in tight commercial structures with all the members of a company having their own functions and defined positions within the group. Moreover, the condition of the existence of chains of command can also be considered as fulfilled, as according to the documented reports and commentators, PMSCs usually have defined leaders and commanders bringing hierarchy and order in a group. The contractors employed by PMSCs do not act on their own, but operate on instructions and orders of their superiors. The requirement of having a set of rules should also be perceived as being met: PMSC members organized in hierarchically structured groups with own chain of command do have a system of written and unwritten internal rules, such as codes of conduct, that they need to comply with. It is also frequently the case that private contractors have certain symbols of authority, such as their specific dress code, uniforms, or weapons they usually carry. It can be argued that the element of openly carrying weapons by PMSCs might be the most obvious sign that private contractors have a certain degree of authority, for in the most countries merely members of the regular armed forces are entitled to use automatic weapons – usually associated with a person or an entity exercising elements of governmental authority.

While it is practically impossible to conduct solid research into the actual operation of PMSCs on the ground and their relationships with States, due to the lack of information regarding contracts concluded between firms and governments and general secrecy around the whole industry, one can argue that PMSCs often show unity of goals with States. Under certain circumstances – when PMSCs are legally established on the territory of the hiring State itself – those companies can even be said to have a unity of ethnicity and ideology with the hiring and home State in question. Reportedly, PMSCs carry out their services in cooperation with the regular armed forces of States, frequently sharing information, coordinating missions,

---

346 Dieter Fleck, (2007) "Individual and State Responsibility for Intelligence Gathering", *Michigan Journal of International Law*, Volume 28, p. 696.

and resolving conflicts during operations.[347] Furthermore, PMSCs often have some leeway to choose necessary means and tactics according to the contracts concluded between them and hiring States. Nevertheless, they often pursue the same goals as the hiring or home States and participate in their common strategies.[348]

Moreover, it is worth noting that substantial differences between the circumstances in the *Nicaragua* case and a variety of situations in which PMSCs operate can lead to the conclusion that one should refrain from assessment on the basis of the "effective control" test and lean towards the application of the control theory as developed in the *Tadić* case.[349] PMSCs often function on the territory of a host State with its explicit approval of the presence of foreign forces and private contractors. In addition, there are agreements reached between host States and hiring States, meaning that PMSCs used by the hiring States do not employ their activities against the will of a host State or without it knowing about their operation, which was a case in the *Nicaragua* judgment. This fact may justify the application of a more flexible "overall control" theory. However, if PMSCs carry out their military and security tasks without a form of explicit acceptance of a host State or without it even being aware that such operations are conducted, one may assume that it is necessary to refer to the "effective control" test in order to determine a degree of control exercised by a hiring or home State over private contractors. Applying the same line of reasoning, it can be stated that if the conduct of PMSCs occurs not on the territory of hiring States, which in practice may be the territory of host States other than States employing PMSCs, while hiring States are distantly removed from the action theatre, the same high degree of control must be proved. The ICTY mentioned in the *Tadić* case that:

> "... if, as in Nicaragua, the controlling State is not the territorial State where the armed clashes occur or where at any rate the armed units perform their acts, more extensive and compelling evidence is required to show that the State is genuinely in control of

347 Dutch Advisory Council on International Affairs, *Employing Private Military Companies: A Question of Responsibility* (Dutch: *De Inhuur van Private Militaire Bedrijven: Een Kwestie van Verantwoordelijkheid*), No. 59, December 2007, The Hague, p. 21, <http://www.aiv-advies.nl/ContentSuite/upload/aiv/doc/webversie_AIV_59eng(1).pdf>, accessed on May 9, 2011; Chia Lehnardt, "Private Military Companies and State Responsibility", in Simon Chesterman and Chia Lehnardt (Eds.), (2007) *From Mercenaries to Market: The Rise and Regulation of Private Military Companies*, Oxford: Oxford University Press, p. 150.

348 Jackson N. Maogoto and Benedict Sheehy, (2009) "Private Military Companies & International Law: Building New Ladders of Legal Accountability & Responsibility", *Cardozo Journal of Conflict Resolution*, Volume 11, p. 113.

349 Chia Lehnardt, "Private Military Companies and State Responsibility", in Simon Chesterman and Chia Lehnardt (Eds.), (2007) *From Mercenaries to Market: The Rise and Regulation of Private Military Companies*, Oxford: Oxford University Press, p. 151.

> the units or groups not merely by financing and equipping them, but also by *generally directing* or *helping plan* their actions."[350] (emphasis added)

The ICTY held further that the test of control does not have a high threshold and can easily be satisfied:

> "... where the controlling State in question is an adjacent State with territorial ambitions on the State where the conflict is taking place, and the controlling State is attempting to achieve its territorial enlargement through the armed forces which it controls."[351]

In other words, it may be assumed that the degree of control exercised by a State over PMSCs that is necessary to attribute their illegal acts to the State in question decreases when the State gets closer to the actual field of PMSC operation. If a State finds itself on the territory of a host State where PMSCs carry out their missions or if PMSCs employed by a State act on the territory of that State, one may acknowledge that the analysis of the *Tadić* case indicates that the "overall control" test would be the most appropriate tool to determine the extent of State control and to attribute illegal activities to the State. It is, however, arguable whether this determination should be based solely on the criterion of physical distance of a State, which might be a hiring State, a home State, or a combination of both, to a host State, as the modern means of communication can enable any State to exercise a particularly high degree of control over their troops or civilian personnel operating abroad. This means that the "overall control" test may be sufficient to attribute illegal acts of PMSCs to the States that control their actions over long distances.

The application of the "overall control" test is associated with a lower threshold of imputing internationally wrongful acts of private persons and entities, such as PMSCs, to States when compared to the "effective control" test and may be very helpful in addressing recent trends in the State use of individuals and entities. The set of trends identified by Antonio Cassese includes the State's support of military and paramilitary groups or armed bands that fight against other States or rebels and secessionist movements, the State's control and support of terrorist groups, and the use by international organizations of the national armed forces in peacekeeping and other military operations.[352] To the list of such trends, one can add the increasing reliance of States on PMSCs that carry out a wide range of tasks on their behalf and may be used to achieve State interests that are not always legitimate. The phenomenon of PMSCs may lead to serious threats to international peace and security and therefore

---

350 International Criminal Tribunal for the former Yugoslavia, *Prosecutor v. Tadić*, Appeals Chamber, Judgment, July 15, 1999, ICTY-IT-94-1-A, para. 138.

351 Ibidem, para. 140.

352 Antonio Cassese, (2007) "The Nicaragua and Tadić Tests Revisited in Light of the ICJ Judgment on Genocide in Bosnia", *The European Journal of International Law*, Volume 18, Issue 4, pp. 665-667.

needs to be addressed from the perspective of the doctrine of State responsibility. In this regard, the theory of overall control may prove to be a more helpful and more desirable option to hold States accountable for the violations of international law and to prevent those breaches from occurring rather than the strict test of effective control. The latter is simply not suitable for capturing the actual relationship between States and private contractors and by often making the attribution ex Article 8 DARS impossible results in the situation of impunity.[353] In other words, this test appears to be rather impractical.[354] One might understand that most cases of PMSCs engaged in conflict areas can be brought under the scope of Article 8 DARS with its essential notion of control. It might be proposed that the test of effective control should be relaxed so that States would not be able to escape international responsibility for the misconduct of these companies and their employees. Against this backdrop, the "overall control" test would be the preferred theory and this opinion is shared by the author of the present study.

One might wonder what would be the outcome of using both tests in a real world scenario involving a PMSC whose personnel commit violations of international law. The Nisour Square shooting can serve as a good example of an incident that is to be scrutinized on the basis of Article 8 DARS and two control theories. In this regard, it must essentially be established whether Blackwater and its personnel acted as "auxiliaries" outside the official State structure.[355] Taking into account the factual circumstances of the Nisour Square incident, it is apparent that no instructions were given by the US or, more specifically, the US Department of State as the hiring entity to engage in the commission of unlawful acts. The position that the US made the decision to kill and to injure Iraqi civilians and to destroy their property cannot be maintained, nor the claim that Blackwater or, to be more precise, the Raven 23 team implemented or made attempts to implement this decision. There is no evidence of the US authorizing the unlawful conduct of the PMSC, meaning that it did not give instructions to the company to use force and to engage in the shooting. Also, it is not obvious that Blackwater had been operating under direction of the US and that the latter directed the unlawful conduct. Blackwater was operating on the basis of a contract with the Department of State and was involved in the shooting as part of its mission to provide security to another Blackwater team. The question of whether the US exercised a certain degree of control in this specific instance is a more interesting and complicated one from the legal point of view.

---

[353] Corinna Seiberth, (2013) *Private Military and Security Companies in International Law – A Challenge for Non-binding Norms: the Montreux Document and the International Code of Conduct for Private Security Service Providers*, Cambridge, Antwerp, Portland: Intersentia, p. 120.

[354] Amanda Tarzwell, (2009) "In Search of Accountability: Attributing the Conduct of Private Security Contractors to the United States Under the Doctrine of State Responsibility", *Oregon Review of International Law*, Volume 11, Issue 1, pp. 201-202.

[355] Commentary to DARS as adopted in 2001, p. 47, Article 8, para. 2.

According to the Commentaries to the ILC's Draft Articles, "such conduct will be attributable to the State only if it directed or controlled the specific operation and the conduct complained of was an integral part of that operation. The principle does not extend to conduct which was only incidentally or peripherally associated with an operation and which escaped from the State's direction or control."[356] The "effective control" test, introduced in the ICJ *Nicaragua* case and confirmed in the ICJ *Bosnian Genocide* case, is a leading criterion that should be used in order to examine the degree of control. In order to satisfy this test, Blackwater and its employees must specifically have been charged by the US government giving certain instructions to commit unlawful acts or to carry out certain activities inconsistent with international law.[357] It is not necessary to show that certain State organs of the US and its State agents have physically been engaged in the commission of unlawful acts: what is required in essence, is the exercise of effective control by the State in question in relation to every operation in which the alleged breaches took place and thus instructions and "practical support"[358] for the realization of specific unlawful activities. Thus, it must be shown that the US as a State hiring Blackwater in practice exercised effective control over all activities of the company during the Nisour Square incident and, in fact, authorized those illegal acts by giving instructions for their commission.[359]

Considering the current scenario involving Blackwater and the United States, it comes to mind that the United States is, in fact, both the hiring and the home State of the corporation. While as a hiring State the US is known for outsourcing a range of activities to PMSCs,[360] in its capacity as a home State the given State is thus a country that has a framework of licensing and certification mechanisms that are aimed at the companies of its nationality. It does not, however, suffice to say that the United States exercises effective control over every operation of a PMSC it employs. In practice, the lack of elaborate oversight and monitoring instruments makes it impossible for the US to effectively exercise control over PMSCs and their activities. Closely investigating the Nisour Square shooting, it cannot be proved that the US had effective control over the Raven 23 team, which acted on its own. Responding to the message concerning the explosion of a vehicle-borne improvised explosive device, the convoy took up positions at Nisour Square to protect another Blackwater team providing

356 Ibidem, p. 47, Article 8, para. 3.

357 Separate Opinion of Judge M. Ago in International Court of Justice, *Case Concerning Military and Paramilitary Activities in and against Nicaragua (Nicaragua v. United States of America)*, Judgment (Merits), June 27, 1986, ICJ Reports 1986, paras. 16-17.

358 Lindsey Cameron and Vincent Chetail, (2013) *Privatizing War: Private Military and Security Companies under Public International Law*, Cambridge: Cambridge University Press, p. 213.

359 Brigitte Stern, "The Elements of An Internationally Wrongful Act", in James Crawford, Alain Pellet, and Simon Olleson (Eds.), (2010) *The Law of International Responsibility*, Oxford: Oxford University Press, p. 206.

360 Deborah D. Avant, (2005) *The Market for Force: The Consequences of Privatizing Security*, Cambridge: Cambridge University Press, p. 147.

security to the US officials. The further chain of events has been elaborated upon above, but essentially it indicates that Blackwater guards operated on their own initiative and engaged in the shooting with no instructions from the US Department of State or any other US agencies or officials. As some experts tend to think, it is often merely a certain degree of "informal co-ordination" between PMSCs and the United States military that is meant to organize meetings for the purposes of sharing information and coordinating and resolving conflicts that may arise during the operations.[361] In Iraq, there is a lack of command and control exercised by the US armed forces over PMSCs and they operate together on the basis of "cooperation and coordination of activities and the desire to work from a common operating picture".[362] In fact, PMSCs employed by various US agencies in Iraq fall outside the scope of the military chain of command.[363] Only when PMSC employees find themselves in the US military installations do military commanders have authority over them.[364] Dr. Hammes, a US colonel, provided his written statement to the US House Committee on Oversight and Government Reform indicating the following:

> "When serving within the combat zone, particularly during a counterinsurgency, contractors create a number of significant problems from the tactical to the strategic level. Three primary characteristics of contractors, particularly armed contractors, create problems for the government. First, the government does not control the quality of the personnel the contractor hires. Second, unless it provides a government officer or NCO for each convoy, personal security detail or facilities protection unit, it *does not control* their daily interactions with the local population. Finally, the population holds the government responsible for everything the contractors do or fail to do."[365]

---

361 US Government Accountability Office, *Rebuilding Iraq: Actions Needed to Improve Use of Private Security Providers*, Report to Congressional Committees, July 2005, pp. 20-21, <http://www.gao.gov/assets/250/247252.pdf>, accessed on October 18, 2013; Chia Lehnardt, "Private Military Companies and State Responsibility", in Simon Chesterman and Chia Lehnardt (Eds.), (2007) *From Mercenaries to Market: The Rise and Regulation of Private Military Companies*, Oxford: Oxford University Press, p. 150.

362 US Government Accountability Office, *Rebuilding Iraq: Actions Needed to Improve Use of Private Security Providers*, Report to Congressional Committees, July 2005, pp. 20-21, <http://www.gao.gov/assets/250/247252.pdf>, accessed on October 18, 2013.

363 Chia Lehnardt, "Private Military Companies and State Responsibility", in Simon Chesterman and Chia Lehnardt (Eds.), (2007) *From Mercenaries to Market: The Rise and Regulation of Private Military Companies*, Oxford: Oxford University Press, p. 50.

364 US Government Accountability Office, *Rebuilding Iraq: Actions Needed to Improve Use of Private Security Providers*, Report to Congressional Committees, July 2005, p. 21, <http://www.gao.gov/assets/250/247252.pdf>, accessed on October 18, 2013.

365 Written Statement of Dr. T.X. Hammes, Senior Research Fellow, Provided to the House Committee on Oversight and Government Reform, Subcommittee on National Security and Foreign Affairs, Hearing on June 22, 2010, <http://oversight.house.gov/wp-content/uploads/2012/01/20100622Hammes.pdf>, accessed on October 1, 2013.

In this regard, it is safe to say that the US did not exercise effective control over the activities of the Raven 23 team at Nisour Square, which means that the "effective control" test cannot be successfully used in order to make attribution ex Article 8 DARS possible. For the purposes of the "effective control" theory, the only fact of States financially controlling activities of PMSCs by means of having contracts with them cannot be used to establish that the effective control is exercised.[366] It also appears that the contractual relationship between Blackwater and the US as a real link between the company and the hiring State and the fact that the latter could terminate the contract and cease to pay Blackwater for its services are not sufficient to conclude that attribution under this specific provision to the US as both the hiring and the home State would be successful.[367]

If, however, the line of reasoning of the ICTY in the *Tadić* case is followed and some merit in the points raised by Antonio Cassese is seen, one might be inclined to make use of the "overall control" test developed by the Tribunal. Such is the position adopted in the current study. Due to the fact that PMSCs are not solely private individuals or unorganized groups of persons but can be seen as organized and hierarchically structured business entities, a different degree of control might suffice to establish that a State exercises control within the meaning of Article 8 DARS. The "overall control" test might also seem to be the most appropriate theory considering the fact that Blackwater operated in Iraq with the consent of the host government allowing American presence on its territory.[368] The closer the hiring State is to the actual place where the activities of PMSCs employed by it are carried out, the less strict control test can be more appropriate.[369] Applied to the Nisour Square incident, the "overall control" test can be met if it is indicated that the US plays a role in organizing, coordinating, or planning of the operations of Blackwater besides financing, training, equipping, and providing other forms of operational support. It is thus not required to establish the effective control of the State in question in relation to the operational activities of the Raven 23 team during which unlawful acts were committed. The requirements of financing, training, and equipping can be partially or, under certain circumstances, fully satisfied, given that the operations of Blackwater are in essence financed by the US government that concluded the contract with the corporation for the provision of a certain array of security services.[370] Taking into

---

[366] Lindsey Cameron and Vincent Chetail, (2013) *Privatizing War: Private Military and Security Companies under Public International Law*, Cambridge: Cambridge University Press, p. 215.

[367] Chia Lehnardt, "Private Military Companies and State Responsibility", in Simon Chesterman and Chia Lehnardt (Eds.), (2007) *From Mercenaries to Market: The Rise and Regulation of Private Military Companies*, Oxford: Oxford University Press, p. 149.

[368] Ibidem, p. 151.

[369] Ibidem, p. 152.

[370] Amanda Tarzwell, (2009) "In Search of Accountability: Attributing the Conduct of Private Security Contractors to the United States Under the Doctrine of State Responsibility", *Oregon Review of International Law*, Volume 11, Issue 1, p. 203.

account the fact that the terms of the actual contract between the State and the company are not made public, it can only be speculated whether it included provisions on training and equipping that are to be performed by the US. Usually, it is not the case, and PMSCs are, in fact, indirectly equipped and trained by means of financial remuneration received from hiring States.[371] However, when dealing with the criteria of providing other forms of support and having a role in organizing, coordinating, or planning of the activities of Blackwater, one may find that there is no sufficient evidence to conclude what the actual involvement of the US Department of State or other agencies in the operations of Blackwater and, more specifically, the Raven 23 team was. From the facts made available to the general public, it can be deduced that the State did play a certain role in the organization, coordination, and planning of the activities of Blackwater as a whole in Iraq, but as such it was not involved in the operations of this specific team. While it can be speculated that the requirements of the "overall control" test have been met in this case, more evidence might be needed to conclude what the actual relationship was between the US and Blackwater and whether the State had overall control over this firm.

In order to sum up, one needs to recall that the extent of control exercised by States over PMSCs varies depending on the particular circumstances of a case.[372] It should be noted that under certain circumstances the "overall control" test appears to be the most suitable one, while States, especially those hiring PMSCs, are likely to easily satisfy the set of criteria for the application of this test. It will automatically lead to the attribution of their unlawful conduct to the States concerned if the reasoning of the ICTY positioning its control theory as realistic[373] is followed. Particular situations need, however, to be subjected to a specific analysis with regard to detailed circumstances of each case, which may indicate that a State is required to exercise the effective control over PMSCs in order to make the attribution possible.

As has already become clear, if the narrow "effective control" test is taken as a starting point of assessment of situations, in which PMSCs currently operate, the attribution of the unlawful conduct of private contractors violating international law to States is highly unlikely, as the *Nicaragua* control test transforms it into a

---

[371] Lindsey Cameron and Vincent Chetail, (2013) *Privatizing War: Private Military and Security Companies under Public International Law*, Cambridge: Cambridge University Press, p. 219.

[372] Dissenting Opinion of the Vice-President of the International Court of Justice A.S. Al-Khasawneh in International Court of Justice, *Case Concerning the Application of the Convention on the Prevention and Punishment of the Crime of Genocide (Bosnia and Herzegovina v. Serbia and Montenegro)*, Judgment (Merits), February 26, 2007, ICJ Reports 2007, para. 37; Chia Lehnardt, "Private Military Companies and State Responsibility", in Simon Chesterman and Chia Lehnardt (Eds.), (2007) *From Mercenaries to Market: The Rise and Regulation of Private Military Companies*, Oxford: Oxford University Press, p. 152.

[373] Tal Becker, (2008) *Terrorism and the State: Rethinking the Rules of State Responsibility*, Oxford and Portland: Hart Publishing, p. 71.

mere theoretical possibility.[374] This means that most States will be able to escape the application of the doctrine of State responsibility for those breaches of international law. Taking into account the fact that both control tests are based on customary international law and judicial practice, it is worth emphasizing that the degree of control exercised by States over PMSCs is to be assessed in accordance with the "overall control" test that most probably will be broadly accepted in the future. In any case, it was one of the strong points of Vice-President of the ICJ Awn Shawkat Al-Khasawneh in his dissenting opinion and these opinions often tend to reflect the future developments of international law.

6.4 Article 11 DARS

6.4.1 General application of the provision

Obviously, when persons, groups of persons, or entities do not constitute State organs, do not exercise elements of governmental authority, and do not function under instructions, direction, or control of a State, their unlawful acts violating rules and principles of international law will not be attributed to the State in question, meaning that no international responsibility of that State will be engaged.[375] However, the State may decide to go over to an *ex post facto* acknowledgment or adoption of illegal acts as its own, in accordance with Article 11 DARS, providing that:

> "Conduct which is not attributable to a State under the preceding articles shall nevertheless be considered an act of that State under international law if and to the extent that the State *acknowledges* and *adopts* the conduct in question as its own." (emphasis added)

Following the idea expressed in the Commentaries, if a certain unlawful conduct of private persons or entities violating international law and constituting an internationally wrongful act was not or may not have been attributable to a State at the time of its commission, the State in question may subsequently acknowledge and adopt it as its own.[376] In the view of the ILC, this provision is based on the main principle underlying its essence that private acts in their pure sense cannot be imputed to a State under international law if individuals and entities do not act on behalf of a State.[377] In this regard, one should point out that a State must not merely support or endorse those activities, but specifically recognize and identify the conduct in question and

---

374 Ibidem, p. 69.

375 Martin Dixon, (2007) *International Law*, Sixth Edition, Oxford: Oxford University Press, p. 250.

376 Commentary to DARS as adopted in 2001, p. 52, Article 11, paras. 1-2.

377 Ibidem, p. 52, Article 11, paras. 2-3.

acknowledge and adopt it.[378] Finally, the State's act of acknowledgment and adoption needs to be clear and unequivocal, but also express or inferable from the conduct of that State.[379]

6.4.2 Application with regard to PMSCs

According to Article 11 DARS, if the unlawful conduct of PMSCs and their personnel is not attributable to a State under other provisions of the Draft Articles discussed above, it will be regarded as an act of that State under international law if it acknowledges and adopts that conduct as its own. In this case, certain illegal actions of PMSCs constituting violations of international humanitarian and human right law will be considered "internationally wrongful acts" and will lead to State responsibility.[380]

Theoretically speaking, all three types of States, i.e. hiring, home, and host States, are in a position to acknowledge and to adopt the misconduct of PMSCs as their own. As pointed out above, however, it is an established practice that States often tend to deny any connection to PMSCs, indicating that they may not be held responsible for their unlawful actions violating international law. Therefore, it might be assumed that States will remain highly reluctant to acknowledge and to adopt those illegal acts as their own, as they are normally unwilling to be associated with violations of international law committed by PMSCs.[381]

After the Abu Ghraib scandal came to light, the US as a hiring and home State did not go as far as to acknowledge and to adopt the abuses committed in Abu Ghraib as its own. On the contrary, what happened was the denial on the part of the US that unlawful acts committed at the detention facility constituted torture. In May 2004, Secretary of Defense Donald H. Rumsfeld made the violations committed by the US military personnel and members of PMSCs seem less serious than they were in reality during the press conference. He noted:

> "My impression is that what has been charged thus far is abuse, which I believe technically is different from torture... I don't know if the – it is correct to say what you just said, that torture has taken place, or that there's been a conviction for torture. And therefore I'm not going to address the torture word."[382]

---

378 Ibidem, p. 53, Article 11, para. 6; Tal Becker, (2008) *Terrorism and the State: Rethinking the Rules of State Responsibility*, Oxford and Portland: Hart Publishing, p. 72.

379 Commentary to DARS as adopted in 2001, p. 53, Article 11, paras. 8-9.

380 Ibidem, p. 53, Article 11, paras. 6-7.

381 James Crawford, (2013) *State Responsibility: The General Part*, Cambridge Studies in International and Comparative Law, No. 100, Cambridge: Cambridge University Press, p. 182.

382 US Defense Department Operational Update Briefing, May 4, 2004, <http://www.defense.gov/transcripts/transcript.aspx?transcriptid=2973>, accessed on November 10, 2013.

In the beginning, the US government attempted to avoid the use of the word torture by all means and was not willing to go as far as an expression of regret or a formal apology. On May 6, 2004, however, the US President George W. Bush said that he was "sorry for the humiliation suffered by the Iraqi prisoners and the humiliation suffered by their families".[383] But at the same time he was also "equally sorry that people seeing these pictures didn't understand the true nature and heart of America". This statement cannot be seen as a formal apology to Iraq and Iraqi citizens who suffered under detention at the Abu Ghraib prison. The US did not adopt the illegal acts of the Abu Ghraib guards and interrogators as its own and did not provide any form of reparation to Iraq required by customary international law.

One may conclude that Article 11 DARS will probably have a little practical application in reality and will not be applicable to the situations when PMSCs breach international law. States are and will most probably remain reluctant to bear responsibility pursuant this provision.

7. Breaches of International Obligations of States

Having established that the unlawful conduct of PMSCs can be attributed to States in certain situations, it must be considered in accordance with Articles 2 and 12 DARS whether this conduct constitutes violations of international law. These must be breaches of international obligations of States. The findings presented below are based on the analysis of different types of misconduct included in the introductory Chapter.

Indiscriminate shootings and civilian casualties

The indiscriminate shootings involving PMSCs that lead to civilian casualties appear to be quite a persistent pattern of operation of many private contractors abroad. The companies sometimes conduct their own quick investigations of the incidents, while no official investigation is carried out by the hiring and home States. The next step PMSCs take is firing the employees in question and sending them back to their country of origin.[384] In the end, contractors do not face prosecution by the State authorities and may freely be employed by other interested firms.[385]

---

[383] Susan Sontag, "Regarding the Torture of Others", *The New York Times*, May 23, 2004, <http://www.nytimes.com/2004/05/23/magazine/regarding-the-torture-of-others.html>, accessed on October 17, 2013.

[384] Human Rights First, *Private Security Contractors at War: Ending the Culture of Impunity*, Report, 2008, New York, p. 16.

[385] Kateri Carmola, (2010) *Private Security Contractors and New Wars: Risk, Law, and Ethics*, Oxon and New York: Routledge, p. 115.

In general, it must be noted that the excessive use of force employed by PMSC contractors and their indiscriminate shootings at civilians of the States in which they operate constitute breaches of the Fourth Geneva Convention of 1949, nowadays recognized as customary international law.[386] Therefore, if the illegal acts of private guards in the meaning of the above-mentioned Convention could be attributed to the hiring and perhaps other States, then it would mean that those States could be held responsible for those atrocities.

These crimes committed by contractors can also be labeled as human rights abuses. In fact, those persons killing or injuring civilians act in violation of one of the most essential human rights, the right to life, which can be said to be included in the right to security.[387] This right to security aims to protect persons from becoming victims of genocide, war crimes, crimes against humanity, torture, and other acts. This has been laid down in Article 3 UDHR, Article 9 ICCPR, Article 5 ECHR, Article 7 ACHR, and Article 6 ACHPR. The right to life is to be found in Article 3 UDHR, Article 6 ICCPR, Article 2 ECHR, Article 4 ACHR, Article 4 ACHPR, and Article 6 CRC. When these human rights treaties have an extraterritorial reach, their provisions are applicable to the operations of a State, its organs, and persons under its control or exercising its governmental authority on the territory of foreign countries.

At the same time, the illegal activities falling under this category can be said to amount to war crimes or crimes against humanity,[388] which are to be regarded as serious violations of international humanitarian law and can give rise to individual criminal responsibility of respective individuals. After the Nisour Square incident involving the horrific acts by Blackwater guards, the representative of the UN Mission Said Arikat stated that "when you kill 17 people like that, it is a crime against humanity if it is proven that it was done in cold blood".[389] It is quite obvious, considering that, following the general idea of Article 7(1)(a) and (b) of the Statute of International Criminal Court (hereinafter: ICC), shootings at civilians as a grave breach of the Fourth Geneva Convention of 1949 might be perceived as murder and

---

386 David S. Mitchell, (2005) "The Prohibition of Rape in International Humanitarian Law as a Norm of Jus Cogens: Clarifying the Doctrine", *Duke Journal of Comparative and International Law*, Volume 15, p. 244.

387 Amol Mehra, (2010) "Bridging Accountability Gaps – The Proliferation of Private Military and Security Companies and Ensuring Accountability for Human Rights Violations", *The Pacific McGeorge Global Business & Development Law Journal*, Volume 22, pp. 326-327; Human Rights Advocates, *Holding Private Military and Security Companies and Mercenaries Accountable for Human Rights Violations*, Report, p. 4, <http://www.humanrightsadvocates.org/wp-content/uploads/2010/05/Arellano2008-Mercenaries-and-Corporate-Accountability-Long.pdf>, accessed on October 10, 2012.

388 Daniel Graeber, "U.N. Report Cites War Crimes by Private Contractors", October 12, 2007, <http://warcrimes.foreignpolicyblogs.com/2007/10/12/un-reports-cites-war-crimes-by-private-contractors/>, accessed on October 10, 2012.

389 Katarina Kratovac, "U.N.: Shootings by Guards Might Well Be War Crimes", *The Star-Ledger*, October 12, 2007.

extermination forming part of a widespread or systematic attack against any civilian population, which must qualify as a crime against humanity.

According to the definition of war crimes provided in Article 8(2)(a)(i) and (iii) of the Statute of the ICC, the same shootings can be considered a war crime, taking into account the fact that they can be described as willful killing, willfully causing great suffering, or serious injury to body or health. Moreover, in case of an international armed conflict, as stipulated in Article 8(2)(b)(i), those indiscriminate shootings at civilian population can be considered as other serious violations of the laws and customs applicable in international armed conflict, such as intentionally directing attacks against the whole civilian population or its individual representatives who do not take part in hostilities. In this regard, indiscriminate shootings at civilians are also, in fact, an intentional launching of an attack knowing that it will cause incidental loss of life or injury to civilians, following the reasoning of Article 8(2)(b)(iv). In case of a conflict of a non-international character according to Article 8(2)(c)(i), Common Article 3 of the Geneva Conventions of 1949 is applicable, meaning that causing violence to life of civilians, including all forms of murder and mutilation, will also be regarded a war crime. Finally, Article 8(2)(e)(i) provides that intentionally directing attacks against the whole civilian population or its individual representatives also constitutes a war crime in a non-international armed conflict.

Applying this knowledge to a specific case of misconduct, one can trace what rules and principles of international law have been breached. Taking the Nisour Square shooting involving employees of Blackwater, the question that bears asking in this regard is whether the conduct of Blackwater guards imputable to the US qualifies to be wrongful in the sense of the non-compliance by the given State with international law. This determination rests on the process of establishing whether there is an incompatibility with the obligations of the US with conventions in force or with other applicable rules of international law.

Before going into the substance of discussion concerning the existence of a breach of an international obligation on the United States, it must be observed that the US fiercely denied the extraterritorial applicability of some human rights treaties. As one of the major States often performing extraterritorial activities, the US did not accept the extraterritorial reach of the ICCPR. During the considerations of the initial report[390] of the US to the Human Rights Committee, the US representatives made clear that the Covenant was not regarded as a human rights convention having extraterritorial application.[391] There is the dual requirement laid down in Article 2(1) ICCPR stating that each State Party to the treaty "undertakes to respect and to ensure to all individuals within its territory and subject to its jurisdiction the rights

---

390 UN Human Rights Committee, *Initial Reports of State Parties due in 1993: United States of America*, August 24, 1994, UN Doc. CCPR/C/81/Add.4 (State Party Report).

391 UN Human Rights Committee, *Summary Record of the 1405th Meeting: United States of America*, April 24, 1995, UN Doc. CCPR/C/SR.1405 (Summary Record), para. 20.

recognized in the present Covenant". This provision appears to limit the scope of application of the Convention to persons residing within United States territory and being under United States jurisdiction. The same position was taken in the second and third periodic reports submitted by the US in one document to the Human Rights Committee.[392] The latter, however, did not agree with the position taken by the State and made clear in its concluding observations that the Covenant does have extraterritorial applicability and that the reasoning used in General Comment No. 31 must be the leading approach accepted by States Parties.[393]

In the incident of the Nisour Square killings and injuries, a number of human rights treaties can be said to have been breached. In the first place, there are violations of the right to security, as incorporated in Article 3 UDHR, Article 9 ICCPR, and Article 7 ACHR. In addition, these instances of misconduct are breaches of the right to life that can be found in a number of human rights instruments, such as Article 3 UDHR, Article 6 ICCPR, Article 4 ACHR, and Article 6 CRC. From this list of human rights Declarations and Conventions, the US has only ratified the ICCPR, on June 8, 1992.[394] Article 6(1) ICCPR provides that "every person has the right to have his life respected". Further to that, the provision adds that "this right shall be protected by law and, in general, from the moment of conception. No one shall be arbitrarily deprived of his life". In addition, Article 3 UDHR provides that "everyone has the right to life". A large number of Iraqi civilians were killed at Nisour Square: their right to life was clearly breached by private actors acting on behalf of the State.

In addition to the breach of human rights law, international humanitarian law applicable in armed conflicts has been violated. The incident took place in September 2007 when the conflict in Iraq could be characterized as a non-international armed conflict. This type of conflict is associated with the applicability of Common Article 3 to the 1949 Geneva Conventions and customary international humanitarian law. The United States ratified the Four Geneva Conventions on August 2, 1955, but these legal instruments do not apply with regard to the non-international armed conflict in question. The US is not a party to the 1977 First Additional Protocol or the 1977 Second Additional Protocol: they were signed in December 1977, but not yet ratified. Notwithstanding this, certain rules of international humanitarian law included in these instruments have a customary law status and are binding upon the US.

---

392 UN Human Rights Committee, *Consideration of Reports Submitted by States Parties under Article 40 of the Covenant, Third Periodic Reports of States Parties due in 2003, United States of America*, November 28, 2005, UN Doc. CCPR/C/USA/3, Annex I.

393 UN Human Rights Committee, *Concluding Observations of the Human Rights Committee: United States of America*, April 6, 1995, UN Doc. CCPR/C/79/Add.50, para. 19; UN Human Rights Committee, *Consideration of Reports Submitted by States Parties under Article 40 of the Covenant, Concluding Observations of the Human Rights Committee, United States of America*, September 15, 2006, UN Doc. CCPR/C/USA/CO/3, para. 10.

394 For the current ratification status of various human rights treaties with regard to the United States, see <http://tbinternet.ohchr.org/_layouts/TreatyBodyExternal/Treaty.aspx>.

First and foremost, the crucial principle of distinction has been breached. This implies that in the conduct of armed hostilities a clear line must be maintained between combatants and civilians. In this respect, rules and principles of customary international humanitarian law specify that civilians are persons that are not characterized as members of the armed forces and that they are protected as long as they do not take direct part in hostilities.[395] In the Nisour Square incident, members of the Raven 23 team discharged their weapons and killed and injured local Iraqi civilians who did not qualify as members of Iraqi armed forces or troops of other States. In non-international armed conflicts, such as the level of hostilities in Iraq, these civilians are protected by Article 13(2) of the 1977 Second Additional Protocol, which stipulates that the civilian population cannot be made the object of attack. This principle of distinction is, in fact, a rule of customary international law requiring that "the parties to the conflict must at all times distinguish between civilians and combatants", that "attacks may only be directed against combatants", and that "attacks must not be directed against civilians".[396] This norm is applicable in both international and non-international armed conflicts and is recognized by many States, including the United States, in the military manuals.[397] In its case law, the ICJ maintained that this is in essence not only one of "the cardinal principles" of the law of armed conflict, but also of the "intransgressible principles of international customary law".[398] In addition, other rules of international humanitarian law having customary character and associated with the principle of distinction have been violated: the prohibition of direct attacks against civilians and civilian objects.[399] Linked to these, the customary international humanitarian law prohibition of indiscriminate attacks has been violated.[400] These are the attacks which are not directed at a specific military objective, which employ certain means or methods of combat that simply cannot be directed at a certain military objective, or which employ means or methods of combat the effects of which cannot be limited pursuant to the law of armed conflict.[401] In this respect, the attacks have such a character as to strike military objectives and civilians or civilian objects without distinction. By directly targeting Iraqi civilians and civilian objects at Nisour Square, Blackwater guards performed actions that in essence constitute a breach of the international obligation arising out of the body of customary international humanitarian law. It is not relevant that the US is not a party to the Additional

---

395 Jean-Marie Henckaerts and Louise Doswald-Beck, (2005) *Customary International Humanitarian Law, Volume I: Rules*, ICRC, Cambridge: Cambridge University Press, pp. 17, 19.

396 Ibidem, p. 3.

397 Ibidem, p. 4.

398 International Court of Justice, *Legality of the Threat or Use of Nuclear Weapons*, Advisory Opinion, July 8, 1996, ICJ Reports 1996, p. 257, paras. 78-79.

399 Jean-Marie Henckaerts and Louise Doswald-Beck, (2005) *Customary International Humanitarian Law, Volume I: Rules*, ICRC, Cambridge: Cambridge University Press, pp. 19-29, 32-36.

400 Ibidem, pp. 37-45.

401 Ibidem, p. 40.

Protocols to the 1949 Geneva Conventions: this obligation is a rule of customary international law and must be complied with by not only the US, but also any other State as a member of the international community.

Intentional killings

Intentional killings on behalf of States can also be seen as breaches of international law. Some reports issued in the past indicated that Blackwater contractors often engage in tactical military operations of the US military.[402] This means that they actively participate in the US military actions and therefore can be targeted by the enemy forces. In this regard, contractors do not merely provide defensive services, but can also be in a position to actually kill or to wound individuals, which may be enemy combatants and civilians, fighting offensively. In addition, as has been pointed out above, sometimes governments can hire private firms in order to eliminate specific individuals, such as terrorist leaders, and thus to end someone's life. All the above-stated facts might demonstrate that these actions can be accepted as serious human rights abuses that violate the right to security covering the right to life. Therefore, intentional killings may be considered breaches of the right to security, as provided in Article 3 UDHR, Article 9 ICCPR, Article 5 ECHR, Article 7 ACHR, and Article 6 ACHPR and the right to life laid down in Article 3 UDHR, Article 6 ICCPR, Article 2 ECHR, Article 4 ACHR, Article 4 ACHPR, and Article 6 CRC.

Property damage

Damage to the property of civilian population caused by shooting incidents, in which PMSC personnel can participate, qualifies as a violation of international humanitarian law in accordance with Article 53 of the Fourth Geneva Convention of 1949. In this regard, PMSCs must be regarded as belonging to the occupying power and their acts must be attributable to it, as the destruction of real or personal property by the occupying power that belongs individually or collectively to private persons is prohibited. That type of misconduct can also be seen as a human rights breach and a violation of the right to property laid down in Article 17 UDHR and Article 1 of the First Protocol to the ECHR.

Touching upon the concept of individual criminal responsibility for those offenses, one could argue that they can be acknowledged to be war crimes in the sense of Article 8 of the Statute of the International Criminal Court. Being a grave breach of the Fourth Geneva Convention of 1949, property damage caused by the actions

---

402 US House of Representatives, Committee on Oversight and Government Reform, "Additional Information About Blackwater USA", Memorandum, October 1, 2007, p. 8, <http://graphics8.nytimes.com/packages/pdf/national/20071001121609.pdf>, accessed on September 16, 2014.

of PMSCs might be concluded to be an extensive destruction and appropriation of property carried out unlawfully and wantonly without any military necessity, as laid down in Article 8(2)(a)(iv). It may also constitute other serious violations of the laws and customs applicable in international armed conflicts, as that type of damage is often caused by intentional attacks on civilian objects and attacks launched with the knowledge that they will cause damage to civilian objects, as articulated in Article 8(2)(b)(iv).

Torture and other cruel, inhuman, or degrading treatment or punishment

In many cases, PMSCs have close links with the governmental structures of the States they work for. For many years, Blackwater was involved in rendition flights of the CIA and this practice of employing private contractors helped provide security on rendition flights transporting detainees after 2001. A former top CIA officer once stated: "There was a feeling that Blackwater eventually became an extension of the agency."[403] Other PMSCs closely cooperate with governments and operate in the highly secretive fields of national security and the provision of intelligence services.[404] DynCorp International is, for instance, another contractor that considers itself to be an extension of the US interests and policies and that is highly dependent on the US being its home State and its main client.[405] Thus, one can imagine that working closely with governmental agencies under their control, direction, and instructions, but also exercising governmental authority, PMSCs might be involved in the commission of torture and other cruel, inhuman, or degrading treatment or punishment, which most probably could be attributed to the employing States.

From the perspective of international humanitarian law, these crimes must qualify as breaches of the Third and Fourth Geneva Conventions of 1949, as they refer in the Common Article 3 to the "violence to life and person, in particular murder of all kinds, mutilation, cruel treatment and torture". The acts of torture and other cruel, inhuman, or degrading treatment or punishment are, in fact, also human rights violations of the CAT, the Inter-American Convention to Prevent and Punish Torture, Article 5 UDHR, Article 5(2) ACHR, Article 3 ECHR, and Article 5 ACHPR.[406] In addition, they might be seen as breaches of the right to security, which can be found

---

[403] James Risen and Mark Mazzetti, "Blackwater Guards Tied to Secret C.I.A. Raids", *The New York Times*, December 10, 2009, <http://www.nytimes.com/2009/12/11/us/politics/11blackwater.html>, accessed on October 10, 2013.

[404] Dana Priest and William M. Arkin, "Top Secret America: A Washington Post Investigation", *The Washington Post*, July 20, 2010.

[405] Lou Pingeot, (2012) *Dangerous Partnership: Private Military & Security Companies and the UN*, Report, June 2012, Global Policy Forum and Rosa Luxemburg Foundation: New York and Berlin, p. 14.

[406] Efrain Staino, (2010) "Suing Private Military Contractors for Torture: How to Use the Alien Tort Statute Without Granting Sovereign Immunity-Related Defenses", *Santa Clara Law Review*, Volume 50, Issue 4, p. 1302.

in Article 3 UDHR, Article 9 ICCPR, Article 5 ECHR, Article 7 ACHR, and Article 6 ACHPR. Finally, according to Articles 7(1)(f), 8(2)(a)(ii), and 8(2)(c)(ii) of the ICC Statute, torture and other similar acts committed by private contractors could have a status of crimes against humanity or war crimes.

The actual operations of contractors in Abu Ghraib, as described in the Fay Report, were associated with various accounts of abuse. One civilian contractor working for CACI used physical violence against one of the detainees by grabbing him, forcing him to the ground, and dragging him into an interrogation booth.[407] Another civilian interrogator engaged in other activities, such as using dogs during interrogations and placing a detainee in an unauthorized stress position.[408] In addition, there were instances of detainee humiliation and other types of abuse.[409] If these unlawful acts qualify as torture and other cruel, inhuman, or degrading treatment or punishment, they will constitute a violation of Article 5 UDHR stating that "no one shall be subjected to torture or to cruel, inhuman, or degrading treatment or punishment". Additionally, the commission of torture and related forms of treatment can be regarded as a breach of the ICCPR ratified by the US on June 8, 1992. Similar to the UDHR, Article 7 of this treaty provides that "no one shall be subjected to torture or to cruel, inhuman or degrading treatment or punishment". In the light of the nature of the abuses committed at Abu Ghraib, there can also have been a breach of the CAT. This treaty was ratified by the United States on October 21, 1994 and therefore it has to comply with its obligations to prevent and to punish acts of torture and other forms of related treatment. Article 1(1) CAT defines torture as:

> "... any act by which severe pain or suffering, whether physical or mental, is intentionally inflicted on a person for such purposes as obtaining from him or a third person information or a confession, punishing him for an act he or a third person has committed or is suspected of having committed, or intimidating or coercing him or a third person, or for any reason based on discrimination of any kind, when such pain or suffering is inflicted by or at the instigation of or with the consent or acquiescence of a public official or other person acting in an official capacity."

Further, Article 2(1) CAT contains an obligation on States Parties to the Convention to take a variety of measures, including legislative, administrative, judicial, and other, in order to prevent acts of torture in any territory under their jurisdiction. Article 2(2) CAT maintains that States are not allowed to invoke exceptional circumstances, such as a state of war or a threat of war, to justify any acts of torture. The abuses documented at the Abu Ghraib prison that were committed by the personnel

---

407 Major General George R. Fay, Investigating Officer, Article 15-6 Investigation of the Abu Ghraib Prison and 205th Military Intelligence Brigade, August 2004, Civilian-05, pp. 130-131, <news.findlaw.com/hdocs/docs/dod/fay82504rpt.pdf>, accessed on October 15, 2013.

408 Ibidem, Civilian-11, p. 132.

409 Ibidem, Civilian-21, p. 134.

of PMSCs could clearly fall within the scope of this definition: they not only intentionally inflicted severe mental and physical pain and suffering of inmates, but did so for the purposes of obtaining information, punishing, and intimidating. In addition, these acts were performed with direct and indirect cooperation of the US military personnel and army officials and can be considered to be realized in an official capacity. According to the UN Committee Against Torture:

> "[W]here detention centres are privately owned or run, the Committee considers that personnel are acting in an official capacity on account of their responsibility for carrying out the State function without derogation of the obligation of State officials to monitor and take all effective measures to prevent torture and ill-treatment."[410]

Furthermore, the right to security protects individuals from torture and therefore the above-mentioned acts can be seen as violations of Article 3 UDHR, Article 9 ICCPR, and Article 7 ACHR. In the context of the abuses committed at the US-run Abu Ghraib prison, the US appears to have breached the right of security of detainees protected by Article 3 UDHR and Article 9 ICCPR, which had been ratified by the State.

It is well known that the United States during the armed conflict in Afghanistan judged the captured Taliban and al-Qaeda fighters not to be prisoners of war, because treating terrorists in accordance with the law of armed conflict would put the State in question in a disadvantageous position.[411] The US presence in Iraq and the US perception of its detainees at, for instance, the Abu Ghraib prison, seem to be influenced by these judgments as well. Complying with human rights law and international humanitarian law while handling terrorists and other criminals on Iraqi soil appears to be incompatible with effective engagement in the war on terrorism by the US. In this regard, detainees are seen as sources of valuable information and "targets for better intelligence".[412] It would, however, be entirely incorrect to stipulate that the law of armed conflict is not applicable to the US operations in Iraq and activities of its military personnel and contractors at the Abu Ghraib detention facility. The period between October and December 2003 during which abuses at the Abu Ghraib prison took place can be described as occupation of the US and coalition troops of certain areas of Iraq. With regard to the belligerent occupation, Article 42 of the Hague Regulations provides:

> "Territory is considered occupied when it is actually placed under the authority of the hostile army. The occupation extends only to the territory where such authority has been established and can be exercised."

[410] UN Committee Against Torture, *General Comment No. 2, Implementation of Article 2 by States Parties*, January 24, 2008, UN Doc. CAT/C/GC/2, para. 17.

[411] Deborah D. Avant, (2005) *The Market for Force: The Consequences of Privatizing Security*, Cambridge: Cambridge University Press, pp. 224-225.

[412] Ibidem, p. 225.

This is a rule of the law of armed conflict that together with other provisions of the Hague Regulations belongs to the body of customary international law, as mentioned by the ICJ.[413] It is important to stress that the whole territory of a State does not have to be occupied for the occupation to take effect. Article 2 common to the 1949 Geneva Conventions provides that in addition to all cases of declared war or any other armed conflict arising between two or more States Parties to the Conventions, these treaties are also applicable "to all cases of partial or total occupation of the territory of a High Contracting Party, even if the said occupation meets with no armed resistance". What is certain, is that the area of Baghdad and Abu Ghraib was occupied by the coalition troops at the time that the US decided to take over the prison and realized its decision in the period between May and June 2003. The same is true for the period in which abuses at the above-mentioned prison were committed. This means that during the occupation of these areas of Iraq by the US and coalition troops, international humanitarian law regulates the actual conduct of hostilities, treatment of prisoners of war, and other matters.

The United States ratified the 1949 Geneva Conventions on August 2, 1955 and is bound by not only these Conventions, but also customary international law. Moreover, the Fay Report mentions that both members of the US armed forces and the civilians who accompany them are bound by the 1949 Geneva Conventions and must comply with their provisions.[414] The nature of abuses committed at the Abu Ghraib prison creates an impression that certain significant rules of the law of armed conflict have been breached: those rules relate to the prohibition of torture and the treatment of prisoners of war and civilians during occupation.[415] It is obvious that, in addition to prohibition by human rights instruments, torture is also prohibited by the 1949 Geneva Conventions. This prohibition can be found in Common Article 3 to the 1949 Geneva Conventions, Article 12 of the First and Second Conventions, Articles 13, 17, and 87 of the Third Convention, and Articles 27 and 32 of the Fourth Convention. Additionally, the First, Second, Third, and Fourth Conventions respectively indicate that torture and other related types of treatment constitute grave breaches of the Geneva law by referring to "torture or inhuman treatment, including biological experiments, willfully causing great suffering or serious injury to body or health".[416]

---

413 International Court of Justice, *Legality of the Threat or Use of Nuclear Weapons*, Advisory Opinion, July 8, 1996, ICJ Reports 1996, p. 256, para. 75; International Court of Justice, *Legal Consequences of the Construction of a Wall in the Occupied Palestinian Territory*, Advisory Opinion, July 9, 2004, ICJ Reports 2004, p. 172, para. 89.

414 Major General George R. Fay, Investigating Officer, Article 15-6 Investigation of the Abu Ghraib Prison and 205th Military Intelligence Brigade, August 2004, pp. 12-13, <news.findlaw.com/hdocs/docs/dod/fay82504rpt.pdf>, accessed on October 15, 2013.

415 Antenor Hallo de Wolf, (2006) "Modern Condottieri in Iraq: Privatizing War from the Perspective of International Human Rights Law", *Indiana Journal of Global Legal Studies*, Volume 13, Issue 2, p. 338.

416 See Article 50 of the First Geneva Convention, Article 51 of the Second Geneva Convention, Article 130 of the Third Geneva Convention, and Article 147 of the Fourth Geneva Convention.

The prohibition of torture activities that cannot be justified under any circumstances is also a *jus cogens* norm of international law and can be distilled from the rules and principles having customary international character.[417] Rule 90 of customary international humanitarian law applicable in both international and non-international armed conflicts suggests:

> "Torture, cruel or inhuman treatment and outrages upon personal dignity, in particular humiliating and degrading treatment, are prohibited."[418]

It is worth examining torture, inhuman treatment, and outrages upon personal dignity for the purpose of classifying the instances of unlawful conduct of PMSC members at the Abu Ghraib prison as these particular categories of treatment. In the study of the ICRC on customary international humanitarian law, references are made to the Elements of Crimes for the ICC. They provide that torture implies infliction of "severe physical or mental pain or suffering upon one or more persons" protected by the 1949 Geneva Conventions that is done for such purposes as "obtaining information or a confession, punishment, intimidation or coercion or for any reason based on discrimination of any kind".[419] In its case law, the ICTY held that "the definition of torture under international humanitarian law does not comprise the same elements as the definition of torture generally applied under human rights law".[420] The Trial Chamber of the Tribunal was of the opinion that "the presence of a state official or of any other authority-wielding person in the torture process is not necessary for the offence to be regarded as torture under international humanitarian law". Apparently, there is a notable contrast between the concept of torture under international human rights law and international humanitarian law leading to the situation that the conduct of PMSC personnel and other individuals engaging in acts inflicting severe pain or suffering can more easily meet the set standard of torture under the law of armed conflict.

Another possibility is that the misconduct of PMSC employees would qualify as inhuman treatment. In line with the Elements of Crimes for the ICC, this act takes places when the perpetrator inflicts "severe physical or mental pain or suffering upon one or more persons" that are protected under the 1949 Geneva Conventions.[421] Inhu-

---

417 International Criminal Tribunal for the former Yugoslavia, *Prosecutor v. Delalic and Others*, Trial Chamber, Judgment, November 16, 1998, ICTY-IT-96-21-T, paras. 452-454; International Criminal Tribunal for the former Yugoslavia, *Prosecutor v. Furundžija*, Trial Chamber, Judgment, December 10, 1998, ICTY-IT-95-17/1-T, paras. 139, 143.

418 Jean-Marie Henckaerts and Louise Doswald-Beck, (2005) *Customary International Humanitarian Law, Volume I: Rules*, ICRC, Cambridge: Cambridge University Press, p. 315.

419 ICC, Elements of Crimes, Article 8(2)(a)(ii)-1, p. 35, <http://www.icc-cpi.int/nr/rdonlyres/336923d8-a6ad-40ec-ad7b-45bf9de73d56/0/elementsofcrimeseng.pdf>, accessed on October 17, 2013.

420 International Criminal Tribunal for the former Yugoslavia, *Prosecutor v. Kunarac and Others*, Trial Chamber, Judgment, February 22, 2001, ICTY-IT-96-23-T & ICTY-IT-96-23/1-T, para. 496.

421 ICC, Elements of Crimes, Article 8(2)(a)(ii)-2, p. 35.

man treatment obviously has a lower threshold than torture and does not require that the treatment to be inflicted for a specific purpose.[422] The ICTY found that "inhuman treatment is an intentional act or omission, that is an act which, judged objectively, is deliberate and not accidental, which causes serious mental or physical suffering or injury or constitutes a serious attack on human dignity".[423] Further, it observed that it is a form of intentional treatment that is not in conformity with the fundamental principle of humanity and falls under the grave breaches of the 1949 Geneva Conventions.

What was done by the CACI and Titan employees, can also be seen as "outrages upon personal dignity, in particular humiliating and degrading treatment". This is the case if the perpetrators "humiliated, degraded or otherwise violated the dignity of one or more persons".[424] The severity of such treatment must be of a certain degree to be generally considered as an outrage upon personal dignity. According to the Elements of Crimes for the ICC, this element takes various relevant aspects of the cultural background of persons into account. The ethical and religious backgrounds of Abu Ghraib detainees is thus of significance, as many of the abuses could have been seen as violating important norms and values of the victims.

In addition to the commission of torture and other related types of treatment, other rules and principles of the law of armed conflict could fall under the category of those breached by civilian contractors. If detainees of the Abu Ghraib prison belong to two groups of individuals – prisoners of war and civilians –, they must be treated in a certain way prescribed by the law of armed conflict. The first possibility is that the Iraqi inmates fall under the definition of prisoners of war under Article 4 Paragraph A mentioning members of the armed forces, members of militia and volunteer corps forming part of such troops, and other types of individuals addressed in the previous Chapters. Additionally, Article 4 Paragraph B of the Third Geneva Convention can also apply, which states that included in the category of prisoners of war must be considered "persons belonging, or having belonged, to the armed forces of the occupied country, if the occupying Power considers it necessary by reason of such allegiance to intern them, even though it has originally liberated them while hostilities were going on outside the territory it occupies, in particular where such persons have made an unsuccessful attempt to rejoin the armed forces to which they belong and which are engaged in combat, or where they fail to comply with a summons made to them with a view to internment".[425] The 1949 Third Geneva Convention establishes a certain standard of treatment that prisoners of war must enjoy. It requires States Parties

[422] Jean-Marie Henckaerts and Louise Doswald-Beck, (2005) *Customary International Humanitarian Law, Volume I: Rules*, ICRC, Cambridge: Cambridge University Press, p. 318.
[423] International Criminal Tribunal for the former Yugoslavia, *Prosecutor v. Delalic and Others*, Trial Chamber, Judgment, November 16, 1998, ICTY-IT-96-21-T, para. 543.
[424] ICC, Elements of Crimes, p. 27, Article 8(2)(b)(xxi).
[425] Article 4(B)(1) of the Third Geneva Convention.

to humanely treat prisoners of war at all times and to protect them against possible acts of violence, intimidation, insults, and public curiosity.[426] Further, it is provided that seriously endangering the health of prisoners of war constitutes a serious breach of the Convention.[427] These detainees must also be treated in all circumstances with respect for their persons and their honor.[428] It is thus clear that individuals disregarding these provisions are in a violation of the Geneva law.

The second possibility is that Iraqi civilians are held at the Abu Ghraib prison for criminal offenses and other categories of similar misconduct and should be perceived and treated as civilians, protected by the 1949 Fourth Geneva Convention. In the situation of occupation, persons protected by this treaty must be in the hands of the occupying power of which they are not nationals.[429] Further, the Geneva law specifies that nationals of a State that is not a party to the Fourth Convention are not protected by its provisions.[430] Due to the fact that the US is an occupying State that captured some Iraqi civilians and that Iraq had ratified the 1949 Geneva Conventions on February 14, 1956, these detainees are protected by the Fourth Geneva Convention. This Convention is applicable after the occupation took place and ceases to apply one year after the close of military operations.[431] In general, its Article 5 stipulates that individuals detained in an occupied territory as persons who are suspected of activities hostile to the occupying power must be treated with humanity and have the full package of rights and privileges conferred on them by the Convention.[432] The level of protection afforded to detainees of Abu Ghraib is reflected in the treaty provisions as follows:

> "Protected persons are entitled, in all circumstances, to respect for their persons, their honour, their family rights, their religious convictions and practices, and their manners and customs. They shall at all times be humanely treated, and shall be protected especially against all acts of violence or threats thereof and against insults and public curiosity."[433]

It is generally clear that inmates placed in the Abu Ghraib prison may not be subjected to physical or moral coercion, which can be used to obtain all sorts of information from them or third parties.[434] It is also not allowed to take measures leading to physical suffering or extermination of the protected individuals finding themselves

---

426 Article 13 of the Third Geneva Convention.
427 Ibidem.
428 Article 14 of the Third Geneva Convention.
429 Article 4 of the Fourth Geneva Convention.
430 Ibidem.
431 Article 6 of the Fourth Geneva Convention.
432 Article 5 of the Fourth Geneva Convention.
433 Article 27 of the Fourth Geneva Convention.
434 Article 31 of the Fourth Geneva Convention.

in the hands of the United States.[435] This prohibition applies to "murder, torture, corporal punishment, mutilation and medical or scientific experiments not necessitated by the medical treatment of a protected person" and "any other measures of brutality whether applied by civilian or military agents".[436] Importantly, the Convention mentions that States in which hands protected persons are, are responsible for the treatment they receive from the State agents.[437] The US did not comply with a number of its obligations under the Fourth Geneva Convention and therefore an argument can be made that it is to be held responsible for these breaches of the law of armed conflict.

Sex trafficking, sex slavery, and running prostitution networks

Considering sex trafficking, sex slavery, and running prostitution networks by PMSCs from the perspective of international humanitarian law, one needs to observe that the prohibition of sexual violence is to be perceived as having a status of a *jus cogens* norm within the international legal system that is prohibited at all times, especially during wars, as it may be employed to serve as a distinct form of warfare.[438] Therefore, private guards breaching that norm act in violation of the law of armed conflict. Moreover, it can be said that trafficking in human beings is a crime falling within the scope of application of the Council of Europe Convention on Action against Trafficking in Human Beings of 2005 and might be seen as a human rights abuse. With regard to the issue of individual criminal responsibility, this kind of offense relating to sexual violence falls under the category of crimes against humanity under Article 7(1)(g) ICC Statute, but also war crimes in accordance with Article 8(2)(b)(xxii) and (e)(vi) ICC Statute.

Drug running, profiteering in transnational criminal networks, and weapons violations

The next type of misconduct includes drug running, profiteering in transnational criminal networks, and weapons violations. It must be stressed that not all misconduct attributed to PMSC contractors can be said to amount to violations of international humanitarian and human rights law. Some illegal actions having a less serious character fall rather within the scope of national criminal legislation, regional treaties, and the UN legal framework. They do not always constitute human rights abuses or violations of the law of armed conflict. Drug running cannot be easily said to

---

435 Article 32 of the Fourth Geneva Convention.

436 Ibidem.

437 Article 29 of the Fourth Geneva Convention.

438 David S. Mitchell, (2005) "The Prohibition of Rape in International Humanitarian Law as a Norm of Jus Cogens: Clarifying the Doctrine", *Duke Journal of Comparative and International Law*, Volume 15, p. 225.

amount to a human rights abuse or a breach of international humanitarian law, but is to be punished in accordance with national criminal laws of the respective States. Apparently, profiteering in transnational criminal networks is also not to be regarded as constituting a violation of international humanitarian or human rights law. Most probably, weapons violations committed in breach of UN resolutions as such also cannot qualify as a violation of such a character. However, if PMSCs make use of illegal weapons, such as cluster bombs prohibited under international humanitarian law, more specifically in the Convention on Cluster Munitions of December 3, 2008 and customary international law with regard to indiscriminate attacks to be found in the First Additional Protocol to the Geneva Conventions of 1949, it will be found to be a breach of the law of armed conflict that in theory could be attributed to States.

Mercenary activities

Finally, mercenary activities of PMSCs can possibly constitute a violation of international law. It needs to be pointed out that PMSCs can be used by States in order to intervene in the internal affairs and politics of other nations. Most importantly, unlawful activities carried out by private contractors and their use of force can threaten not only national sovereignty and territorial integrity of States, but also violate the right of self-determination, as one of the most essential human rights listed in Article 1 of both the International Covenant on Civil and Political Rights and International Covenant on Economic, Social and Cultural Rights.[439] In the UN General Assembly Resolution 2625 of October 24, 1970, the right of self-determination with regard to mercenary activities was defined as follows:

> "All people have the right freely to determine, without external interference, their political status and to pursue their economic, social and cultural development, and that by the context of mercenary activities, the right to self-determination was and that every State has the duty to respect this right in accordance with the [UN] Charter."

The phenomenon of mercenarism is regulated by the 1989 UN International Convention on Mercenaries and the 1977 OAU Convention on Mercenarism. Moreover, the Resolution of the UN General Assembly 63/164 states that States should "exercise the utmost vigilance against any kind of recruitment, training, hiring or financing of mercenaries, including nationals, by private companies offering international

---

[439] Human Rights Advocates, *Holding Private Military and Security Companies and Mercenaries Accountable for Human Rights Violations*, Report, p. 3, <http://www.humanrightsadvocates.org/wp-content/uploads/2010/05/Arellano2008-Mercenaries-and-Corporate-Accountability-Long.pdf>, accessed on October 10, 2012; Amol Mehra, (2010) "Bridging Accountability Gaps – The Proliferation of Private Military and Security Companies and Ensuring Accountability for Human Rights Violations", *The Pacific McGeorge Global Business & Development Law Journal*, Volume 22, pp. 326-327.

military consultancy and security services, as well as to impose a specific ban on such companies intervening in armed conflicts or actions to destabilize constitutional regimes". Because mercenaries regularly participate in armed conflicts and can easily violate rules and principles of international humanitarian law, this notion of a mercenary has also been mentioned in the law of armed conflict. However, one should bear in mind that it is not a violation of the 1949 Geneva Conventions or 1977 Additional Protocols to be a mercenary, while mercenarism as such does not lead to international criminal responsibility of individuals.[440] Article 47 of the 1977 First Additional Protocol to the 1949 Geneva Conventions indicates that mercenaries do not have the right to be combatants and cannot benefit from prisoner-of-war status when captured. It is, however, not very probable that PMSCs will meet all the requirements listed in the above-mentioned provision.

Most importantly, experts seem to agree on the fact that the issue of State responsibility does not arise when PMSC employees, just as other individuals, fall under the definition of a mercenary under Article 47 of the 1977 First Additional Protocol to the 1949 Geneva Conventions.[441] However, the two other international Conventions assessed above contain legal rules stipulating that States may be held responsible for the use of mercenaries: this reasoning can be derived from Articles 5, 6, and 7 of the UN International Convention on Mercenaries and Article 5 and 6 of the OAU Convention on Mercenarism. Legal experts share the view that both Conventions create additional legal bases for the international responsibility of States Parties to them, when members of the PMSC personnel hired by them fall within the definition provided in those international legal instruments and participate in mercenary activities.[442]

States might be in breach of these two Conventions and bear international responsibility if they recruit, use, finance, train mercenaries, or in other possible manners participate in their activities, or do not prevent the use of mercenaries. The UN International Convention on Mercenaries requires in Article 5(1) that States refrain from recruiting, using, financing, or training mercenaries and forbids such activities. In the same vein, according to Article 5(2) they are not allowed to recruit, use, finance, or train mercenaries for the purpose of violating the legitimate exercise of the right of peoples to self-determination and must take all necessary measures to prevent these activities. Furthermore, Articles 6 and 7 provide that States are required to cooperate

---

440 Lindsey Cameron, (2006) "Private Military Companies: Their Status under International Humanitarian Law and Its Impact on Their Regulation", *International Review of the Red Cross*, Volume 88, Issue 863, p. 577.

441 University Centre for International Humanitarian Law, *Expert Meeting on Private Military Contractors: Status and State Responsibility for Their Actions*, Report, Geneva, August 29-30, 2005, p. 28, <http://www.geneva-academy.ch/docs/expert-meetings/2005/2rapport_compagnies_privees.pdf>, accessed on October 10, 2014.

442 Ibidem.

in the prevention of offenses related to the use of mercenaries and in the process of taking necessary measures for the implementation of the Convention. Finally, there are additional State obligations, such as a duty to communicate any relevant information about mercenary activity to the States Parties affected that can be found in Article 8 and a duty laid down in Article 9 to establish jurisdiction over any offenses set forth in the Convention when they are committed in a territory of a State or on board of its ship or aircraft, or when they are committed by its nationals.

The second major treaty in this field, the OAU Convention on Mercenarism, specifies in Article 1(2) what the crime of mercenarism implies by stressing that it can also be committed by States and includes three categories of acts. In the first place, these acts are sheltering, organizing, financing, assisting, equipping, training, promoting, supporting, or in other ways employing mercenaries. Secondly, they comprise such activities as enlisting, and enrolling or trying to enroll mercenaries. Finally, to the crime of mercenarism also belong the activities from the first category when they are allowed to be performed in a territory under States' jurisdiction or in a place under their control or when States provide facilities for transit, transport, or operations of mercenaries. When a State is accused of committing such a crime, other States Parties may invoke provisions of the Convention in their relations with that State and before any competent international or other tribunal or body, as provided in Article 5(2). Additionally, under the OAU Convention on Mercenarism, States Parties have certain obligations that they can disregard and thus act in violation of the treaty provisions. In accordance with Article 6, States must prevent their nationals or foreigners on their territory from engaging in mercenary activities; prevent entry into or passage through their territory of mercenaries and their equipment; prohibit on their territory activities of persons and entities using mercenaries against other States of the Organization of African Unity or the people of Africa fighting for liberation; communicate to other Member States of the Organization of African Unity any information about the use of mercenaries and their activities; prohibit on their territory any recruitment, training, financing, and equipment of mercenaries, and other activities promoting mercenarism; and take all possible legislative and other measures for the immediate entry into force of the OAU Convention. Unfortunately, both the UN International Convention on Mercenaries and the OAU Convention on Mercenarism have not widely been ratified and their importance for tackling the problem of mercenaries and mercenarism remains rather limited.[443]

From the point of view of international human rights law, the activities of PMSCs can in theory constitute breaches of that field of international law. According to some experts, there are still many threats posed by mercenaries and PMSCs to the right of

[443] As of July 2014, there are 33 States that became party to the UN International Convention on Mercenaries through ratification or accession, see <https://treaties.un.org/>; there are 31 ratifications of the OAU Convention on Mercenarism as of March 2013, see <http://www.au.int/en/treaties>.

self-determination of peoples.[444] Extreme nationalism and ethnic and religious intolerance appear to have intensified the interests of States and other entities in using those firms or individuals. The United States, China, and the United Kingdom could be seen to be willing to gain more influence in the regions of strategic and economic importance, pursuing their own aims and interfering in the affairs of less powerful nations.

8. CONCLUSION

Various types of misconduct by the employees of PMSCs can qualify as different breaches of international law. They can be perceived as violations of international humanitarian law or human rights and lead to international responsibility of hiring, host, and home States in case of successful attribution. The ILC's Draft Articles contain four main modes of attribution that are of the utmost importance for the application with regard to the State use of PMSCs. The first type of attribution laid down in Article 4 DARS is highly unlikely to result in attaching international responsibility for the commission of internationally wrongful acts by PMSCs to States, as States remain unwilling to incorporate those companies into their traditional armed forces and therefore they cannot qualify as official State organs. It is, however, possible that they will be able to be determined as *de facto* State organs considering their possible complete dependence on certain States and the degree of State control. The other mode of attribution, also less relevant for attaching certain types of misconduct of PMSCs to States, is to be found in Article 11 DARS. It must be stated that States are and will always be reluctant to acknowledge and to adopt the unlawful conduct of private persons and entities violating international law as their own, diminishing the practical relevance of this provision.

The most significant provisions containing rules on attribution for the concept of PMSCs are without doubt Article 5 DARS and Article 8 DARS. The former focuses on the illegal activities of PMSCs empowered by the law of a State to exercise elements of governmental authority, and the latter aims at addressing situations when those firms and their employees operate in accordance with instructions or under the direction and control of a State. Certain PMSCs can be said to exercise elements of governmental powers when the situations are examined on the basis of criteria assessed above. Regarding the degree of control that a State needs to have over a PMSC, no clear-cut answer can be given which particular control test must be applied. However, customary international law and international judicial practice suggest that the "overall control" test is the most appropriate theory under normal

---

[444] Human Rights Advocates, *Holding Private Military and Security Companies and Mercenaries Accountable for Human Rights Violations*, Report, p. 3, <http://www.humanrightsadvocates.org/wp-content/uploads/2010/05/Arellano2008-Mercenaries-and-Corporate-Accountability-Long.pdf>, accessed on October 10, 2012.

circumstances of PMSC operation. It is a practical test to establish the State control over private contractors that can address challenges of State employment of PMSCs in conflict areas, which the international community has to deal with.

CHAPTER VI
STATE RESPONSIBILITY FOR NON-COMPLIANCE WITH POSITIVE INTERNATIONAL LAW OBLIGATIONS

Who will be responsible for any repressive excesses that the security companies may commit against the civilian population...? Who will take responsibility for any violations of international humanitarian law or human rights they may commit?

Report of the Special Rapporteur on the question of the use of mercenaries[1]

1. INTRODUCTION

Examination of the ILC's Draft Articles on State Responsibility reveals that sole reliance on these provisions is not an effective tool for holding States responsible for the illegal activities committed by PMSCs. In determining State responsibility, the central focus of examination is the behavior of certain government officials acting on their own or in cooperation with others.[2] In principle, the State is responsible for the conduct of those individuals, which exercise "its machinery of power and authority".[3] The purpose of the attribution on the basis of criteria established by international law is to determine whether a certain conduct, such as the illegal activities performed by individuals, is to be perceived as an act of State. This determination is required due to the fact that the State itself remains responsible for its own acts.[4] The comparison of PMSCs with regular armed forces and the assessment of different types of attribu-

---

1 UN Commission on Human Rights, *Report on the Question of the Use of Mercenaries as a Means of Violating Human Rights and Impeding the Exercise of the Right of Peoples to Self-Determination, Submitted by Mr. Enrique Bernales Ballesteros, Special Rapporteur*, February 20, 1997, UN Doc. E/CN.4/1997/24, para. 93.

2 Gordon A. Christenson, (1991) "Attributing Acts of Omission to the State", *Michigan Journal of International Law*, Volume 12, p. 312.

3 Ibidem, p. 322.

4 James Crawford, (2002) *The International Law Commission's Articles on State Responsibility: Introduction, Text and Commentaries*, Cambridge: Cambridge University Press, p. 92; Hannah Tonkin, (2011)

tion indicate that the question of attribution of the unlawful conduct of PMSCs is not clear cut. An impression is created that States are less likely to be held responsible for violations of international law committed by private contractors than for crimes committed by their military personnel because the attribution of those violations realized by private actors to States proves to be more problematic. The practice of employing PMSCs by governments seems to indicate that it is not possible to attribute PMSC conduct that falls beyond the scope of State control or is *ultra vires* to States.[5] The same goes for the attributability of illegal activities of the personnel of private contractors that do not belong to the official State military apparatus or exercise elements of governmental authority while acting off duty.[6]

This assessment reveals a responsibility gap that can be exploited by States using private contractors in order to escape international responsibility. States hiring PMSCs and other States involved can find themselves in the position of being able to avoid being held accountable on the basis of the principle that has been called in the past "direct responsibility"[7] connected with the concept of attribution. However, a set of obligations of States from the fields of international humanitarian and human rights law can be helpful in narrowing and ultimately closing that responsibility gap when the misconduct of private persons and entities cannot be attributed to a State under given circumstances. Rules and principles of those areas of international law might create positive obligations, which set up a basis for incurring international responsibility in case of violations and are without doubt to be taken into account and respected by States. In this regard, Olivier De Schutter, mentioning the primary State duty to protect and to promote human rights, correctly points out:

> "[T]he international responsibility of States in the context of economic globalization, where TNCs [transnational corporations] have gained an almost unprecedented influence and bargaining power vis-à-vis the States in which they operate, may have to be expanded even further."[8]

Under international law, it might be possible to hold States responsible for the illegal conduct of private contractors violating international law not only by means of

---

State Control over Private Military and Security Companies in Armed Conflict, Cambridge: Cambridge University Press, p. 56.

5 See Chapter V, Section 6.3; see also Carsten Hoppe, (2008) "Passing the Buck: State Responsibility for Private Military Companies", *The European Journal of International Law*, Volume 19, Issue 5, pp. 992-993.

6 See Chapter V, Sections 6.1 and 6.2.

7 Jan. A. Hessbruegge, (2004) "The Historical Development of the Doctrines of Attribution and Due Diligence in International Law", *New York University Journal of International Law and Politics*, Volume 36, p. 268.

8 Olivier De Schutter, "The Challenge of Imposing Human Rights Norms on Corporate Actors", in Olivier De Schutter (Ed.), (2006) *Transnational Corporations and Human Rights*, Oxford and Portland, Oregon: Hart Publishing, p. 37.

attribution of those illegal activities to States, but also by means of their non-compliance with a variety of positive obligations under international humanitarian and human rights law.[9] In this Chapter, particular attention will be devoted to the notion of positive obligations in the fields of international humanitarian and human rights law and State responsibility for non-compliance with these branches of international law that might lead to the international responsibility of hiring, host, and home States.

2. Indirect Responsibility

Back in 1975, the ILC, considering a provision on the conduct of persons or groups of persons not acting on behalf of a State, stressed in the Commentary the following:

> "The strictly negative conclusion reached regarding the attribution to the State of the acts of private natural and legal persons and of the other persons... does not imply that, however, that the State cannot incur international responsibility for such acts *on other grounds*... The point of departure for its [the ILC] analysis was the discovery that a State has often been held internationally responsible on the occasion of *acts or omissions* whose material perpetrator was *a private natural or legal person* who, on that occasion, was *not acting on the State's behalf*."[10] (emphasis added)

In general, States have undertaken many positive obligations under international law and in principle should not be able to avoid them by transferring these obligations to private actors, such as PMSCs or other groups and individuals.[11] For the purpose of examining the main research question, positive obligations might constitute an alternative route to engage international responsibility of States if the illegal acts committed by PMSCs as private actors cannot be attributed to them.[12] This is the

---

[9] Carsten Hoppe, (2008) "Passing the Buck: State Responsibility for Private Military Companies", *The European Journal of International Law*, Volume 19, Issue 5, pp. 989-990; Chia Lehnardt, "Private Military Companies and State Responsibility", in Simon Chesterman and Chia Lehnardt (Eds.), (2007) *From Mercenaries to Market: The Rise and Regulation of Private Military Companies*, Oxford: Oxford University Press, p. 152; Olivier De Schutter, "The Responsibility of States", in Simon Chesterman and Angelina Fisher (Eds.), (2009) *Private Security, Public Order: The Outsourcing of Public Services and Its Limits*, Oxford: Oxford University Press, p. 34; Olga Martin-Ortega, (2013) "Human Rights Due Diligence for Corporations: From Voluntary Standards to Hard Law at Last?", *Netherlands Quarterly of Human Rights*, Volume 31, Issue 4, pp. 53-54.

[10] International Law Commission, *Report of the International Law Commission on the Work of Its Twenty-Seventh Session, 5 May – 25 July 1975*, Official Records of the General Assembly, Thirtieth Session, Supplement No. 10, (1975) Yearbook of the International Law Commission, Volume II, UN Doc. A/10010/Rev.1, p. 71, paras. 4-5.

[11] Louise Doswald-Beck, "Private Military Companies under International Humanitarian Law", in Simon Chesterman and Chia Lehnardt (Eds.), (2007) *From Mercenaries to Market: The Rise and Regulation of Private Military Companies*, Oxford: Oxford University Press, pp. 131-132.

[12] Nicola Jägers, (2002) *Corporate Human Rights Obligations. In Search of Accountability*, School of Human Rights Research Series, Volume 17, Antwerp: Intersentia, pp. 144-145.

so-called principle of "indirect responsibility" or State responsibility due to the failure to comply with positive obligations of international law.[13] These positive obligations will be analyzed in the light of the division of States in three main categories as outlined in the previous Chapters – the hiring States, the host States, and the home States of PMSCs. All these different types of States might bear responsibility for breaching different positive obligations applicable to their conduct and the misconduct of PMSCs.

In general, positive obligations are to be distinguished from negative obligations arising out of the body of international law and resting upon States.[14] While the latter are in essence duties to abstain from the commission of internationally wrongful acts, such as torture and crimes against humanity, the former require States to take certain positive actions prescribed by international law. The ILC's Draft Articles maintain that a breach of an international obligation exists when an act of a State is not in conformity with what is required of that State by that specific obligation, irrespective of its origin and character.[15] Remarkably, the language of the Draft Articles avoids a distinction between negative or positive State obligations. The Commentaries stress, however, that a breach of an international obligation lies in the disconformity between the conduct required of a State by a certain obligation and the conduct actually performed by this State.[16] It is to be found in the difference between the requirements prescribed by international law and the actual facts of a State's conduct. International obligations may not only expect a specifically defined conduct from States, but also set a minimum standard according to which States need to act.[17] Rules of international law can contain references to the expected conduct of States that includes acts, omissions, or combinations of both and the term "not in conformity with" used in the Draft Articles includes different manners of specifying an obligation and different types of its possible breaches.[18]

Several human rights bodies have so far addressed a variety of positive obligations in their case law, elaborating on the essence of these duties. In the case *Joaquín David Herrera Rubio et al. v. Colombia,* the UN Human Rights Committee did not specifically mention the due diligence standard but noted that the State failed to comply with its positive obligations under the International Covenant on Civil and Political

---

13 Jan A. Hessbruegge, (2004) "The Historical Development of the Doctrines of Attribution and Due Diligence in International Law", *New York University Journal of International Law and Politics*, Volume 36, pp. 268-269.

14 Hannah Tonkin, (2011) *State Control over Private Military and Security Companies in Armed Conflict*, Cambridge: Cambridge University Press, p. 59.

15 Article 12 DARS.

16 Commentary to DARS as adopted in 2001, p. 54, Article 12, para. 2.

17 Ibidem, p. 55, Article 12, para. 2.

18 Ibidem; International Law Commission, *Second Report on State Responsibility, by Roberto Ago, Special Rapporteur – the Origin International Responsibility*, (1970) Yearbook of the International Law Commission, Volume II, UN Doc. A/CN.4/233, p. 188, para. 35.

Rights, which automatically implied its indirect responsibility. In his individual complaint, Joaquín David Herrera Rubio claimed to have been tortured by members of the armed forces and stated that his parents were killed by a group of individuals in military uniforms.[19] It was not clear whether members of the armed forces participated in the violations, and Colombia claimed that this was not the case.[20] The Committee found, however, that the State had failed to comply with its positive obligation of taking "effective measures to remedy the violations that Mr. Herrera Rubio has suffered and further to investigate said violations, to take action thereon as appropriate and to take steps to ensure that similar violations do not occur in the future".[21]

The Inter-American Court of Human Rights played an important role in the development of the concept of positive obligations in relation to the illegal activities of private actors. In its case law, the Court came to the conclusion that:

> "An illegal act which violates human rights and which is *initially not directly imputable to a State* (for example, because it is the act of a private person or because the person responsible has not been identified) can lead to *international responsibility* of the State, not because of the act itself, but because of *the lack of due diligence* to prevent the violation or to respond to it as required by the Convention... What is decisive is whether a violation of the rights recognized by the Convention has occurred with the support or the acquiescence of the government, or whether the State has allowed the act to take place without taking measures to prevent it or to punish those responsible... The State has *a legal duty* to take *reasonable steps to prevent* human rights violations and to use the means at its disposal to carry out *a serious investigation* of violations committed within its jurisdiction, to identify *those responsible*, to impose *the appropriate punishment* and to ensure the victim *adequate compensation*."[22] (emphasis added)

In general, the Human Rights Committee is also convinced of the significance of positive human rights for the compliance of States with the ICCPR, and is of the opinion that:

> "... *the positive obligations* on States Parties to ensure Covenant rights will only be fully discharged if individuals are protected by the State, not just against violations of Covenant rights by its agents, but also against *acts committed by private persons or entities* that would impair the enjoyment of Covenant rights in so far as they are amenable to application between private persons or entities. There may be circumstances in which a failure to ensure Covenant rights as required by article 2 would give rise to violations by States Parties of those rights, as a result of States Parties'

---

19 UN Human Rights Committee, *Joaquín David Herrera Rubio et al. v. Colombia*, November 2, 1987, Communication No. 161/1983, UN Doc. CCPR/C/OP/2 at 192 (1990), paras. 1.2, 1.5.

20 Ibidem, para. 10.2.

21 Ibidem, para. 12.

22 Inter-American Court of Human Rights, *Velasquez Rodriguez v. Honduras*, Judgment, July 29, 1988, Series C, No. 4 (1988), paras. 172-174.

> *permitting or failing to take appropriate measures or to exercise due diligence to prevent, punish, investigate or redress the harm caused by such acts by private persons or entities...* The Covenant itself envisages in some articles certain areas where there are positive obligations on States Parties to address the activities of private persons or entities."[23] (emphasis added)

Finally, the European Court of Human Rights provided insight into these positive State duties with regard to the right to a private life in its judgment in the case *X and Y v. The Netherlands* from 1985, holding that:

> "... although the object of Article 8 (art. 8) is essentially that of protecting the individual against arbitrary interference by the public authorities, it does not merely compel the State to abstain from such interference: in addition to this primarily negative undertaking, there may be *positive obligations* inherent in an effective respect for private or family life. These obligations may involve the adoption of measures designed to secure respect for private life even in the sphere of the relations of individuals between themselves."[24] (emphasis added)

For the purposes of the current investigation, positive State obligations can further be classified into obligations of result and obligations of diligent conduct.[25] The same distinction between obligations of result and obligations of conduct is drawn in the Commentaries to the ILC's Draft Articles.[26] Therein, the ILC specifies that the first group of positive obligations concerns duties to achieve a certain result irrespective of the chosen means. The second group is based on the idea of "*une obligation de s'efforcer*" – a duty to strive to obtain a specific result. These are "best efforts" obligations resting upon States that are reasonably expected to take all possible measures lying within their power in order to reach a required outcome.

These two categories of positive obligations can be distinguished by the factor of the different risks involved. The fulfillment of the object and purpose of the obligations of result is not associated with a particular risk.[27] In the case of obligations of diligent conduct, however, there are more uncertainties in the process of attaining

---

23 UN Human Rights Committee, *General Comment 31*, May 26, 2004, UN Doc. CCPR/C/21/Rev.1/Add.13, para. 8.

24 European Court of Human Rights, *X and Y v. The Netherlands*, Judgment, March 26, 1985, Application No. 8978/80, ECHR Series A, para. 23.

25 Pierre-Marie Dupuy, (1999) "Reviewing the Difficulties of Codification: On Ago's Classification of Obligations of Means and Obligations of Result in Relation to State Responsibility", *The European Journal of International Law*, p. 375; Riccardo Pisillo-Mazzeschi, (1992) "The Due Diligence Rule and the Nature of the International Responsibility of States", *German Yearbook of International Law*, Volume 35, pp. 46-49.

26 Commentary to DARS as adopted in 2001, p. 56, Article 12, para. 11.

27 Riccardo Pisillo-Mazzeschi, (1992) "The Due Diligence Rule and the Nature of the International Responsibility of States", *German Yearbook of International Law*, Volume 35, p. 48.

the purpose of these duties. For their realization, some efforts must be made by the State obliged to do so, and circumstances of a particular case must also be such as not to render the performance impossible, while a number of objective risk factors are auspicious for the fulfillment of State obligations.[28] According to the ILC, the distinction between two types of positive obligations can be useful in establishing whether a breach of an international obligation has occurred.[29] However, it is not exclusive – according to the ICJ, one can speak of obligations of conduct, performance, and result[30] – and does not lead to specific consequences in relation to the application of the Draft Articles.

In the *Bosnian Genocide* case from 2007, the ICJ elaborated upon the distinction of two types of positive obligations and explained what they actually meant. When assessing the State's duty to prevent genocide laid down in Article 1 of the Genocide Convention, the Court pointed out that:

> "[I]t is clear that the obligation in question is one of conduct and not one of result, in the sense that a State cannot be under an obligation to succeed, whatever the circumstances, in preventing the commission of genocide: the obligation of States parties is rather to employ all means reasonably available to them, so as to prevent genocide so far as possible."[31]

At the outset, it is worth stressing that international responsibility is at stake if States fail to achieve a specific outcome and thus to comply with their positive obligations of result. The obligations of diligent conduct are quite different in that regard. The ICJ indicated that in case of non-compliance with this form of positive duty, States are held responsible not because they do not succeed in the achievement of a certain result, but because they manifestly fail to take all measures to prevent the acts of genocide that are within their power and can reasonably contribute to preventing genocide. In this respect, the notion of "due diligence" is, according to the ICJ, of crucial significance and will be discussed in the following Sections.[32]

The mentioned dichotomy between obligations of result and conduct will further be utilized as an instrument in assessing a spectrum of positive obligations resting upon States to some extent involved in the practice of hiring PMSCs: contracting States, host States, and home States. With regard to the obligations of result, it is clear what the result should be that must be realized by States and it can relatively easily be

---

28 Ibidem.

29 Commentary to DARS as adopted in 2001, p. 56, Article 12, para. 11.

30 International Court of Justice, *Gabčíkovo-Nagymaros Project (Hungary v. Slovakia)*, Judgment, September 25, 1997, ICJ Reports 1997, para. 135.

31 International Court of Justice, *Case Concerning the Application of the Convention on the Prevention and Punishment of the Crime of Genocide (Bosnia and Herzegovina v. Serbia and Montenegro)*, Judgment (Merits), February 26, 2007, ICJ Reports 2007, para. 430.

32 Ibidem.

established whether a breach of such obligations has taken place. The question, however, of whether a State has complied with its obligations of conduct requires more elaboration and assessment of all relevant circumstances of a particular case. There must be a clear failure on the part of that State to adopt all measures aimed at achieving a certain outcome. Making a distinction between duties of result and conduct enables an orderly and comprehensive presentation of the array of State obligations arising from the fields of international humanitarian and human rights law and allows a better understanding of their nature and content.

3. Positive State obligations of result

3.1 General analysis

When a State is placed under an obligation of result posed by international law, it has to act in a certain way to attain a specific result in accordance with the required standard. Failure on the part of the State to comply with this obligation to achieve a particular goal or goals consistent with human rights and international humanitarian law will lead to its international responsibility and respective legal consequences.[33] It is not important whether the State has taken all possible measures to make sure that the obligation is met: by acting according to the standard set by an international positive obligation of result or ensuring that an act is performed in this manner, the State is obliged to realize the expected outcome, which must be derived from its conduct.[34] In the words of the ICJ, a State is then "under an obligation to succeed, whatever the circumstances".[35] The assessment of the State's compliance or non-compliance with its result obligations is therefore made solely on the basis of State's achievements in practice and not the steps taken to meet some objectives in question.

There are various positive obligations of result, which can be arrived at through the analysis of the body of international humanitarian law. The 1949 Geneva Conventions contain certain provisions, such as those relating to the treatment of detained prisoners of war and civilian internees, incorporating positive State obligations of result. Human rights law is also familiar with the concept of this type of positive obligation and may be useful in relation to the practice of employing PMSCs by States. These duties will form the core of the examination made below.

---

33 Hannah Tonkin, (2011) *State Control over Private Military and Security Companies in Armed Conflict*, Cambridge: Cambridge University Press, p. 60.

34 Riccardo Pisillo-Mazzeschi, (1992) "The Due Diligence Rule and the Nature of the International Responsibility of States", *German Yearbook of International Law*, Volume 35, pp. 47-48.

35 International Court of Justice, *Case Concerning the Application of the Convention on the Prevention and Punishment of the Crime of Genocide (Bosnia and Herzegovina v. Serbia and Montenegro)*, Judgment (Merits), February 26, 2007, ICJ Reports 2007, para. 430.

3.2 Application to hiring States, host States, and home States

Positive obligations of result constitute duties that are to be fulfilled by States, which are in a position to choose any means possible to achieve a required result. PMSCs as non-State actors become relevant in the context of complying with and breaching this type of positive obligation if States opt to hire these corporations to act in accordance with this category of duties. Therefore, the only group of States subject to the analysis that may potentially breach positive obligations of result through the acts performed by PMSCs is hiring States. Note must be taken of the fact that these might be initially home States where PMSCs are established, host States where PMSCs actually operate, or third States having no such relationships with private firms, which by employing PMSCs become hiring States as a separate identified group. In this regard, they engage into contractual relationships with PMSCs to provide a large variety of military and security services and need to ensure that rules of international humanitarian and human rights law are observed and respected. By taking positive actions of their choice, these hiring States bound by international law must seek to comply with the applicable international duties of result.

International humanitarian law

It is true that States Parties to the 1949 Geneva Conventions and Additional Protocols must act in accordance with the rules and principles laid down in the Geneva law, but are also expected to take positive actions in order to achieve some specifically prescribed goals. Most, if not all, State obligations set out in international humanitarian law are obligations of result.[36] In the first instance, States that are party to a conflict are obliged to take a number of positive actions with regard to the protected groups of individuals. For example, Articles 16 and 19 of the First and Second Geneva Conventions respectively provide that they need to record as soon as possible details of identification for each sick, wounded, dead, or shipwrecked at sea person of the adverse party that falls into their hands. Such records must include, *inter alia*, designation on the power on which an individual depends; his or her army, regimental, personal, or serial number; surname and first name; date of birth; and other essential information. States are obliged to send these details to an official Information Bureau for prisoners of war, as mentioned in Article 122 of the Third Geneva Convention, which will then forward this information to the adversarial power on which persons in question might depend.

---

[36] Louise Doswald-Beck, "Private Military Companies under International Humanitarian Law", in Simon Chesterman and Chia Lehnardt (Eds.), (2007) *From Mercenaries to Market: The Rise and Regulation of Private Military Companies*, Oxford: Oxford University Press, p. 133.

Furthermore, the Third Geneva Convention contains a number of standards that must be met by a detaining power. Section II on the internment of prisoners of war provides that a detaining power must guarantee proper conditions for the detention of prisoners of war. It needs to take into account a set of requirements that must be fulfilled in accordance with the law of armed conflict. With regard to the standards of good hygiene and health of the detainees, the Third Geneva Convention stipulates that "prisoners of war may be interned only in premises located on land and affording every guarantee of hygiene and healthfulness".[37] In addition, "the basic daily food rations shall be sufficient in quantity, quality and variety to keep prisoners of war in good health and to prevent loss of weight or the development of nutritional deficiencies". There is also a provision regarding the clothing of the prisoners of war stating that "clothing, underwear and footwear shall be supplied to prisoners of war in sufficient quantities by the Detaining Power, which shall make allowance for the climate of the region where the prisoners are detained".[38]

Another array of positive obligations of result that is worth mentioning can be found in the Fourth Geneva Convention. Section IV concerning the treatment of civilian internees incorporates a number of standards that must be complied with by a detaining power. For instance, it "shall not set up places of internment in areas particularly exposed to the dangers of war" and "shall give the enemy Powers, through the intermediary of the Protecting Powers, all useful information regarding the geographical location of places of internment".[39] Also, the provisions of the Fourth Geneva Convention state that the "food rations for internees shall be sufficient in quantity, quality and variety to keep internees in a good state of health and prevent the development of nutritional deficiencies".[40] Furthermore, the detaining power must give all facilities to internees "to provide themselves with the necessary clothing, footwear and change of underwear, and later on, to procure further supplies if required".[41]

In addition to the duty to take certain measures in relation to some groups of persons, there is a separate broad obligation to train and to specifically instruct.[42] This duty applies to military and civilian personnel of States and is only applicable in international armed conflicts.[43] Specific rules with regard to persons that do not belong to the armed forces are in the Third and Fourth Geneva Conventions.

---

[37] Article 22 of the Third Geneva Convention.

[38] Article 27 of the Third Geneva Convention.

[39] Article 83 of the Fourth Geneva Convention.

[40] Article 89 of the Fourth Geneva Convention.

[41] Article 90 of the Fourth Geneva Convention.

[42] University Centre for International Humanitarian Law, *Expert Meeting on Private Military Contractors: Status and State Responsibility for Their Actions*, Report, Geneva, August 29-30, 2005, pp. 43-44, <http://www.geneva-academy.ch/docs/expert-meetings/2005/2rapport_compagnies_privees.pdf>, accessed on October 10, 2014.

[43] Hannah Tonkin, (2011) *State Control over Private Military and Security Companies in Armed Conflict*, Cambridge: Cambridge University Press, p. 198.

Article 127(2) of the Third Geneva Convention refers to "any military or other authorities", who in times of war need to fulfill States' duties towards prisoners of war, must not only have the actual text of the respective Geneva Convention, but also be instructed to apply its provisions. Article 144(2) of the Fourth Geneva Convention deals with "any civilian, military, police or other authorities", who in times of war must protect civilian persons, and are required to have the text of the Convention and to receive instructions on its application. Persons referred to in both Articles must "assume responsibilities" towards two groups of protected people – prisoners of war and civilians – while the provisions specifically mention "other authorities" having these responsibilities. The drafters of the Geneva Conventions did not have PMSCs in mind in the process of creating these legal rules, but it can be assumed that they are applicable to private contractors performing military and security tasks and undertaking these responsibilities of States.[44] Read in conjunction with Common Article 1, the duty to train applicable to PMSCs can further be extended to the obligations to vet employees of PMSCs and to issue clear rules of engagement in accordance with international humanitarian law.[45] It can be argued that the necessity of the training cannot be underestimated. Many employees of PMSCs have a military background and may be inclined to engage in conflict areas from the position of lawful combatants.[46] Therefore, governments relying on the deployment of these individuals must ensure that they have an adequate level of training in the matters of international law and use force within the limits of the Geneva law.

In this context, a duty to disseminate closely connected to the obligation to train and to instruct also plays an essential role. This is of crucial importance for the deployment of PMSCs in armed conflicts in general, for the law of armed conflict must be known to individuals participating in armed conflicts. As the Commentaries to the Geneva Conventions put it, "one of the worst enemies of the Geneva Conventions is ignorance".[47] Therefore, States must ensure that their national armed forces are familiar with them and apply rules and principles of international humanitarian law in situations of armed conflict. This goal can be achieved by disseminating the

---

44 Louise Doswald-Beck, "Private Military Companies under International Humanitarian Law", in Simon Chesterman and Chia Lehnardt (Eds.), (2007) *From Mercenaries to Market: The Rise and Regulation of Private Military Companies*, Oxford: Oxford University Press, p. 132.

45 Carsten Hoppe, (2008) "Passing the Buck: State Responsibility for Private Military Companies", *The European Journal of International Law*, Volume 19, Issue 5, p. 993.

46 Lindsey Cameron and Vincent Chetail, (2013) *Privatizing War: Private Military and Security Companies under Public International Law*, Cambridge: Cambridge University Press, p. 270.

47 Jean S. Pictet, (1952) *Commentary, I Geneva Convention for the Amelioration of the Condition of the Wounded and Sick in Armed Forces in the Field*, Geneva: International Committee of the Red Cross, p. 348; Jean S. Pictet, (1960) *Commentary, II Geneva Convention for the Amelioration of the Condition of the Wounded, Sick and Shipwrecked Members of Armed Forces at Sea*, Geneva: International Committee of the Red Cross, p. 257; Jean S. Pictet, (1958) *Commentary, IV Geneva Convention Relative to the Protection of Civilian Persons in Time of War*, Geneva: International Committee of the Red Cross, p. 581.

texts of the relevant Conventions. The State obligation to disseminate can be found in Article 47 of the First Geneva Convention, Article 48 of the Second Geneva Convention, Article 127 of the Third Geneva Convention, and Article 144 of the Fourth Geneva Convention. Furthermore, there is Article 83 of the First Additional Protocol that reaffirms these provisions and obligations laid down therein.[48] The obligation concerning dissemination set out in the Geneva law appears to be general and absolute in nature.[49] This particular State duty is applicable in not only times of war, but also in peace, and States are required to disseminate the texts of the treaties "as widely as possible". This needs to be done in the respective countries. If PMSCs belong to State armed forces, States hiring them will be obliged to instruct them in international humanitarian law in accordance with these legal rules.

While States are not specifically required to ensure that international humanitarian law is in practice known to all civilians,[50] the above-mentioned provisions indicate that States are encouraged to make the Geneva law known to the civilian population as well. Article 47 of the First Geneva Convention, having similar wording as Article 48 of the Second Geneva Convention and Article 83 of the First Additional Protocol, specifically refers to "the entire population, in particular the armed fighting forces, the medical personnel and the chaplains". In addition, Article 83(2) of the First Additional Protocol maintains that not only military, but also civilian authorities assuming responsibilities in relation to the application of the Geneva law in armed conflicts must be acquainted with the text of the Conventions and the Protocol in question.

Thus, States are required to ensure that both their military personnel and civilians have knowledge of the law of armed conflict. Taking into account the general spirit of the provisions in question read in conjunction with Common Article 1 to the 1949 Geneva Conventions, it can be said that PMSCs often hired by States to perform essential State functions on the battlefield might be considered as falling under this category of persons that must be educated in international humanitarian law and its application.[51] Such education would be required in order to avoid non-compliance

---

[48] Claude Pilloud and Others, (1987) *Commentary on the Additional Protocols of 8 June 1977 to the Geneva Conventions of 12 August 1949*, International Committee of the Red Cross, Geneva: Martinus Nijhoff Publishers, p. 960, para. 3369.

[49] Jean S. Pictet, (1952) *Commentary, I Geneva Convention for the Amelioration of the Condition of the Wounded and Sick in Armed Forces in the Field*, Geneva: International Committee of the Red Cross, p. 348; Jean S. Pictet, (1960) *Commentary, II Geneva Convention for the Amelioration of the Condition of the Wounded, Sick and Shipwrecked Members of Armed Forces at Sea*, Geneva: International Committee of the Red Cross, p. 257; Jean S. Pictet, (1958) *Commentary, IV Geneva Convention Relative to the Protection of Civilian Persons in Time of War*, Geneva: International Committee of the Red Cross, p. 581.

[50] University Centre for International Humanitarian Law, *Expert Meeting on Private Military Contractors: Status and State Responsibility for Their Actions*, Report, Geneva, August 29-30, 2005, p. 43, <http://www.geneva-academy.ch/docs/expert-meetings/2005/2rapport_compagnies_privees.pdf>, accessed on October 10, 2014.

[51] Hannah Tonkin, (2011) *State Control over Private Military and Security Companies in Armed Conflict*, Cambridge: Cambridge University Press, p. 198; Carsten Hoppe, (2008) "Passing the Buck: State

with the law of armed conflict by PMSCs and other non-State armed groups capable of using force contrary to this branch of international law due to the absence of necessary knowledge.[52] Although the obligation to disseminate is not taken very seriously by some States,[53] many of them go as far as to emphasize the need to educate their judiciary, police, and prison personnel on matters of international humanitarian law and this is consistent with the idea of instructing PMSCs hired by States to operate in armed conflicts.[54] When PMSCs are active in armed conflicts, not only hiring States, but also host and home States are required to make sure that their personnel are aware of the Geneva law.

In sum, a State might have an established practice of employing PMSCs or hire them occasionally, when a particular need arises, to perform specific services connected to their obligations under the First and Second Geneva Conventions and be present on a battlefield to collect and to care for the sick, wounded, shipwrecked, and dead. It is also highly possible that these corporations can be entrusted with various tasks related to the provision of security, detention, and guarding, which are carried out in the facilities where prisoners of war or civilian internees are held. Such was clearly the case in the detention facilities at the US-run Abu Ghraib prison in Iraq. Also, host States where PMSCs are active or home States where they are incorporated need to be aware of the presence of these entities on their territory and within their jurisdiction. In these scenarios, a spectrum of standards laid down in the four 1949 Geneva Conventions in the form of positive obligations of result must be met by the PMSCs dealing with the groups of protected individuals. Thus, the respective State should be aware of the fact that the failure of the PMSC to treat prisoners of war, civilian internees, or other protected persons according to the required standards can lead to its international responsibility and be prepared to deal with its legal consequences.[55] Additionally, all States in all circumstances must ensure that PMSCs are familiar with the law of armed conflict by disseminating the Geneva law, instructing,

---

Responsibility for Private Military Companies", *The European Journal of International Law*, Volume 19, Issue 5, p. 993.

[52] Nicolas Lamp, (2011) "Conceptions of War and Paradigms of Compliance: The 'New War' Challenge to International Humanitarian Law", *Journal of Conflict & Security Law*, Volume 16, Issue 2, pp. 241-242.

[53] University Centre for International Humanitarian Law, *Expert Meeting on Private Military Contractors: Status and State Responsibility for Their Actions*, Report, Geneva, August 29-30, 2005, p. 43, <http://www.geneva-academy.ch/docs/expert-meetings/2005/2rapport_compagnies_privees.pdf>, accessed on October 10, 2014.

[54] Louise Doswald-Beck, "Private Military Companies under International Humanitarian Law", in Simon Chesterman and Chia Lehnardt (Eds.), (2007) *From Mercenaries to Market: The Rise and Regulation of Private Military Companies*, Oxford: Oxford University Press, pp. 132-133.

[55] University Centre for International Humanitarian Law, *Expert Meeting on Private Military Contractors: Status and State Responsibility for Their Actions*, Report, Geneva, August 29-30, 2005, pp. 44-45, <http://www.geneva-academy.ch/docs/expert-meetings/2005/2rapport_compagnies_privees.pdf>, accessed on October 10, 2014.

or training them. Hiring States in particular will have this whole range of obligations due to the close link between them and the PMSCs they hire. In this regard, the possibility of attributing a failure to comply with these positive obligations to a State in question in accordance with the rules on attribution, as laid down in the Draft Articles assessed above or efforts made by the State to comply with the respective obligation are not relevant. If the expected result has not been achieved and a set of standards of the 1949 Geneva Conventions has not been met, the State under these positive obligations will be held responsible and will have to face the legal consequences.

International human rights law

Positive obligations of result are a concept developed in the field of international humanitarian law, and according to the same logic can also be found in the context of international human rights law. In addition to the 1949 Geneva Conventions, an array of human rights conventions contain provisions with State obligations of this nature. As an example, Hannah Tonkin cites the right to a fair trial imposing positive obligations of result.[56] Article 6(1) ECHR clearly stresses that:

> "In the determination of his civil rights and obligations or of any criminal charge against him, everyone is entitled to a fair and public hearing within a reasonable time by an independent and impartial tribunal established by law."

This legal rule might be understood to imply that if the desired outcome is not realized, it is not significant to look at the efforts made in order to find out whether a particular State has fulfilled its positive obligation. What counts, is the achieved result. The European Court of Human Rights confirmed this idea and further elaborated upon the obligation in question as follows:

> "The Court recalls that the Contracting States are under the obligation to organise their legal systems 'so as to ensure compliance with the requirements of Article 6 para. 1'... The Court's task is to determine whether the Contracting States have achieved the result called for by the Convention, not to indicate the particular means to be utilised."[57]

Human rights of a similar nature can also be found in Article 14 ICCPR, Article 8 ACHR, and Article 7 ACHPR. It is not entirely beyond the realm of possibility that PMSCs – just as other private entities – might be hired to ensure that the right to a fair trial laid down in the above-mentioned provisions is guaranteed and that the

---

[56] Hannah Tonkin, (2011) *State Control over Private Military and Security Companies in Armed Conflict*, Cambridge: Cambridge University Press, p. 62.

[57] European Court of Human Rights, *De Cubber v. Belgium*, Judgment (Merits), October 26, 1984, Application No. 9186/80, para. 35.

legal systems of the hiring States are constructed in such a way as to ensure this particular human right. For instance, if PMSCs are contracted to participate in criminal investigations, they need to presume that somebody is innocent until he or she has been proved to be guilty in accordance with law.[58] In that respect, States Parties to the European Convention on Human Rights and other human rights treaties are under an obligation to achieve the result of ensuring the right to a fair trial and to meet the standards set in the respective Conventions.

Positive duties of result also rest upon States when they need to comply with positive obligations of conduct, such as those relating to the human right to life. States need, for instance, to train the members of the police in such a manner that they will not violate the right to life. If PMSCs are used to provide training to these individuals, they must achieve an appropriate level of awareness and knowledge of the human rights among the personnel of the police. By employing PMSCs for such purposes, States can thus fail to comply with their obligations of result if private contractors do not provide a sufficient training in relation to the prevention of human rights abuses. As will be observed in the following parts of the study, positive obligations of result may be necessary to prevent human rights violations. In this regard, States are expected to achieve a certain result, for instance, by establishing national laws and regulations, that will enable them to take an array of positive measures for the prevention of human rights violations.

From the above considerations, it follows that States employing PMSCs can be held responsible under international law when PMSCs are used as an instrument to guarantee the States' compliance with their positive obligations of result and fail to achieve the required standards by acting improperly or employing unlawful activities breaching these positive State obligations that arise out of international humanitarian and human rights law. Those duties seek to ensure that States take all appropriate measures and necessarily meet the array of established standards, while their breach means the engagement of the responsibility of States concerned.

4. Positive State Obligations of Diligent Conduct

4.1 General analysis

Perhaps of more pertinence to the State practice of using PMSCs is another particular type of positive obligation: that of diligent conduct. As pointed out by Chia Lehnardt, Carsten Hoppe, Olivier De Schutter, Hannah Tonkin, and other experts, a great deal of attention needs to be paid to this issue, which may give rise to the responsibility of

---

58 Article 6(2) ECHR, Article 14(2) ICCPR, Article 8(2) ACHR, and Article 7(1)(b) ACHPR.

States.[59] This is often referred to as the concept of due diligence obligations constituting an alternative instrument to incur international responsibility if the attribution of unlawful conduct of PMSCs and other private actors to States under public international law is not possible. By referring to the *Velasquez Rodriguez* case as the first case where the notion of due diligence was used, the Inter-American Commission on Human Rights observed what it implies when a State is under an obligation to exercise due diligence for the prevention of human rights abuses and that the non-compliance with this duty can result in State responsibility. The Commission stressed that:

> "This means that, even where conduct may not initially be directly imputable to a state (for example, because the actor is unidentified or not a state agent), a violative act may lead to state responsibility 'not because of the act itself, but because of the lack of due diligence to prevent the violation or respond to it as the Convention requires'."[60]

The due diligence standard has been developed as a part of the State responsibility doctrine regarding the protection of aliens from the misconduct of private actors.[61] Nowadays, due diligence obligations can be found in various fields of international law, including, *inter alia*, environmental law, international investment law, and, importantly, international humanitarian and human rights law.[62] Although some argue that there are currently no clearly articulated and widely accepted definitions of this concept, it has been used and explained in different areas where it applies.[63] Essen-

---

[59] Chia Lehnardt, "Private Military Companies and State Responsibility", in Simon Chesterman and Chia Lehnardt (Eds.), (2007) *From Mercenaries to Market: The Rise and Regulation of Private Military Companies*, Oxford: Oxford University Press, pp. 152-155; Carsten Hoppe, (2008) "Passing the Buck: State Responsibility for Private Military Companies", *The European Journal of International Law*, Volume 19, Issue 5; Olivier De Schutter, "The Responsibility of States", in Simon Chesterman and Angelina Fisher (Eds.), (2009) *Private Security, Public Order: The Outsourcing of Public Services and Its Limits*, Oxford: Oxford University Press, pp. 27-28; Willem van Genugten, Nicola Jägers, and Marie-José Van der Heijden, "Protecting the Victims of the Privatization of War", in Rianne Letschert and Jan Van Dijk (Eds.), (2011) *The New Faces of Victimhood*, Dordrecht: Springer; Riccardo Pisillo-Mazzeschi, (1992) "The Due Diligence Rule and the Nature of the International Responsibility of States", *German Yearbook of International Law*, Volume 35; Hannah Tonkin, (2011) *State Control over Private Military and Security Companies in Armed Conflict*, Cambridge: Cambridge University Press, pp. 63-64.

[60] Inter-American Commission on Human Rights, *Report on the Situation of Human Rights in Brazil*, September 29, 1997, OEA/Ser.L/V/II.97, Doc. 29 rev. 1, Chapter VIII, F.Conclusions, para. 30, <http://www.cidh.org/countryrep/brazil-eng/index%20-%20brazil.htm>, accessed on January 14, 2014.

[61] Stephanie Farrior, (1998) "State Responsibility for Human Rights Abuses by Non-state Actors", *American Society of International Law Proceedings*, Volume 92, p. 302.

[62] See Robert p. Barnidge, Jr., (2006) "The Due Diligence Principle Under International Law", *International Community Law Review*, Volume 8; Olga Martin-Ortega, (2013) "Human Rights Due Diligence for Corporations: From Voluntary Standards to Hard Law at Last?", *Netherlands Quarterly of Human Rights*, Volume 31, Issue 4, p. 53.

[63] Stephanie Farrior, (1998) "State Responsibility for Human Rights Abuses by Non-state Actors", *American Society of International Law Proceedings*, Volume 92, p. 302; Chia Lehnardt, "Private Military Companies and State Responsibility", in Simon Chesterman and Chia Lehnardt (Eds.), (2007) *From*

tially, it is used to indicate the level of effort made by States in order to comply with their obligations under treaty law[64] and can be employed as an instrument for assessing State responsibility for violations of international law committed by non-State actors, such as PMSCs. In general, it is true that these particular duties are not something extraordinary in the law of armed conflict.[65] Dealing with this field of international law, the Geneva Conventions and their Additional Protocols do not specifically refer to due diligence duties and it may be questioned whether there is indeed such a notion in humanitarian law.[66] Article 91 of the First Additional Protocol, however, mentions what this concept entails. This provision concerns the responsibility of States and suggests that States participating in an armed conflict and violating provisions of the Geneva Conventions and their Additional Protocols will be liable to pay compensation.[67] They will be responsible for acts and omissions of individuals belonging to their armed forces. Additionally and more importantly, the ICRC Commentary refers to due diligence obligations of States with regard to the conduct of private individuals and emphasizes that international responsibility is engaged if States parties to armed conflicts fail to comply with their duties to prevent and to punish. It stresses the following:

> "In international law the conduct of any organ of the State, whether military or civilian, constitutes an act of State, provided that it acted in its official capacity, regardless of its position, whether superior or subordinate... It can be imputed not only for acts committed by a person or persons who form part of the armed forces, as this provision lays down, but also for possible omissions. As regards damages which may be caused by *private individuals*, i.e., by persons who are not members of the armed forces (nor of any other organ of the State), legal writings and case-law show that the responsibility of the State is involved if it has not taken *such preventive or repressive measures* as

---

Mercenaries to Market: The Rise and Regulation of Private Military Companies, Oxford: Oxford University Press, p. 154; Olga Martin-Ortega, (2013) "Human Rights Due Diligence for Corporations: From Voluntary Standards to Hard Law at Last?", *Netherlands Quarterly of Human Rights*, Volume 31, Issue 4, p. 53.

64 Tineke Lambooy, (2010) "Corporate Due Diligence As a Tool to Respect Human Rights", *Netherlands Quarterly of Human Rights*, Volume 28, Issue 3, p. 418.

65 Hannah Tonkin, (2009) "Common Article 1: A Minimum Yardstick for Regulating Private Military and Security Companies", *Leiden Journal of International Law*, Volume 22, Issue 4, p. 791.

66 University Centre for International Humanitarian Law, *Expert Meeting on Private Military Contractors: Status and State Responsibility for Their Actions*, Report, Geneva, August 29-30, 2005, p. 42, <http://www.geneva-academy.ch/docs/expert-meetings/2005/2rapport_compagnies_privees.pdf>, accessed on October 10, 2014; Corinna Seiberth, (2013) *Private Military and Security Companies in International Law – A Challenge for Non-binding Norms: the Montreux Document and the International Code of Conduct for Private Security Service Providers*, Cambridge, Antwerp, Portland: Intersentia, p. 93.

67 University Centre for International Humanitarian Law, *Expert Meeting on Private Military Contractors: Status and State Responsibility for Their Actions*, Report, Geneva, August 29-30, 2005, p. 42, <http://www.geneva-academy.ch/docs/expert-meetings/2005/2rapport_compagnies_privees.pdf>, accessed on October 10, 2014.

> could reasonably be expected to have been taken in the circumstances. In other words, responsibility is incurred if the Party to the conflict has not acted with *due diligence to prevent* such acts from taking place, or *to ensure their repression* once they have taken place."[68] (emphasis added)

In order to distill positive State obligations of diligent conduct, individual provisions of the Geneva law will be analyzed in order to provide clarity regarding the scope of State duties to adopt positive measures under the law of armed conflict. Examination of the law of armed conflict reveals a set of due diligence obligations applicable to certain States. Common Article 1 of the 1949 Geneva Conventions and Article 1 of the First Additional Protocol impose a general positive obligation upon States Parties to comply with the law of armed conflict. Both provisions require the High Contracting Parties to "undertake to respect and to ensure respect" for the Conventions and the Protocol in all circumstances. The positive character of the obligation creates the impression that no specific duties rest upon States and that it is a rather general policy established with regard to the Geneva Conventions and the First Additional Protocol.[69] However, the wording of the provision is misleading. The words "to ensure respect" stipulate that it requires States to take positive measures to ensure compliance of PMSCs and other non-State actors with international humanitarian law.

In international human rights law, the due diligence obligations appear to constitute a well-established idea that is traceable in primary State obligations laid down in various human rights treaties. Of particular importance for the practice of employing PMSCs are the obligations stemming from the field of human rights law to "ensure" and to "secure" human rights within the jurisdiction of States.[70] These duties can be found in practically every human rights instrument.[71] They have been incorporated into Article 2 ICCPR, Article 1 ECHR, Article 1(1) ACHR, Article 25 ACHPR, Article 2(1) CRC, and Articles 2 and 3 CEDAW. These obligations require duty-holding States to exercise due diligence to prevent, investigate, punish, and redress human rights violations and to take appropriate measures to comply with them.[72] In addi-

---

68 Claude Pilloud and Others, (1987) *Commentary on the Additional Protocols of 8 June 1977 to the Geneva Conventions of 12 August 1949*, International Committee of the Red Cross, Geneva: Martinus Nijhoff Publishers, pp. 1057-1058, para. 3660.

69 Chia Lehnardt, "Private Military Companies and State Responsibility", in Simon Chesterman and Chia Lehnardt (Eds.), (2007) *From Mercenaries to Market: The Rise and Regulation of Private Military Companies*, Oxford: Oxford University Press, p. 153.

70 Hannah Tonkin, (2011) *State Control over Private Military and Security Companies in Armed Conflict*, Cambridge: Cambridge University Press, p. 65.

71 Stephanie Farrior, (1998) "State Responsibility for Human Rights Abuses by Non-state Actors", *American Society of International Law Proceedings*, Volume 92, p. 301.

72 Hannah Tonkin, (2011) *State Control over Private Military and Security Companies in Armed Conflict*, Cambridge: Cambridge University Press, p. 65; Stephanie Farrior, (1998) "State Responsibility for Human Rights Abuses by Non-state Actors", *American Society of International Law Proceedings*, Volume 92, p. 301.

tion to these provisions, international and regional human rights bodies have elaborated upon the concept of due diligence obligations. Besides the reasoning of the Inter-American Court of Human Rights in the *Velasquez Rodriguez* case, the UN Human Rights Committee provided its views on the notion.[73] The UN Committee on the Elimination of Discrimination Against Women,[74] the UN Committee on the Rights of the Child that addressed due diligence duties with respect to the business sector,[75] and the European Court of Human Rights also looked into the matter.[76]

In the terminology of the Ruggie Framework, the due diligence obligations of States are expressed in the State duty to protect, constituting a standard of conduct.[77] What the notion of due diligence entails with regard to transnational corporations and other non-State actors is stated, for instance, in the Maastricht Guidelines that specifically focus on violations of economic, social, and cultural rights. The Guidelines clearly recall the State obligation to protect, and state with respect to the acts of non-State entities breaching the above-mentioned rights that:

> "The obligation to protect includes the State's responsibility to ensure that *private entities or individuals, including transnational corporations* over which they exercise jurisdiction, do not deprive individuals of their economic, social and cultural rights. States are *responsible* for violations of economic, social and cultural rights that result from their *failure to exercise due diligence* in controlling the behaviour of such non-state actors."[78] (emphasis added)

States are thus responsible for taking appropriate measures and exercising due diligence in relation to the activities of private contractors.[79] Even if unlawful acts of

---

[73] UN Human Rights Committee, *General Comment 31*, May 26, 2004, UN Doc. CCPR/C/21/Rev.1/Add.13, para. 8.

[74] UN Committee on the Elimination of Discrimination Against Women, *General Recommendation 19*, January 30, 1992, UN Doc. A/47/38 (1993), para. 9.

[75] UN Committee on the Rights of the Child, *General Comment No. 16 on State Obligations Regarding the Impact of the Business Sector on Children's Rights*, April 17, 2013, UN Doc. CRC/C/GC/16, paras. 62-65.

[76] See, for instance, European Court of Human Rights, *Bevacqua and S. v. Bulgaria*, Judgment, June 12, 2008, Application No. 71127/01.

[77] Report of the Special Representative of the Secretary-General on the Issue of Human Rights and Transnational Corporations and Other Business Enterprises, John Ruggie, *Business and Human Rights: Towards Operationalizing the 'Protect, Respect and Remedy' Framework*, UN Human Rights Council, Eleventh Session, April 22, 2009, UN Doc. A/HRC/11/13, p. 7, para. 14; Antenor Hallo de Wolf, (2013) "Human Rights and the Regulation of Privatized Essential Services", *Netherlands International Law Review*, Volume 60, Issue 2, pp. 166, 174.

[78] UN Committee on Economic, Social and Cultural Rights, Maastricht Guidelines on Violations of Economic, Social and Cultural Rights, Twenty-Fourth Session, November 13 – December 1, 2000, UN Doc. E/C.12/2000/13, Guideline 18.

[79] UN Human Rights Council, *Report of the Working Group on the Use of Mercenaries as a Means of Violating Human Rights and Impeding the Exercise of the Right of People to Self-Determination*, January

private actors do not fall under qualification of acts of a State as such or due to the involvement of State organs in their commission, the inaction of a State will engage international responsibility.[80] It is important to note that these are not the private acts themselves triggering State responsibility, but the lack of due diligence on the part of a State unable to prevent those illegal activities or to punish responsible individuals.[81] This does not mean that a State has to commit an unlawful act itself, it actually implies that the State fails to comply with its obligation to employ all necessary means to prevent or to punish the commission of certain acts and is thus in violation of international law. While in general the due diligence notion is known for its vagueness, one should not be hesitant to examine it in the context of the operation of PMSCs, as it may prove to be pertinent as a legal concept for holding States internationally responsible for the breaches of international law committed by private contractors.[82]

Nowadays, good governments are expected to take an appropriate level of care in certain cases when they pursue a range of specific activities, especially by outsourcing some of their functions to private entities and individuals. Some provisions of international humanitarian and human rights law might impose upon States obligations to take all reasonable actions in order to comply with the standards required by international law. In this regard, the facts of particular cases and respective circumstances may indicate that an omission or a failure on the part of States to act leads to the realization of a certain conduct prohibited by an international legal mechanism. When there is sufficient evidence that the State conduct has not met standards of behavior set in specific rules of international law constituting positive obligations of diligent conduct, a breach of these obligations must be considered to be established.[83] As a result, it triggers the international responsibility of States in question and involves respective legal consequences, even when illegal activities concerned are of a private nature and cannot directly be attributed to these States. In this vein, the notion of due diligence serves a significant purpose as "a tool for promoting greater State responsibility".[84]

9, 2008, UN Doc. A/HCR/7/7.

[80] Antenor Hallo de Wolf, (2012) *Reconciling Privatization with Human Rights*, School of Human Rights Research Series, Volume 49, Cambridge, Antwerp and Portland: Intersentia, p. 230.

[81] Nicola Jägers, (2002) *Corporate Human Rights Obligations. In Search of Accountability*, School of Human Rights Research Series, Volume 17, Antwerp: Intersentia, pp. 144-145.

[82] Charlotte Beaucillon, Julian Fernandez, and Hélène Raspail, "State Responsibility for Conduct of Private Military and Security Companies Violating Ius ad Bellum", in Francesco Francioni and Natalino Ronzitti (Eds.), (2011) *War by Contract: Human Rights, Humanitarian Law, and Private Contractors*, Oxford: Oxford University Press, p. 413.

[83] Pierre-Marie Dupuy, (1999) "Reviewing the Difficulties of Codification: On Ago's Classification of Obligations of Means and Obligations of Result in Relation to State Responsibility", *The European Journal of International Law*, p. 382.

[84] Carin Benninger-Budel, "Introduction", in Carin Benninger-Budel (Ed.), (2008) *Due Diligence and Its Application to Protect Women From Violence*, Leiden: Koninklijke Brill NV, p. 15; Tineke Lambooy,

For a long time, international law has been familiar with the concept of obligations of conduct and this notion for holding States responsible in addition to the attribution doctrine has been applied in a variety of ways over centuries.[85] This category of positive obligations comprises various States duties that can be classified as the duty to prevent and the duty to punish.[86] Ian Brownlie refers to three claims brought before the General Claims Commission[87] and indicates that there is a State duty to exercise due diligence in case of the misconduct by private individuals.[88] He notes that it is associated with the obligation to take adequate steps to protect, apprehend, and punish. In his report, the Special Representative of the Secretary-General sheds some light on the issue of due diligence obligations and sets the State duty to protect apart from the question of direct attribution of corporate misconduct to States. He notes that this duty implies that:

> "... States are not held responsible for corporate-related human rights abuse per se, but may be considered in breach of their obligations where they fail to take appropriate steps to prevent it and to investigate, punish and redress it when it occurs."[89]

Both duties to prevent and to punish represent different ideas or views on the nature of positive actions to be taken by States, but seem to be connected to the same general duty to provide protection to individuals and have the same aim. Special Rapporteur Roberto Ago put it as follows:

> "*Prevention* and *punishment* are simply two aspects of the same obligation to provide protection and have a common aim, namely to discourage potential attackers of

---

(2010) "Corporate Due Diligence As a Tool to Respect Human Rights", *Netherlands Quarterly of Human Rights*, Volume 28, Issue 3, p. 425.

[85] Jan A. Hessbruegge, (2004) "The Historical Development of the Doctrines of Attribution and Due Diligence in International Law", *New York University Journal of International Law and Politics*, Volume 36, p. 266.

[86] Hannah Tonkin, (2011) *State Control over Private Military and Security Companies in Armed Conflict*, Cambridge: Cambridge University Press, p. 63; Carsten Hoppe, (2008) "Passing the Buck: State Responsibility for Private Military Companies", *The European Journal of International Law*, Volume 19, Issue 5, p. 994.

[87] *Thomas H. Youmans (United States of America v. United Mexican States)*, Reports of International Arbitral Awards, Volume IV, p. 110; *Laura M. B. Janes et al. (United States of America v. United Mexican States)*, Reports of International Arbitral Awards, Volume IV, p. 82; *Gertrude Parker Massey (United States of America v. United Mexican States)*, April 15, 1927, Reports of International Arbitral Awards, Volume IV, p. 155.

[88] Ian Brownlie, (1983) *System of the Law of Nations: State Responsibility Part I*, Oxford: Clarendon Press, p. 161.

[89] Report of the Special Representative of the Secretary-General on the Issue of Human Rights and Transnational Corporations and Other Business Enterprises, John Ruggie, *Business and Human Rights: Towards Operationalizing the 'Protect, Respect and Remedy' Framework*, UN Human Rights Council, Eleventh Session, April 22, 2009, UN Doc. A/HRC/11/13, p. 7, para. 14.

> protected persons from carrying out such attacks. The system of protection that the State must provide therefore includes not only the adoption of *measures* to avoid certain acts being committed but also provision for, and application of, *sanctions* against the authors of acts which the implementation of preventive measures has failed to avert."[90] (emphasis added)

To provide greater clarity on the possibility of incurring international responsibility for private acts by means of attribution, the ILC stated that it had no intention to go into the content, scope, and specifics of "the international obligations of States which may be breached by acts, or more often, omissions on the part of organs of the State in relation to the acts of individuals".[91] It merely stressed that "it is enough to point out that such obligations exist"[92] and gave a number of examples of such duties and their possible violations. According to the ILC:

> "There is no doubt that, to an extent which varies from case to case, the State must afford protection to foreign States, their official representatives and their ordinary nationals against any attacks by individuals. If, in a particular situation, the State or the entities mentioned in article 7 [organs of a territorial governmental entity within a State or of an entity which is not part of the formal structure of a State or a territorial governmental entity, but is empowered to exercise elements of the governmental authority] *failed to take adequate protective measures* and an attack by individuals took place, there would be an act of the State related to the acts of individuals – an act of the State constituting a breach of an international obligation of that State. The same would be true if an individual made an attack and the organs of the State *failed... to discharge an international obligation to punish or extradite* that individual."[93] (emphasis added)

That the duty to punish might be understood as a broad international obligation to legislate, investigate, prosecute, punish, and provide redress appears to be quite clear. It is connected to the human right to an effective remedy and other significant human rights, such as the right to life and freedom from torture and other cruel, inhuman, or degrading treatment.[94] In general, States are free to decide how they comply with their obligations to prevent and to punish, and can opt for the use of legislative,

---

[90] International Law Commission, *Fourth Report on State Responsibility, by Roberto Ago, Special Rapporteur – the Internationally Wrongful Act of the State, Source of International Responsibility (Continued)*, (1972) Yearbook of the International Law Commission, Volume II, UN Doc. A/CN.4/264 and Add.1, p. 98, para. 67.

[91] International Law Commission, *Report of the International Law Commission on the Work of Its Twenty-Seventh Session, 5 May – 25 July 1975*, Official Records of the General Assembly, Thirtieth Session, Supplement No. 10, (1975) Yearbook of the International Law Commission, Volume II, UN Doc. A/10010/Rev.1, p. 71, para. 4.

[92] Ibidem.

[93] Ibidem.

[94] Carsten Hoppe, "Positive Human Rights Obligations of the Hiring State in Connection with the Provision of 'Coercive Services' by a Private Military or Security Company", in Francesco Francioni and

administrative, and judicial measures in accordance with international human rights law.[95] Against this background, the duty to punish is a positive obligation of an international character imposed upon States to take adequate measures in order to create legislative acts in accordance with respective international legal instruments, to investigate the committed violations of international law, and to prosecute and to punish the accused offenders. For instance, a failure on the part of a State to punish an individual who successfully managed to attack another individual and to incur physical and other damage to him or her is a violation of the obligation to punish, which is no less serious than the obligation to prevent.[96] The latter is mainly an obligation imposed on a State to take all possible measures to prevent breaches of international law from occurrence. The ILC's Draft Articles deal with this particular positive obligation of prevention in Article 14 DARS focusing on the extension in time of the breach of an international obligation. Article 14(3) DARS reads:

> "The breach of an international obligation requiring a State to prevent a given event occurs when the event occurs and extends over the entire period during which the event continues and remains not in conformity with that obligation."

Going into the nature of this particular type of positive obligation, the Commentaries to the Draft Articles further explain that there is indeed a specific category of violations of international obligations to prevent the occurrence of a certain event.[97] These obligations of prevention are, in fact, "best efforts obligations" that require States to take all reasonable or necessary measures within their power to prevent a certain event from occurring. States are, however, not under a duty to warrant that that particular event will not occur: they must merely do whatever is in their power to prevent it.[98] The positive measures to be taken by States may include the duty to intervene when a violation of international law is likely to occur and to regulate the activities of private actors in order to prevent breaches of international humanitarian

---

Natalino Ronzitti (Eds.), (2011) *War by Contract: Human Rights, Humanitarian Law, and Private Contractors*, Oxford: Oxford University Press, pp. 117-118.

95 Report of the Special Representative of the Secretary-General on the Issue of Human Rights and Transnational Corporations and Other Business Enterprises, John Ruggie, *Business and Human Rights: Towards Operationalizing the 'Protect, Respect and Remedy' Framework*, UN Human Rights Council, Eleventh Session, April 22, 2009, UN Doc. A/HRC/11/13, p. 7, para. 14.

96 International Law Commission, *Fourth Report on State Responsibility, by Roberto Ago, Special Rapporteur – the Internationally Wrongful Act of the State, Source of International Responsibility (Continued)*, (1972) Yearbook of the International Law Commission, Volume II, UN Doc. A/CN.4/264 and Add.1, p. 98, para. 67.

97 Commentary to DARS as adopted in 2001, p. 62, Article 14, para. 14.

98 Monica Hakimi, (2010) "State Bystander Responsibility", *The European Journal of International Law*, Volume 21, Issue 2, pp. 371-372.

and human rights law. In the ICJ *Corfu Channel* case,[99] the Court held that Albania was responsible under international law for the incurred damage to British warships struck by mines in Albanian waters and the loss of human life. Albania faced international responsibility given that the minelaying could not have been realized without the knowledge of Albania and that it failed to notify the existence of the minefield and to warn the British warship of the approaching danger. In other words, the ICJ found that Albania did not do everything it could to prevent the disaster and that these "grave omissions" involve the international responsibility of this particular State.[100] In addition to the World Court, the due diligence concept has been frequently used by a large number of international arbitral tribunals assessing the failures on the part of States to prevent or to punish private unlawful conduct and this jurisprudence will be used in the present Chapter to analyze the scope of due diligence obligations of States.

It makes sense to emphasize that to a certain extent all three types of States – hiring, host, and home States – have positive obligations to prevent and to legislate, investigate, prosecute, and punish.[101] These obligations can be found in both international humanitarian and human rights law. When PMSCs engage in unlawful conduct violating international law, all these States involved to a certain degree in their employment and use can in theory face international responsibility for failing to meet their due diligence obligations.

According to Hannah Tonkin, the analysis of the positive State duties needs to be performed in a number of steps.[102] In the first place, primary obligations to prevent and to punish certain misconduct of PMSCs must be identified. Under human rights treaties in particular, State obligations come into play when provisions of these conventions are applicable. In situations when PMSCs operate abroad, it must thus be determined whether human rights instruments apply in an extraterritorial manner to particular States: this will be the case when States exercise effective control over foreign territories and persons, as noted in Chapter III. Secondly, it needs to be determined what degree of State knowledge of illegal private activities is required for the obligation in question to be activated. Some primary obligations may contain requirements of fault that should be identified in the light of those particular obligations.[103] They may provide that a certain level of State knowledge or "constructive

---

[99] International Court of Justice, *Corfu Channel (United Kingdom v. Albania)*, Judgment (Merits), April 9, 1949, ICJ Reports 1949.

[100] Ibidem, p. 23.

[101] University Centre for International Humanitarian Law, *Expert Meeting on Private Military Contractors: Status and State Responsibility for Their Actions*, Report, Geneva, August 29-30, 2005, p. 34, <http://www.geneva-academy.ch/docs/expert-meetings/2005/2rapport_compagnies_privees.pdf>, accessed on October 10, 2014.

[102] Hannah Tonkin, (2011) *State Control over Private Military and Security Companies in Armed Conflict*, Cambridge: Cambridge University Press, p. 64.

[103] Ibidem, p. 67.

knowledge of the prohibited activity" is required for determining whether a State is under an obligation to take a positive action.[104] Thirdly, positive actions that must be taken by States obliged to comply with the due diligence standard are to be identified. The due diligence standard has "undoubtedly an elastic and relative nature"[105] revealing a certain degree of flexibility and subjectivity.[106] On the one hand, States must create tools necessary for the fulfillment of these duties, such as legislative acts, administrative frameworks, and other regulations and policies.[107] On the other hand, these instruments must be used to prevent and to punish. Some crucial objective factors may determine the level of diligence to be exercised by States in a particular case.[108] The first factor is the degree of effectiveness of control exercised by a State over its territory and thus its capacity to have influence on private individuals. The second element proposed by Hannah Tonkin is the availability of State resources to comply with positive obligations. Riccardo Pisillo-Mazzeschi, however, believes that the interest to be protected should prevail as the second variable. The third factor concerns the risks or the degree of predictability that activities of private individuals will cause harm and will result in violations of international law. Fourthly, the focus must be on the causal connection or nexus between the unlawful conduct of PMSCs and the State's non-compliance with its due diligence obligations that must be derived from primary obligations.[109] Finally, the question of what type of damage – moral or material – that must be suffered by the claimant according to primary rules in order to successfully hold a State responsible must be answered.[110]

In the following part, an array of positive obligations of diligent conduct originating from both international humanitarian law and human rights law will be examined. For the purpose of the analysis, the division is made between three types of States – hiring, host, and home States – that can have different positive duties of

104 Ibidem, p. 69.

105 Riccardo Pisillo-Mazzeschi, (1992) "The Due Diligence Rule and the Nature of the International Responsibility of States", *German Yearbook of International Law*, Volume 35, p. 44.

106 Hannah Tonkin, (2011) *State Control over Private Military and Security Companies in Armed Conflict*, Cambridge: Cambridge University Press, p. 69; Riccardo Pisillo-Mazzeschi, (1992) "The Due Diligence Rule and the Nature of the International Responsibility of States", *German Yearbook of International Law*, Volume 35, pp. 44-45.

107 Hannah Tonkin, (2011) *State Control over Private Military and Security Companies in Armed Conflict*, Cambridge: Cambridge University Press, p. 70.

108 Riccardo Pisillo-Mazzeschi, (1992) "The Due Diligence Rule and the Nature of the International Responsibility of States", *German Yearbook of International Law*, Volume 35, p. 44; Hannah Tonkin, (2011) *State Control over Private Military and Security Companies in Armed Conflict*, Cambridge: Cambridge University Press, p. 70.

109 Hannah Tonkin, (2011) *State Control over Private Military and Security Companies in Armed Conflict*, Cambridge: Cambridge University Press, p. 73.

110 Ibidem, p. 75.

diligent conduct under international law depending on their role in the deployment of PMSCs.[111] The approach used to examine these obligations is derived from the structure of the Montreux Document dealing in its Part One with pertinent legal obligations of States in relation to PMSCs. Starting with the most elaborate duties of hiring States, the analysis will shift to host and home States respectively. It is to be noted that States under examination are representatives of one of the three categories of States involved in the deployment of PMSCs. If they belong to two or three categories at the same time, the spectrum of their obligations and responsibilities might be extended.[112] Thus, if a home State of a PMSC is also a hiring State of that corporation, its positive obligations can comprise the duties of home and hiring States. The Montreux Document is also referred to in order to distill obligations of States and suggested good practices and to establish the degree of due diligence they need to exercise.[113]

4.2 Application to hiring States

It can be argued that at the international level a spectrum of international standards exist to hold States responsible for their failure to meet positive obligations of diligent conduct when these States conclude contracts with PMSCs and hire their employees for deployment in conflict zones. These duties are to be derived from the bodies of international humanitarian and human rights law that require States to take positive actions to prevent, legislate, investigate, prosecute, and punish the misconduct of PMSCs active in conflict zones. The main focus will be on the primary rules of international law, which are capable of imposing a due diligence norm on States, rather than the set of secondary rules containing provisions on State responsibility, as laid down in the Draft Articles.[114] These norms will essentially impose on hiring or contracting States a set of obligations to take positive measures in relation to PMSCs

---

[111] Corinna Seiberth, (2013) *Private Military and Security Companies in International Law – A Challenge for Non-binding Norms: the Montreux Document and the International Code of Conduct for Private Security Service Providers*, Cambridge, Antwerp, Portland: Intersentia, p. 94; University Centre for International Humanitarian Law, *Expert Meeting on Private Military Contractors: Status and State Responsibility for Their Actions*, Report, Geneva, August 29-30, 2005, p. 34, <http://www.geneva-academy.ch/docs/expert-meetings/2005/2rapport_compagnies_privees.pdf>, accessed on October 10, 2014.

[112] Francesco Francioni, "The Role of the Home State in Ensuring Compliance with Human Rights by Private Military Contractors", in Francesco Francioni and Natalino Ronzitti (Eds.), (2011) *War by Contract: Human Rights, Humanitarian Law, and Private Contractors*, Oxford: Oxford University Press, pp. 94-95.

[113] Corinna Seiberth, (2013) *Private Military and Security Companies in International Law – A Challenge for Non-binding Norms: the Montreux Document and the International Code of Conduct for Private Security Service Providers*, Cambridge, Antwerp, Portland: Intersentia, pp. 239-240.

[114] Robert p. Barnidge, Jr., (2008) *Non-State Actors and Terrorism: Applying the Law of State Responsibility and the Due Diligence Principle*, The Hague: T.M.C. Asser Press, p. 62.

under international humanitarian and human rights law. These States appear to have the most extensive due diligence obligations, which will be explored in detail below.[115]

4.2.1 Positive State obligations under international humanitarian law

The Montreux Document gives some guidance with regard to the existing humanitarian law obligations of hiring States by restating the pertinent State obligations under this part of international law:

> "Contracting States have an obligation, within their power, to ensure respect for international humanitarian law by PMSCs they contract, in particular to:
>
> a) ensure that PMSCs that they contract and their personnel are aware of their obligations and trained accordingly;
> b) not encourage or assist in, and take appropriate measures to prevent, any violations of international humanitarian law by personnel of PMSCs;
> c) take measures to suppress violations of international humanitarian law committed by the personnel of PMSCs through appropriate means, such as military regulations, administrative orders and other regulatory measures as well as administrative, disciplinary or judicial sanctions, as appropriate."[116]

Common Article 1 of the 1949 Geneva Conventions

In the first place, with regard to international humanitarian law applicable in international armed conflicts it must be observed that there is a general duty of States Parties to the 1949 Geneva Conventions to ensure respect for that body of law "in all circumstances" laid down in Common Article 1 of the 1949 Geneva Conventions and Article 1 of the First Additional Protocol. Common Article 1 of the Geneva Conventions reads:

> "The High Contracting Parties undertake to respect and to ensure respect for the present Convention in all circumstances."

This is arguably the most significant provision of the law of armed conflict in relation to the responsibility of States employing PMSCs and is considered part of customary international law.[117] It contains a notion that is similar to the concept of due diligence

---

[115] Montreux Document, Part One "Pertinent International Legal Obligations Relating to Private Military and Security Companies", pp. 11-12.
[116] Ibidem, p. 11, para. 3.
[117] Louise Doswald-Beck, "Private Military Companies under International Humanitarian Law", in Simon Chesterman and Chia Lehnardt (Eds.), (2007) *From Mercenaries to Market: The Rise and Regulation of Private Military Companies*, Oxford: Oxford University Press, p. 131; Hannah Tonkin, (2009)

under human rights law.[118] Under this provision, States must take all necessary measures in order to ensure that private individuals and groups respect international humanitarian law.[119] The term "to ensure respect" emphasizes the responsibility of States Parties to the 1949 Geneva Conventions[120] and implies that this is a specific duty of conduct and not an obligation of result. The notion "in all circumstances" essentially means that the duty to respect and to ensure respect applies not only to international armed conflicts, but also to non-international armed conflicts when they fall within the scope of Common Article 3.[121] The application of Common Article 1 containing this State obligation is not territorially limited and is applicable to the conduct of PMSCs functioning outside the State territory of States employing them.[122] It means that States hiring PMSCs to perform their activities abroad are bound by the positive obligation of Common Article 1 to the same extent as when these firms operate on their own territory.

Importantly, Common Article 1 of the Geneva Conventions is binding upon all hiring States in all circumstances and not only those contracting States party to an armed conflict.[123] According to the ICRC Commentary:

> "... the Government must of necessity prepare in advance, that is to say in peace-time, the legal, material or other means of ensuring the faithful enforcement of the Convention when the occasion arises. This applies to the respect of each individual State for the Convention, but that is not all: in the event of a Power failing to fulfil its obligations, each of the other Contracting Parties (neutral, allied or enemy) should endeavour to bring it back to an attitude of respect for the Convention."[124]

"Common Article 1: A Minimum Yardstick for Regulating Private Military and Security Companies", *Leiden Journal of International Law*, Volume 22, Issue 4, p. 799.

[118] University Centre for International Humanitarian Law, *Expert Meeting on Private Military Contractors: Status and State Responsibility for Their Actions*, Report, Geneva, August 29-30, 2005, p. 41, <http://www.geneva-academy.ch/docs/expert-meetings/2005/2rapport_compagnies_privees.pdf>, accessed on October 10, 2014.

[119] Lindsey Cameron and Vincent Chetail, (2013) *Privatizing War: Private Military and Security Companies under Public International Law*, Cambridge: Cambridge University Press, p. 246.

[120] Jean S. Pictet, (1958) *Commentary, IV Geneva Convention Relative to the Protection of Civilian Persons in Time of War*, Geneva: International Committee of the Red Cross, p. 16.

[121] Laurence Boisson de Chazournes and Luigi Condorelli, (2000) "Common Article 1 of the Geneva Conventions Revisited: Protecting Collective Interests", *International Review of the Red Cross*, Volume 82, Issue 837, pp. 68-69.

[122] Hannah Tonkin, (2011) *State Control over Private Military and Security Companies in Armed Conflict*, Cambridge: Cambridge University Press, p. 188.

[123] University Centre for International Humanitarian Law, *Expert Meeting on Private Military Contractors: Status and State Responsibility for Their Actions*, Report, Geneva, August 29-30, 2005, p. 41, <http://www.geneva-academy.ch/docs/expert-meetings/2005/2rapport_compagnies_privees.pdf>, accessed on October 10, 2014; Hannah Tonkin, (2011) *State Control over Private Military and Security Companies in Armed Conflict*, Cambridge: Cambridge University Press, pp. 190, 191.

[124] Jean S. Pictet, (1960) *Commentary, III Geneva Convention Relative to the Treatment of Prisoners of War*, Geneva: International Committee of the Red Cross, Article 1, p. 18.

In 2004, the ICJ supported this broad interpretation of Common Article 1 in the *Construction of a Wall* Advisory Opinion by referring to its Advisory Opinion on the *Legality of the Threat or Use of Nuclear Weapons* from 1996. It stated that a large number of rules of international humanitarian law applicable in armed conflicts are "so fundamental to the respect of the human person and 'elementary considerations of humanity' that the Hague and Geneva Conventions have enjoyed a broad accession".[125] The Court appeared to be of the opinion that these fundamental rules and principles of the law of armed conflict are to be perceived as "intransgressible principles of international customary law"[126] and are to be observed by all States, irrespective of whether they have ratified the Conventions in question. Further, it added that:

> "It follows from that provision [Common Article 1] that every State party to that Convention, whether or not it is a party to a specific conflict, is under an obligation to ensure that the requirements of the instruments in question are complied with."[127]

This reasoning implies that States party to an armed conflict, such as the United States in Iraq in 2003, which opt to make use of services provided by PMSCs are under a positive obligation to comply with Common Article 1 of the Geneva Conventions. Additionally, third States that do not participate in an armed conflict and, for instance, hire PMSCs to contribute to a peacekeeping mission or to protect persons and State assets on the territory of other States must take reasonable steps to ensure that these firms act in accordance with international humanitarian law.[128]

An important condition for the application of Common Article 1 has been incorporated into the requirement that States hiring PMSCs must be in a position to influence PMSCs effectively.[129] Only if a State can exercise effective control over these corporations might it be expected to take positive actions to control the behavior of the firms in question and their employees.[130] Thus, international responsibility will not arise if a State did not have control over the actions of a particular PMSC or its control was ineffective. In its *Bosnian Genocide* case, the ICJ looked into the State obligation to prevent genocide indicating that a number of parameters had to be taken

---

125 International Court of Justice, *Legality of the Threat or Use of Nuclear Weapons*, Advisory Opinion, July 8, 1996, ICJ Reports 1996, p. 257, para. 79.

126 Ibidem.

127 International Court of Justice, *Legal Consequences of the Construction of a Wall in the Occupied Palestinian Territory*, Advisory Opinion, July 9, 2004, ICJ Reports 2004, pp. 67-68, para. 158.

128 Jean-Marie Henckaerts and Louise Doswald-Beck, (2005) *Customary International Humanitarian Law, Volume I: Rules*, ICRC, Cambridge: Cambridge University Press, pp. 509-510.

129 Birgit Kessler, (2001) "The Duty to 'Ensure Respect' under Common Article 1 of the Geneva Conventions: Its Implications on International and Non-International Armed Conflicts", *German Yearbook of International Law*, Volume 44, p. 505.

130 Hannah Tonkin, (2011) *State Control over Private Military and Security Companies in Armed Conflict*, Cambridge: Cambridge University Press, p. 189.

into account. The first parameter was one "which varies greatly from one State to another, is clearly the capacity to influence effectively the action of persons likely to commit, or already committing, genocide".[131] Interestingly, this capacity appears to be dependent, *inter alia*, on "the geographical distance of the State concerned from the scene of the events, and on the strength of the political links, as well as links of all other kinds, between the authorities of that State and the main actors in the events".[132] Because States must act in accordance with rules and principles of international law, legal criteria should also be taken into consideration. Therefore, a State's capacity to influence the actions of non-State actors, including PMSCs, also varies in accordance with a particular legal position of a State in question *vis-à-vis* the situations and individuals faced with the danger of violations of the law of armed conflict.[133] The degree of effectiveness of such influence, however, must be assessed on an individual basis in every instance of States resorting to the use of PMSCs. In most cases, hiring States might be said to be capable of controlling the PMSCs they contract and influence their behavior, even if firms are employed to function abroad outside the territory of contracting States. Thus, Common Article 1 appears to impose an obligation upon hiring States to adopt positive measures in order to realize the compliance of PMSCs with international humanitarian law. The same obligation is mentioned in the Montreux Document as a pertinent international legal obligation on States contracting PMSCs to ensure that international humanitarian law is respected by these companies and their employees.[134]

If members of PMSC personnel engage in a violation of international humanitarian law, the hiring State in question will face international responsibility for this particular breach. In this scenario, the State would fail to exercise due diligence towards a PMSC and to control its behavior. It would be unable to comply with its positive duty to take all possible measures within its power to ensure that the law of armed conflict is respected by a PMSC and to prevent its violation. This particular provision, Common Article 1, constitutes an additional possibility to hold States liable for violations of international humanitarian law committed by PMSCs that cannot qualify as State actors acting on behalf of States. It might be used to close a responsibility gap emerging due to many difficulties of incurring State responsibility by means of attributing unlawful activities of PMSCs and other non-State actors to hiring States. Hannah Tonkin sees it as a "minimum standard of conduct for the hiring State"[135] that

[131] International Court of Justice, *Case Concerning the Application of the Convention on the Prevention and Punishment of the Crime of Genocide (Bosnia and Herzegovina v. Serbia and Montenegro)*, Judgment (Merits), February 26, 2007, ICJ Reports 2007, para. 430.

[132] Ibidem.

[133] Ibidem.

[134] Montreux Document, Part One "Pertinent International Legal Obligations Relating to Private Military and Security Companies", p. 11, para. 3.

[135] Hannah Tonkin, (2011) *State Control over Private Military and Security Companies in Armed Conflict*, Cambridge: Cambridge University Press, p. 197.

can prove its utility in the years to come, as all hiring States must act in accordance with its requirements in all circumstances.

State intervention

States have an obligation to take all appropriate measures to prevent private persons and groups from violating international law. In the *Tehran Hostages* case, the ICJ shed some light on the conditions for holding States responsible for the failure to prevent such conduct. The World Court observed that the authorities of Iran:

> "(a) were fully aware of their obligations under the convention in force to take appropriate steps to protect the premises of the United States Embassy and its diplomatic and consular staff from any attack and from any infringement of their inviolability, and to ensure the security of such other persons as might be present on the said premises;
> (b) were fully aware, as a result of the appeals for help made by the United States Embassy, of the urgent need for action on their part;
> (c) had the means at their disposal to perform their obligations;
> (d) completely failed to comply with these obligations."[136]

Thus, when a hiring State is fully aware that there is a need for intervening in case PMSCs are likely to engage in a violation of the law of armed conflict and has all necessary means at its disposal to comply with its obligation to prevent under international law, it must take appropriate steps to comply with this obligation. When international law is breached by PMSCs active on the territory under control of a particular State, it does not mean that the State in question knew or should have known of this breach. The ICJ stated in the *Corfu Channel* case:

> "... it cannot be concluded from the mere fact of the control exercised by a State over its territory and waters that that State necessarily knew, or ought to have known, of any unlawful act perpetrated therein, nor yet that it necessarily knew, or should have known, the authors."[137]

In these scenarios, the knowledge of a State can be derived from "inferences of fact and circumstantial evidence", if they leave "no room for reasonable doubt".[138] Thus, a violation can be said to be predictable when a State knows or should know that it

---

[136] International Court of Justice, *United States Diplomatic and Consular Staff in Tehran (United States of America v. Iran)*, Judgment, May 24, 1980, ICJ Reports 1980, para. 68.
[137] International Court of Justice, *Corfu Channel (United Kingdom v. Albania)*, Judgment (Merits), April 9, 1949, ICJ Reports 1949, p. 18.
[138] Ibidem.

can occur.[139] If a violation is foreseeable in the light of particular circumstances of a case and individuals can fall victim of PMSC activities, hiring States are expected to exercise due diligence and to prevent these acts from materializing.[140]

Certain operations of PMSCs acting on behalf of contracting States and using armed force increase the likelihood of a breach of international humanitarian law by these private actors and therefore States are under an obligation to intervene. This duty becomes even more pertinent when there is more information available to a State regarding the predictability of a breach. It might be the case, for instance, when private contractors participate in a military operation with a high risk of escalation of violence.

Protection of certain groups of people

In international armed conflicts, hiring States have positive duties to protect specific groups of individuals present on the battlefield. States are obliged to take positive measures in order to control and to coordinate the behavior of both their State organs and officials and non-State actors, such as the personnel of PMSCs.[141] Several provisions of the 1949 Geneva Conventions grant protection to certain groups and contain State duties to ensure this protection. Some of these duties can be explored in more detail. Article 12 of the First Geneva Convention regarding members of the armed forces and other persons who are wounded and sick stresses that these individuals must be respected and protected in all circumstances. In this context, "respect" means "to spare, not to attack" while "protect" implies "to come to someone's defence, to lend help and support".[142] This obligation extends to not only combatants, but also civilians, including employees working for PMSCs.[143] Among other things, Article 15 requires States Parties to an international armed conflict to take all possible measures to search for wounded and sick individuals and to collect them. In this regard, States need to protect these persons against pillage and ill-treatment and to ensure an adequate level of care. Under Article 18, civilian populations and relief societies collecting and caring for the wounded and sick must respect them and abstain from using violence against these individuals. Thus, PMSCs hired by States are required to act in accordance with this obligation, while contracting States have a duty to

[139] Lindsey Cameron and Vincent Chetail, (2013) *Privatizing War: Private Military and Security Companies under Public International Law*, Cambridge: Cambridge University Press, p. 257.
[140] Ibidem, pp. 257-258.
[141] Hannah Tonkin, (2011) *State Control over Private Military and Security Companies in Armed Conflict*, Cambridge: Cambridge University Press, p. 137.
[142] Jean S. Pictet, (1952) *Commentary, I Geneva Convention for the Amelioration of the Condition of the Wounded and Sick in Armed Forces in the Field*, Geneva: International Committee of the Red Cross, pp. 134-135.
[143] Ibidem, p. 135.

ensure that they do so.[144] Moreover, Article 19 provides that fixed establishments and mobile medical units may not be attacked and must be respected and protected by States, similar to transport of the wounded, sick, and medical equipment, as provided in Article 35. The Second Geneva Convention contains provisions having the same spirit as those of the First Geneva Convention. Article 12 stipulates that members of the armed forces and other persons who are wounded, sick, or shipwrecked must be protected and respected in all circumstances. Articles 22 and 36 mention the obligation to protect and to respect military hospital ships and their personnel. According to Article 13 of the Third Geneva Convention, prisoners of war need to be protected at all times, especially against acts of violence and intimidation, insults and public curiosity. The Fourth Geneva Convention in Article 16 also emphasizes the duty to protect and to respect the wounded, sick, infirm, and expectant mothers and Articles 18, 20, and 21 regarding civilian hospitals, hospital staff, and land and sea transport of the wounded, sick, infirm, and expectant mothers respectively. Finally, Article 10(1) of the First Additional Protocol requires States to respect and to protect all wounded, sick, and shipwrecked. This duty is also applicable to civilian civil defense organizations and their personnel in accordance with Article 62(1) and members of the armed forces and military units assigned to these civil defense organizations, as provided in Article 67(1).

Furthermore, States also have an important positive obligation under the law of armed conflict to protect the civilian population and several groups of civilians in international armed conflicts, particularly women and children. This specific duty can be found in Article 27 of the Fourth Geneva Convention and Articles 76 and 77 of the First Additional Protocol. The first provision requires that persons protected by the Fourth Geneva Convention must be treated humanely at all times and protected "against all acts of violence or threats thereof and against insults and public curiosity". The second and third provisions refer to two particular protected groups indicating that women need to be protected by States "against rape, forced prostitution and any other form of indecent assault" and children must be protected against "any form of indecent assault". Children must also be provided with any form of care and aid they may need. The obligation to protect civilians is an effective one requiring States that employ PMSCs to exercise due diligence towards these corporations and their employees and to control their activities. States must take positive steps in order to ensure that the civilian population of the country of their operation, particularly women and children, are not threatened by their functioning and are not mistreated.[145]

---

144 Lindsey Cameron and Vincent Chetail, (2013) *Privatizing War: Private Military and Security Companies under Public International Law*, Cambridge: Cambridge University Press, p. 242.

145 University Centre for International Humanitarian Law, *Expert Meeting on Private Military Contractors: Status and State Responsibility for Their Actions*, Report, Geneva, August 29-30, 2005, pp. 35-36, <http://www.geneva-academy.ch/docs/expert-meetings/2005/2rapport_compagnies_privees.pdf>, accessed on October 10, 2014.

This duty, also applicable to off-duty PMSC employees, requires hiring States to regulate the activities of these firms and the exercise of military and security services, to properly vet and train these individuals, to create and apply disciplinary measures, and to instigate criminal proceedings in relation to PMSC personnel engaging in unlawful activities.[146]

In the context of non-international armed conflicts, Article 3 common to the 1949 Geneva Conventions contains no positive duty resting upon States that would require them to protect persons who are considered *hors de combat*.[147] There are, however, positive obligations of States scattered throughout the Second Additional Protocol. Article 7(1) of the Protocol states that the wounded, sick, and shipwrecked must be respected and protected. Article 8 sets out a State duty to take all necessary measures to search for these individuals mentioned in the previous provision, to protect them against pillage and ill-treatment, and to ensure an adequate care for them. In addition, Article 9(1) and Article 11(1) provide that medical and religious personnel, medical units, and transport must be respected and protected.

The above analysis illustrates that States that are parties to armed conflicts have a range of positive duties with regard to PMSCs they employ. They need to respect and to protect a large number of groups of persons from possible instances of PMSC misconduct and are thus required to take positive measures to comply with their obligations under the law of armed conflict.

Suppressing and repressing violations of the law of armed conflict

The next positive obligation under the law of armed conflict with respect to the employment of PMSCs is that of suppressing and repressing violations of international humanitarian law in international armed conflicts imposed by the 1949 Geneva Conventions. While the duty to prevent violations of the law of armed conflict may not be completely clear, the obligations to suppress and to repress pose few difficulties.[148] The Montreux Document can be used to clarify further positive obligations of hiring States. It provides that contracting States are under an obligation to ensure

---

146 Carsten Hoppe, (2008) "Passing the Buck: State Responsibility for Private Military Companies", *The European Journal of International Law*, Volume 19, Issue 5, p. 993; Hannah Tonkin, (2011) *State Control over Private Military and Security Companies in Armed Conflict*, Cambridge: Cambridge University Press, p. 200.

147 Lindsey Cameron and Vincent Chetail, (2013) *Privatizing War: Private Military and Security Companies under Public International Law*, Cambridge: Cambridge University Press, pp. 243-244; Hannah Tonkin, (2011) *State Control over Private Military and Security Companies in Armed Conflict*, Cambridge: Cambridge University Press, p. 139.

148 University Centre for International Humanitarian Law, *Expert Meeting on Private Military Contractors: Status and State Responsibility for Their Actions*, Report, Geneva, August 29-30, 2005, p. 45, <http://www.geneva-academy.ch/docs/expert-meetings/2005/2rapport_compagnies_privees.pdf>, accessed on October 10, 2014.

respect for the law of armed conflict by taking a variety of appropriate measures meant to suppress violations of international humanitarian law.[149] These measures can be "military regulations, administrative orders and other regulatory measures as well as administrative, disciplinary or judicial sanctions, as appropriate".

In non-international armed conflicts, an obligation to suppress can be derived from Common Article 1 to the Geneva Conventions ensuring respect for international humanitarian law.[150] In the situations of international armed conflicts, this obligation is to be derived from the provisions of the Geneva Conventions and Additional Protocols. More specifically, the duty to repress related to the commission of grave breaches of the Geneva Conventions is laid down in Article 49(1) of the First Geneva Convention, Article 50(1) of the Second Geneva Convention, Article 129(1) of the Third Geneva Convention, and Article 146(1) of the Fourth Geneva Convention. The grave breaches in question involve the following crimes committed against individuals or property protected by the Geneva law: willful killing, torture or inhuman treatment, willfully causing great suffering or serious injury to body or health, and other acts.[151] Under these provisions, States are obliged to enact all possible legislative acts for the provision of effective legislative sanctions against PMSCs committing, or ordering to be committed, grave breaches described above. In addition, States must take measures to find individuals alleged to have committed grave breaches and to bring them to justice before their own courts or to hand them over to other States having a *prima facie* case against them. Also, States have a duty to take all necessary measures to suppress violations of the law of armed conflict other than grave breaches and to prevent them from being materialized or repeated. These breaches are mentioned in Articles 49(3) of the First Geneva Convention, Article 50(3) of the Second Geneva Convention, Article 129(3) of the Third Geneva Convention, and Article 146(3) of the Fourth Geneva Convention.[152] It also implies a duty not only to criminalize other breaches of the Conventions in the national legal system,[153] but also to suppress them by taking other positive steps, such as properly vetting and training PMSC staff, monitoring their activities in the field, and creating contractual clauses for imposing penalties for breaches of international humanitarian law.[154]

---

149 Montreux Document, Part Two "Good Practices Relating to Private Military and Security Companies", p. 11, para. 3(c).

150 Hannah Tonkin, (2011) *State Control over Private Military and Security Companies in Armed Conflict*, Cambridge: Cambridge University Press, p. 201.

151 Article 50 of the First Geneva Convention, Article 51 of the Second Geneva Convention, Article 130 of the Third Geneva Convention, and Article 147 of the Fourth Geneva Convention.

152 Jean S. Pictet, (1958) *Commentary, IV Geneva Convention Relative to the Protection of Civilian Persons in Time of War*, Geneva: International Committee of the Red Cross, Article 146, p. 594.

153 Ibidem.

154 Hannah Tonkin, (2011) *State Control over Private Military and Security Companies in Armed Conflict*, Cambridge: Cambridge University Press, p. 201.

Obligations of occupying powers

The last type of positive obligation from the field of the law of armed conflict relates to the duties of hiring States in situations of military occupation when they qualify as occupying powers. In this context, Article 43 of the Hague Convention IV provides:

> "The authority of the legitimate power having in fact passed into the hands of the occupant, the latter shall take all the measures in his power to restore, and ensure, as far as possible, public order and safety, while respecting, unless absolutely prevented, the laws in force in the country."

The due diligence standard to be used in this category of situations has been looked at by the ICJ in the *Case Concerning Armed Activities on the Territory of the Congo* from 2005.[155] The Court concluded that Uganda was to be regarded as an occupying power in Ituri and therefore under an obligation to comply with Article 43 of the Hague Convention IV, which entails "the duty to secure respect for the applicable rules of international human rights law and international humanitarian law, to protect the inhabitants of the occupied territory against acts of violence, and not to tolerate such violence by any third party".[156] According to the ICJ, Uganda's international responsibility was engaged because its armed forces violated international obligations of this State and due to the lack of vigilance in preventing breaches of international humanitarian law and human rights on the occupied territory realized by other actors active on the occupied territory, such as rebel groups operating on their own.[157] In other words, States are obliged to exercise vigilance or due diligence in the prevention of violations of the law of armed conflict and human rights that might be committed by third actors, such as PMSCs. Notably, Part One of the Montreux Document on pertinent international legal obligations of States with regard to their employment of PMSCs contains a provision confirming this particular legal duty. It states:

> "If they [contracting States] are occupying powers, they have an obligation to take all measures in their power to restore, and ensure, as far as possible, public order and safety, i.e. exercise *vigilance* in preventing violations of international humanitarian law and human rights law."[158] (emphasis added)

---

[155] International Court of Justice, *Case Concerning Armed Activities on the Territory of the Congo (Democratic Republic of the Congo v. Uganda)*, Judgment (Merits), December 19, 2005, ICJ Reports 2005.

[156] Ibidem, para. 178.

[157] Ibidem, para. 179.

[158] Montreux Document, Part One "Pertinent International Legal Obligations Relating to Private Military and Security Companies", p. 11, para. 1.

Therefore, a conclusion may be drawn that both Article 43 of the Hague Convention IV and Common Article 1 impose a duty on the hiring States that can be considered occupying powers to always exercise due diligence towards PMSCs they employ. This obligation concerns not only PMSCs hired by the occupying powers, but also all other private contractors active within the occupied territory.[159] Failure to comply with the due diligence standard by the occupying State will necessarily trigger international responsibility of the State concerned. This particular obligation on contracting States acting as occupying powers laid down in Article 43 concerns a duty to ensure that contractors of PMSCs providing coercive services are not left unsupervised when they are off-duty.[160] Therefore, States are required to ensure that there are monitoring instruments in place to control the conduct of off-duty personnel of PMSCs. They are also obliged to make sure that violations of international humanitarian law are reported and the perpetrators are subjected to disciplinary, judicial, and other sanctions.[161]

Lindsey Cameron and Vincent Chetail note that, on the one hand, the employment of PMSCs in occupied territories can lead to international responsibility for the failure to exercise due diligence when these companies and their personnel engage in various types of unlawful conduct.[162] This use of private contractors can be counterproductive in ensuring respect for international humanitarian and human rights law, granting protection to the local population of the occupied territories from possible acts of violence, and not tolerating violence by third parties. On the other hand, hiring States might claim that PMSCs are employed in order to implement their obligations following from Article 43 of the Hague Convention IV.[163] The deployment of PMSCs in the role of security guards and other functions, like those involving reconstruction work, is necessary for restoring and ensuring public order and safety and is better suited to achieving these goals than the use of regular armed forces.

Specific positive measures to comply with positive obligations

Thus, what do contracting States have to do to meet the above-mentioned obligations? From the more general obligation of Common Article 1, one can arrive at various specific positive duties of hiring States under the law of armed conflict.[164] With this

---

[159] Lindsey Cameron and Vincent Chetail, (2013) *Privatizing War: Private Military and Security Companies under Public International Law*, Cambridge: Cambridge University Press, p. 240.

[160] Carsten Hoppe, (2008) "Passing the Buck: State Responsibility for Private Military Companies", *The European Journal of International Law*, Volume 19, Issue 5, p. 994.

[161] Ibidem, pp. 993-994.

[162] Lindsey Cameron and Vincent Chetail, (2013) *Privatizing War: Private Military and Security Companies under Public International Law*, Cambridge: Cambridge University Press, p. 239.

[163] Ibidem.

[164] Carsten Hoppe, (2008) "Passing the Buck: State Responsibility for Private Military Companies", *The European Journal of International Law*, Volume 19, Issue 5, p. 994.

in mind, they must carefully choose which PMSCs they employ and have appropriate procedures for selection and contracting.[165] The main hiring and home State, the US, has, for example, a complex system for contracting and selecting PMSCs that aims at addressing a variety of issues of the industry.[166] Additionally, States need to have clear contracts with contractual clauses and requirements ensuring respect for the law of armed conflict and establishing the rules of engagement in the same manner.[167] The contract is in this context perhaps the most effective instrument.[168] Moreover, hiring States have a duty not to contract PMSCs for the performance of tasks that according to the law of armed conflict should be assigned to State agents or authorities.[169] They must train contractors and educate them on matters of international humanitarian law.[170] Hiring States can be considered to have a duty to set up a regulatory framework addressing the PMSCs they employ.[171] These States must create monitoring mechanisms to control the conduct of off-duty employees of PMSCs. For example, the system of monitoring PMSC conduct in the US is an elaborate one and is based on many recommendations made in the Montreux Document.[172] There is also a need to have reporting instruments and procedures and the policies by which the perpetrators are subject to disciplinary, judicial, and other sanctions.[173]

4.2.2 Positive State obligations under international human rights law

State obligations under human rights law originate from various conventions. Some human rights laid down in those treaties are likely to be violated by PMSCs active in areas affected by conflicts: those are the right to life and the prohibition of torture. Of course, violations of these rights are the most serious breaches of human rights law that can be committed by PMSCs providing coercive services, however, it is also

---

165 Montreux Document, Part Two "Good Practices Relating to Private Military and Security Companies", pp. 17-18, paras. 2-13.

166 Rebecca DeWinter-Schmitt (Ed.), *Montreux Five Years On: An Analysis of State Efforts to Implement Montreux Document Legal Obligations and Good Practices*, Report, 2013, pp. 8, 56-57.

167 Montreux Document, Part Two "Good Practices Relating to Private Military and Security Companies", pp. 18-19, paras. 14-18.

168 Michael Cottier, (2006) "Elements for Contracting and Regulating Private Security and Military Companies", *International Review of the Red Cross*, Volume 88, Issue 863, pp. 638-640.

169 Montreux Document, Part One "Pertinent International Legal Obligations Relating to Private Military and Security Companies", p. 11, para. 2.

170 University Centre for International Humanitarian Law, *Expert Meeting on Private Military Contractors: Status and State Responsibility for Their Actions*, Report, Geneva, August 29-30, 2005, p. 41, <http://www.geneva-academy.ch/docs/expert-meetings/2005/2rapport_compagnies_privees.pdf>, accessed on October 10, 2014.

171 Ibidem, p. 44.

172 Rebecca DeWinter-Schmitt (Ed.), *Montreux Five Years On: An Analysis of State Efforts to Implement Montreux Document Legal Obligations and Good Practices*, Report, 2013, pp. 8, 76-79.

173 Carsten Hoppe, (2008) "Passing the Buck: State Responsibility for Private Military Companies", *The European Journal of International Law*, Volume 19, Issue 5, pp. 993-994.

possible that private contractors violate other human rights and fundamental freedoms. Importantly for the practice of deploying PMSCs in conflict-affected zones, the due diligence obligations under international human rights law have a clear extraterritorial dimension, meaning that States can be held responsible for the extraterritorial activities of private corporations taking place beyond their borders.[174] However, "envisaging the existence of a state obligation to prevent and to punish corporate human rights violations committed abroad is not uncontroversial."[175] Hiring States can be regarded as having both the duty to prevent and the duty to punish under international human rights law. According to the set of international legal duties referred to in the Montreux Document:

> "Contracting States are responsible to implement their obligations under international human rights law, including by adopting such legislative and other measures as may be necessary to give effect to these obligations. To this end they have the obligation, in specific circumstances, to take appropriate measures to prevent, investigate and provide effective remedies for relevant misconduct of PMSCs and their personnel."[176]

4.2.2.1 Obligation to prevent

At this stage, an answer should be found to the question as to which specific positive duties applicable to hiring States can be derived from the general obligation to prevent. In order to comply with the duty to prevent, contracting States have additional duties they have to comply with and are required to take certain measures.

State intervention

First, States must employ all appropriate means to prevent private persons and groups from violating human rights law. With regard to the right to life, the UN Human Rights Committee provided that States Parties to the Covenant cannot ignore known threats to the life and personal security of non-detained individuals under their jurisdiction.[177]

---

174 Robert McCorquodale and Penelope Simons, (2007) "Responsibility Beyond Borders: State Responsibility for Extraterritorial Violations by Corporations of International Human Rights Law", *Modern Law Review*, Volume 70, Issue 4, pp. 618-619.

175 Nadia Bernaz, (2013) "Enhancing Corporate Accountability for Human Rights Violations: Is Extraterritoriality the Magic Potion?", *Journal of Business Ethics*, Volume 117, Issue 3, p. 503. A view that international human rights law does impose these positive obligations on States with regard to the conduct of non-State actors, such as PMSCs, can be found in Cedric M.J. Ryngaert and Jean D'Aspermont, *Third Report on the Obligations and Responsibility of Non-State Actors in International Law*, April 2014, International Law Association, Committee on Non-State Actors, pp. 21-22.

176 Montreux Document, Part One "Pertinent International Legal Obligations Relating to Private Military and Security Companies", p. 11, para. 4.

177 UN Human Rights Committee, *William Eduardo Delgado Páez v. Colombia*, July 12, 1990, Communication No. 195/1985. UN Doc. CCPR/C/39/D/195/1985 (1990), para. 5.5.

As indicated in relation to the same obligation under the law of armed conflict, a violation is foreseeable in the light of particular circumstances of a case when a hiring State knows or should know that it is likely to occur, meaning that human rights of individuals can be breached as the result of PMSC activities.[178] In these cases, hiring States are required to exercise due diligence and to intervene in order to make sure that a breach does not take place. This duty can be complied with by means of the control exercised by hiring States and other positive measures, such as inquiries into the functioning of particular companies. In general, States can be said to be under an obligation to take positive measures, for example, intervention, in order to avoid recurrence of human rights violations committed by possible sources of danger and harm.[179] Impunity for various breaches of international law may be an important characteristic in the recurrence of breaches that might be committed by PMSCs, such as torture.[180] In addition, States must aim to control individuals that may be dangerous to States and their population.[181] Similar to the UN Human Rights Committee,[182] the ECtHR noted:

> "Article 2 may also imply in certain well-defined circumstances a positive obligation on the authorities to take preventive operational measures to protect an individual whose life is at risk from the criminal acts of another individual."[183]

It is obvious that when PMSCs provide a variety of coercive services they are in a position to violate the most significant human rights, for instance, the right to life, and to engage in the commission of torture or other forms of related treatment. Coercive services always have "an element of compelling individuals or groups by force or authority" and include active combat participation, provision of security and protection to persons and property, detention, and other activities.[184] It might be evident that

---

178 Lindsey Cameron and Vincent Chetail, (2013) *Privatizing War: Private Military and Security Companies under Public International Law*, Cambridge: Cambridge University Press, pp. 257-258.

179 UN Human Rights Committee, *General Comment 31*, May 26, 2004, UN Doc. CCPR/C/21/Rev.1/Add.13, para. 17.

180 Ibidem, para. 18.

181 European Court of Human Rights, *Mastromatteo v. Italy*, Judgment, October 24, 2002, Application No. 37703/97.

182 Christine Bakker, "Duties to Prevent, Investigate, and Redress Human Rights Violations by Private Military and Security Companies: The Role of the Host State", in Francesco Francioni and Natalino Ronzitti (Eds.), (2011) *War by Contract: Human Rights, Humanitarian Law, and Private Contractors*, Oxford: Oxford University Press, pp. 137-138.

183 European Court of Human Rights, *Mastromatteo v. Italy*, Judgment, October 24, 2002, Application No. 37703/97, para. 67.

184 Carsten Hoppe, "Positive Human Rights Obligations of the Hiring State in Connection with the Provision of 'Coercive Services' by a Private Military or Security Company", in Francesco Francioni and Natalino Ronzitti (Eds.), (2011) *War by Contract: Human Rights, Humanitarian Law, and Private Contractors*, Oxford: Oxford University Press, p. 111.

PMSCs and their employees exercising military and security services and having the potential to use violence and to cause death of persons must be considered as entities and persons that hiring States need to exercise control over in order to prevent human rights violations. When a State has knowledge that the engagement of PMSCs in certain missions would likely lead to the use of force or cause human rights violations, it has to intervene and to take positive measures to prevent these illegalities.[185] Under the ICCPR, this duty of intervention is crucial when a State knows about possible threats to the life of individuals under its jurisdiction posed by PMSCs and other actors.[186] In the words of the Human Rights Committee, States cannot:

> "… ignore known threats to the life of persons under their jurisdiction, just because that he or she is not arrested or otherwise detained. States parties are under an obligation to take *reasonable* and *appropriate measures to protect* them."[187] (emphasis added)

The European Commission of Human Rights also found that States need to take preventive measures which must seek to protect a person's right to life.[188] The ECtHR stated that Article 2 ECHR may imply:

> "… in certain well-defined circumstances a positive obligation on the authorities to take *preventive operational measures* to protect an individual whose life is at risk from the criminal acts of another individual."[189] (emphasis added)

In this regard, when a State allegedly fails to comply with its positive obligations to protect the right to life and to act in accordance with its duty to prevent by omitting to protect persons against possible offenses of PMSCs and other individuals, it must be determined:

---

185 University Centre for International Humanitarian Law, *Expert Meeting on Private Military Contractors: Status and State Responsibility for Their Actions*, Report, Geneva, August 29-30, 2005, p. 37, <http://www.geneva-academy.ch/docs/expert-meetings/2005/2rapport_compagnies_privees.pdf>, accessed on October 10, 2014.

186 Carsten Hoppe, "Positive Human Rights Obligations of the Hiring State in Connection with the Provision of 'Coercive Services' by a Private Military or Security Company", in Francesco Francioni and Natalino Ronzitti (Eds.), (2011) *War by Contract: Human Rights, Humanitarian Law, and Private Contractors*, Oxford: Oxford University Press, p. 112.

187 UN Human Rights Committee, *William Eduardo Delgado Páez v. Colombia*, July 12, 1990, Communication No. 195/1985, UN Doc. CCPR/C/39/D/195/1985 (1990), para. 5.5.

188 European Commission of Human Rights, *W v. United Kingdom*, Judgment, February 28, 1983, Application No. 9348/81, Series A, No. 121; European Court of Human Rights, *McCann and Others v. United Kingdom*, Judgment, September 27, 1995, Application No. 18984/91, Series A, No. 324.

189 European Court of Human Rights, *Osman v. United Kingdom*, Judgment, October 28, 1998, Application No. 23452/94, ECHR Reports 1998-VIII, para. 115; see also European Court of Human Rights, *Kiliç v. Turkey*, Judgment, March 28, 2000, Application No. 22492/93, ECHR Reports 2000-III, para. 62.

> "... that the authorities *knew or ought to have known* at the time of the existence of *a real and immediate risk* to the life of *an identified individual or individuals* from the criminal acts of a third party and that they failed to take measures within the scope of their powers which, judged reasonably, might have been expected to avoid that risk."[190] (emphasis added)

Importantly, a State is under an obligation to intervene when there is an identifiable victim or a group of victims who are threatened by PMSCs as third parties, as otherwise this obligation might impose an impossible burden on States to comply with.[191] Therefore, States are not obligated to prevent possible violations of the right to life of civilian population in general when PMSC employees are off duty.[192]

The IACtHR had a similar standpoint when it indicated that States failing to exercise due diligence necessary to prevent violations of the Convention by means of taking all reasonable measures are regarded as breaching positive obligations of the Convention to prevent.[193] When a State is aware that there is a real and imminent risk or danger posed to the right to life of certain persons or groups of persons and it is possible to prevent or to avoid these, the State has a duty to protect those individuals and to prevent a breach of the human right to life. The IACtHR emphasized that State obligations to prevent and to protect persons in their relationships with each other "are conditioned by the awareness of a situation of real and imminent danger for a specific individual or group of individuals and to the reasonable possibilities of preventing or avoiding that danger".[194] Thus, when PMSCs carry out coercive tasks and the hiring States have knowledge of a real and imminent risk to the life of civilian population, they have to intervene and to prevent possible misconduct of private contractors and their personnel.[195]

In line with the reasoning of the IACtHR and the ECtHR, the African Commission maintained that States Parties to the ACHPR are obliged to ensure human rights

---

[190] Ibidem, para. 116; ibidem, para. 63.

[191] European Court of Human Rights, *Mahmut Kaya v. Turkey*, Judgment, March 28, 2000, Application No. 22535/93, ECHR Reports 2000-III, para. 86.

[192] Carsten Hoppe, "Positive Human Rights Obligations of the Hiring State in Connection with the Provision of 'Coercive Services' by a Private Military or Security Company", in Francesco Francioni and Natalino Ronzitti (Eds.), (2011) *War by Contract: Human Rights, Humanitarian Law, and Private Contractors*, Oxford: Oxford University Press, p. 116.

[193] Inter-American Court of Human Rights, *Velasquez Rodriguez v. Honduras*, Judgment, July 29, 1988, Series C, No. 4 (1988), paras. 172, 175-177.

[194] Inter-American Court of Human Rights, *Pueblo Bello Massacre v. Colombia*, Judgment, January 31, 2006, Series C, No. 140 (2006), para. 123; see also Inter-American Court of Human Rights, *Valle Jaramillo et al. v. Colombia*, Judgment, November 27, 2008, Series C, No. 192 (2008), para. 78.

[195] Carsten Hoppe, "Positive Human Rights Obligations of the Hiring State in Connection with the Provision of 'Coercive Services' by a Private Military or Security Company", in Francesco Francioni and Natalino Ronzitti (Eds.), (2011) *War by Contract: Human Rights, Humanitarian Law, and Private Contractors*, Oxford: Oxford University Press, p. 114.

laid down in the Charter.[196] International responsibility can be triggered when they do not comply with the standard of due diligence by failing to prevent human rights violations.[197] The extent of this responsibility should be considered on a case-by-case basis.[198] States have a duty to ensure that third parties do not interfere with human rights of individuals under their jurisdiction[199] and need to protect their citizens from damage and harm emanating from private persons by taking positive actions for the fulfillment of this obligation.[200] This "minimum conduct" is to be expected from governments.[201] If a State neglects this duty, it will be in violation of this particular human rights treaty even if the State itself or its agents have not caused the misconduct. Addressing the situation in Chad, the Commission underlined the idea that a State has "a responsibility to secure the safety and the liberty of its citizens" and must intervene to prevent serious human rights violations, for example, assassination of individuals or groups.[202]

In addition to PMSC employees using force and discharging their weapons, human rights violations might also take place when private contractors are employed to provide detention and interrogation services and might be expected to commit torture and other related acts according to the information available to a State. There is a State duty to protect the right to life and physical integrity of detainees. When a hiring State performs tasks of a detaining power and uses PMSCs to carry out detention services and other related activities, such as the two PMSCs Titan and CACI at the Abu Ghraib prison, it is under an obligation to take positive steps to prevent detainees being killed or injured by the representatives of the State apparatus, PMSCs, or other detainees. A contracting State running detention facilities must have knowledge of possible violations of the right to life or to humane treatment under detention.[203] States remain responsible for detention establishments and have a duty to guarantee these important rights to prisoners. In this regard, a hiring State being aware of PMSCs posing threat to human rights in the area of their operation must exercise control over them and take other appropriate measures to ensure that these private contractors do not abuse detainees and detainees are protected from possible abuses

---

196 African Commission on Human and Peoples' Rights, *Commission Nationale des Droits de l'Homme et des Libertes v. Chad*, Communication No. 74/92 (1995), para. 20.

197 African Commission on Human and Peoples' Rights, *Zimbabwe Human Rights NGO Forum v. Zimbabwe*, May 15, 2006, Communication No. 245/02 (2006), paras. 143, 147.

198 Ibidem, para. 155.

199 Ibidem, para. 154.

200 African Commission on Human and Peoples' Rights, *The Social and Economic Rights Action Center and the Center for Economic and Social Rights v. Nigeria*, Communication No. 155/96 (2001), para. 57.

201 Ibidem, para. 58.

202 African Commission on Human and Peoples' Rights, *Commission Nationale des Droits de l'Homme et des Libertes v. Chad*, Communication No. 74/92 (1995), para. 22.

203 Inter-American Court of Human Rights, *Neira Alegria et al. v. Peru*, Judgment, January 19, 1995, Series C, No. 21 (1995), para. 60.

by other detainees. In the General Comment 20, the Human Rights Committee specifically notes that States are required to take positive measures in the context of their duty to prevent in order to ensure that torture and other forms of related treatment do not take place. It specifically refers to a variety of "legislative, administrative, judicial and other measures... to prevent",[204] which must be taken by State authorities. The ECtHR also looked into the prohibition of Article 3 ECHR and declared that States need to ensure that the right of life of detainees is guaranteed and that they are responsible for the actions of both own State organs and third persons.[205] This duty can be derived from Article 3 read in conjunction with Article 1 and includes a positive obligation to protect individuals from violations of the right not to be tortured by PMSCs and other non-State actors.[206] The IACtHR also took a position that there is an obligation on the part of States to prevent "any situations... that might lead, by action or omission, to suppression of inviolability of the right to life" of detainees by taking positive measures.[207] Finally, Article 2(2) CAT requires States to take effective measures in order to prevent torture in any territory under their jurisdiction.

In conclusion, a hiring State is obliged to adopt certain measures in order to comply with its duty to prevent and to intervene when there is a likelihood of a human rights violation. In this respect, it is obliged to assess the situation of the PMSC operation on the ground and to take positive steps, including control over PMSCs, that must be even more far-reaching when particular private contractors are known for their violent behavior and there were instances of human rights breaches and occurrence of illegalities in the past.

As argued by Hannah Tonkin, in general hiring States must employ all possible means to control and to carefully plan the operations of PMSCs in order to minimize the risk to human life and eventually to eliminate that risk.[208] This is a structural necessity in the context of military and other operations that involve the use of force, which might be used against individuals and cause death. In the ECtHR case *McCann and Others v. UK*, the Court emphasized a State obligation to plan and to control operations carried out by State security forces by connecting it to the States' duty to protect life ex Article 2(1) of the European Convention. The case dealt

---

[204] UN Human Rights Committee, *General Comment 20 Replacing General Comment 07*, UN Doc. HRI/GEN/1/Rev.1 (1994), para. 8.

[205] European Court of Human Rights, *Gezici v. Turkey*, Judgment, March 17, 2005, Application No. 34594/97, paras. 49-54.

[206] Carsten Hoppe, "Positive Human Rights Obligations of the Hiring State in Connection with the Provision of 'Coercive Services' by a Private Military or Security Company", in Francesco Francioni and Natalino Ronzitti (Eds.), (2011) *War by Contract: Human Rights, Humanitarian Law, and Private Contractors*, Oxford: Oxford University Press, p. 117.

[207] Inter-American Court of Human Rights, *Juan Humberto Sánchez v. Honduras*, Judgment, June 7, 2003, Series C, No. 99 (2003), para. 111.

[208] Hannah Tonkin, (2011) *State Control over Private Military and Security Companies in Armed Conflict*, Cambridge: Cambridge University Press, pp. 216-217.

with an anti-terrorist operation carried out by the British forces in Gibraltar during which a number of terrorist suspects were killed. The Court was of the opinion that it had to consider "whether the anti-terrorist operation was planned and controlled by the authorities so as to minimize, to the greatest extent possible, recourse to lethal force".[209] In this regard, it emphasized the necessity of considering "not only the actions of the agents of the State who actually administer the force but also all the surrounding circumstances including such matters as the planning and control of the actions under examination".[210] In this case, it was also important to establish whether the violence used by the State forces was "strictly proportionate" to the main aim of conducting the military operation, which is to protect persons from unlawful violence.[211] In addition, the question regarding the proportionality of the employed measures, such as the use of force, to the aims of the operation was also raised in the case law of the IACtHR and the Human Rights Committee.[212]

The case of *McCann and Others v. UK* concerned the conduct of State security forces acting as State organs whose actions and omissions are directly attributable to the State. Another case, *Andronicou and Constantinou v. Cyprus*, contained references to the matters of planning and control exercised by the State over operations of the officers of a special police unit, but also concerned the conduct of State organs.[213] Therefore, the reasoning used in these judgments cannot directly be used with regard to the private conduct of PMSCs, as indicated by Carsten Hoppe.[214] In the case *Avsar v. Turkey*, the ECtHR dealt with the situation of civilian volunteers providing quasi-police functions with the official security forces of Turkey.[215] This scenario might be closely related to the employment of PMSCs for the provision of services of similar nature. The Court relied on its own reasoning from the *Ergi v. Turkey* case from 1998 to indicate that the international responsibility of States might be incurred if States cannot ensure the right to life of persons failing to comply with their obligation

---

209 European Court of Human Rights, *McCann and Others v. United Kingdom*, Judgment, September 27, 1995, Application No. 18984/91, Series A, No. 324, para. 194.
210 Ibidem, para. 150.
211 Ibidem, para. 194.
212 Inter-American Court of Human Rights, *Neira Alegria et al. v. Peru*, Judgment, January 19, 1995, Series C, No. 21 (1995); UN Human Rights Committee, *Suarez de Guerrero v. Colombia*, March 31, 1982, UN Doc. Supp. No. 40 (A/37/40).
213 European Court of Human Rights, *Andronicou and Constantinou v. Cyprus*, Judgment, October 9, 1997, Application No. 25052/94, ECHR 1997-VI, para. 171.
214 Carsten Hoppe, "Positive Human Rights Obligations of the Hiring State in Connection with the Provision of 'Coercive Services' by a Private Military or Security Company", in Francesco Francioni and Natalino Ronzitti (Eds.), (2011) *War by Contract: Human Rights, Humanitarian Law, and Private Contractors*, Oxford: Oxford University Press, p. 116.
215 European Court of Human Rights, *Avsar v. Turkey*, Judgment, July 10, 2001, Application No. 25657/94.

to plan and to control security operations.[216] This might be the case even if the loss of life cannot be attributed to the State in question. In the *Avsar v. Turkey* decision, the ECtHR observed that Turkey did not manage to provide proper training to volunteers who were active in a variety of official operations, and to ensure that they would be subjected to discipline measures, which constituted a violation of the Convention's right to life.

Moreover, hiring States must intervene and take positive measures in order to offer protection to women and children in areas of conflict, more specifically armed conflicts and peace operations, because these two vulnerable groups of people are usually in need of protection in these particularly difficult situations. The abuses that might be committed by PMSCs, just as by regular armed forces in conflict zones, concern sexual activities aimed at women and children. In this regard, women are protected by Article 3 CEDAW obliging States to protect women and stating that:

> "States Parties shall take in all fields, in particular in the political, social, economic and cultural fields, all appropriate measures, including legislation, to ensure the full development and advancement of women, for the purpose of guaranteeing them the exercise and enjoyment of human rights and fundamental freedoms on a basis of equality with men."

Article 1 of the CEDAW provides a definition of the term "discrimination against women" that includes gender-based violence, which is directed against women because they belong to this particular group and disproportionally affects them.[217] Gender-based violence concerns all possible acts that "inflict physical, mental or sexual harm or suffering, threats of such acts, coercion and other deprivations of liberty".[218] Importantly for the operation of PMSCs, Article 2 CEDAW stipulates that States Parties to the Convention condemn discrimination against women and agree to immediately adopt appropriate means in order to pursue a policy of eliminating this form of discrimination. To realize this goal, according to Article 2(e) of the CEDAW they must, *inter alia*, take all appropriate measures to combat discrimination against women by "any person, organization or enterprise". Thus, States have to act to protect women from all possible assaults by non-State actors, including PMSCs. The Committee on the Elimination of Discrimination Against Women underlines the idea of the due diligence obligations of States in this respect and proclaims that:

> "Under general international law and specific human rights covenants, States may also be responsible for *private acts* if they fail to act with *due diligence* to prevent violations

[216] European Court of Human Rights, *Ergi v. Turkey*, Judgment, July 28, 1998, Application No. 23818/94, ECHR 1998-IV, No. 81.

[217] UN Committee on the Elimination of Discrimination Against Women, *General Recommendation 19*, January 30, 1992, UN Doc. A/47/38, para. 6.

[218] Ibidem.

of rights or to investigate and punish acts of violence, and for providing compensation."[219] (emphasis added)

In addition to women, children are also protected by Article 2 of the CRC, which provides that States Parties to the Convention must respect and ensure the rights laid down herein and are under an obligation to employ all appropriate means in order to guarantee that children are protected against all forms of discrimination or punishment on the basis of the status, beliefs, opinions, and activities of their parents, family members, or legal guardians. More concrete duties can be found in Article 35 of the Convention emphasizing that States need to adopt all appropriate measures to prevent the abduction, sale, and trafficking of children. Crucial to the reality of PMSC operation and the history of the crimes committed by them in the past is, however, Article 34 stating the following:

"States Parties undertake to protect the child from all forms of sexual exploitation and sexual abuse. For these purposes, States Parties shall in particular take all appropriate national, bilateral and multilateral measures to prevent:

a) The inducement or coercion of a child to engage in any unlawful sexual activity;
b) The exploitative use of children in prostitution or other unlawful sexual practices;
c) The exploitative use of children in pornographic performance and materials."

The assessed provisions of the two Conventions relating to the protection of women and children clearly impose strong due diligence obligations on hiring States to ensure that these groups are not discriminated against, abused, or sexually exploited.

Preventive regulation

The Montreux Document refers to an obligation imposed on hiring States to regulate the activities of PMSCs and to protect against their misconduct. It specifically mentions that these States are free to adopt a variety of measures that might include legislative and other measures.[220] It is important to note that the duty to regulate is common to all States that are required to prevent human rights abuses by means of taking certain measures. This is an obligation of result, which is essential in the context of the duty to prevent. It seeks to ensure that States effectively employ regulatory and other tools to have a regulatory framework in place for the prevention of human rights abuses.

Article 2(2) ICCPR contains an obligation on States Parties to adopt positive measures in order to comply with their duties under the Covenant. It reads as follows:

---

219 Ibidem, para. 9.

220 Montreux Document, Part One "Pertinent International Legal Obligations Relating to Private Military and Security Companies", p. 11, para. 4.

> "Where not already provided for by existing legislative or other measures, each State Party to the present Covenant undertakes to take the necessary steps, in accordance with its constitutional processes and with the provisions of the present Covenant, to adopt such laws or other measures as may be necessary to give effect to the rights recognized in the present Covenant."

The UN Human Rights Committee emphasized that under Article 2 ICCPR States Parties are required to use not only legal, but also judicial, administrative, educative, and other means to fulfill their human rights obligations.[221] This is necessary for raising the awareness of both public officials and State agents and the population at large about the human rights standards of the ICCPR.[222] The State duty to take positive measures from the regulatory field to prevent a recurrence of a human rights violation is crucial to the ICCPR.[223] In order to avoid such a recurrence, States are required to adjust their domestic regulatory frameworks by changing national laws and practices concerning PMSCs.

Regarding the right to life, the ECtHR emphasizes the duty of States to secure this right by establishing:

> "... an appropriate legal and administrative framework to deter the commission of offences against the person, backed up by law-enforcement machinery for the prevention, suppression and punishment of breaches of such provisions."[224]

Further, the ECtHR mentions that positive State obligations meant to ensure respect for private and family life may involve the adoption of effective measures, such as domestic legislation that can go as far as regulating the sphere of private relations among individuals.[225] This is of crucial importance in relation to the industry of PMSCs and the activities of these actors.

Under Article 2 ACHR, there is an obligation imposed on States Parties to adopt legislative and other measures that are required for giving effect to the rights and freedoms that they are obligated to respect and the free and full exercise of which must be ensured to all individuals within their jurisdiction. The Inter-American Court of Human Rights articulated that:

---

221 UN Human Rights Committee, *General Comment 31*, May 26, 2004, UN Doc. CCPR/C/21/Rev.1/Add.13, para. 7.

222 Ibidem.

223 Ibidem, para. 17.

224 European Court of Human Rights, *Makaratzis v. Greece*, Judgment, December 20, 2004, Application No. 50385/99, Reports 2004-XI, para. 57.

225 European Court of Human Rights, *X and Y v. The Netherlands*, Judgment, March 26, 1985, Application No. 8978/80, Series A, No. 91, para. 23.

"... the duty to prevent includes all those means of a legal, political, administrative and cultural nature that promote the protection of human rights."[226]

Article 1 of the ACHPR provides that States Parties must give effect to the rights, duties, and freedoms laid down in the Charter by undertaking to adopt legislative and other measures. Referring to international and regional human rights standards, the African Commission maintained that States are required to regulate the activities of non-State actors by adopting effective measures to prevent possible human rights breaches.[227] It was of the opinion that establishing legislation is one of the necessary components of the State compliance with the duty to protect their citizens from the misconduct of other subjects.[228] This protection offered by governments "generally entails the creation and maintenance of an atmosphere or framework of an effective interplay of laws and regulations so that individuals will be able to freely realize their rights and freedoms".[229] In addition, there is a duty to promote the enjoyment of human rights strongly connected to the obligation to protect. In this respect, States can use a large number of means, like raising awareness and building infrastructures, to comply with this duty, which might require regulation on the part of governments.[230]

Hence, hiring States are required to adopt domestic regulatory frameworks aimed at PMSCs and their activities and regulating the conduct of private contractors. These obligations need to be further reinforced in order to tackle the problems of the industry.[231] Various positive measures are needed to prevent possible human rights violations that can especially be committed by the personnel of private contractors who are capable of using force and thus can breach human rights law, for instance, the right to life. These rules and regulations can imply several means of controlling PMSCs, monitoring their activities and establishing oversight mechanisms, providing supervision over their operations, and other measures. As noted by some scholars,

---

226 Inter-American Court of Human Rights, *Velasquez Rodriguez v. Honduras*, Judgment, July 29, 1988, Series C, No. 4 (1988), para. 175.

227 African Commission on Human and Peoples' Rights, *Zimbabwe Human Rights NGO Forum v. Zimbabwe*, May 15, 2006, Communication No. 245/02 (2006), para. 147.

228 African Commission on Human and Peoples' Rights, *The Social and Economic Rights Action Center and the Center for Economic and Social Rights v. Nigeria*, Communication No. 155/96 (2001), paras. 46, 57.

229 African Commission on Human and Peoples' Rights, *Zimbabwe Human Rights NGO Forum v. Zimbabwe*, May 15, 2006, Communication No. 245/02 (2006), para. 152; see also African Commission on Human and Peoples' Rights, *The Social and Economic Rights Action Center and the Center for Economic and Social Rights v. Nigeria*, Communication No. 155/96 (2001), para. 46.

230 African Commission on Human and Peoples' Rights, *The Social and Economic Rights Action Center and the Center for Economic and Social Rights v. Nigeria*, Communication No. 155/96 (2001), para. 46.

231 Alexandra Bohm, Kerry Senior, and Adam White, "The United Kingdom", in Christine Bakker and Mirko Sossai (Eds.), (2012) *Multilevel Regulation of Military and Security Contractors: The Interplay Between International, European and Domestic Norms*, Oxford and Portland, Oregon: Hart Publishing, p. 326.

the duties to establish adequate legal and administrative frameworks for regulating PMSCs are essentially obligations of result and not of conduct.[232] In order to comply with the duty to intervene and to control activities of private contractors, States are obliged to make use of a variety of these instruments.

Specific positive measures to comply with the obligation to prevent

As established above, contracting States must act pro-actively and are required to prevent human rights violations that might be committed by the employees of private contractors working in conflict zones and often providing coercive services that might cause death and destruction. They are under a duty to take "all feasible precautions" in form of positive actions to avoid human rights abuses.[233] In order to comply with their obligation of prevention, hiring States have to take a spectrum of positive measures with regard to the conduct of PMSCs. These can relate to both the duty to intervene and the duty to regulate and aim at having a more extensive control over PMSCs and their operations. There are various measures that can be adopted by these States.

Hiring States are required to employ only those private contractors that will operate in accordance with international law, more specifically the law of armed conflict and human rights law.[234] Contracting States must examine which particular services can be outsourced to PMSCs and which can better be left to official armed forces or other State organs.[235] It has been demonstrated in the previous Chapters that the US, for instance, has extensive regulations on the types of services that may be outsourced by the US Department of Defense.[236] These regulatory instruments, however, are quite confusing in relation to the use of the vague term of "inherently governmental functions" that has been interpreted by various governmental agencies differently.[237] Furthermore, there is a need for the clear criteria for the supervised selection that must be transparent, and the careful selection procedures of the most reliable PMSCs

---

232 Lindsey Cameron and Vincent Chetail, (2013) *Privatizing War: Private Military and Security Companies under Public International Law*, Cambridge: Cambridge University Press, p. 258.

233 Carsten Hoppe, "Positive Human Rights Obligations of the Hiring State in Connection with the Provision of 'Coercive Services' by a Private Military or Security Company", in Francesco Francioni and Natalino Ronzitti (Eds.), (2011) *War by Contract: Human Rights, Humanitarian Law, and Private Contractors*, Oxford: Oxford University Press, pp. 113-114, 116-117.

234 Lindsey Cameron and Vincent Chetail, (2013) *Privatizing War: Private Military and Security Companies under Public International Law*, Cambridge: Cambridge University Press, p. 267.

235 Montreux Document, Part Two "Good Practices Relating to Private Military and Security Companies", p. 16, para. 1; Hannah Tonkin, (2011) *State Control over Private Military and Security Companies in Armed Conflict*, Cambridge: Cambridge University Press, p. 217.

236 The FAIR Act, FAR, OMB Circular, US Department of Defense Instruction 1100.22, and US Department of Defense Instruction 3020.50.

237 Rebecca DeWinter-Schmitt (Ed.), *Montreux Five Years On: An Analysis of State Efforts to Implement Montreux Document Legal Obligations and Good Practices*, Report, 2013, pp. 41-43.

that are capable of carrying out military and security services in conformity with national legislation of respective States and international law.[238] In this regard, all relevant information, for example, the performance of a company in the past, references from its clients, its superior personnel, and relations with sub-contractors, must be taken into account. If a company is likely to engage in violations of international law, the State must refrain from hiring it. In addition, PMSC personnel must receive sufficient training with regard to the application of relevant national and international law and States need to ensure that the required standard of training is met.[239] A contract as an important legal tool for controlling PMSC conduct in hands of hiring States must also be set up with utmost care. It should contain clauses and performance requirements aimed to ensure that hired PMSCs comply with national laws and international humanitarian and human rights law.[240] Besides specifying obligations under domestic and international law, States could also include selection criteria and standards of training and vetting in the contracts.[241] In this respect, governments are required to allocate necessary resources to monitoring compliance of PMSCs with contracts and to oversee the actual performance on the ground.[242] Another important set of positive measures to be taken by contracting States involves establishing effective monitoring mechanisms to ensure the compliance of PMSCs with contracts and the accountability of companies and their personnel for instances of unlawful conduct.[243] One recent US legislative act[244] aims at achieving exactly this goal by ensuring better oversight and accountability mechanisms.[245] This is an essential step towards effective compliance with the duty to prevent and constitutes a measure to comply with the obligation of result.

---

238 Montreux Document, Part Two "Good Practices Relating to Private Military and Security Companies", p. 17, paras. 2, 4-5.
239 Ibidem, p. 18, para. 10.
240 Ibidem, p. 18, para. 14.
241 Michael Cottier, (2006) "Elements for Contracting and Regulating Private Security and Military Companies", *International Review of the Red Cross*, Volume 88, Issue 863, p. 642.
242 Laura A. Dickinson, "Contract as a Tool for Regulating Private Military Companies", in Simon Chesterman and Chia Lehnardt (Eds.), (2007) *From Mercenaries to Market: The Rise and Regulation of Private Military Companies*, Oxford: Oxford University Press, pp. 223-224.
243 Montreux Document, Part Two "Good Practices Relating to Private Military and Security Companies", p. 20, para. 21.
244 Ike Skelton National Defense Authorization Act for Fiscal Year 2011, 111th Congress of the United States of America, Second Session, H.R. 6523, <http://www.gpo.gov/fdsys/pkg/BILLS-111hr6523enr/pdf/BILLS-111hr6523enr.pdf>, accessed on July 10, 2014.
245 Lindsey Cameron and Vincent Chetail, (2013) *Privatizing War: Private Military and Security Companies under Public International Law*, Cambridge: Cambridge University Press, p. 579.

4.2.2.2 Obligation to punish

Reactive regulation

As has been emphasized above, in addition to the State duty to prevent there is a general obligation of hiring States to punish, covering an array of sub-obligations: a duty to legislate in conformity with human rights instruments, and the duties to investigate, prosecute, and punish the perpetrators. One of the constitutive duties comprising the obligation to punish is the duty to legislate, which is an obligation of result that is to be achieved by a State. This means that States are required to not only adopt positive measures aiming at achieving certain goals, but also to realize those results in practice.

Under the ICCPR, Article 2(3) sets out a series of obligations with regard to the availability of effective remedies in case of human rights violations that can be committed by private individuals and groups, including PMSCs. This provision in combination with Article 2(1) establishes a duty to provide an effective and accessible remedy to a victim of a human rights breach within State jurisdiction.[246] The obligation covers the process of taking "appropriate measures" or exercising "due diligence to prevent, punish, investigate or redress the harm caused" not only by State agents, but also non-State actors.[247] In particular, the UN Human Rights Committee states that in this regard a variety of appropriate administrative and judicial mechanisms must be established in order to address possible claims of human rights violations.[248] The means by which the national judiciary can assure the enjoyment of these rights can be "direct applicability of the Covenant, application of comparable constitutional or other provisions of law, or the interpretive effect of the Covenant in the application of national law".[249] Administrative instruments must be employed to comply with the duty to investigate human rights violations by means of independent and impartial bodies, such as national human rights institutions.[250] It is necessary that a State adopts these positive measures by creating new legislative acts, administrative frameworks, and other instruments, but also revises and modifies the existing regulatory framework. In its concluding observations, the UN Human Rights Committee declared with regard to the use of certain interrogation techniques by a State Party to the ICCPR the following:

---

[246] UN Human Rights Committee, *General Comment 31*, May 26, 2004, UN Doc. CCPR/C/21/Rev.1/Add.13, paras. 8, 15.

[247] Ibidem, para. 8.

[248] Ibidem, para. 15.

[249] Ibidem.

[250] Ibidem.

> "The State party should ensure that any revision of the Army Field Manual only permits interrogation techniques consistent with the international understanding of the scope of the prohibition contained in article 7 of the Covenant; it should ensure that the current techniques or any revised techniques are binding on all agencies of government and others acting for them; it should ensure that there are effective means of recourse against abuses committed by agencies operating outside the military structure..."[251]

The ECHR contains Article 13, providing for a right to an effective remedy, which is of importance in the context of these positive obligations, including duties to legislate and to regulate.[252] There is, however, no specific duty resting upon States to legislate in accordance with the Convention, but the obligations to investigate, prosecute, and punish, which will be examined below.

As has been pointed out above, the ACHR specifically mentions the obligation of States to legislate in conformity with its provisions in Article 2 on the domestic legal effect requiring them "to adopt, in accordance with their constitutional processes and the provisions of this Convention, such legislative or other measures as may be necessary to give effect to those rights or freedoms". These measures include domestic regulation, such as legislation adopted in accordance with the Convention, for the purpose of compliance with the State duty to punish.[253]

Similarly, under Article 1 ACHPR, States are obligated to use legislative and other means for giving effect to the rights, duties, and freedoms incorporated into the treaty. The African Commission confirmed this obligation imposed on States Parties.[254] It stressed that they are required to comply with the due diligence standard by taking effective positive measures aimed at the regulation of the conduct of private actors.[255]

Investigation, prosecution, and punishment

In the context of the ICCPR, the UN Human Rights Committee noted the State duty to "punish, investigate or redress the harm" caused by the conduct of private persons

---

251 UN Human Rights Committee, *Concluding Observations on the Second and Third U.S. Reports to the Committee*, Eighty-Seventh Session, July 10-28, 2006, para. 13.

252 University Centre for International Humanitarian Law, *Expert Meeting on Private Military Contractors: Status and State Responsibility for Their Actions*, Report, Geneva, August 29-30, 2005, p. 39, <http://www.geneva-academy.ch/docs/expert-meetings/2005/2rapport_compagnies_privees.pdf>, accessed on October 10, 2014.

253 Carsten Hoppe, "Positive Human Rights Obligations of the Hiring State in Connection with the Provision of 'Coercive Services' by a Private Military or Security Company", in Francesco Francioni and Natalino Ronzitti (Eds.), (2011) *War by Contract: Human Rights, Humanitarian Law, and Private Contractors*, Oxford: Oxford University Press, p. 123.

254 African Commission on Human and Peoples' Rights, *The Social and Economic Rights Action Center and the Center for Economic and Social Rights v. Nigeria*, Communication No. 155/96 (2001), para. 46.

255 African Commission on Human and Peoples' Rights, *Zimbabwe Human Rights NGO Forum v. Zimbabwe*, May 15, 2006, Communication No. 245/02 (2006), para. 147.

and entities.[256] States Parties to the ICCPR must ensure that individuals have "accessible and effective remedies" in order to be able to vindicate the human rights covered by the ICCPR.[257] These remedies must be established in such a manner as to take into account the vulnerability of certain groups of people, like children, and to address the cessation of an ongoing human rights violation.[258] When a violation takes place, the Committee observes that States Parties are required to make reparation to individuals who fell victim to a human rights violation.[259] In accordance with customary international law, reparations can take form of compensation, restitution, rehabilitation, and satisfaction.[260]

States Parties are obliged to establish appropriate administrative mechanisms required for the compliance with their obligation to investigate allegations of human rights violations under the Covenant and must ensure that "those responsible are brought to justice".[261] The investigations must be independent and impartial, and prompt and effective, and must be conducted with regard to State agents and contractors, such as PMSCs.[262] If a State fails to investigate a human rights breach, this failure can lead to a separate breach of the ICCPR.[263] Similarly, a failure to prosecute responsible individuals and entities can be regarded as a separate violation of the treaty.[264] According to the Committee, States Parties are expected to assist each other in prosecuting those suspected of violating human rights laid down in the Covenant.[265]

With respect to the human right to life, the ECtHR declares that the State duty to protect this right laid down in Article 2 of the Convention is strongly connected to the duty to investigate:

> "The obligation to protect the right to life under this provision, read in conjunction with the State's general duty under Article 1 of the Convention to 'secure to everyone within their jurisdiction the rights and freedoms defined in [the] Convention', requires by implication that there should be some form of *effective official investigation* when

[256] UN Human Rights Committee, *General Comment 31*, May 26, 2004, UN Doc. CCPR/C/21/Rev.1/Add.13, para. 8.
[257] Ibidem, para. 15.
[258] Ibidem.
[259] Ibidem, para. 16.
[260] Ibidem, para. 15.
[261] Ibidem, paras. 15, 18.
[262] UN Human Rights Committee, *Concluding Observations on the Second and Third U.S. Reports to the Committee*, Eighty-Seventh Session, July 10-28, 2006, para. 14.
[263] UN Human Rights Committee, *General Comment 31*, May 26, 2004, UN Doc. CCPR/C/21/Rev.1/Add.13, para. 15.
[264] Ibidem, para. 18.
[265] Ibidem.

individuals have been killed as a result of the use of force by, inter alios, agents of the State."[266] (emphasis added)

States must commence effective and independent investigations on their own initiative.[267] In case of human rights violations, the authorities are obliged to carry out an "impartial and careful examination of the circumstances" of the misconduct.[268] They must take all reasonable steps to ensure that evidence is collected and secured.[269] Possible conclusions made as a result of investigations need to be based on a "thorough, objective and impartial analysis of all relevant elements" and the authorities are required to use a standard, such as "no more than absolutely necessary" in Article 2(2) of the Convention.[270] It is important to note that the duty to start investigations cannot be set aside even when a violent armed struggle takes place within the State and there is a high rate of fatalities.[271] The ECtHR further examines the scope of the right to an effective remedy in the European human rights context – which can be found in Article 13 ECHR – and holds that investigations of violations of Convention's rights – Article 2 in this regard – must eventually result in the identification and punishment of responsible individuals.[272]

As mentioned above, obligations of States Parties to the ACHR laid down in Article 2 ACHR include the duty to legislate and to take other measures that include duties to investigate and to sanction individuals responsible for a particular human rights violation. States are obligated to investigate every instance of human rights violations that are committed.[273] The duty to conduct investigations, which must have an objective and must be initiated by the States themselves, has to be taken very

[266] European Court of Human Rights, *McCann and Others v. United Kingdom*, Judgment, September 27, 1995, Application No. 18984/91, Series A, No. 324, para. 161.

[267] European Court of Human Rights, *Ilhan v. Turkey*, Judgment, June 27, 2000, Application No. 22277/93, Reports 2000-VII, para. 63; European Court of Human Rights, *Nachova and Others v. Bulgaria*, Judgment, July 6, 2005, Application Nos. 43577/98 and 43579/98, Reports 2005-VII, para. 111.

[268] European Court of Human Rights, *McCann and Others v. United Kingdom*, Judgment, September 27, 1995, Application No. 18984/91, Series A, No. 324, para. 162.

[269] European Court of Human Rights, *Nachova and Others v. Bulgaria*, Judgment, July 6, 2005, Application Nos. 43577/98 and 43579/98, Reports 2005-VII, para. 113.

[270] Ibidem.

[271] European Court of Human Rights, *Kaya v. Turkey*, Judgment, February 19, 1998, Application No. 22729/93, Report 1998-I, para. 91.

[272] European Court of Human Rights, *Assenov and Other v. Bulgaria*, Judgment, October 28, 1998, Application No. 24760/94, Reports 1998-VIII, para. 102; European Court of Human Rights, *Ogur v. Turkey*, Judgment, May 20, 1999, Application No. 21954/93, Reports 1999-III, para. 88; European Court of Human Rights, *McKerr v. United Kingdom*, Judgment, May 4, 2001, Application No. 28883/95, Reports 2001-III, para. 121.

[273] Inter-American Court of Human Rights, *Velasquez Rodriguez v. Honduras*, Judgment, July 29, 1988, Series C, No. 4 (1988), para. 176.

seriously and should not be seen as a formality.[274] These investigations need to be complete, impartial, and effective and continue as long as there is uncertainty about the facts of a particular case, especially the fate of disappeared persons.[275] Just as the duty to prevent is an obligation of conduct, it is not necessarily breached only when investigations do not lead to a satisfactory result given that there may be scenarios which are difficult to investigate.[276]

Article 25 ACHR provides for the right to judicial protection before "a competent court or tribunal for protection against acts that violate his fundamental rights recognized by the constitution or laws of the state concerned or by this Convention". In the *Velásquez Rodríguez* case, the Inter-American Court of Human Rights declared:

> "Under the Convention, States Parties have an obligation to provide effective judicial remedies to victims of human rights violations (Art. 25), remedies that must be substantiated in accordance with the rules of due process of law (Art. 8(1)), all in keeping with the general obligation of such States to guarantee the free and full exercise of the rights recognized by the Convention to all persons subject to their jurisdiction (Art. 1)."[277]

The IACtHR states that States Parties cannot refrain from complying with their duty to ensure the exercise of the human rights laid down in the Convention and must act in every situation of a human rights breach. The inaction of States will result in their failure to meet this obligation:

> "If the State apparatus acts in such a way that the violation goes unpunished and the victim's full enjoyment of such rights is not restored as soon as possible, the State has failed to comply with its duty to ensure the free and full exercise of those rights to the persons within its jurisdiction. The same is true when the State allows private persons or groups to act freely and with impunity to the detriment of the rights recognized by the Convention."[278]

---

[274] Ibidem, para. 177; Inter-American Court of Human Rights, *Godínez Cruz v. Honduras*, Judgment, January 20, 1989, Series C, No. 5 (1989), para. 188; Inter-American Court of Human Rights, *Villagran-Morales et al. v. Guatemala*, Judgment, November 19, 1999, Series C, No. 63 (1999), para. 226.

[275] Inter-American Court of Human Rights, *Ituango Massacres v. Colombia*, Judgment, July 1, 2006, Series C, No. 148 (2006), para. 319; Inter-American Court of Human Rights, *Velasquez Rodriguez v. Honduras*, Judgment, July 29, 1988, Series C, No. 4 (1988), para. 181.

[276] Inter-American Court of Human Rights, *Velasquez Rodriguez v. Honduras*, Judgment, July 29, 1988, Series C, No. 4 (1988), para. 177.

[277] Inter-American Court of Human Rights, *Velasquez Rodriguez v. Honduras*, Preliminary Objections, Judgment, June 26, 1989, Series C, No. 1 (1994), para. 91.

[278] Inter-American Court of Human Rights, *Velasquez Rodriguez v. Honduras*, Judgment, July 29, 1988, Series C, No. 4 (1988), para. 176.

What States Parties to the ACHR need to prevent is the situation of impunity leading to the repetition of unlawful acts that has been defined by the Court as "the overall failure to investigate, pursue, capture, prosecute and convict those responsible".[279] The IACtHR has further characterized this as:

> "... the total lack of investigation, prosecution, capture, trial and conviction of those responsible for violations of the rights protected by the American Convention, in view of the fact that the State has the obligation to use all the legal means at its disposal to combat that situation, since impunity fosters chronic recidivism of human rights violations, and total defenselessness of victims and their relatives."[280]

States are not allowed to create this impunity by means of domestic legal rules and their institutions that may hinder investigations and punishment which must be undertaken by States.[281] These *de facto* and *de jure* obstacles, such as immunity granted to private contractors operating in situations of armed conflict and peace operations,[282] have to be removed within a reasonable period of time.[283]

The African Commission explicitly mentioned that a failure on the part of the States to comply with the due diligence standard by not taking necessary steps to provide victims with reparations can lead to their responsibility when human rights are violated by private individuals. According to the Commission:

> "A failure to exercise due diligence to prevent or remedy violation, or failure to apprehend the individuals committing human rights violations gives rise to State responsibility even if committed by private individuals."[284]

It considered States Parties to the Charter as being under an obligation to conduct investigations into human rights violations, such as murder, when perpetrated by

---

279 Inter-American Court of Human Rights, *Blanco-Romero et al. v. Venezuela*, Judgment, November 28, 2005, Series C, No. 138 (2005), para. 94.

280 Inter-American Court of Human Rights, *Paniagua Morales et al. v. Guatemala*, Judgment, March 8, 1998, Series C, No. 37 (1998), para. 173.

281 Inter-American Court of Human Rights, *Bulacio v. Argentina*, Judgment, September 18, 2003, Series C, No. 100 (2003), paras. 117-118; Inter-American Court of Human Rights, *Ituango Massacres v. Colombia*, Judgment, July 1, 2006, Series C, No. 148 (2006), para. 402.

282 Carsten Hoppe, "Positive Human Rights Obligations of the Hiring State in Connection with the Provision of 'Coercive Services' by a Private Military or Security Company", in Francesco Francioni and Natalino Ronzitti (Eds.), (2011) *War by Contract: Human Rights, Humanitarian Law, and Private Contractors*, Oxford: Oxford University Press, pp. 124-125.

283 Inter-American Court of Human Rights, *Montero-Aranguren et al. v. Venezuela*, Judgment, July 5, 2006, Series C, No. 150 (2006), para. 138; Inter-American Court of Human Rights, *Servellón-García et al. v. Honduras*, Judgment, September 21, 2006, Series C, No. 150 (2006), para. 195.

284 African Commission on Human and Peoples' Rights, *Zimbabwe Human Rights NGO Forum v. Zimbabwe*, May 15, 2006, Communication No. 245/02 (2006), para. 147.

private actors.[285] Additionally, they are required to protect their citizens by adopting legislation and providing effective enforcement.[286]

Specific positive measures to comply with the obligation to punish

In sum, it is underlined that in the first place hiring States employing PMSCs have a duty to legislate, meaning that they need to ensure that contracts concluded between them and PMSCs are clear and precise and there are legal frameworks available to react and to respond to possible violations of human rights. They must have legislation allowing for prosecution of PMSC personnel that commit illegal acts on their own territory or on the territory of host States.[287] The United States can be used as an example of a State that has adopted legislation concerning criminal jurisdiction over PMSC personnel.[288] The US MEJA is suitable for such purposes. Hiring States must not only enact these national laws forming the basis of criminal jurisdiction over the members of PMSCs, but also need to provide for non-criminal accountability mechanisms, such as contractual sanctions, referral of the cases of violations to competent investigative authorities, and instruments for civil liability.[289] Laura Dickinson even argues that it is desirable that hiring States allow third-party claims under private law on the basis of contracts for the non-compliance of PMSCs with international humanitarian and human rights law.[290] Such claims could be addressed through a variety of judicial or non-judicial mechanisms to be established by contracting States. States are required to create external and internal monitoring and oversight mechanisms in order to have insight into and information about the compliance of PMSCs with contracts and their accountability.[291] Such systems at company level could pro-

---

[285] African Commission on Human and Peoples' Rights, *Commission Nationale des Droits de l'Homme et des Libertes v. Chad*, Communication No. 74/92 (1995), para. 22.

[286] African Commission on Human and Peoples' Rights, *The Social and Economic Rights Action Center and the Center for Economic and Social Rights v. Nigeria*, Communication No. 155/96 (2001), para. 57; African Commission on Human and Peoples' Rights, *Zimbabwe Human Rights NGO Forum v. Zimbabwe*, May 15, 2006, Communication No. 245/02 (2006), para. 152.

[287] Carsten Hoppe, "Positive Human Rights Obligations of the Hiring State in Connection with the Provision of 'Coercive Services' by a Private Military or Security Company", in Francesco Francioni and Natalino Ronzitti (Eds.), (2011) *War by Contract: Human Rights, Humanitarian Law, and Private Contractors*, Oxford: Oxford University Press, p. 123.

[288] Lindsey Cameron and Vincent Chetail, (2013) *Privatizing War: Private Military and Security Companies under Public International Law*, Cambridge: Cambridge University Press, p. 579.

[289] Montreux Document, Part Two "Good Practices Relating to Private Military and Security Companies", pp. 19-20, paras. 19-20.

[290] Laura A. Dickinson, "Contract as a Tool for Regulating Private Military Companies", in Simon Chesterman and Chia Lehnardt (Eds.), (2007) *From Mercenaries to Market: The Rise and Regulation of Private Military Companies*, Oxford: Oxford University Press, pp. 224-225.

[291] Montreux Document, Part Two "Good Practices Relating to Private Military and Security Companies", p. 20, para. 21; Michael Cottier, (2006) "Elements for Contracting and Regulating Private Security and Military Companies", *International Review of the Red Cross*, Volume 88, Issue 863, p. 644.

vide avenues for "whistleblowers" to access an internal compliance system.[292] Also, reporting obligations on PMSCs can be included in contracts. Given that governments rarely make use of termination clauses laid down in the contracts, these agreements between States and PMSCs should contain provisions directed at States and enabling them to "take over" the contracts when PMSCs and their personnel violate international law.[293]

In addition, contracting States need to carry out proper investigations in order to identify potential perpetrators if violations of human rights occur, and to ensure that these persons are subjected to disciplinary, administrative, and criminal sanctions. One can think of the investigations that took place after the Nisour Square incident and the Abu Ghraib abuses by the US authorities as examples that have been described in the previous parts of the study. When necessary, hiring States are under an obligation to combine their efforts with host and home States in matters of common concern regarding investigation and regulation of PMSC activities.[294] In the case of hiring States, the investigations may not always lead to actual prosecution and punishment, as these States must have criminal jurisdiction over responsible individuals. PMSC employees, often being considered civilians accompanying armed forces and not the military personnel of a State, would normally fall under the criminal jurisdiction of a State on which territory they function, a host State. However, private contractors might have immunity from criminal jurisdiction of local authorities due to the special arrangements and agreements between host States and hiring States. The former type of State can also be incapable of initiating criminal investigations because of the lack of necessary State structures or because it is militarily occupied. The Montreux Document's good practices suggest, however, that contracting States need to cooperate with host States in negotiating these immunity agreements, taking into account their impact on the compliance of PMSCs with national legislation and regulations and providing for appropriate civil, administrative, and criminal remedies for possible instances of misconduct.[295] In the case concerning the incident in Bosnia involving the PMSC DynCorp contracted by the US, not a single individual involved in the misconduct was criminally prosecuted and the company itself faced no consequences. Unfortunately, this indicates that hiring States can be reluctant or unable to punish members of PMSC personnel for their violations of international law.

---

292 Michael Cottier, (2006) "Elements for Contracting and Regulating Private Security and Military Companies", *International Review of the Red Cross*, Volume 88, Issue 863, p. 644.

293 Laura A. Dickinson, "Contract as a Tool for Regulating Private Military Companies", in Simon Chesterman and Chia Lehnardt (Eds.), (2007) *From Mercenaries to Market: The Rise and Regulation of Private Military Companies*, Oxford: Oxford University Press, p. 224.

294 Montreux Document, Part Two "Good Practices Relating to Private Military and Security Companies", p. 20, para. 23.

295 Ibidem, p. 20, para. 22.

4.3 Application to host States

An analysis will now be made of the positive obligations of host States or territorial States arising out of international humanitarian and human rights law. The focus will be on those States on which territory PMSCs operate and which cannot be perceived as States employing private contractors or as home States where they are established as private firms. As an example of a territorial State, one could think of Iraq where a large number of PMSCs have been operating on behalf of the United States and the United Kingdom, being on many occasions both contracting and home States.

It must be noted that host States usually face two main difficulties when they have to comply with positive obligations from the field of international law. In the first place, they are often unable to act in accordance with these obligations due to the lack of institutional capacities.[296] Secondly, host States might fail to meet positive obligations when they are occupied by another State or when a third State exercises effective control over their territory.[297] These circumstances might lead to the inability of host States to discharge their positive duties under international law. Additionally, it should be underlined that, in contrast to hiring States, territorial States – in the strict sense of the notion – do not have an effective instrument of contract to control and to oversee activities of PMSCs.[298] Therefore, in the absence of such an important factual and legal link between PMSCs and host States, national regulation and legislation of these States must be seen as a tool to comply with certain duties under the law of armed conflict and human rights law, and to regulate the operations of PMSCs on their soil.

4.3.1 Positive State obligations under international humanitarian law

Observing all positive duties of host States under international humanitarian law, the Montreux Document notes that these States have to comply with the following obligations:

> "Territorial States have an obligation, within their power, to ensure respect for international humanitarian law by PMSCs operating on their territory, in particular to:
>
> a) disseminate, as widely as possible, the text of the Geneva Conventions and other relevant norms of international humanitarian law among PMSCs and their personnel;

---

296 Christine Bakker, "Duties to Prevent, Investigate, and Redress Human Rights Violations by Private Military and Security Companies: The Role of the Host State", in Francesco Francioni and Natalino Ronzitti (Eds.), (2011) *War by Contract: Human Rights, Humanitarian Law, and Private Contractors*, Oxford: Oxford University Press, p. 131.

297 Ibidem, pp. 132-135.

298 Lindsey Cameron and Vincent Chetail, (2013) *Privatizing War: Private Military and Security Companies under Public International Law*, Cambridge: Cambridge University Press, p. 280.

b) not encourage or assist in, and take appropriate measures to prevent, any violations of international humanitarian law by personnel of PMSCs;
c) take measures to suppress violations of international humanitarian law committed by the personnel of PMSCs through appropriate means such as military regulations, administrative orders and other regulatory measures as well as administrative, disciplinary or judicial sanctions, as appropriate."[299]

Common Article 1 of the 1949 Geneva Conventions

In the first place, there is the same general duty of States Parties to the 1949 Geneva Conventions to ensure respect for that body of law "in all circumstances". This has been laid down in Common Article 1 of the Geneva Conventions addressed with respect to the positive obligations of hiring States. Following the visible trend of its broad interpretation and application to all States, it is clear that this provision is applicable to host States to the same extent as hiring States.[300] In this regard, all States need to take every possible positive step in their power to ensure that international humanitarian law is observed and respected. Rule 144 of the ICRC's study on customary international law indicates:

> "States may not encourage violations of international humanitarian law by parties to an armed conflict. They must exert their influence, to the degree possible, to stop violations of international humanitarian law."[301]

This rule has been established by State practice as a norm of customary international law and applies to all States in both international and non-international armed conflicts.[302] The Montreux Document, supported by a large number of States, also includes provisions confirming the view that host States indeed need to ensure respect for international humanitarian law by private contractors operating on their territory.[303]

Common Article 1, the scope of application of which is not territorially limited, obliges host States to take adequate measures to ensure that PMSCs present on their

---

299 Montreux Document, Part One "Pertinent International Legal Obligations Relating to Private Military and Security Companies", p. 12, para. 9.
300 University Centre for International Humanitarian Law, *Expert Meeting on Private Military Contractors: Status and State Responsibility for Their Actions*, Report, Geneva, August 29-30, 2005, p. 41, <http://www.geneva-academy.ch/docs/expert-meetings/2005/2rapport_compagnies_privees.pdf>, accessed on October 10, 2014.
301 Jean-Marie Henckaerts and Louise Doswald-Beck, (2005) *Customary International Humanitarian Law, Volume I: Rules*, ICRC, Cambridge: Cambridge University Press, p. 509.
302 Ibidem; Hannah Tonkin, (2011) *State Control over Private Military and Security Companies in Armed Conflict*, Cambridge: Cambridge University Press, p. 128.
303 Montreux Document, Part One "Pertinent International Legal Obligations Relating to Private Military and Security Companies", p. 12, para. 9.

sovereign territory comply with the rules and principles of the law of armed conflict. As addressed in relation to hiring States, host States must have a certain capacity to influence the conduct of PMSCs, which serves as a link between the obligation incorporated in Common Article 1 and the State duty to take positive actions with regard to private contractors.[304] In general, host States on whose territory PMSCs perform their services and can potentially violate domestic and international laws have a special interest in controlling these entities.[305] When PMSCs are active in the sovereign territory of a host State and the latter has control over its territory, this State might be considered to be in a position of controlling PMSCs and regulating their conduct. In practice, however, host States will often have no effective control over their territory and will lack the capacity to effectively influence PMSC conduct.[306] Host States that are not home States to PMSCs or hiring States do not usually finance these corporations and provide no other forms of support. If such States are occupied by other States, obligations under international humanitarian law will be limited in scope to the areas where the authorities in question are able to exercise effective control.[307] It goes without saying that if a territorial State can have effective influence on the actions of PMSCs, it will have to use all positive measures to ensure that international humanitarian law is observed and respected by these firms and their personnel.

State intervention

Similar to hiring States, host States of PMSC operations are required to intervene when there is the likelihood that a violation of international humanitarian law will be committed by PMSCs operating on their territory. Host States can be expected to know whether a violation is predictable and to protect individuals in danger. Under the law of armed conflict, they are obliged to protect civilians in international armed conflicts, particularly women and children. This duty stems from Article 27 of the Fourth Geneva Convention and Articles 76 and 77 of the First Additional Protocol. Similar to hiring States, territorial States must ensure that PMSC activities and conduct of their personnel do not threaten women and children. PMSCs exercising coercive services can constitute danger to these two protected groups. Therefore, host States must take reasonable measures within their power to protect women and

---

[304] Hannah Tonkin, (2011) *State Control over Private Military and Security Companies in Armed Conflict*, Cambridge: Cambridge University Press, p. 131.

[305] Emanuela-Chiara Gillard, (2006) "Business Goes to War: Private Military / Security Companies and International Humanitarian Law", *International Review of the Red Cross*, Volume 88, Issue 863, p. 558.

[306] Hannah Tonkin, (2011) *State Control over Private Military and Security Companies in Armed Conflict*, Cambridge: Cambridge University Press, pp. 131-132.

[307] Montreux Document, Part One "Pertinent International Legal Obligations Relating to Private Military and Security Companies", p. 13, para. 13.

children from violence, sexual abuse, and other forms of harm that might emanate from private contractors.[308]

Protection of certain groups of people

As has already been indicated in relation to hiring States, States participating in armed conflicts, having both international and non-international character, are obligated to respect and to protect specific groups of individuals. These obligations, laid down in the 1949 Geneva Conventions and their Additional Protocols, are also imposed upon host States on whose territory armed conflicts usually take place. Territorial States are required to protect persons who are wounded, sick, and shipwrecked; prisoners of war; and civilians, including women and children.

Suppressing and repressing violations of the law of armed conflict

Host States are also under a duty to suppress and to repress violations of international humanitarian law in international armed conflicts. Just as for the contracting States, grave breaches of the Geneva Conventions require host States to enact legislation and to prosecute and to punish responsible individuals.[309] These violations are mentioned in Article 50 of the First Geneva Convention, Article 51 of the Second Geneva Convention, Article 130 of the Third Geneva Convention, and Article 147 of the Fourth Geneva Convention. The non-grave breaches are to be found in Article 49(3) of the First Geneva Convention, Article 50(3) of the Second Geneva Convention, Article 129(3) of the Third Geneva Convention, and Article 146(3) of the Fourth Geneva Convention. They also must be suppressed by the host States, which must adopt positive measures to realize this undertaking. In non-international armed conflicts, the duty to suppress can be derived from Common Article 1.[310]

Obligations of occupying powers

It has already been established that contracting States have a duty to act diligently in accordance with Article 43 of the Hague Convention IV and Common Article 1. These provisions set out a State obligation for occupying States to exercise due diligence with regard to private actors, like PMSCs active in the occupied territory. This is necessary to prevent violations of international humanitarian law from being

---

308 Hannah Tonkin, (2011) *State Control over Private Military and Security Companies in Armed Conflict*, Cambridge: Cambridge University Press, p. 138.

309 Article 49(1) of the First Geneva Convention, Article 50(1) of the Second Geneva Convention, Article 129(1) of the Third Geneva Convention, and Article 146(1) of the Fourth Geneva Convention.

310 Hannah Tonkin, (2011) *State Control over Private Military and Security Companies in Armed Conflict*, Cambridge: Cambridge University Press, p. 140.

committed by individuals. This particular duty concerns all private persons, groups of persons, and entities that are present in the territory occupied by a State. An occupying power can be a host State to PMSC operations without actually employing these corporations. In this respect, they must take positive measures seeking to control private contractors and their conduct, for example, by creating oversight and monitoring mechanisms and procedures. In addition, these States need to ensure that breaches of the law of armed conflict are investigated and possible perpetrators are prosecuted and punished. Similar to the situation with hiring States, the non-compliance of a host State with this obligation will incur their State responsibility.

Specific positive measures to comply with positive obligations

In order to comply with the above-mentioned duties, host States are expected to take positive measures from different fields, such as regulation. Under the obligations arising out of Common Article 1, this particular type of State can be required to establish a licensing regime for PMSCs and to ensure that they apply for authorization to provide services on the territory of host States.[311] They must carefully assess the capacity of PMSCs to operate on their territory, in conformity with national laws and international humanitarian and human rights law, and create quality criteria for issuing a license.[312] Host States are also expected to have national rules and policies on the use of force by PMSCs and their possession and use of firearms.[313] In addition, there is a need to establish monitoring instruments and accountability mechanisms to evaluate the compliance of PMSCs with contracts and national and international law and to respond to possible violations.[314] It should be noted, however, that the failure to establish a regulatory framework as a measure to comply with Common Article 1 applicable to all States does not in itself constitute a breach of that provision under the current body of international humanitarian law.[315] These States are, however, expected to enact necessary legislation for providing effective penal sanctions for persons committing grave breaches of the Geneva Conventions and the First Additional Protocol and to investigate these violations, searching for individuals committing them or ordering them to be prosecuted.[316] In some instances, host States must

---

311 Montreux Document, Part Two "Good Practices Relating to Private Military and Security Companies", p. 21, para. 25.

312 Ibidem, pp. 21-23, paras. 28, 30-39.

313 Ibidem, pp. 23-24, paras. 43-44.

314 Ibidem, p. 24, paras. 46-52.

315 University Centre for International Humanitarian Law, *Expert Meeting on Private Military Contractors: Status and State Responsibility for Their Actions*, Report, Geneva, August 29-30, 2005, p. 45, <http://www.geneva-academy.ch/docs/expert-meetings/2005/2rapport_compagnies_privees.pdf>, accessed on October 10, 2014.

316 Montreux Document, Part One "Pertinent International Legal Obligations Relating to Private Military and Security Companies", p. 13, paras. 11, 12.

avoid taking regulatory measures, such as immunity agreements with contracting States, when the latter cannot guarantee that they will comply with the duty to punish.[317] Iraq as a host State could have been expected to clarify the Order 17 of the Coalition Provisional Authority granting immunity to PMSCs active on its territory and, if needed, to modify it. The new SOFA agreement from 2009 that was negotiated with the US abolished immunity of some foreign PMSC employees[318] and can be seen as a measure to comply with this particular positive duty.

4.3.2 Positive State obligations under international human rights law

Host States have primary responsibility for protecting against human rights violations by PMSCs operating on their soil, given that contractors' activities take place in the territory or under the jurisdiction of these States.[319] In theory, positive obligations to prevent and to punish also pertain to territorial States, as to hiring States. The practice of PMSC involvement in armed conflicts and peace operations, however, shows that they are often unable to comply with these duties.[320] In this respect, it might be correct to point out that the obligations to prevent and to punish of territorial States are suspended when they are occupied by foreign States or when third States exercise effective control over their territory or certain areas and facilities, such as prisons.[321] PMSCs appear to be used in similar situations when third States have effective control or exercise State agent authority over individuals residing in the territory of host States.[322] In other cases, however, States in which private contractors perform their tasks might still be obligated to comply with their positive duties under international human rights law. The Montreux Document seeks to clarify the scope of responsibilities of host States under human rights law and offers some guidance by emphasizing that:

---

[317] Ibidem, Part Two "Good Practices Relating to Private Military and Security Companies", p. 24, para. 51.

[318] Rebecca DeWinter-Schmitt (Ed.), *Montreux Five Years On: An Analysis of State Efforts to Implement Montreux Document Legal Obligations and Good Practices*, Report, 2013, p. 116.

[319] Smita Narula, "International Financial Institutions, Transnational Corporations and Duties of States", in Malcolm Langford, Wouter Vandenhole, Martin Scheinin, and Willem van Genugten, (2014) *Global Justice, State Duties: The Extraterritorial Scope of Economic, Social and Cultural Rights in International Law*, Cambridge: Cambridge University Press, p. 127.

[320] Francesco Francioni, "The Role of the Home State in Ensuring Compliance with Human Rights by Private Military Contractors", in Francesco Francioni and Natalino Ronzitti (Eds.), (2011) *War by Contract: Human Rights, Humanitarian Law, and Private Contractors*, Oxford: Oxford University Press, pp. 104-105.

[321] Christine Bakker, "Duties to Prevent, Investigate, and Redress Human Rights Violations by Private Military and Security Companies: The Role of the Host State", in Francesco Francioni and Natalino Ronzitti (Eds.), (2011) *War by Contract: Human Rights, Humanitarian Law, and Private Contractors*, Oxford: Oxford University Press, p. 146.

[322] Ibidem, p. 147.

> "Territorial States are responsible to implement their obligations under international human rights law, including by adopting such legislative and other measures as may be necessary to give effect to these obligations. To this end they have the obligation, in specific circumstances, to take appropriate measures to prevent, investigate and provide effective remedies for relevant misconduct of PMSCs and their personnel."[323]

4.3.2.1 Obligation to prevent

Assessing the obligation to prevent, one must pay attention to a number of duties belonging to this category and applicable to the conduct of PMSCs on the territory of host States. These States might often be in close proximity to the events involving PMSCs, exercise a certain degree of control over their territory, and have jurisdiction over individuals concerned. Therefore, depending on circumstances of a particular case, they are expected to comply with the positive obligation to prevent human rights violations.

In this respect, one of the fundamental principles mentioned in the report of John Ruggie, the Special Representative of the Secretary-General on the issue of human rights and transnational corporations and other business enterprises, is the duty of territorial States to protect human rights. It reads as follows:

> "States must protect against human rights abuse within their territory and/or jurisdiction by third parties, including business enterprises. This requires taking appropriate steps to prevent, investigate, punish and redress such abuse through effective policies, legislation, regulations and adjudication."[324]

In general, this provision requires that host States comply with their international human rights obligations to respect, protect, and fulfill human rights of persons within their jurisdiction and/or territory. These duties are considered to include an obligation to protect against human rights violations committed by third parties, for instance, PMSCs and other business enterprises.[325] Much has been written on businesses and human rights by legal scholars.[326] As such, host States are not responsible

---

[323] Montreux Document, Part One "Pertinent International Legal Obligations Relating to Private Military and Security Companies", p. 12, para. 10.

[324] Report of the Special Representative of the Secretary-General on the Issue of Human Rights and Transnational Corporations and Other Business Enterprises, John Ruggie, *Guiding Principles on Business and Human Rights: Implementing the United Nations 'Protect, Respect and Remedy' Framework*, UN Human Rights Council, Seventeenth Session, March 21, 2011, UN Doc. A/HRC/17/31, pp. 6-7.

[325] Ibidem, p. 7.

[326] Nicola Jägers, (2002) *Corporate Human Rights Obligations. In Search of Accountability*, School of Human Rights Research Series, Volume 17, Antwerp: Intersentia; Andrew Clapham, (2006) *Human Rights Obligations of Non-State Actors*, Oxford: Oxford University Press; Steven R. Ratner, (2001) "Corporations and Human Rights: A Theory of Legal Responsibility", *Yale Law Journal*, Volume 111; Peter T. Muchilinski, (2007) *Multinational Enterprises & The Law*, Oxford: Oxford University Press.

for the human rights breaches by private contractors and other private actors, but, in fact, these States violate international human rights law where human rights abuses "can be attributed to them, or where they fail to take appropriate steps to prevent, investigate, punish and redress private actors' abuse".[327]

State intervention

As with hiring States, host States must take positive measures to target known sources of danger and harm, including PMSCs and their employees. This is imperative for the compliance of territorial States with their human rights duties to protect their population from threats posed by PMSCs that are often capable of using force and exercising violence.[328] Therefore, they have an obligation to take preventive measures targeting all PMSCs operating on their territory and especially those that are known for the involvement in various forms of unlawful conduct or the potential to do so. As has been assessed with regard to contracting States, host States are obligated to intervene and to exercise control over PMSCs in order to prevent human rights abuses.

Host States are obliged to take positive measures in order to protect individuals whose life might be at risk and to seek to minimize this risk. It is especially the case in situations of armed conflict and peace operations, during which PMSCs are active on their territory. When a host State has knowledge that certain individuals are in danger and might potentially become victims of PMSCs, it needs to act and to guarantee that it will not take place. If a host State knows or ought to know that there is a real and immediate risk that human rights violations will be committed by PMSCs in a certain place or that a particular group of individuals is at risk of being targeted by these corporations and their employees, it must take preventive steps to avoid or to minimize that risk.[329] To such vulnerable groups might belong women and children, who are protected by a number of human rights instruments. The sources of positive human rights obligations with regard to these two groups are Article 3 of the CEDAW and Article 34 of the CRC, which are equally applicable to host States as to hiring States.

---

[327] Report of the Special Representative of the Secretary-General on the Issue of Human Rights and Transnational Corporations and Other Business Enterprises, John Ruggie, *Guiding Principles on Business and Human Rights: Implementing the United Nations 'Protect, Respect and Remedy' Framework*, UN Human Rights Council, Seventeenth Session, March 21, 2011, UN Doc. A/HRC/17/31, p. 7.

[328] Hannah Tonkin, (2011) *State Control over Private Military and Security Companies in Armed Conflict*, Cambridge: Cambridge University Press, p. 155.

[329] European Court of Human Rights, *Osman v. United Kingdom*, Judgment, October 28, 1998, Application No. 23452/94, ECHR Reports 1998-VIII, para. 116; Hannah Tonkin, (2011) *State Control over Private Military and Security Companies in Armed Conflict*, Cambridge: Cambridge University Press, pp. 156-157.

Preventive regulation

Obligations of hiring States to adopt positive measures – legislative acts and other measures – aimed at the prevention of human rights violations that can be found in the analyzed human rights frameworks are also applicable to territorial States. It has been argued by some experts that this duty to regulate has a more extensive basis in human rights law than in the law of armed conflict.[330] Not only do contracting States have the duty to effectively regulate PMSC activities, but also States where they operate, and this national regulation is indeed an important instrument to ensure control over PMSCs and their compliance with international law.[331] As already indicated, there is one notable difference between hiring and host States. Although it may be assumed that host States have a sufficiently high degree of control over operations of private corporations, like PMSCs, active on their territory,[332] they are often unable to comply with this particular duty. Host States can also be said to be obliged to adopt administrative and other measures to have a minimum degree of control over the activities of PMSCs on their territory and over their selection.[333] In addition, under the ECHR regime, territorial States should not conclude agreements granting immunity to PMSC personnel with hiring States, given that they hinder the establishment of an effective legal framework for preventing human rights abuses and punishing responsible individuals.[334]

Specific positive measures to comply with the obligation to prevent

Obviously, in order to discharge these obligations, host States must take certain positive measures, such as adopting licensing schemes and establishing authorization regimes enabling them to allow on their territory only those companies that conduct

---

330 University Centre for International Humanitarian Law, *Expert Meeting on Private Military Contractors: Status and State Responsibility for Their Actions*, Report, Geneva, August 29-30, 2005, p. 57, <http://www.geneva-academy.ch/docs/expert-meetings/2005/2rapport_compagnies_privees.pdf>, accessed on October 10, 2014.

331 Alexandra Bohm, Kerry Senior, and Adam White, "The United Kingdom", in Christine Bakker and Mirko Sossai (Eds.), (2012) *Multilevel Regulation of Military and Security Contractors: The Interplay Between International, European and Domestic Norms*, Oxford and Portland, Oregon: Hart Publishing, p. 326; Michael Cottier, (2006) "Elements for Contracting and Regulating Private Security and Military Companies", *International Review of the Red Cross*, Volume 88, Issue 863, pp. 639-640.

332 Nicola Jägers, (2002) *Corporate Human Rights Obligations: in Search of Accountability*, School of Human Rights Research Series, Volume 17, Antwerpen, Oxford, New York: Intersentia, p. 172.

333 Christine Bakker, "Duties to Prevent, Investigate, and Redress Human Rights Violations by Private Military and Security Companies: The Role of the Host State", in Francesco Francioni and Natalino Ronzitti (Eds.), (2011) *War by Contract: Human Rights, Humanitarian Law, and Private Contractors*, Oxford: Oxford University Press, p. 137.

334 Ibidem, p. 140.

themselves and carry out operations properly.[335] When PMSCs get involved in human rights abuses, territorial States have to withdraw an authorization or license of a company whose employees were engaged in the misconduct.[336] The host State Iraq has such an authorization and licensing regime in place that allowed the government to revoke the license of Blackwater after the Nisour Square shooting in 2007.[337] The creation of monitoring mechanisms and other similar measures can be used to increase the overall transparency of representatives of the private military and security industry active on the territory of host States and to give to these States more control over PMSCs.[338] In practice, host States torn apart by armed conflict are frequently unable to effectively regulate activities of PMSCs. In these cases, they can be expected to notify hiring States using private contractors on their territory about their inability to prevent human rights violations by means of regulation and to require that contracting States apply their own national requirements and safeguards for the operation of PMSCs and supervise their activities.[339]

4.3.2.2 Obligation to punish

Reactive regulation

Finally, we assess the positive duty to punish, which requires host States to undertake a number of activities to legislate, investigate, prosecute, punish, and provide redress. These obligations laid down in Article 2(1) and 2(3) ICCPR, Article 13 ECHR, Articles 2 and 25 ACHR, and Article 1 ACHPR are also applicable to host States. As observed in relation to the duties of hiring States, States need to adopt legislation and other forms of regulation in order to protect their population from human rights violations by non-State actors, in particular PMSCs. National criminal codes adopted by territorial States must criminalize conduct breaching such essential human rights as the right to life and the prohibition of torture and slavery.[340]

---

335 Ibidem, p. 137.
336 Ibidem, p. 138.
337 Jordi Palou-Loverdos and Leticia Armendáriz, *The Privatization of Warfare, Violence and Private Military & Security Companies: A Factual and Legal Approach to Human Rights Abuses by PMSC in Iraq*, Report, 2011, Nova-Institute for Active Non-violence, p. 90, <http://nova.cat/wp-content/uploads/2011/12/Informe_PMSC_Iraq_Nova_ok.pdf>; "Iraq Shooting Firm Loses Licence", *BBC*, September 17, 2007, <http://news.bbc.co.uk/2/hi/middle_east/6998788.stm>, accessed on June 10, 2014.
338 Hannah Tonkin, (2011) *State Control over Private Military and Security Companies in Armed Conflict*, Cambridge: Cambridge University Press, p. 158.
339 Christine Bakker, "Duties to Prevent, Investigate, and Redress Human Rights Violations by Private Military and Security Companies: The Role of the Host State", in Francesco Francioni and Natalino Ronzitti (Eds.), (2011) *War by Contract: Human Rights, Humanitarian Law, and Private Contractors*, Oxford: Oxford University Press, pp. 137-138.
340 Ibidem, p. 137.

Investigation, prosecution, and punishment

States are also required to actively engage in the investigations into committed crimes and prosecution and punishment of the offenders. In this regard, territorial States must take all appropriate measures to comply with this obligation.[341] Especially under the UN Convention Against Torture, host States have to take positive actions and to prosecute or to extradite members of PMSC personnel that are on the territory of their jurisdiction and are accused of the commission of torture and related forms of treatment, according to Article 7 of the Convention. Time and again, it must be acknowledged that host States can frequently be unable to comply with this duty to punish, due to the immunity agreements concluded with hiring States and practical difficulties of having criminal jurisdiction over responsible individuals.

Specific positive measures to comply with the obligation to punish

The positive actions which host States must take in order to comply with the above human rights duties of prevention and punishment include a number of crucial steps. Host States must adopt legislative acts and other measures, such as rules on provision of PMSC services ensuring compliance of private contractors with human rights standards.[342] Their authorities need to determine which services may be carried out by PMSCs, to set up an authorization system for PMSCs, for example, a licensing regime, and to collect all relevant information regarding the activities of every single PMSC willing to function on their territory.[343] Host States must be clear about quality criteria for authorization and its terms aimed at ensuring that PMSCs comply with international human rights law.[344] It is also necessary to create monitoring and accountability mechanisms to be able to assess the compliance of PMSCs with contracts and national and international law and to address possible human rights vio-

---

[341] Lindsey Cameron and Vincent Chetail, (2013) *Privatizing War: Private Military and Security Companies under Public International Law*, Cambridge: Cambridge University Press, p. 259; Christine Bakker, "Duties to Prevent, Investigate, and Redress Human Rights Violations by Private Military and Security Companies: The Role of the Host State", in Francesco Francioni and Natalino Ronzitti (Eds.), (2011) *War by Contract: Human Rights, Humanitarian Law, and Private Contractors*, Oxford: Oxford University Press, p. 144.

[342] Montreux Document, Part One "Pertinent International Legal Obligations Relating to Private Military and Security Companies" and Part Two "Good Practices Relating to Private Military and Security Companies", pp. 12, 23-24, paras. 10, 43-45.

[343] Ibidem, Part Two "Good Practices Relating to Private Military and Security Companies", p. 21, paras. 24-29; University Centre for International Humanitarian Law, *Expert Meeting on Private Military Contractors: Status and State Responsibility for Their Actions*, Report, Geneva, August 29-30, 2005, pp. 41-42, <http://www.geneva-academy.ch/docs/expert-meetings/2005/2rapport_compagnies_privees.pdf>, accessed on October 10, 2014.

[344] Montreux Document, Part Two "Good Practices Relating to Private Military and Security Companies", pp. 21-23, paras. 30-42.

lations. Host States are also encouraged to cooperate with hiring States and home States in such matters as investigations and regulation of the activities of private contractors.[345] When Blackwater employees killed and injured Iraqi civilians in the Nisour Square incident, the Iraqi National Police started investigations. It means that the Iraqi government made an immediate attempt to comply with its obligation to punish under international human rights law. In addition, the operating license of the company was withdrawn.

4.4 Application to home States

The final part of the analysis will be on the so-called home States where PMSCs are registered or incorporated. In general, international law does not impose broad obligations on home States for regulating or controlling the activities of private corporations operating abroad.[346] As such, home States will not automatically be responsible for the unlawful activities of PMSCs committed on the territory of third States only because there is a territorial connection in terms of the place of the PMSC establishment between those States and private firms.[347] In practice, home States seldom control operations of PMSCs established on their soil when they function abroad.[348] These States, however, can be said to have extensive control and jurisdiction over PMSCs based in their territory.[349] Nicola Jägers makes an argument that home States do exercise control over companies, especially large multinational corporations, such as various PMSCs, due to the fact that there is a range of connections between two actors: political, legal, and economic.[350] Relationships between transnational corporations and home States are currently more interdependent than they were in the past.[351] These States often appear to financially and politically support activities of

---

345 Ibidem, p. 24, para. 52.

346 Ian Brownlie, (1983) *System of the Law of Nations: State Responsibility*, Oxford: Oxford University Press, p. 165; Nicola Jägers, (2002) *Corporate Human Rights Obligations: in Search of Accountability*, School of Human Rights Research Series, Volume 17, Antwerpen, Oxford, New York: Intersentia, p. 166.

347 Hannah Tonkin, (2011) *State Control over Private Military and Security Companies in Armed Conflict*, Cambridge: Cambridge University Press, p. 229.

348 Kathrin Herbst, "Private Security Companies and Civil-Military Cooperation", in Thomas Jäger and Gerhard Kümmel (Eds.), (2007) *Private Military and Security Companies: Chances, Problems, Pitfalls and Prospects*, Wiesbaden: VS Verlag für Sozialwissenschaften, p. 282.

349 Francesco Francioni, "The Role of the Home State in Ensuring Compliance with Human Rights by Private Military Contractors", in Francesco Francioni and Natalino Ronzitti (Eds.), (2011) *War by Contract: Human Rights, Humanitarian Law, and Private Contractors*, Oxford: Oxford University Press, p. 104.

350 Nicola Jägers, (2002) *Corporate Human Rights Obligations: in Search of Accountability*, School of Human Rights Research Series, Volume 17, Antwerpen, Oxford, New York: Intersentia, p. 171.

351 Smita Narula, "International Financial Institutions, Transnational Corporations and Duties of States", in Malcolm Langford, Wouter Vandenhole, Martin Scheinin, and Willem van Genugten, (2014) *Global Justice, State Duties: The Extraterritorial Scope of Economic, Social and Cultural Rights in International Law*, Cambridge: Cambridge University Press, p. 139.

transnational companies and are thus to a high degree involved in their functioning abroad.[352] Therefore, PMSCs as legally established businesses incorporated under the jurisdiction of home States are closely connected to the latter. This relationship, in addition to territorial competences of home States and effective control exercised by them, enables these States to comply with their positive obligations under international law.[353] It is thus not wrong to assume that under international humanitarian and human rights law there are a number of positive obligations applicable to home States that might lead to their international responsibility if they fail to comply with these duties.

4.4.1 Positive State obligations under international humanitarian law

The Montreux Document contains a provision concerning the obligations of home States under international humanitarian law and provides:

> "Home States have an obligation, within their power, to ensure respect for international humanitarian law by PMSCs of their nationality, in particular to:
>
> a) disseminate, as widely as possible, the text of the Geneva Conventions and other relevant norms of international humanitarian law among PMSCs and their personnel;
> b) not encourage or assist in, and take appropriate measures to prevent, any violations of international humanitarian law by personnel of PMSCs;
> c) take measures to suppress violations of international humanitarian law committed by the personnel of PMSCs through appropriate means such as administrative or other regulatory measures as well as administrative, disciplinary or judicial sanctions, as appropriate."[354]

Common Article 1 of the 1949 Geneva Conventions

As has previously been explored, Common Article 1 to the Geneva Conventions and customary international law impose an obligation on all States to take positive measures in order to ensure respect for international humanitarian law by State agents and private actors, including PMSCs, which can be influenced or controlled by them. This obligation applies in all situations: in times of war and peace. States of PMSC incorporation do not always participate in armed conflicts or peace operations in which

---

352 Ibidem.

353 Francesco Francioni, "The Role of the Home State in Ensuring Compliance with Human Rights by Private Military Contractors", in Francesco Francioni and Natalino Ronzitti (Eds.), (2011) *War by Contract: Human Rights, Humanitarian Law, and Private Contractors*, Oxford: Oxford University Press, p. 104.

354 Montreux Document, Part One "Pertinent International Legal Obligations Relating to Private Military and Security Companies", p. 13, para. 14.

PMSCs established on their territory operate. It has been observed by an expert that this provision cannot be interpreted to imply obligations of result for States that are not parties to a conflict.[355] In these scenarios, Common Article 1 is thus an obligation of diligent conduct. It can generally be assumed that in practice home States can exercise certain influence over private contractors established on their territory and be under an obligation to guarantee that these corporations comply with the law of armed conflict.[356] They might be expected to adopt an array of positive measures to comply with this duty to ensure respect of international humanitarian law.

State intervention

It might be the case that in the territorial sense, home States are at a remove from an armed conflict or a peace mission in which PMSCs are deployed. Nevertheless, just as hiring and host States, these States also have to intervene if there is information that private contractors established on their territory and their employees are likely to engage in violations of international law. These States may not simply disregard their obligations under Common Article 1 and remain silent, failing to prevent a possible breach of the law of armed conflict.

Suppressing and repressing violations of the law of armed conflict

In international armed conflicts, home States are obliged to investigate the commission of grave breaches of international humanitarian law and to prosecute and to extradite responsible individuals according to Article 50 of the First Geneva Convention, Article 51 of the Second Geneva Convention, Article 130 of the Third Geneva Convention, and Article 147 of the Fourth Geneva Convention. They also must suppress non-grave breaches in conformity with Articles 49(3) of the First Geneva Convention, Article 50(3) of the Second Geneva Convention, Article 129(3) of the Third Geneva Convention, and Article 146(3) of the Fourth Geneva Convention. In addition, home States must facilitate the creation of legislative, administrative, and judicial arrangements to deal with possible violations and perpetrators.[357]

---

355 University Centre for International Humanitarian Law, *Expert Meeting on Private Military Contractors: Status and State Responsibility for Their Actions*, Report, Geneva, August 29-30, 2005, p. 45, <http://www.geneva-academy.ch/docs/expert-meetings/2005/2rapport_compagnies_privees.pdf>, accessed on October 10, 2014.

356 Hannah Tonkin, (2011) *State Control over Private Military and Security Companies in Armed Conflict*, Cambridge: Cambridge University Press, p. 254.

357 Montreux Document, Part Two "Good Practices Relating to Private Military and Security Companies", p. 27, paras. 71-72.

Specific positive measures to comply with positive obligations

Given that PMSCs are established on their territory in accordance with their legislation and are subject to these laws, States can be regarded as being able to effectively control these companies. Thus, as for contracting States and host States, in order to "ensure respect" for the law of armed conflict, home States are required to regulate the activities of PMSCs and to possess domestic legal frameworks protecting individuals from violations of the law of armed conflict.[358] This duty follows from the general obligation laid down in Common Article 1. Home States need to assess which services carried out by PMSCs can be exported, and to establish an appropriate authorization system for the operation of these companies.[359] This system can ensure that PMSCs comply with certain requirements concerning vetting and training and will not violate the law of armed conflict.[360] If the unlawful conduct takes place, licenses can easily be revoked. In the US, for instance, this duty is complied with. Criteria concerning the level of sufficient training on international humanitarian law, the use of force, and other matters do not form part of the selection process, but they are included in the contracting procedures.[361]

Louise Doswald-Beck argues that the lack of regulation of PMSCs or their insufficient training does not lead to a violation of the obligations of home States under Common Article 1 according to current State practice.[362] Nevertheless, complying with the duty to regulate is an appropriate positive measure that States are required to take in order to meet the more general obligation. It becomes especially pertinent when the State is aware that PMSCs incorporated on its territory are engaged in armed conflicts: in these situations, they need to be properly trained and educated in the matters of the law of armed conflict. A failure to do so might lead to a breach of the obligation ex Common Article 1. It goes without saying that when a home State is also a contracting State, this duty must be complied with – an omission to

---

[358] Lindsey Cameron and Vincent Chetail, (2013) *Privatizing War: Private Military and Security Companies under Public International Law*, Cambridge: Cambridge University Press, p. 256.

[359] Montreux Document, Part Two "Good Practices Relating to Private Military and Security Companies", p. 25, paras. 53-56.

[360] University Centre for International Humanitarian Law, *Expert Meeting on Private Military Contractors: Status and State Responsibility for Their Actions*, Report, Geneva, August 29-30, 2005, pp. 41-42, <http://www.geneva-academy.ch/docs/expert-meetings/2005/2rapport_compagnies_privees.pdf>, accessed on October 10, 2014.

[361] Rebecca DeWinter-Schmitt (Ed.), *Montreux Five Years On: An Analysis of State Efforts to Implement Montreux Document Legal Obligations and Good Practices*, Report, 2013, p. 59.

[362] Louise Doswald-Beck, "Private Military Companies under International Humanitarian Law", in Simon Chesterman and Chia Lehnardt (Eds.), (2007) *From Mercenaries to Market: The Rise and Regulation of Private Military Companies*, Oxford: Oxford University Press, p. 134; University Centre for International Humanitarian Law, *Expert Meeting on Private Military Contractors: Status and State Responsibility for Their Actions*, Report, Geneva, August 29-30, 2005, p. 45, <http://www.geneva-academy.ch/docs/expert-meetings/2005/2rapport_compagnies_privees.pdf>, accessed on October 10, 2014.

train PMSCs and to disseminate knowledge about international humanitarian law will engage its international responsibility, as has been observed above.

4.4.2 Positive State obligations under international human rights law

Following the reasoning incorporated in the Montreux Document, home States have an obligation to comply with the duties to prevent and to punish. The Montreux Document stresses:

> "Home States are responsible to implement their obligations under international human rights law, including by adopting such legislative and other measures as may be necessary to give effect to these obligations. To this end they have the obligation, in specific circumstances, to take appropriate measures to prevent, investigate and provide effective remedies for relevant misconduct of PMSCs and their personnel."[363]

Thus, the Montreux Document, restating current international legal State obligations, clearly considers home States to be under a duty to take necessary positive measures in order to comply with the obligations to prevent and to punish. These States might be able to effectively address numerous concerns raised with regard to PMSCs and their involvement in human rights violations.

As has been stated with respect to hiring States, the human rights obligation to ensure rights appears to be binding on States within their own jurisdiction, which is a territorially limited concept. Therefore, no general obligation resting upon home States can be found in the body of international human rights law that requires these States to take positive measures in order to prevent PMSCs from committing human rights violations abroad.[364] If a home State is, in fact, a State contracting PMSCs or it exercises effective control over the territory where PMSCs operate or over particular individuals, then it might have due diligence obligations to ensure compliance of PMSCs with human rights standards. However, this is not usually the case and, as such, home States where PMSCs are established are considered as being outside the scope of application of this due diligence standard.

According to the report of John Ruggie from 2009, it cannot be denied that the extraterritorial application of the duty to protect under international law is still not settled.[365] It is, however, true that a variety of close links between home States and

---

363 Montreux Document, Part One "Pertinent International Legal Obligations Relating to Private Military and Security Companies", p. 13, para. 15.

364 Hannah Tonkin, (2011) *State Control over Private Military and Security Companies in Armed Conflict*, Cambridge: Cambridge University Press, p. 256.

365 Report of the Special Representative of the Secretary-General on the Issue of Human Rights and Transnational Corporations and Other Business Enterprises, John Ruggie, *Business and Human Rights: Towards Operationalizing the 'Protect, Respect and Remedy' Framework*, UN Human Rights Council, Eleventh Session, April 22, 2009, UN Doc. A/HRC/11/13, p. 7, para. 15.

PMSCs in combination with their territorial competences and effective control exercised over corporations within their jurisdiction can be said to trigger their positive obligations. The reasoning of the ECtHR in one of its judgments is of importance for home States. The ECtHR held that the term "jurisdiction" has no limitations based on the national territory of States Parties to the ECHR.[366] It further stressed that the responsibility of States can be engaged when acts of their authorities produce effects outside their own territory.[367] It seems that when a home State is essentially capable of exercising effective control over private contractors established in its jurisdiction, this decision can also be applied to the activities of PMSCs. In many situations, home States where PMSCs are registered or incorporated can effectively control these companies operating in accordance with their laws and policies. In this respect, these States are not allowed to tolerate the commission of international law violations by these companies and possessing knowledge about their operations are required to comply with the obligations to prevent and to punish.

4.4.2.1 Obligation to prevent

State intervention

Home States might be required under human rights law to intervene when there is a likelihood that a human rights violation is about to be committed by PMSCs domiciled in their territory. These States can be said to be under an obligation to exercise control over PMSCs. Olivier De Schutter observes in this regard:

> "The obligations of States towards internationally recognized human rights may imply that States should, insofar as legally and practically possible, control the activities abroad of the corporations which are incorporated under their jurisdiction..."[368]

If the conduct of PMSCs might lead to human rights abuses, a home State can be obligated to protect human rights and to employ a variety of positive measures in order to exercise control over these corporations.[369] In addition to human rights treaties, this duty can be derived from Article 56 UN Charter and general principles of law.[370]

---

[366] European Court of Human Rights, *Drozd and Janousek v. France and Spain*, Judgment, June 26, 1992, Application No. 12747/87, para. 91.

[367] Ibidem.

[368] Olivier De Schutter, "The Challenge of Imposing Human Rights Norms on Corporate Actors", in Olivier De Schutter (Ed.), (2006) *Transnational Corporations and Human Rights*, Oxford and Portland, Oregon: Hart Publishing, p. 37.

[369] Olivier De Schutter, "The Responsibility of States", in Simon Chesterman and Angelina Fisher (Eds.), (2009) *Private Security, Public Order: The Outsourcing of Public Services and Its Limits*, Oxford: Oxford University Press, p. 34.

[370] Ibidem.

Member States of the United Nations are required to take joint or separate actions in cooperation with this international organization in order to achieve universal respect for and observance of human rights and fundamental freedoms listed in Article 55 UN Charter.

Preventive regulation

In the report of John Ruggie from 2011, it is stressed that "at present States are not generally required under international human rights law to regulate the extraterritorial activities of businesses domiciled in their territory and/or jurisdiction".[371] The report acknowledges, however, the fact that States are generally not prohibited from regulating the extraterritorial activities of these businesses. In practice, many States choose to set up regulatory frameworks for the operations of such companies and their extraterritorial activities.[372] It stands to reason that home States can be considered to be able to regulate the conduct of PMSCs, given that these companies are located in their territory and these States have a wide range of tools to effectively tackle various issues of the PMSC industry.[373] Due to this close link between PMSCs and home States allowing their incorporation on the State territory, some even argue that the duty to regulate under human rights law, particularly the European human rights system, is more prominent for these States.[374] This particular connection between home States and PMSCs might appear to be stronger than a mere contract in the relationship between hiring States and private contractors. In this context, domestic regulation by home States is capable of establishing control over activities of private contractors, increasing their accountability, and ensuring their compliance with human rights law.[375]

It is recommended that home States create an expectation that corporations must respect human rights abroad, especially when these States support the business

---

371 Report of the Special Representative of the Secretary-General on the Issue of Human Rights and Transnational Corporations and Other Business Enterprises, John Ruggie, *Guiding Principles on Business and Human Rights: Implementing the United Nations 'Protect, Respect and Remedy' Framework*, UN Human Rights Council, Seventeenth Session, March 21, 2011, UN Doc. A/HRC/17/31, p. 7.

372 Lindsey Cameron and Vincent Chetail, (2013) *Privatizing War: Private Military and Security Companies under Public International Law*, Cambridge: Cambridge University Press, p. 282.

373 Francesco Francioni, "The Role of the Home State in Ensuring Compliance with Human Rights by Private Military Contractors", in Francesco Francioni and Natalino Ronzitti (Eds.), (2011) *War by Contract: Human Rights, Humanitarian Law, and Private Contractors*, Oxford: Oxford University Press, p. 105.

374 University Centre for International Humanitarian Law, *Expert Meeting on Private Military Contractors: Status and State Responsibility for Their Actions*, Report, Geneva, August 29-30, 2005, p. 35, <http://www.geneva-academy.ch/docs/expert-meetings/2005/2rapport_compagnies_privees.pdf>, accessed on October 10, 2014.

375 Michael Cottier, (2006) "Elements for Contracting and Regulating Private Security and Military Companies", *International Review of the Red Cross*, Volume 88, Issue 863, pp. 639-640.

activities of these companies or are involved in them.[376] This strategy can help home States to distance themselves from the possible abuses committed by businesses and at the same time to provide necessary support to host States often struggling to create effective regulatory mechanisms on their own.[377] By adopting legislative acts with an extraterritorial reach, home States may be required to exercise control and to have oversight over PMSCs incorporated under their jurisdiction but conducting business abroad.[378] It may be argued that this obligation to regulate extends even to the situations when a PMSC is employed by another private corporation.[379] It is also necessary to establish regulatory mechanisms, which may take form of licensing and authorization regimes, in order to ensure PMSC compliance with international law and to prevent possible human rights abuses.[380] Home States should also create tools of "systematic monitoring" in order to have oversight over the conduct of PMSCs and to make sure that their activities are not in breach of international law.[381]

Specific positive measures to comply with the obligation to prevent

Home States need to clearly set out the lines within the array of services exported by PMSCs established in their territory. They are expected to determine which services

---

376 Report of the Special Representative of the Secretary-General on the Issue of Human Rights and Transnational Corporations and Other Business Enterprises, John Ruggie, *Guiding Principles on Business and Human Rights: Implementing the United Nations 'Protect, Respect and Remedy' Framework*, UN Human Rights Council, Seventeenth Session, March 21, 2011, UN Doc. A/HRC/17/31, p. 7.

377 Report of the Special Representative of the Secretary-General on the Issue of Human Rights and Transnational Corporations and Other Business Enterprises, John Ruggie, *Business and Human Rights: Towards Operationalizing the 'Protect, Respect and Remedy' Framework*, UN Human Rights Council, Eleventh Session, April 22, 2009, UN Doc. A/HRC/11/13, pp. 7-8, para. 16.

378 Olivier De Schutter, "The Challenge of Imposing Human Rights Norms on Corporate Actors", in Olivier De Schutter (Ed.), (2006) *Transnational Corporations and Human Rights*, Oxford and Portland, Oregon: Hart Publishing, p. 37; Emanuela-Chiara Gillard, (2006) "Business Goes to War: Private Military / Security Companies and International Humanitarian Law", *International Review of the Red Cross*, Volume 88, Issue 863, p. 559.

379 Olivier De Schutter, "The Responsibility of States", in Simon Chesterman and Angelina Fisher (Eds.), (2009) *Private Security, Public Order: The Outsourcing of Public Services and Its Limits*, Oxford: Oxford University Press, p. 44.

380 Emanuela-Chiara Gillard, (2006) "Business Goes to War: Private Military / Security Companies and International Humanitarian Law", *International Review of the Red Cross*, Volume 88, Issue 863, pp. 559-560; Francesco Francioni, "The Role of the Home State in Ensuring Compliance with Human Rights by Private Military Contractors", in Francesco Francioni and Natalino Ronzitti (Eds.), (2011) *War by Contract: Human Rights, Humanitarian Law, and Private Contractors*, Oxford: Oxford University Press, p. 106.

381 Francesco Francioni, "The Role of the Home State in Ensuring Compliance with Human Rights by Private Military Contractors", in Francesco Francioni and Natalino Ronzitti (Eds.), (2011) *War by Contract: Human Rights, Humanitarian Law, and Private Contractors*, Oxford: Oxford University Press, p. 107.

may or may not be performed abroad.[382] Some activities, such as direct participation in armed conflicts, should and can be prohibited.[383] Furthermore, exporting States could consider establishing an authorization system with appropriate rules on accountability and export and return of weapons and ammunition by PMSCs to ensure that PMSCs are properly selected and comply with national and international law.[384] This system and decisions taken within its framework are to be harmonized with authorization mechanisms of other States.[385] Home States can scrutinize activities of PMSCs in order to determine whether they have engaged in human rights violations in the past and revoke their license or authorization if necessary.[386] As an example, the US has taken some specific positive measures to comply with the duty to prevent by setting up a licensing scheme, as assessed above. South Africa also has a special committee granting licenses to PMSCs but the licensing regime itself appears to be "only skeletal in form".[387] Furthermore, home States should constantly monitor the compliance of PMSCs with the granted authorizations and impose sanctions on companies failing to comply.[388] The "systematic monitoring" by home States can take place by requiring that PMSCs include social impact reports in their annual financial reports and by making use of, for instance, diplomatic missions in the territory of host States.[389] It is interesting to note that the UK, as one of the main home States, has failed so far to take this positive measure of monitoring due to the lack of national regulation aimed at British PMSCs operating overseas.[390] Finally, govern-

---

[382] Montreux Document, Part Two "Good Practices Relating to Private Military and Security Companies", p. 25, para. 53.

[383] Emanuela-Chiara Gillard, (2006) "Business Goes to War: Private Military / Security Companies and International Humanitarian Law", *International Review of the Red Cross*, Volume 88, Issue 863, p. 560.

[384] Montreux Document, Part Two "Good Practices Relating to Private Military and Security Companies", p. 25, paras. 54-55; see also Francesco Francioni, "The Role of the Home State in Ensuring Compliance with Human Rights by Private Military Contractors", in Francesco Francioni and Natalino Ronzitti (Eds.), (2011) *War by Contract: Human Rights, Humanitarian Law, and Private Contractors*, Oxford: Oxford University Press, p. 107.

[385] Montreux Document, Part Two "Good Practices Relating to Private Military and Security Companies", p. 25, para. 56.

[386] Francesco Francioni, "The Role of the Home State in Ensuring Compliance with Human Rights by Private Military Contractors", in Francesco Francioni and Natalino Ronzitti (Eds.), (2011) *War by Contract: Human Rights, Humanitarian Law, and Private Contractors*, Oxford: Oxford University Press, p. 107.

[387] Lindsey Cameron and Vincent Chetail, (2013) *Privatizing War: Private Military and Security Companies under Public International Law*, Cambridge: Cambridge University Press, p. 577.

[388] Montreux Document, Part Two "Good Practices Relating to Private Military and Security Companies", p. 27, paras. 68-69.

[389] Francesco Francioni, "The Role of the Home State in Ensuring Compliance with Human Rights by Private Military Contractors", in Francesco Francioni and Natalino Ronzitti (Eds.), (2011) *War by Contract: Human Rights, Humanitarian Law, and Private Contractors*, Oxford: Oxford University Press, p. 107.

[390] Rebecca DeWinter-Schmitt (Ed.), *Montreux Five Years On: An Analysis of State Efforts to Implement Montreux Document Legal Obligations and Good Practices*, Report, 2013, p. 79.

ments of home States can cooperate with authorities of host States in establishing effective monitoring instruments in the territory of PMSC operation.[391]

4.4.2.2 Obligation to punish

Reactive regulation

For the purpose of complying with the duty to punish, home States are required to adopt regulatory measures enabling them to react on possible instances of PMSC misconduct. They need to have instruments of "reactive monitoring" in place, which are necessary for conducting investigations into human rights violations, prosecuting possible perpetrators, and providing remedies.[392]

Investigation, prosecution, and punishment

Home States can be said to have a duty "to investigate and, as required by international law, or otherwise as appropriate, prosecute, extradite or surrender persons suspected of having committed other crimes under international law, such as torture or hostage taking".[393] In practice, a positive obligation to punish resting upon home States may arise if employees working for PMSCs engage in the commission of torture and return to their home country.[394] The provisions of the CAT require that home States take positive steps to prosecute or to extradite members of PMSC personnel who have committed torture and related forms of treatment, according to Article 7 of the Convention. Also, Article 14 CAT creates a duty for home States to establish national mechanisms, which can be used by the victims of torture to obtain redress and to have a right to compensation. The scope of this provision is not territorially limited and covers an obligation of home States to create necessary procedures for the victims of torture that can be committed anywhere around the globe.[395] In other instances of human rights violations, the duties to investigate and to punish are dependent upon the gravity of human rights violations.[396]

---

[391] Montreux Document, Part Two "Good Practices Relating to Private Military and Security Companies", p. 27, para. 70.

[392] Francesco Francioni, "The Role of the Home State in Ensuring Compliance with Human Rights by Private Military Contractors", in Francesco Francioni and Natalino Ronzitti (Eds.), (2011) *War by Contract: Human Rights, Humanitarian Law, and Private Contractors*, Oxford: Oxford University Press, p. 107.

[393] Montreux Document, Part One "Pertinent International Legal Obligations Relating to Private Military and Security Companies", p. 13, para. 17.

[394] Hannah Tonkin, (2011) *State Control over Private Military and Security Companies in Armed Conflict*, Cambridge: Cambridge University Press, p. 257.

[395] Ibidem.

[396] Francesco Francioni, "The Role of the Home State in Ensuring Compliance with Human Rights by Private Military Contractors", in Francesco Francioni and Natalino Ronzitti (Eds.), (2011) *War by*

Specific positive measures to comply with the obligation to punish

In conclusion, if home States are to follow the good practices suggested by the Montreux Document they will have to adopt certain positive measures. Criminal and non-criminal accountability mechanisms should be in place, which can be used if PMSCs fail to comply with terms of authorization or its personnel commit human rights violations, in particular torture.[397] Depending on the character of a human rights breach, home States can be required to conduct investigations and to prosecute responsible individuals, to make sure that these individuals are relocated, and to revoke a license or authorization of involved PMSCs.[398] The United States has sought to comply with its duty to punish in the previously examined cases of Blackwater and CACI/Titan when it commenced investigations into the incidents as not only a hiring, but also as a home State of these corporations. The question of how effective these efforts were and whether the expected outcome was reached remains open. Finally, in order to address human rights issues of the PMSC industry, international cooperation is crucial.[399] In this vein, home States should also cooperate with host and hiring States in matters of regulation and investigation.[400]

5. State Compliance with Positive Obligations

Under international law, States must comply with their positive duties arising out the law of armed conflict and human rights law. Generally speaking, many States fail to meet their obligations to prevent violations of international law and to punish responsible individuals. In its report, the IACommHR looked into the emergence of private security companies in the Americas and concluded that there is an obvious lack of procedures to sufficiently control PMSCs.[401] In addition, it expressed its concerns about complaints regarding violations of individual rights committed by PMSC employees.[402]

Contract: Human Rights, Humanitarian Law, and Private Contractors, Oxford: Oxford University Press, p. 108.

397 Montreux Document, Part Two "Good Practices Relating to Private Military and Security Companies", p. 27, paras. 70-72; ibidem, Part One "Pertinent International Legal Obligations Relating to Private Military and Security Companies", p. 13, para. 17.

398 Francesco Francioni, "The Role of the Home State in Ensuring Compliance with Human Rights by Private Military Contractors", in Francesco Francioni and Natalino Ronzitti (Eds.), (2011) *War by Contract: Human Rights, Humanitarian Law, and Private Contractors*, Oxford: Oxford University Press, p. 108.

399 Ibidem, pp. 108-109.

400 Montreux Document, Part Two "Good Practices Relating to Private Military and Security Companies", p. 27, para. 73.

401 Inter-American Commission on Human Rights, *Report on Citizen Security and Human Rights*, December 31, 2009, OEA/Ser.L/V/II, p. 28, para. 71, <http://www.cidh.oas.org/pdf%20files/SEGURIDAD%20CIUDADANA%202009%20ENG.pdf>, accessed on May 20, 2014.

402 Ibidem.

One may think about the Nisour Square incident from September 2007 as an illustration of State compliance with positive obligations. In this particular situation, the United States as a hiring and home State of the company failed to prevent not only breaches of the law of armed conflict, but also international human rights law from occurring. It did not intervene in the operations of Blackwater, had no appropriate level of control over their activities, and also failed to protect civilians present in the area. In addition, the compliance of the US government with their duty to suppress and to repress violations of international humanitarian law and obligation to punish under human rights has clearly not been sufficient. The question can legitimately be asked as to whether the already mentioned Order 17 of the Coalition Provisional Authority in Iraq did not violate the due diligence obligation of the United States to take adequate measures for prosecuting and punishing breaches of human rights and humanitarian law.[403] As an occupying power, this State provided immunity to PMSCs from local prosecution in the context of uncertainty regarding the possibilities to prosecute and to punish under the jurisdiction of contracting and home States. While in theory the home State could exercise its jurisdiction in this regard, it significantly limited the capacity of the host State to comply with its due diligence obligations and thus contributed to the situation when PMSCs and their personnel were able to escape responsibility. In September and October 2007, the US Department of State's Diplomatic Security Service and the FBI started investigating the incident, but the question is how effective this investigation and especially prosecution of alleged suspects was. As rightly mentioned by Lindsey Cameron and Vincent Chetail, the US domestic legal system faced some difficulties in the process of prosecuting employees of Blackwater.[404] As discussed in Chapter I, the case against Blackwater guards was closed and reopened again and the trial will finally take place in June 2014. This means that the US authorities needed nearly seven years to bring individuals responsible for violations of international law to justice and it is still unclear whether they will face criminal responsibility. Similarly, it was indicated in Chapter I that in the torture-related incident the cases against CACI and Titan had been dismissed while employees of DynCorp involved in another scandal were not prosecuted at all, but merely sent back home. Therefore, it might be assumed that the United States did not meet its duty to punish after the incidents discussed above: it did not adequately investigate various abuses in which PMSC members had been implicated and failed to prosecute and to punish them. Unfortunately, this particular State is not alone in this regard.

With respect to the proliferation of PMSCs, the Inter-American Commission stressed in its report that the responsibility of States for protecting human rights is a

403 Lindsey Cameron and Vincent Chetail, (2013) *Privatizing War: Private Military and Security Companies under Public International Law*, Cambridge: Cambridge University Press, p. 240.
404 Ibidem, p. 260.

non-delegable obligation and PMSCs cannot be used "to augment or supplant" State duties in the field of citizen security. It listed a set of obligations of State authorities in relation to private contractors.[405] States still hold their monopoly on the legitimate use of force and when many of the governmental tasks are outsourced to private firms, they need to step up and to adopt positive measures meant to prevent, deter, and suppress crimes and violence by PMSC personnel. The operations of PMSCs on the territory of States must be properly regulated. This domestic regulation must focus on the activities that may be performed by PMSCs, categories of weapons they are allowed to use, oversight and monitoring mechanisms, a system of licensing, and reporting. Additionally, governments are required to observe compliance of PMSC employees with selection and training requirements.

There appears to be a discrepancy between the obligations of States under international law and the level of their compliance with these duties. It might be instructive to examine whether they meet the pertinent international legal obligations imposed upon them, especially those requiring positive measures to be taken that can be found in the Montreux Document. Analyzing the possibility that this instrument would evolve into customary international law, Corinna Seiberth touched upon a spectrum of domestic measures some States have adopted so far in accordance with it. This examination reveals the compliance of States with positive duties of international humanitarian and human rights law, given that the Montreux Document is a restatement of pertinent obligations of States under international law. The conclusion she draws is that currently no uniform and consistent State practice of implementing the good practices suggested by this document among States can be traced.[406] These good practices form the most effective positive measures that must be adopted by States for the compliance with due diligence duties.[407] They concern the determination of services that may or may not be performed by PMSCs; procedures with regard to their selection, contracting, and authorization; monitoring; accountability; and other important matters.[408]

As has been noted, States have an obligation to regulate in order to control PMSC activities under the law of armed conflict and are required to adopt regulatory measures for compliance with the duty to prevent and to punish derived from international human rights law. For practical reasons, it is more tangible and efficient to investigate

---

405 Inter-American Commission on Human Rights, *Report on Citizen Security and Human Rights*, December 31, 2009, OEA/Ser.L/V/II, p. 28, paras. 72-73, <http://www.cidh.oas.org/pdf%20files/SEGURIDAD%20CIUDADANA%202009%20ENG.pdf>, accessed on May 20, 2014.

406 Corinna Seiberth, (2013) *Private Military and Security Companies in International Law – A Challenge for Non-binding Norms: the Montreux Document and the International Code of Conduct for Private Security Service Providers*, Cambridge, Antwerp, Portland: Intersentia, p. 255.

407 Lindsey Cameron and Vincent Chetail, (2013) *Privatizing War: Private Military and Security Companies under Public International Law*, Cambridge: Cambridge University Press, p. 266.

408 See Montreux Document, Part Two "Good Practices Relating to Private Military and Security Companies", pp. 16-27.

the level of State compliance with these specific obligations and this exercise might prove to be illustrative of the willingness of a number of States – out of 50 countries being part of the Swiss Initiative – to comply with their duties under international law and to follow good practices laid down in the Montreux Document. If States opt to act in accordance with these practices, they will essentially show that they act with an appropriate level of due diligence required by the standards of international law.[409] A failure to do so might be concluded to be a breach of international obligations of these States resulting in their responsibility.

So far, only a relatively small group of the 50 States have endorsed the Montreux Document. It has been pointed out by the experts that in Latin America and the Carribbean there is practically no interest among States for the Montreux initiative.[410] PMSCs active in this region, are, however, the most armed contractors worldwide.[411] The countries that have joined the initiative can be divided in three groups: those that have not regulated PMSC industry so far, those that are in the process of adopting some regulatory measures, and those that have quite elaborate regulation focusing on PMSCs.

One State that has decided not to regulate the industry of PMSCs is Germany.[412] Its government is a supporter of compliance with international obligations restated in the Montreux Document and believes that it acts in accordance with these standards without any need to introduce national legislation on the matter. In addition, there are States taking a step further and considering the adoption of regulatory measures, such as Switzerland and France.[413] Finally, some States adhere to the Montreux Document and recognize their pertinent legal obligations, but fail to comply with all good practices. The United Kingdom, as one of the main home and hiring States of PMSCs, supports the Montreux Document, but has adopted a rather limited scope of measures to comply with its provisions. It has chosen the path of self-regulation of the industry and developing and adhering to international norms and standards. The set of these norms and standards includes the Montreux Document, the ICoC, and the 2009/2010 consultation document of the UK's Foreign and Commonwealth Office specifying

---

409 Lindsey Cameron and Vincent Chetail, (2013) *Privatizing War: Private Military and Security Companies under Public International Law*, Cambridge: Cambridge University Press, p. 286.

410 Rebecca DeWinter-Schmitt (Ed.), *Montreux Five Years On: An Analysis of State Efforts to Implement Montreux Document Legal Obligations and Good Practices*, Report, 2013, p. 33.

411 Ibidem; Graduate Institute of International and Development Studies, *Small Arms Survey 2011: States of Security*, Geneva, 2011, p. 114, <http://www.smallarmssurvey.org/publications/by-type/yearbook/small-arms-survey-2011.html>, accessed on July 10, 2014.

412 Corinna Seiberth, (2013) *Private Military and Security Companies in International Law – A Challenge for Non-binding Norms: the Montreux Document and the International Code of Conduct for Private Security Service Providers*, Cambridge, Antwerp, Portland: Intersentia, p. 229.

413 Ibidem, pp. 247-248.

international sanctions and victim recompense.[414] On the one hand, the UK strongly supports the Swiss Initiative and requires PMSCs it contracts with to adhere to the ICoC, but has not adopted domestic laws on certification and licensing of these companies, which are required for its compliance with the duty to regulate forming part of the duty to prevent human rights violations and its duty to control PMSCs it hires under international humanitarian law. At the moment, the existing national legislation cannot be said to be sufficient or effective for these purposes,[415] but there is a very positive development in form of the application of the ANSI/ASIS PSC.1-2012 standard to UK-based contractors providing security services abroad.[416] The United States as another major home and hiring State of PMSCs has put some regulatory instruments in place to address these corporations. For instance, it has extensive contracting and selection rules and procedures that clearly reflect many of the Montreux Document's good practices.[417] Nevertheless, there are some areas in its legislation concerning authorization, licensing, selecting, and contracting that require significant improvement. The ITAR does not mention international humanitarian and human rights law or their norms, training on rules on the use of force, or cultural and religious matters.[418] There is a lack of regulation, DoD instructions, and similar instruments making explicit references to training on international human rights law.[419]

In conclusion, it may be observed that there is no broad compliance of States with the Montreux Document. What is clear, however, is that "a beginning process of internalisation" of some of its recommendations is taking place, especially regarding the duty to regulate.[420] A fairly low degree of State compliance with the Montreux Document time and again reinforces the importance of the UN Guiding Principles that might be a promising framework for ensuring that States act in accordance with their positive obligations. In this regard, national action plans prepared by some governments can reflect traces of evidence for State practice that serve as an indication

---

414 Alexandra Bohm, Kerry Senior, and Adam White, "The United Kingdom", in Christine Bakker and Mirko Sossai (Eds.), (2012) *Multilevel Regulation of Military and Security Contractors: The Interplay Between International, European and Domestic Norms*, Oxford and Portland, Oregon: Hart Publishing, p. 324.

415 Ibidem, p. 312; Rebecca DeWinter-Schmitt (Ed.), *Montreux Five Years On: An Analysis of State Efforts to Implement Montreux Document Legal Obligations and Good Practices*, Report, 2013, pp. 9, 61-65.

416 UK Foreign and Commonwealth Office, "Private Security Companies", Written Ministerial Statement, December 17, 2012, <http://www.publications.parliament.uk/pa/cm201213/cmhansrd/cm121217/wmstext/121217m0001.htm>, accessed on June 24, 2014.

417 Rebecca DeWinter-Schmitt (Ed.), *Montreux Five Years On: An Analysis of State Efforts to Implement Montreux Document Legal Obligations and Good Practices*, Report, 2013, pp. 8, 55-60.

418 Ibidem, p. 56.

419 Ibidem, p. 60.

420 Corinna Seiberth, (2013) *Private Military and Security Companies in International Law – A Challenge for Non-binding Norms: the Montreux Document and the International Code of Conduct for Private Security Service Providers*, Cambridge, Antwerp, Portland: Intersentia, p. 256.

of the process of internalization of the UN Guiding Principles. They can be used to understand the position of States towards some of their positive due diligence obligations under international law.

The European Commission has called upon the Member States of the EU to develop national action plans to implement the Guiding Principles by the end of 2012.[421] The EU Council also invited these States to do so in 2013.[422] At the moment of writing, only four States have produced a national action plan: the United Kingdom, The Netherlands, Italy, and Denmark.[423] Switzerland and Finland are in the process of publishing their plans of action. The national action plan of the UK, as the first country to deliver its contribution, is a valuable document reflecting the policies of one of the main home States to PMSCs and also a major contracting State. In its plan, the government emphasizes various policies to further implement its State obligations to protect human rights within jurisdiction of the UK where businesses operate.[424] It holds a view that there is no general requirement to regulate extraterritorial activities of companies established under its jurisdiction, for human rights duties apply only within the territory of a State and/or within its jurisdiction. The UK is required to do so only in a number of limited circumstances, such as under some human rights regimes, and may consider regulating extraterritorial activities of its businesses. The government underlined its role in the establishment of the ICoC and made clear that it sought to address human rights abuses by means of public procurement rules allowing public bodies to exclude certain tenderers from competing for a contract when there are indications that companies have been involved in violations of human rights. Importantly, the UK aims at regulating the PMSC industry by introducing a certification process and working towards creation of an international monitoring and oversight mechanism within the ICoC. Only certified companies that participate in the ICoC will be allowed to contract with State and non-State clients. Additionally, in order to ensure access to remedy for human rights abuses, the government puts much emphasis on its judicial remedy possibilities and non-judicial

---

421 European Commission, A Renewed EU Strategy 2011-2014 for Corporate Social Responsibility, Communication from the Commission to the European Parliament, the Council, the European Economic and Social Committee and the Committee of the Regions, October 25, 2011, COM(2011) 681, p. 14, <http://eur-lex.europa.eu/LexUriServ/LexUriServ.do?uri=COM:2011:0681:FIN:EN:PDF>, accessed on May 30, 2014.

422 Council of the European Union, EU Strategic Framework and Action Plan on Human Rights and Democracy, June 25, 2012, <http://www.consilium.europa.eu/uedocs/cms_data/docs/pressdata/EN/foraff/131181.pdf>, accessed on May 30, 2014.

423 OHCHR, "State National Action Plans", <http://www.ohchr.org/EN/Issues/Business/Pages/NationalActionPlans.aspx>, accessed on May 30, 2014.

424 The Government of the United Kingdom, *Good Business: Implementing the UN Guiding Principles on Business and Human Rights*, September 2013, 1.Introduction, <https://www.gov.uk/government/uploads/system/uploads/attachment_data/file/236901/BHR_Action_Plan_-_final_online_version_1_.pdf>, accessed on May 30, 2014.

grievance mechanisms that will be established in cooperation with companies and local authorities.

The Netherlands submitted its national action plan a few months after the United Kingdom.[425] Although the plan does not explicitly refer to the obligations of the State with regard to the conduct of PMSCs, it contains a number of duties taken into account in relation to businesses in general. The Dutch government clearly understands the importance of a proactive stance that it must take in order to comply with the duty to protect.[426] It has a role to play in raising awareness of the need for due diligence by offering companies information and other forms of aid.[427] In addition, it must also apply the standard of due diligence in relation to its own activities, for example, financing businesses abroad, export credit insurance, and trade missions.[428]

The Netherlands stresses that State-controlled companies and businesses, in which the government invests, are expected to comply with the Guiding Principles.[429] Under the national sustainable procurement policy, companies that supply the government with goods or provide other services are required to comply with human rights standards.[430] Also, transparency and reporting procedures need to be improved. Large companies will be obliged to report on human rights related and other issues.[431] The government seeks to extend the scope for remedy, including judicial and non-judicial grievance mechanisms, and established the ACCESS Facility to build up knowledge and to facilitate effective dispute settlement between companies and communities at the local level.[432] It did not, however, accept the idea of effectiveness of legislation with extraterritorial application for initiating criminal and civil law proceedings against foreign companies in order to prevent human rights abuses.[433] The Netherlands Institute for Human Rights indicated in its advisory report that the government had rejected the necessity of such legislation regulating extraterritorial activities of corporations and that there is sufficient support for it in international case law and literature.[434]

---

425 The Government of the Netherlands, *National Action Plan on Business and Human Rights*, <http://www.netherlandsmission.org/binaries/content/assets/postenweb/v/verenigde_staten_van_amerika/the-permanent-mission-to-the-un/actionplanbhr.pdf>, accessed on May 30, 2014.

426 Ibidem, pp. 3-4.

427 Ibidem, p. 7.

428 Ibidem, p. 8.

429 Ibidem, p. 4.

430 Ibidem, p. 5.

431 Ibidem, p. 9.

432 Ibidem, pp. 10-13.

433 Ibidem, p. 13.

434 Netherlands Institute for Human Rights, *Response to the National Action Plan on Business and Human Rights "Knowing and Showing"* (Dutch: *Reactie op het Nationaal Actieplan Bedrijfsleven en Mensenrechten "Knowing and Showing")*, Advisory Report, February 2014, pp. 10-11, <http://zoekservice.mensenrechten.nl/StippWebDLL/Resources/Handlers/DownloadBestand.ashx?id=2236>, accessed on July 10, 2014.

The Danish government also came up with a national action plan to implement the UN Guiding Principles.[435] It does not mention PMSCs, but addresses State policies to comply with the Ruggie Framework with respect to business in general. Among other things, the Danish government enhances the level of transparency and implements reporting procedures with regard to the measures taken by the companies to respect human rights.[436] The government also struggled with the adoption of extraterritorial application and sought to encourage more discussion on the matter.[437] Like Denmark, Italy does not discuss the question of proliferation of PMSCs and its positive measures taken to comply with the duties under international law in relation to these non-State actors. It devotes significant attention to the first and third pillars of the Guiding Principles and focuses for the major part on agricultural policies.[438]

One can conclude that only a handful of States have submitted their national action plans on the implementation of the UN Guiding Principles. While Denmark aims at attaining a higher level of understanding of the extraterritorial application of legislation for granting access to justice, The Netherlands does not see a potential in this endeavor. The United Kingdom considers itself as not being under an obligation to establish legislation regulating the activities of private contractors abroad and may only consider doing so. Interestingly, the UK government changed its position with respect to the regulation method of its PMSC industry. It takes necessary steps towards establishing a certification process that will be helpful in preventing international law violations by PMSCs and will require that only certified companies participating in the ICoC are hired. Generally speaking, the governments seem to realize that they need to take an active position and to require that companies comply with human rights under their procurement policies. Unfortunately, no extensive and coherent practice can be established here: only a small number of States have shared their action plans. There is, however, a positive development to be perceived that States are willing to comply with an essentially non-binding set of principles and actively adjust their policies to comply with it. In this regard, their obligations to prevent and to punish human rights violations in the context of private businesses and their operations clearly have a prominent place in the course of State action based on a variety of positive measures to be adopted. In this vein, the United Kingdom, being one of the main home State to PMSCs, made its decision to regulate the industry, but

---

435 The Government of Denmark, *Danish National Action Plan – Implementation of the UN Guiding Principles on Business and Human Rights*, March 2014, <http://www.ohchr.org/Documents/Issues/Business/NationalPlans/Denmark_NationalPlanBHR.pdf>, accessed on May 30, 2014.

436 Ibidem, p. 18.

437 Ibidem, p. 15.

438 The Government of Italy, *The Foundations of the Italian Action Plan on the United Nations "Guiding Principles on Business and Human Rights"*, pp. 3-59, 65-75, <http://www.ohchr.org/Documents/Issues/Business/NationalPlans/NationalPlanActionItaly.pdf>, accessed on July 10, 2014.

is not willing to go as far as addressing extraterritorial activities of companies domiciled in its territory.

6. Conclusion

It is sometimes said that positive obligations arising out of international law and their relevance for the State practice of using non-State entities and private persons, such as PMSCs, are not clear enough and cannot be considered effective.[439] In the bodies of international human rights law and the law of armed conflict, however, they form an important source of State duties with regard to PMSCs and other private persons and entities and non-compliance with them might lead to State responsibility. Chia Lehnardt appears to be quite optimistic when she argues that if the personnel of PMSCs violate rules and principles of international humanitarian and human rights law, such conduct can generate the international responsibility of hiring States or host States for the mere failure to prevent or to respond to those illegal acts.[440] This might also be the case even if there is no clarity regarding the actual participation of State organs in breaching those rules. Lindsey Cameron and Vincent Chetail are also positive about the possibility of engaging State responsibility for failure to comply with positive obligations, more specifically the duties of due diligence, that are available in situations when the private conduct cannot be attributed to States.[441] Hannah Tonkin believes that positive obligations from international humanitarian and human rights law may impose not only duties to prevent, but also duties to punish on all three types of States, including home States, and that States failing to meet these standards will face international responsibility.[442]

Having made an analysis of the concepts and issues associated with them, it cannot be denied that the array of these positive obligations can indeed be very useful in the process of holding States responsible for violations of international humanitarian and human rights law by private contractors. If the illegal acts of PMSCs cannot be attributed to States according to the rules of attribution laid down in the Draft Articles, States can also be held responsible for those violations due to their failure to

---

[439] Carsten Hoppe, (2008) "Passing the Buck: State Responsibility for Private Military Companies", *The European Journal of International Law*, Volume 19, Issue 5, pp. 1012-1014; Chia Lehnardt, "Private Military Companies and State Responsibility", in Simon Chesterman and Chia Lehnardt (Eds.), (2007) *From Mercenaries to Market: The Rise and Regulation of Private Military Companies*, Oxford: Oxford University Press, p. 153.

[440] Chia Lehnardt, "Private Military Companies and State Responsibility", in Simon Chesterman and Chia Lehnardt (Eds.), (2007) *From Mercenaries to Market: The Rise and Regulation of Private Military Companies*, Oxford: Oxford University Press, pp. 154-155.

[441] Lindsey Cameron and Vincent Chetail, (2013) *Privatizing War: Private Military and Security Companies under Public International Law*, Cambridge: Cambridge University Press, p. 286.

[442] Hannah Tonkin, (2011) *State Control over Private Military and Security Companies in Armed Conflict*, Cambridge: Cambridge University Press, pp. 63-65.

meet a set of positive obligations arising out of international law, divided into obligations of result and diligent conduct. In particular, the doctrine of due diligence obligations is of crucial importance for the State practice of using PMSCs in conflict areas and is also applicable with regard to States employing PMSCs, host States where these companies operate, and home States where they are established.[443] While hiring States can be regarded as having the largest number of positive obligations under both international humanitarian and human rights law, host States and home States having less obvious duties also need to act in a certain way by taking positive steps to ensure the compliance of PMSCs with these significant branches of international law. It is conceivable that one particular State can be seen as three distinct types of States and bear a wide spectrum of responsibilities, however, it is not often the case and these groups of States are usually responsible for meeting their positive obligations separately. If they fail to comply with positive obligations imposed by international humanitarian and human rights law, this will lead to the international responsibility even if the unlawful conduct of PMSCs in question cannot be attributed to them under the classic doctrine of State responsibility, which is a significant development that will undoubtedly promote and ensure the compliance of States with international law.

Finally, the analysis of State compliance with international obligations has been performed on the basis of the Montreux Document and the UN Guiding Principles. A number of States took into account the good practices offered by the Montreux Document and implemented them to a certain extent. Also, some governments adhering to the Ruggie Framework issued national action plans that served as an indication to their compliance or willingness to comply with some positive duties under international law. A brief examination of these cases revealed that there is no consistent level of compliance among States with the positive obligations under international humanitarian law and human rights treaties. In addition, these are merely a few States, meaning that it is not possible to draw credible conclusions regarding the overall compliance with the above-mentioned duties among the rest of the international community. There is, however, a positive step forward to be perceived here: some States appear to realize the importance of complying with the duties to prevent and to punish human rights abuses and positive duties under the law of armed conflict, and modifying their policies in accordance with these standards.

---

443 University Centre for International Humanitarian Law, *Expert Meeting on Private Military Contractors: Status and State Responsibility for Their Actions*, Report, Geneva, August 29-30, 2005, p. 34, <http://www.geneva-academy.ch/docs/expert-meetings/2005/2rapport_compagnies_privees.pdf>, accessed on October 10, 2014.

CHAPTER VII
SUMMARY, OVERALL CONCLUSIONS, AND FINAL OBSERVATIONS

The price of greatness is responsibility.

Sir Winston Churchill[1]

INTRODUCTION

The State use of private contractors providing military and security services is an increasingly popular practice that has notably found its expression in various areas around the world in the era of globalization. In conflict zones, PMSCs have been deployed to carry out numerous tasks ranging from the provision of training and advice to the interrogation of prisoners and actual participation in hostilities. What has come to light, are several violations of international law, including international humanitarian and human rights law, that have been committed by the personnel of these corporations. These breaches reveal the impotence of not only domestic and transnational legal frameworks, but also international law to deal with the issues of control, oversight, and accountability with respect to PMSCs. The main aim of the present study is to investigate the possibility of application of the State responsibility doctrine to PMSC operations in conflict-affected areas. Having analyzed the phenomenon of privatization of war and security and the possibility of invoking international responsibility of hiring, host, and home States for the unlawful conduct of these legal entities and their employees breaching international law, the main focus of this final Chapter is on providing a summary, drawing some general conclusions, and making final observations with regard to the posed research questions. The main question of the study is:

> "To what extent should and can States be held responsible for the misconduct of PMSCs, taking place in conflict areas and constituting breaches of international humanitarian and human rights law?"

---

1 A speech given by Sir Winston Churchill on September 6, 1943 at Harvard, Massachusetts, United States, where he received an honorary degree. See Winston S. Churchill, (1993) "The Price of Greatness is Responsibility", *Finest Hour*, No. 80, p. 25, <http://www.winstonchurchill.org/images/finesthour/Vol.01%20No.80.pdf>, accessed on May 3, 2014.

The investigation concerned the extent and circumstances of attribution of the misconduct of PMSCs and their personnel under the law of State responsibility and the scope and circumstances of the application of positive State duties under international humanitarian and human rights law to the deployment of private contractors in conflict zones.

Misconduct and Moral Implications

It has been established that currently States rely heavily on the employment of PMSCs and outsource a large variety of tasks to these non-State entities in times of war and peace. As noted by many experts, these businesses and their personnel have frequently engaged in numerous illegal activities that can be put in a number of categories: indiscriminate shootings and civilian casualties; intentional killings on behalf of governments; property damage; torture and other cruel, inhuman, or degrading treatment or punishment; sexual abuses, including sex trafficking, sex slavery, and running prostitution networks; drug running; profiteering in transnational criminal networks; weapons violations; and mercenary activities. In practice, it is not an easy task to start criminal or civil law suits against individual companies and their personnel for these violations due to various legal and other constraints. Therefore, the pertinent question that bears asking is: should States be held responsible for this misconduct? Before the legal questions are responded to, it should be observed that there are two main moral implications of the use of private military and security contractors by States in the context of war: their lack of attachment to a cause and reduced level of democratic control caused by their involvement in armed conflicts and peace operations. From a legal perspective, the key issue is that States have the monopoly over the use of force, but often outsource important State functions and other activities to private actors. By adopting this course of action, they have the possibility to turn a blind eye to the instances of misconduct in which companies and individuals hired by them might be implicated. This practice enables States to pursue their goals on the battlefield and in other scenarios without having to face international responsibility. In the light of these considerations, it is concluded that the current situation needs to be changed and the reassessment of the rules and principles on State responsibility leads to the acknowledgment that States should be held responsible for violations of international law, from which they must not be allowed to disassociate themselves.

The Phenomenon of PMSCs

In the course of human history, private actors offering military and security expertise to a large number of customers appear to be commonplace. Following the emergence of the nation-State, some remarkable changes in the perception of individuals participating in conflicts and earning money were introduced: mercenaries were certainly no longer the norm. After the end of the Cold War, a massive explosion in the proliferation of PMSCs had taken place and the world was shaken under the boots of "corporate soldiers of fortune". While employees working for PMSCs show some notable similarities with regular mercenaries, in most cases these individuals cannot qualify as such. Without applying a legal definition of a mercenary under international law to PMSC employees, one might recognize that PMSCs are in essence legally established for-profit businesses having a corporate structure and often operating as international firms. They have the potential to exercise force and perform various military and security tasks on behalf of States, as well as international organizations and other corporations. The diverse and fragmented industry of private military and security contractors, well-known for its transparency and legitimacy issues, keeps maintaining significant levels of growth and there are no signs of slowing down. While under national laws PMSCs are legitimate business entities registered or incorporated in the territory of States under their domestic legislation, at the international level they can be seen as transnational or multinational corporations that often operate as "quasi-State actors". This poses many challenges in the context of international law when these firms operating transnationally are employed by States and the conduct of PMSCs and associated members of their personnel is at the core of discussion.

Legal Framework of PMSC Activities

The existing legal framework of operation of PMSCs as transnational companies hired by governments encompasses in the first place rules and principles of public international law, specifically international humanitarian and human rights law. While the former is applied in all situations of armed conflict, both international and internal, the application of especially the latter field of international law can become a more complicated matter. Although there are human rights having the status of *jus cogens* norms and forming part of customary international law, in principle States must become party to a human rights treaty in order to be bound by it. In addition, human rights treaties contain derogation clauses allowing States to derogate from their duties under human rights law in certain exceptional circumstances. States may not introduce derogation clauses from certain human rights, such as the prohibition of torture and the right to life.[2] In the context of employment of PMSCs, they can lawfully resort to derogations from the right not to be subjected to arbitrary or unlawful detention due to the fact that this right is derogable.[3] This could mean that these particular States would be able to violate human rights while at the same time avoiding

2 See Article 4(2) ICCPR, Article 15(2) ECHR, and Article 27(2) ACHR.

3 Christine Bakker, "Duties to Prevent, Investigate, and Redress Human Rights Violations by Private Military and Security Companies: The Role of the Host State", in Francesco Francioni and Natalino Ronzitti (Eds.), (2011) *War by Contract: Human Rights, Humanitarian Law, and Private Contractors*, Oxford: Oxford University Press, p. 136.

responsibility for such conduct. Finally, the body of international human rights law may not be applicable given the different extent and circumstances of extraterritorial applicability of several human rights treaties. In general, it is clear that when States exercise effective control over territories and persons abroad, they are bound to comply with human rights obligations arising out of the human rights conventions to which they are party.

In addition to the law of armed conflict and human rights law, there are a number of essential international regulatory initiatives on business and human rights and specific legal instruments with the focus on PMSCs, such as the UN Guiding Principles and the Montreux Document. These international legal instruments could provide some inspiration on the scope of State obligations under international humanitarian and human rights law and serve as an important evaluation tool to delineate them. Both might prove to be significantly relevant to the operations of PMSCs in the future. Furthermore, national laws and regulations of some States focus on the industry of private military and security service providers. The extent of this domestic regulation varies in different jurisdictions: only some States have extensively regulated PMSCs so far, with the United States having the most extensive scope of regulatory measures. As has been stressed, national regulation offers rather ineffective mechanisms for holding PMSCs and their employees accountable for breaches of international law when active beyond State borders. Transnational private regulatory regimes also do not present a viable option to achieve justice for the misconduct committed by private contractors. These schemes are voluntary in nature and – among other drawbacks – lack effective monitoring, oversight, and enforcement mechanisms. Finally, PMSC activities are also to a certain extent regulated by the hiring contracts, standard operating procedures, rules of engagement, and status of forces agreements. The status of forces agreements and other orders can often be adopted to protect the personnel working for PMSCs abroad from local prosecution.[4] They notably diminish the capacity of host States to prosecute individuals involved in the commission of violations of international law on their soil and might create an actual image of impunity of civilian contractors.

LEGAL STATUS OF PMSC EMPLOYEES

The legal status of PMSC employees in armed conflicts determines the scope of their rights and duties and is closely connected to the obligations and responsibilities of States towards those who are considered combatants or civilians. These individuals are in essence private persons employed by PMSCs through direct hire or under a contract. In international armed conflicts, employees of these corporations can be

---

[4] Cedric Ryngaert, (2008) "Litigating Abuses Committed by Private Military Companies", *The European Journal of International Law*, Volume 19, Issue 5, pp. 1048-1049.

considered members of the armed forces of a party to an armed conflict only in a very limited number of situations. Similarly, their status as members of militias or volunteer corps forming part of the armed forces also has little practical relevance. It has been observed that under some circumstances they can be seen as members of other militias or volunteer corps. Most probably, the vast majority of members of PMSC personnel have the legal status of civilians accompanying armed forces without actually being members of these forces. The Geneva law protects this particular group of individuals and grants to them the right to be prisoners of war upon capture by an adverse party to an international armed conflict. It is also possible that PMSC employees belong to the group of ordinary civilians or, in a very limited number of scenarios, mercenaries. In armed conflicts of a non-international character, these persons cannot easily meet the required criteria to be regarded as members of the armed forces of a legitimate government, armed opposition groups, or mercenaries. Most likely, and depending on the context of the cases at hand, their status in this specific type of armed conflict is the status of civilians.

THE ILC'S DRAFT ARTICLES AND ATTRIBUTION

This study attempted to closely examine the invocation of State responsibility for the actions and omissions of PMSCs and their employees that resemble violations of international humanitarian and human rights law. The central question of whether States can be held responsible for the breaches of international law must be answered in the affirmative; the degree of this responsibility, however, varies according to particular circumstances of a specific case under analysis. The application of international responsibility doctrine to the deployment of PMSCs in conflict scenarios has been approached from two different perspectives. The first angle of international responsibility was chosen to specifically focus on the concept of direct responsibility and to address the issue of attribution of private acts to States under the ILC's Draft Articles on State Responsibility. The second perspective stems from the notion of indirect State responsibility and the availability of a range of positive State obligations under the law of armed conflict and international human rights law that might substantially contribute to the possibility of holding States accountable for the essentially private conduct of PMSCs.

The general international law rules and principles of State responsibility set out in the Draft Articles deal with such matters as establishment of an internationally wrongful act, implementation of responsibility, circumstances precluding wrongfulness, legal consequences, and other related concepts. The Draft Articles are used to determine whether an internationally wrongful act has been committed. In the first place, the question of whether the conduct of PMSCs, and other natural persons, groups of individuals, and entities, can qualify as an act of State must be examined under international law. This connection between the conduct violating international law and a State is established by the process of attribution.

In the present study, the importance of attribution in relation to the misconduct of PMSC employees and companies themselves has clearly been demonstrated. It has been established that, in theory, State responsibility can be generated for their misconduct violating international law on four main attribution grounds included in Articles 4, 5, 8, and 11 DARS. Article 4 DARS contains a general rule used to attribute breaches of international humanitarian and human rights law to States if PMSCs fall into the category of official State organs and have this status according to an internal State law. This might be the case when they are incorporated into the State apparatus by means of, for instance, possible integration into the regular army and there is a domestic law specifically conferring on them the legal status of members of the armed forces. This provision does not have much practical value, as only in exceptional circumstances PMSCs can be seen as official State organs when they are active in conflict zones.[5] Another attribution provision that is unlikely to be effective in practice is Article 11 DARS. States acknowledging and adopting unlawful activities of PMSCs and their personnel as their own trigger international responsibility for the conduct of private actors ex Article 11 DARS. It does not seem probable that on a large scale States would be willing to comply with the requirements of this particular provision. Therefore, its relevance will largely remain a theoretical one.

Cases brought before the ICJ, international human rights adjudicatory bodies, and even domestic courts against States employing PMSCs or in any other manner connected to the deployment of these companies in conflict zones are most likely to be successful in assigning State responsibility for the misconduct of contractors and their employees in two particular situations. Articles 5 and 8 DARS constitute two modes of attribution of the highest importance in the situations when PMSCs are deployed in conflict areas. Article 5 DARS has the potential to be an effective tool in holding States responsible for the unlawful activities of private contractors exercising elements of governmental authority on the basis of State law. When PMSCs are employed to perform inherently governmental functions, such as direct participation in armed conflicts or guarding of prisons and interrogation of prisoners, and operate under an internal law of a State, the application of Article 5 DARS is activated. If one opts to use a more flexible notion of the domestic law, it might even be acknowledged that employment contracts concluded between States and PMSCs and licensing and certification schemes serve as its substitute when they have the same purpose of delegating State functions to PMSCs or authorizing them to perform governmental activities. This would make the application of Article 5 DARS to PMSCs more realistic. In order to prevent possible instances of abuses to be attributed to States, these States must refrain from using private contractors that exercise elements of governmental

---

[5] Olivier De Schutter, "The Responsibility of States", in Simon Chesterman and Angelina Fisher (Eds.), (2009) *Private Security, Public Order: The Outsourcing of Public Services and Its Limits*, Oxford: Oxford University Press, p. 26.

authority or take necessary measures to ensure that governmental tasks are exercised in accordance with international law. For instance, the Draft International Convention on PMSCs discourages the process of outsourcing the exercise of "fundamental State functions" to private contractors and requires governments to prohibit the use of PMSCs for essentially governmental tasks.[6] In reality, States might still rely on the provision of these services by private contractors and, by triggering the imputation of their conduct, be held responsible.

The most significant attribution provision of the DARS in relation to the misconduct of PMSCs is undoubtedly Article 8. This concerns PMSCs and any other private persons and entities acting under instructions, direction, or control of States. On the one hand, the question of instructions and direction provided by States is rather straightforward and does not introduce a ground for attribution that can be easily applied in practice. While some might argue that there is a high probability of successful imputation of the misconduct of PMSCs to States when the companies act on State instructions,[7] it is rather a difficult standard of proof to meet. In both situations of instructions and direction, it needs to be proved that a State has specifically instructed or directed the commission of violations of international law. On the other hand, the issue of control poses some difficulties of application while at the same time it constitutes the most relevant factual link between States and PMSCs. In this regard, the dictum of Clyde Eagleton may perfectly reflect the situation of a State controlling activities of private actors and being accountable for possible instances of misconduct: "responsibility derives from control".[8]

It has been shown that there are currently two main control theories having a variable degree of relevance for PMSC operations: these have been put forward by the ICJ and the ICTY. In the first place, the "effective control" test developed by the ICJ in the *Nicaragua* case and confirmed in the *Bosnian Genocide* case requires that States exercise effective control over military or other operations in which PMSCs violate international law. In this regard, it is necessary to establish that States provide a high degree of "practical support" to operations or the actual realization of misconduct. This particular control theory creates "an extremely high threshold" for

6 See Articles 4(3) and 9 of the Draft International Convention in UN Human Rights Council, *Draft of a Possible Convention on Private Military and Security Companies (PMSCs) for Consideration and Action by the Human Rights Council*, Prepared by the Working Group on the use of mercenaries as a means of violating human rights and impeding the exercise of the right of peoples to self-determination, May 13, 2011, First Session, Geneva, May 23-27, 2011, UN Doc. A/HRC/WG.10/1/2, pp. 6, 9.

7 Robert McCorquodale and Penelope Simons, (2007) "Responsibility Beyond Borders: State Responsibility for Extraterritorial Violations by Corporations of International Human Rights Law", *Modern Law Review*, Volume 70, Issue 4, p. 610.

8 Clyde Eagleton, (1950) "International Organization and the Law of Responsibility", *Collected Courses of the Hague Academy of International Law (Recueil des Cours)*, Volume 76, p. 385.

attributing the private conduct to States under Article 8 DARS.[9] It has also been proved to be rather impractical in this regard. Therefore, this test is not suitable for establishing the actual link between PMSCs and States and may result in the non-attributability of acts and omissions of the former to the latter.

In contrast to the "effective control" notion, the "overall control" test created by the ICTY in the *Tadić* case significantly lowers the threshold of attribution under Article 8 DARS. The Tribunal found that the degree of control is variable according to the circumstances of a particular case. It stressed that in case of organized and hierarchically structured groups acting on behalf of States, the control may also be of an overall character. If States not only finance, train, and equip or provide operational support to PMSCs, but also play a role in organizing, coordinating, or planning military and other actions of these companies, they might be considered to exercise control over these non-State entities and persons. Although the ICJ appeared reluctant to follow the reasoning of the ICTY in the ICJ's *Bosnian Genocide* case, there is most certainly some merit to the use of the Tribunal's control test in relation to unlawful acts of PMSCs. Currently, PMSCs are deployed in various scenarios, including armed conflicts and peacekeeping missions, and States relying on these companies can easily avoid being held responsible due to the fact that they in practice do not exercise a significantly high level of effective control over them. The adherence to the less strict "overall control" test might be seen as a more realistic approach to State responsibility when well-organized and hierarchically structured PMSCs are controlled by States. This control theory is perfectly in line with the rules and principles of customary international law. This seeks to prevent States from escaping their responsibility for the breaches of international law committed by private individuals and groups and is capable of closing the responsibility gaps left open by the ICJ's reasoning. The "overall control" test fiercely advocated by the late Antonio Cassese is relevant and cannot be simply put aside and disregarded by international judicial bodies, especially the ICJ. The World Court, however, clearly chose the path of the higher degree of control and if this theory is given preference, it will be highly unlikely that States will face responsibility if in reality they have a lower level of control over private contractors than the effective control.

Secondly, besides the successful imputation of the conduct of PMSCs to States, there must be a breach of an international obligation of a State in order to incur its international responsibility. As has been made clear, States have duties under international humanitarian and human rights law that they are required to comply with. In order to establish international responsibility of States, it must be determined whether States are bound by the law of armed conflict and international human rights law

[9] Sigrun I. Skogly, "Causality and Extraterritorial Human Rights Obligations", in Malcolm Langford, Wouter Vandenhole, Martin Scheinin, and Willem van Genugten (Eds.), (2013) *Global Justice, State Duties: The Extraterritorial Scope of Economic, Social and Cultural Rights in International Law*, Cambridge: Cambridge University Press, pp. 241-242.

by participating in armed conflicts and having effective control over territories and persons, even when extraterritorial operations of State agents or PMSCs are at stake. Additionally, it has to be examined whether the misconduct of PMSCs materializes as a breach of international State obligations. It is the case that most instances of misconduct of PMSCs and their personnel analyzed in the study can qualify as violations of international State obligations. These breaches in combination with successful attribution to respective States will result in State responsibility. In this regard, rules and principles of customary international law reflected in the ILC's Draft Articles constitute an essential source of secondary rules to assess and to incur international responsibility.

If a claim is made against States breaching international law through actions and omissions of PMSCs and their employees, these States have the possibility to invoke six circumstances precluding wrongfulness. These are consent, self-defense, countermeasures, *force majeure*, distress, and necessity. In the context of PMSC operations in conflict zones, the most relevant ground for precluding wrongfulness is that of self-defense, which must be exercised in accordance with the UN Charter. While States do defend themselves, they might resort to the use of PMSCs against attacking States. Their self-defense, however, does not preclude the wrongfulness of the conduct of private contractors and their personnel breaching international humanitarian law and non-derogable human rights.[10] If the wrongfulness of the misconduct cannot be precluded and States are held responsible, they will face a number of legal consequences: they will be required to cease the wrongful conduct and to make reparations. Full reparations for moral and material damages include restitution, compensation, and satisfaction. When PMSC employees, for instance, engage in violent shootings and kill innocent civilians or torture detainees, the most appropriate types of reparation will be compensation of victims and their relatives and satisfaction, given that restitution – and thus re-establishment of the situation as it existed before internationally wrongful acts were committed – is impossible in this regard.

Positive State obligations

The second part of the study dealt extensively with the concept of indirect responsibility of States for unlawful acts of private contractors and members of their personnel carrying out military and security tasks. As shown above, there is still a responsibility gap when PMSCs are used by States in conflict areas, as their activities are imputable to States only in rare situations. In effect, this implies that States manage to escape their responsibility for breaches of international law committed by PMSCs and their staff. A wide spectrum of positive obligations, however, present a valuable additional way of holding States responsible under the regime of the ILC's Draft Articles that

---

10 Commentary to DARS as adopted in 2001, p. 74, Article 21, para. 3.

relies heavily on different types of attribution. These positive State obligations stem from the fields of both international humanitarian and human rights law and constitute duties of both result and diligent conduct requiring States to take all necessary measures to achieve a certain result or to prevent and to punish. While positive State duties form an additional instrument for engaging State responsibility, there are certain difficulties especially with applying human rights treaties, given that States may derogate from some of their provisions under a limited number of circumstances and the extraterritorial reach of those conventions can only extend beyond "own" borders of States in the two particular situations explored above.

It has been concluded that under international humanitarian law, hiring, host, and home States have a number of obligations of result. They are required to take a variety of positive measures in relation to certain groups of persons in the situations of armed conflict. Those can relate to, for instance, the duty to identify sick, wounded, dead, or shipwrecked individuals and to guarantee adequate detention conditions of prisoners of war if these tasks are outsourced to private contractors. Furthermore, in addition to an obligation to disseminate the Geneva law among PMSCs, States are bound to train and to instruct PMSCs in them. In the field of human rights, some positive obligations can also be constructed with respect to PMSCs: these concern the right to fair trial and the right to life. In this regard, States guaranteeing both human rights must achieve the result and, if they resort to PMSCs to ensure these rights, positive measures must be adopted.

The most important category of positive obligations comprises due diligence obligations. Even before the famous *Velasquez Rodriguez* decision was made, an expert on the issue of missing and disappeared persons in Chile, Judge Abdoulaye Dieye from Senegal, in his report to the UN General Assembly, referred to the notion of due diligence. He made clear that a State is responsible for acts and omissions with respect to disappearances if, *inter alia*, the State authorities fail to promptly react to reliable reports, there are no relevant legal remedies or they are not effective, the State fails to act in order clarify the situation on the basis of existing evidence, or it takes no action to punish responsible individuals.[11] Currently, the due diligence concept is inseparable from the body of international law.

Under international humanitarian law, all three types of States have certain duties of diligent conduct that apply to them to a different extent. Hiring States appear to have most obligations with the widest reach. From the general obligation "to ensure respect" laid down in Article 1 common to the 1949 Geneva Conventions, one can

---

[11] UN Commission on Human Rights, *Extrajudicial, Summary or Arbitrary Executions*, Report of the Special Rapporteur, Philip Alston, December 22, 2004, UN Doc. E/CN.4/2005/7, para. 73; see also Joanna Bourke-Martignoni, "The History and Development of the Due Diligence Standard in International Law and Its Role in the Protection of Women against Violence", in Carin Benninger-Budel (Ed.), (2008) *Due Diligence and Its Application to Protect Women From Violence*, Leiden: Koninklijke Brill NV, p. 50.

derive a large number of positive measures to be taken by all States. Those measures include establishing domestic regulation of PMSCs and appropriate licensing and certification mechanisms. Furthermore, hiring, host, and home States have a duty to intervene when a breach is likely to happen and an obligation to suppress and to oppress violations. Additionally, some duties might be the exclusive domain of contracting and territorial States, often being parties to an armed conflict. They are obligated to protect certain groups of persons in conflict zones, such as civilians, and when they act as occupying powers they must prevent, suppress, and oppress violations of international humanitarian law.

The body of international human rights law consists of various treaties containing an array of positive obligations of diligent conduct. These are obligations to prevent and to punish. Following the steps made in the analysis, one might observe that both duties contain sub-duties further specifying the course of action that needs to be followed by a State. The duty to prevent comprises two more specific obligations to intervene and to adopt "preventive regulation" regarding PMSCs and their activities. The duty to punish consists of two obligations to investigate, prosecute, and punish, and to adopt "reactive regulation" aimed at *ex post facto* addressing the unlawful conduct. Both duties concerning regulation are admittedly duties of result and are necessary for complying with the duties to prevent and to punish. Given that contracting States have an effective tool of a contract with PMSCs, they have most obligations under this branch of international law and their scope is most extensive. When there is a likelihood that a PMSC or its personnel would commit a human rights abuse, this State is required to act and to control the activities of the company and persons in question in order to prevent a possible violation of human rights law. This control and careful planning of PMSC operations can be said to be necessary even when the State is not aware that these companies and employees would engage in instances of misconduct. Contracting States are also required to protect women and children in armed conflicts. Furthermore, in order to effectively prevent human rights violations, these States must adopt regulatory measures, such as oversight and monitoring mechanisms and administrative frameworks. Finally, the duty to punish can be complied with by establishing legislation and using other means to ensure that States can adequately meet their obligation to investigate, prosecute, and punish.

Although host States are in a close proximity to the conduct of PMSCs, against the background of armed conflicts they are often not able to meet their positive duties under international human rights law. Frequently, they seem to lack the necessary institutional capacity and do not exercise effective control over their own territories in situations of occupation. Having no contracts with PMSCs, these States need to rely on their national regulatory frameworks to meet various positive obligations. Under the duty to prevent, they are required to intervene and to control activities of all PMSCs present in their territory, especially those companies known for their human rights abuses. The protection of individuals whose life might be at risk and women and children also needs to be ensured. In order to establish a sufficient degree of control over PMSCs, territorial States must adopt regulatory measures, such as certification schemes, and refrain from certain arrangements, for example, immunity agreements with other States. Furthermore, the duty to punish requires that these States adopt legislation and other forms of regulation to facilitate investigations, prosecution, and punishment when human rights abuses are committed. This duty is specifically important in situations when torture and related types of treatment occur.

The final group, home States, is the most controversial when it comes to their positive obligations under human rights regimes. These States are not unquestionably responsible for the unlawful conduct committed abroad by the PMSCs domiciled in their territory. They do, however, have a strong capacity to exercise control and jurisdiction over these non-State actors and are often connected to them by means of political, financial, and other ties. Therefore, these States are also bound to comply with certain positive duties. In order to prevent human rights violations, they must intervene and control PMSCs when they are likely to engage in abuses. Given that PMSCs are registered and incorporated in the territory of home States and this connection is more substantial than an instrument of a contract, States in question must adopt regulatory measures, such as authorization and certification regimes, to prevent human rights breaches. They also have an obligation of punishment. They are expected to adopt regulations for investigating, prosecuting, and punishing human rights violations, depending on their gravity, such as instances of torture.

It is clear that the law of armed conflict, more specifically the Geneva law, and international human rights law, obligate States to take a sufficiently wide range of positive measures in relation to PMSCs. Some positive developments can be perceived at the level of a number of States willing to comply with good State practices listed in the Montreux Document and to adhere to the UN Guiding Principles framework. It is notable that the United States, a home State to many PMSCs, has changed its mind and decided to set up a certification system and to adopt other positive measures that specifically concern PMSCs. The number of States taking the lead in this process is still very limited and it seems that many governments simply disregard their duties under international law while contracting these corporations, hosting them in their territory, or allowing them to be registered on their soil.

The way to deal with the uncertainties concerning the spectrum of positive human rights obligations of home States – but also their application to hiring and host States – with regard to transnational operations of PMSCs and other businesses could be the adoption of a new international convention that could explicitly address these duties, determine their scope, and even further extend it.[12] While the Montreux Document

---

[12] Olivier De Schutter, (2006) *Extraterritorial Jurisdiction as a Tool for Improving the Human Rights Accountability of Transnational Corporations*, Report and Background Paper to the seminar organized in collaboration with the Office of the UN High Commissioner for Human Rights in Brussels on November 3-4, 2006, p. 51, <http://www.reports-and-materials.org/Olivier-de-Schutter-report-for-SRSG-re-extraterritorial-jurisdiction-Dec-2006.pdf>, accessed on July 9, 2014.

has some serious weaknesses, the adoption of a new convention focusing on PMSCs could further strengthen the international legal framework on private contractors and set out State obligations under international law.[13] The 2010 Draft International Convention on PMSCs prepared by the UN Working Group on the Use of Mercenaries could be perceived as such a progressive international legal instrument. Its provisions relating to the issue of State responsibility recognize positive State obligations under international humanitarian and human rights law and specifically mention the notion of due diligence that needs to be applied by PMSCs and their staff.[14] It is obvious that such a treaty must have sufficiently strong oversight and accountability mechanisms in order to be truly effective. The UN Draft Convention aims at establishing an oversight and monitoring mechanism that seeks to ensure State compliance with the treaty provisions by means of reporting, inquiry and complaints procedures, individual and group petitions.[15] It is, however, yet to be seen how much interest States will have for supporting the Convention and which States will become a party to it. It seems that in the debates regarding the UN Draft Convention the main States in the field of private military and security service provision – the United States, the United Kingdom, and some States from the EU – were not in favor of such a treaty, while a large number of developing countries on the UN Human Rights Council expressed opposite views.[16]

There is certainly some merit to the position adopted by the Western States belonging to the first group of countries mentioned above. Their preference goes to an alternative regulation system than that envisaged by a possible legally binding UN Convention on PMSCs. Currently, processes of globalization and privatization can require a variety of measures to be taken beyond the scope of State influence. For responding to a myriad of issues associated with the PMSC industry, it is necessary that these corporate stakeholders possessing much knowledge and expertise of their field of activity engage in regulatory processes with States and other stakeholders,

---

13 Corinna Seiberth, (2013) *Private Military and Security Companies in International Law – A Challenge for Non-binding Norms: the Montreux Document and the International Code of Conduct for Private Security Service Providers*, Cambridge, Antwerp, Portland: Intersentia, p. 240.

14 UN Human Rights Council, *Draft of a Possible Convention on Private Military and Security Companies (PMSCs) for Consideration and Action by the Human Rights Council*, Prepared by the Working Group on the use of mercenaries as a means of violating human rights and impeding the exercise of the right of peoples to self-determination, May 13, 2011, First Session, Geneva, May 23-27, 2011, UN Doc. A/HRC/WG.10/1/2, Article 7(1) and (2); see also Article 4 and Parts III and IV.

15 Ibidem, Part V.

16 Nigel D. White, "Regulatory Initiatives at the International Level", in Christine Bakker and Mirko Sossai (Eds.), (2012) *Multilevel Regulation of Military and Security Contractors: The Interplay Between International, European and Domestic Norms*, Oxford and Portland, Oregon: Hart Publishing, pp. 16-18; Corinna Seiberth, (2013) *Private Military and Security Companies in International Law – A Challenge for Non-binding Norms: the Montreux Document and the International Code of Conduct for Private Security Service Providers*, Cambridge, Antwerp, Portland: Intersentia, p. 241.

such as NGOs.[17] These public-private partnerships can attract many financial and other resources from their participants in addition to those solely available to the government and be more effective in taking into account different approaches and utilizing this knowledge.[18] In this sense, traditional international law solutions might indeed be less effective than new governance approaches that form a central theme of many current debates. The notion "new governance" implies a "shift away from a certain traditional mode of governing associated with 'government'".[19] In this context, self-regulation of the industry and adoption of non-binding legal instruments, such as the Montreux Document, are considered to have the potential of filling the existing regulatory void.

As an example, one could discuss the decision of the United Kingdom not to adopt laws addressing PMSCs, but to opt for the route of self-regulation of the industry mainly through the special interest group SCEG open to all UK-based PMSCs that are members to the ICoC. Despite the fact of being a participating State in the Montreux initiative, the UK is not willing to acknowledge obligations of home States of private military and security contractors that are listed in the Montreux Document.[20] Such voluntary schemes reflecting the "protect" and "remedy" sides of the UN Guiding Principles are associated with the danger that States may choose to join them, but remain reluctant to comply with their suggested good practices and recommendations.[21] Furthermore, other self-regulatory arrangements of the industry itself focusing on the "respect" side of the UN Guiding Principles do not necessarily address the question of positive obligations of States and their application with regard to PMSCs, but are concerned with corporate responsibility to act in accordance with human rights standards and international humanitarian law.

Currently, it is not an overstatement to say that in order to address the issues that go hand in hand with the activities of PMSCs in conflict areas around the world, it is of great importance to identify and to recognize a variety of positive State obligations of diligent conduct stemming from the field of international law, leading to a treaty or not.[22] For the realization of these duties, it is necessary to strengthen them at the

---

17 David M. Trubek and Louise G. Trubek, (2007) "New Governance & Legal Regulation: Complementarity, Rivalry, and Transformation", *Columbia Journal of European Law*, Volume 13, Issue 3, p. 542.

18 Anne-Marie Slaughter, "International Law and International Relations Theory: Twenty Years Later", in Jeffrey L. Dunoff and Mark A. Pollack (Eds.), (2012) *Interdisciplinary Perspectives on International Law and International Relations: The State of the Art*, Cambridge: Cambridge University Press, p. 620.

19 Kenneth Armstrong and Claire Kilpatrick, (2007) "Law, Governance, or New Governance? The Changing Open Method of Coordination", *Columbia Journal of European Law*, Volume 13, Issue 3, p. 651.

20 Corinna Seiberth, (2013) *Private Military and Security Companies in International Law – A Challenge for Non-binding Norms: the Montreux Document and the International Code of Conduct for Private Security Service Providers*, Cambridge, Antwerp, Portland: Intersentia, p. 8.

21 Ibidem.

22 Alexandra Bohm, Kerry Senior, and Adam White, "The United Kingdom", in Christine Bakker and Mirko Sossai (Eds.), (2012) *Multilevel Regulation of Military and Security Contractors: The Interplay*

national level by adopting the set of positive measures assessed in the present study and enforcing them against PMSCs themselves.

FINAL OBSERVATIONS

In conclusion, it may be submitted that international law is certainly not an omnipotent legal framework, but – without doubt – its different branches, particularly international humanitarian and human rights law, create a variety of State obligations that governments of not only hiring and host, but also home States must meet when faced with the deployment of PMSCs in conflict areas. When these companies and their employees commit human rights abuses or violate international humanitarian law, all three types of States can be held responsible on the basis of attribution of unlawful activities or on the ground of non-compliance with positive obligations under the law of armed conflict and human rights law. In this regard, State responsibility is a valuable instrument preventing violations of international law and ensuring compliance of States with their international obligations. It addresses abuses committed by PMSCs and their staff, achieves justice, and ensures that these private actors respect international law.

The famous saying attributed to Seneca would perfectly recapture future prospects of the application of the doctrine of State responsibility in relation to the misconduct of PMSCs and their employees breaching international law – *"Per aspera ad astra"*. There is a long way to go, through many difficulties and uncertainties that international legal scholars and policy makers as well as NGOs are currently dealing with, in order to reach the stars where the great are courageous enough to face their responsibilities. The damage incurred by PMSCs must be repaired when these non-State actors are incorporated in the State machinery, are authorized to carry out essential governmental tasks and operate under State control, and when States acknowledge and adopt their conduct as their own. Additionally, reparations must be provided when States disregard a wide spectrum of their positive duties. Importantly, there is a strong need for far-reaching actions on the part of State authorities aimed at preventing human rights abuses and breaches of the law of armed conflict from occurrence and punishing possible perpetrators. In the meantime, it is appropriate to acknowledge that international law offers all the necessary tools to engage State responsibility for the unlawful conduct of PMSCs and their personnel, but only time will tell whether, and if so, how they are going to be effectively used.

Between International, European and Domestic Norms, Oxford and Portland, Oregon: Hart Publishing, p. 327.

DUTCH SUMMARY
NEDERLANDSE SAMENVATTING

In dit boek staan het concept van "de privatisering van oorlog en veiligheid" en het leerstuk van staatsaansprakelijkheid centraal. Tegenwoordig worden de zogenoemde private militaire en veiligheidsbedrijven – "private military and security companies" of PMSCs – door diverse actoren in conflictgebieden ingezet. Internationale organisaties, NGO's, multinationals en staten gebruiken deze bedrijven voor het uitvoeren van een breed scala aan taken, zoals het beveiligen van diplomaten en het bewaken van gevangenissen. In de afgelopen jaren bleken PMSCs bij verschillende schendingen van internationaal recht – internationaal humanitair recht en mensenrechten – betrokken te zijn. Hierbij zijn nationale rechtssystemen, transnationale rechtsraamwerken en internationaal recht vaak niet in staat gebleken om cruciale aspecten van het functioneren van deze ondernemingen en hun personeel, zoals de controle en toezicht over hun operaties en aansprakelijkheid voor hun misdrijven, effectief te reguleren. Het voornaamste doel van dit promotieonderzoek is na te gaan of het mogelijk is om op grond van internationaal recht staten aansprakelijk te stellen voor de illegale activiteiten van PMSCs en hun werknemers. De onderzoeksvraag die dit onderzoeksproject tracht te beantwoorden, luidt:

> "In hoeverre zouden staten op grond van internationaal recht aansprakelijk gesteld moeten worden en daadwerkelijk aansprakelijk gehouden kunnen worden voor het wangedrag van PMSCs dat in conflictgebieden plaatsvindt en als schendingen van internationaal recht aangemerkt kan worden?"

In deze studie wordt de kwestie van staatsaansprakelijkheid onderzocht aan de hand van de toerekening van misdragingen van PMSCs en hun werknemers aan staten en een spectrum aan positieve verplichtingen van staten onder internationaal humanitair recht en mensenrechtenrecht. Meer in het bijzonder wordt er ingegaan op de omstandigheden en de reikwijdte van de toerekening en de toepassing van positieve staatsverplichtingen. De hypothese van het onderzoek is dat staten voor activiteiten van PMSCs en hun personeel aansprakelijk gesteld zouden moeten worden en dat ze in bepaalde gevallen en tot op zekere hoogte daadwerkelijk aansprakelijk kunnen zijn.

Hoofdstuk 1 is in essentie een inleiding waarin de volgende onderwerpen aan bod komen: de problematiek, de methodologie van de studie, de onderzoeksvraag, de hypothese, onderzoeksdoeleinden, typen misdragingen van PMSCs en hun werknemers, morele implicaties van de inzet van PMSCs door staten in gewapende conflicten en de structuur van het boek. Er wordt geconstateerd dat PMSCs momenteel door staten voor een groot aantal taken in conflictgebieden worden gebruikt. Het blijkt dat deze bedrijven en hun personeel zich bij het verrichten van hun werkzaamheden aan verschillende illegale activiteiten schuldig hebben gemaakt in het verleden. Ze waren betrokken bij de volgende incidenten die in de literatuur en in de media belicht zijn: willekeurige aanvallen en beschietingen van burgers, liquidaties namens regeringen, het toebrengen van schade aan persoonlijke eigendommen, folteringen en andere vormen van gerelateerde misdragingen, mensensmokkel, seksueel misbruik, drugsmisdrijven, deelname aan criminele organisaties, misdrijven met betrekking tot wapenbezit en deelname aan gewapende conflicten als huurlingen. Het gebruik van PMSCs door staten in gewapende conflicten kan vanuit een moreel oogpunt problematisch zijn en geeft deze staten een mogelijkheid om hun aansprakelijkheid te ontwijken. Ten eerste zijn PMSCs ondernemingen die voornamelijk een financieel oogmerk hebben en per definitie niet dezelfde belangen nastreven als die van een staat die partij bij een gewapend conflict is. Ten tweede leidt de inzet van PMSCs door regeringen in conflictgebieden tot de verzwakking van de democratische controle over de deelname aan gewapende conflicten en andere essentiële overheidstaken. Er kan dus gesteld worden dat staten voor schendingen van internationaal recht begaan door PMSCs en hun werknemers in conflictgebieden de aansprakelijkheid zouden moeten dragen.

Hoofdstuk 2 richt zich op het fenomeen van "de privatisering van oorlog en veiligheid". Door de eeuwen heen werden huurlingen door diverse personen en entiteiten in talloze conflicten gebruikt. In dit Hoofdstuk wordt een beeld van de grootschalige inzet van deze personen in de Oudheid en de Middeleeuwen geschetst. Vervolgens wordt de situatie in de zeventiende, achttiende en negentiende eeuw besproken. Na het sluiten van de Vrede van Westfalen in 1648 en de totstandkoming van de moderne notie van de staat werden huurlingen niet meer gezien als een norm en hun gebruik nam aanzienlijk af. In de jaren zestig van de vorige eeuw konden ze weer in verschillende delen van de wereld hun activiteiten ontplooien. Na het einde van de Koude Oorlog in het begin van de jaren negentig kwam de industrie van PMSCs, zoals we die nu kennen, tot bloei. Deze ondernemingen werden naar nationaal recht opgericht, hadden een duidelijke bedrijfsstructuur en -cultuur en opereerden dikwijls als internationale bedrijven. Dit zijn de belangrijkste verschillen tussen PMSCs en gewone huurlingen terwijl er ook overeenkomsten zijn, zoals financiële motieven van beide typen particulieren voor het leveren van hun diensten. Enerzijds zijn PMSCs private ondernemingen onder nationaal recht, anderzijds kunnen ze onder internationaal recht als transnationale corporaties gezien worden. In het algemeen kunnen ze gedefinieerd worden als naar nationaal recht opgerichte bedrijven met een bedrijfsstructuur die militaire en veiligheidstaken uitvoeren, vaak als internationale ondernemingen functioneren en tot het gebruik van geweld kunnen overgaan.

Tegenwoordig worden PMSCs ingehuurd door verschillende actoren, zoals bedrijven, internationale en niet-gouvernementele organisaties en staten. Te onderscheiden zijn drie soorten staten die bij het gebruik van PMSCs betrokken zijn: contracterende staten die deze ondernemingen inhuren, gastheerstaten waar ze actief zijn en thuisstaten waar ze gevestigd zijn. Dit onderscheid moet in aanmerking genomen worden bij de analyse van internationaalrechtelijke verplichtingen van staten en de reikwijdte van hun aansprakelijkheid. Zowel de transparantie als de legitimiteit van PMSCs laten te wensen over, maar de industrie blijkt in een constante ontwikkeling te zijn waarbij een groot aantal contracten tussen PMSCs en hun klanten wereldwijd voor zeer hoge en kleinere bedragen gesloten wordt. Helaas zijn deze contracten vaak niet beschikbaar voor een breed publiek of ze worden verborgen achter een reeks van contracten met onderaannemers.

Het is niet correct om te stellen dat de industrie van PMSCs niet gereguleerd is. Er zijn bepaalde reguleringsinstumenten die op PMSCs en staten van toepassing zijn wanneer deze bedrijven in conflictgebieden opereren. Het doel van Hoofdstuk 3 van het boek is om te bespreken wat het rechtsraamwerk is waarin PMSCs in deze context hun taken uitvoeren. Dit raamwerk wordt in de eerste plaats gevormd door de rechtsregels en principes van internationaal publiekrecht, internationaal humanitair recht en mensenrechtenrecht. Deze zijn te vinden in onder andere de ILC-Ontwerpartikelen inzake Staatsaansprakelijkheid, de Conventies van Genève en diverse mensenrechtenverdragen. Vooral de extraterritoriale werking van mensenrechtenverdragen schept obstakels voor de toepassing van mensenrechten op contracterende staten en thuisstaten en de bespreking hiervan komt uitgebreid aan bod in dit Hoofdstuk. Er wordt geconcludeerd dat de meeste mensenrechtenverdragen extraterritoriale toepassing genieten als een staat die partij bij zo'n verdrag is, effectieve controle over een grondgebied van een andere staat uitoefent of macht en autoriteit over individuen heeft. Vervolgens is er een aantal internationale reguleringsmechanismen dat op PMSCs en bedrijven in het algemeen betrekking heeft, zoals het Montreux Document en de UN Guiding Principles. Verder reguleert de nationale wetgeving van sommige staten PMSCs en hun activiteiten en vormen contracten een rechtsbasis voor de inzet van PMSCs. Er zijn ook "standard operating procedures", "rules of engagement" en "status of forces agreements" die op de operaties van deze ondernemingen van toepassing zijn. Tot slot zijn transnationale private reguleringsmechanismen, zoals de International Code of Conduct for Private Security Service Providers, erop gericht om het gedrag van PMSCs in goede banen te leiden. Bij deze initiatieven ontbreken vaak echter niet alleen monitoring- maar ook handhavingsmechanismen wat een zekere mate van effectiviteit wegneemt.

Wanneer PMSCs in gewapende conflicten actief zijn, is de juridische positie van hun personeel dikwijls verre van duidelijk. In Hoofdstuk 4 is de analyse van de rechtsstatus van deze personen te vinden. Deze rechtsstatus wordt toegekend aan bovengenoemde individuen in gewapende conflicten en heeft invloed op niet alleen hun rechten en plichten maar ook op de verplichtingen en verantwoordelijkheden van staten met betrekking tot bepaalde personen. Werknemers van PMSCs worden rechtstreeks of door middel van een contract ingehuurd. In internationale gewapende conflicten kunnen ze slechts in een gelimiteerd aantal gevallen beschouwd worden als leden van de gewapende macht van een partij bij een gewapend conflict en als leden van de militiën en vrijwilligers-korpsen die deel van deze gewapende macht uitmaken. Dit aspect wordt verkend aan de hand van de evaluatie van artikel 43 van het Eerste Aanvullende Protocol bij de Conventies van Genève en artikel 4(A)(1) van de Derde Conventie van Genève. Onder bepaalde omstandigheden kunnen deze personen echter aangemerkt worden als leden van andere militiën en leden van andere vrijwilligers-korpsen, zoals neergelegd in artikel 4(A)(2) van de Derde Conventie van Genève. Naar alle waarschijnlijkheid vallen de meeste leden van het personeel van PMSCs onder de categorie van personen die de gewapende macht volgen zonder dat ze daarvan deel uitmaken, als bedoeld in artikel 4(A)(4) van de Derde Conventie van Genève. Deze burgers worden beschermd door de Conventies van Genève en hebben de status van krijgsgevangenen als ze in een internationaal gewapend conflict gevangen genomen worden. Het is ook mogelijk dat werknemers van PMSCs gezien moeten worden als burgers die onder de Vierde Conventie van Genève vallen. Verder kunnen deze personen als huurlingen bestempeld worden op grond van artikel 47 van het Eerste Aanvullende Protocol bij de Conventies van Genève en relevante bepalingen van de VN- en OAU-Conventies inzake huurlingen. In conflicten die niet van een internationaal karakter zijn, zijn leden van het personeel van PMSCs slechts in een beperkt aantal situaties te kwalificeren als leden van de strijdkrachten van een staat en dissidente strijdkrachten of andere georganiseerde gewapende groepen die in artikel 1(1) van het Tweede Aanvullende Protocol bij de Conventies van Genève genoemd worden. Bovendien is het onwaarschijnlijk dat ze onder de definitie van huurlingen vallen die in de VN- en OAU-Conventies inzake huurlingen is vastgelegd en in niet-internationale gewapende conflicten van toepassing is. Gezien de specifieke omstandigheden van de deelname van PMSCs aan dit type gewapende conflicten hebben hun werknemers meestal de rechtsstatus van burgers.

Hoofdstuk 5 snijdt de problematiek van de staatsaansprakelijkheid aan en is voornamelijk gewijd aan het inroepen van deze aansprakelijkheid voor het doen en nalaten van PMSCs en hun personeel die als schendingen van internationaal humanitair recht en mensenrechten aangemerkt kunnen worden. Staten kunnen aansprakelijk gehouden worden voor schendingen van internationaal recht, maar de mate van hun aansprakelijkheid hangt af van de concrete omstandigheden van elk specifiek geval. Naast staten kunnen ook internationale organisaties aansprakelijk zijn voor deze activiteiten, vooral in de context van vredesmissies. Dit Hoofdstuk betreft de zogenoemde directe aansprakelijkheid van staten die met het concept van toerekening verweven is. Artikel 1 van de ILC-Ontwerpartikelen inzake Staatsaansprakelijkheid bepaalt dat elke internationale onrechtmatige daad die door een staat begaan wordt, tot de aansprakelijkheid van deze staat leidt. Deze aansprakelijkheid brengt bepaalde rechtsgevolgen met zich mee: zo moet de aansprakelijke staat volgens artikelen 30 en 31 van de ILC-Ontwerpartikelen de onrechtmatigheid beëindigen en in rechtsherstel voorzien. Krachtens artikel 2 is er sprake van een internationale onrechtmatige daad als een handelen of nalaten van een staat aan deze staat kan worden toegerekend en een schending van een internationale verplichting van deze staat vormt. In de eerste instantie wordt er onderzocht of de bovengenoemde misdragingen van PMSCs en hun werknemers krachtens de ILC-Ontwerpartikelen aan staten toegerekend kunnen worden. Door middel van de toerekening wordt er vastgesteld of deze handelingen met de handelingen van staten zelf gelijkgesteld kunnen worden. Er zijn vier toerekeningsgronden die in dit opzicht van belang zijn; deze zijn vastgelegd in artikelen 4, 5, 8 en 11 van de Ontwerpartikelen. Artikel 4 is van toepassing als PMSCs op grond van het interne recht van een betrokken staat als officiële staatsorganen optreden. Dit kan het geval zijn als ze in de krijgsmacht van een staat zijn geïncorporeerd en als de interne regels van de staat deze rechtsstatus aan hen toekent. Enkel in een zeer beperkt aantal gevallen kunnen PMSCs en hun personeel in de hoedanigheid van staatsorganen optreden en daarom heeft artikel 4 weinig praktische waarde met betrekking tot the functioneren van deze bedrijven in conflictzones. Een andere bepaling van de Ontwerpartikelen die minder relevant is, is artikel 11 dat vereist dat een staat door middel van erkenning en adoptie van de misdragingen van PMSCs en hun werknemers de verantwoordelijkheid voor hun gedrag neemt. Aangezien staten in het algemeen niet geneigd zijn om onrechtmatige acties van PMSCs en andere private actoren als hun eigen te erkennen en te adopteren, is de toepassing van artikel 11 in dit geval ook niet vanzelfsprekend.

Uit het onderzoek is gebleken dat de meest relevante bepalingen van de ILC-Ontwerpartikelen artikelen 5 en 8 zijn. De eerste toerekeningsgrond vastgelegd in artikel 5 betreft de situaties waarin PMSCs gemachtigd zijn om op grond van het recht van een staat elementen van overheidsgezag uit te oefenen. Wat deze overheidstaken inhouden, hangt van de maatschappij in kwestie, haar geschiedenis en tradities af. Als PMSCs worden ingezet om bijvoorbeeld aan gewapende conflicten deel te nemen en tot het gebruik van geweld over te gaan en daartoe bij wet gemachtigd zijn, dan is hun gedrag aan staten toe te rekenen. Vervolgens bevat artikel 8 de belangrijkste toerekeningsgrond met betrekking tot het functioneren van PMSCs in conflictgebieden. Het betreft situaties waarbij private personen en entiteiten op basis van instructies of onder de leiding of controle van een staat handelen. Instructies en leiding moeten door een staat gegeven worden om schendingen van internationaal recht te bewerkstelligen. In de praktijk zijn deze vormen van staatsinmenging in het gedrag van PMSCs minder relevant omdat ze vaak moeilijk te bewijzen zijn. De kwestie van controle uitgevoerd door een staat over PMSCs is ook niet eenvoudig, maar deze controle is tegelijkertijd de meest relevante feitelijke band tussen deze entiteiten. Er zijn twee controletheorieën die van belang zijn bij het beantwoorden van de vraag of PMSCs onder controle van een staat staan. De theorie van "effective control" of effectieve controle werd als uitgangspunt genomen door het Internationaal Gerechtshof in de *Nicaragua*-zaak in 1986 en bevestigd in de zaak tussen *Bosnië-Herzegovina* en *Servië-Montenegro* in 2007. In overeenstemming met deze theorie moet er worden vastgesteld of een staat effectieve controle uitoefende over militaire en andere operaties van PMSCs waarbij internationaal recht geschonden werd. Hierbij moet er worden beoordeeld of er sprake van een hoge mate van "practical support" oftewel praktische ondersteuning aan de operaties of aan de schendingen was. Deze hoogdrempelige effectieve controle-test draagt in de meeste gevallen niet bij aan het demonstreren van feitelijke banden tussen PMSCs en staten en leidt er niet toe dat misdragingen van deze bedrijven en hun werknemers aan deze staten toegerekend worden.

In tegenstelling tot de "effective control" test heeft de theorie van "overall control" die door het Joegoslavië-Tribunaal in de *Tadić*-zaak in 1999 gebruikt werd, een grotere kans van slagen aangezien deze een lagere drempel heeft. Mits particulieren in georganiseerde en hiërarchisch gestructureerde groepsverbanden opereren, is het voldoende aan te tonen dat een staat een rol heeft gespeeld bij het organiseren, coördineren of plannen van de militaire operaties van zulke groepen en dus niet alleen betrokken was bij het financieren, trainen, uitrusten en leveren van operationele steun. Het toepassen van de "overall control" test op de handelingen van PMSCs lijkt een logische stap te zijn waarbij een realistische benadering van staatsaansprakelijkheid centraal staat. Het zou kunnen verhinderen dat de staten PMSCs inhuren en door middel van hun inzet in conflictgebieden, waarbij internationaal recht geschonden wordt, aansprakelijkheid vermijden. Deze aanpak zou de "responsibility gap" aanzienlijk kunnen verkleinen.

Vervolgens wordt er in Hoofdstuk 5 aandacht besteed aan mogelijke misdragingen van PMSCs en hun personeel om te onderzoeken of ze als schendingen van internationaal recht en dus schendingen van de op staten rustende rechtsplichten beschouwd kunnen worden. In de eerste instantie moet er worden vastgesteld of staten gebonden worden door internationaal humanitair recht en mensenrechtenrecht waarbij zaken, zoals de deelname van staten aan gewapende conflicten en het uitoefenen van effectieve controle over grondgebied en personen, cruciaal zijn. Vervolgens dient er bepaald te worden of er sprake is van een schending van een volkenrechtelijke verplichting. Het onderzoek heeft aangetoond dat de meeste vormen van misdragingen van PMSCs en hun werknemers die in het voorafgaande besproken zijn, als schendingen van internationaalrechtelijke verplichtingen van staten gezien kunnen worden.

Hoofdstuk 6 legt nadruk op de tweede manier om staten aansprakelijk te stellen door middel van hun positieve verplichtingen in het internationaal recht, de zogenoemde indirecte aansprakelijkheid. Aangezien de toerekening van het gedrag van PMSCs en hun personeel aan staten niet altijd mogelijk is, biedt deze indirecte aansprakelijkheid een oplossing voor het wegwerken van de "responsibility gap". De positieve verplichtingen vloeien voort uit zowel het internationaal humanitair recht als het internationaal mensenrechtenrecht en zijn te onderverdelen in resultaats- en inspanningsverplichtingen. Staten zijn namelijk verplicht om alle nodige maatregelen te nemen ten einde een bepaald resultaat te verwezenlijken of schendingen van het internationaal recht te voorkomen en de verantwoordelijken te straffen. De analyse strekt zich in de eerste instantie uit tot een scala aan positieve resultaatsverplichtingen dat in het internationaal humanitair recht kan worden geïdentificeerd en bij de inzet van PMSCs van belang is. Het gaat bijvoorbeeld om rechtsplichten van de gevangennemende mogendheden waar bij de behandeling van krijgsgevangenen aan moet worden voldaan. In het mensenrechtenrecht zijn de resultaatsverplichtingen van staten ten opzichte van PMSCs niet zo talrijk en in theorie zou het recht op een eerlijk proces onder deze categorie kunnen vallen.

"Due diligence"-verplichtingen of inspanningsverplichtingen van staten zijn ook te vinden in zowel het internationaal humanitair recht als het mensenrechtenrecht en komen aan alle drie soorten staten toe. In het oorlogsrecht bestaat een verplichting om "te doen eerbiedigen" die in gemeenschappelijk artikel 1 van de Conventies van Genève is vastgelegd. Deze bepaling vereist dat contracterende staten, gastheerstaten en thuisstaten bepaalde maatregelen nemen, zoals het reguleren van PMSCs. Ze hebben ook een verplichting om te interveniëren als een schending van het internationaal humanitair recht dreigt plaats te vinden en om mogelijke schendingen tegen te gaan. Bovendien moeten contracterende staten en gastheerstaten die vaak partij bij een gewapend conflict zijn, bepaalde groepen mensen in gewapende conflicten beschermen en schendingen van het oorlogsrecht voorkomen en tegengaan als ze als bezettende mogendheden optreden.

In het mensenrechtenrecht zijn inspanningsverplichtingen in principe te onderverdelen in verplichtingen om te voorkomen en verplichtingen om te straffen. Het eerste type verplichtingen houdt in dat staten moeten interveniëren als het mensenrechtenrecht geschonden kan worden en een "preventieve regulering" gericht op PMSCs en hun activiteiten in het leven te roepen. Het tweede type betreft een verplichting om te onderzoeken, te vervolgen en te straffen en een verplichting om een "reagerende regulering" in te voeren die erop moet toezien dat schendingen van mensenrechten gestraft worden. Contracterende staten zijn verplicht om mogelijke mensenrechtenschendingen te voorkomen door op te treden en controle over de activiteiten van PMSCs uit te oefenen. Verder moeten ze vrouwen en kinderen in conflictgebieden beschermen en reguleringsmaatregelen nemen, zoals het instellen van monitoringsmechanismen. Om schendingen van het mensenrechtenrecht te straffen dienen deze staten de nodige wetgeving uit te vaardigen en op andere manieren actie te ondernemen. Gastheerstaten worden ook verwacht soortgelijke maatregelen te nemen als contracterende staten. Ten slotte zijn thuisstaten verplicht om bepaalde activiteiten te verrichten. Om te voldoen aan de verplichting om te voorkomen dienen ze op te treden als er schendingen van mensenrechten dreigen te gebeuren en de controle over PMSCs uit te oefenen. Ze moeten ook nodige reguleringsmechanismen in het leven roepen om zulke mensenrechtenschendingen als foltering te straffen. Het onderzoek toont aan dat sommige staten bereidwillig zijn om overeenkomstig de goede praktijken van het Montreux Document en het UN Guiding Principles-raamwerk te handelen en al bepaalde stappen in deze richting hebben genomen. Een groot aantal staten heeft daarentegen vooralsnog geen actie ondernomen.

Hoofdstuk 7 dient als het sluitstuk van het boek en bevat de samenvatting gecombineerd met de eindconclusies en opmerkingen van de auteur. De voorafgaande analyse toont aan dat de staten die bij de inzet van PMSCs in conflictgebieden betrokken zijn, vaak hun aansprakelijkheid kunnen ontvluchten wanneer deze bedrijven en hun werknemers internationaal recht schenden. Dat betekent dat er geen rechtsgevolgen intreden die vooral voor de personen die door deze schendingen zijn getroffen, van belang zijn. In de praktijk kunnen staten echter in twee gevallen met succes aansprakelijk gesteld worden. Ten eerste is het mogelijk mits elementen van staatsgezag door PMSCs die daartoe bij wet gemachtigd zijn, uitgeoefend worden. In dit geval kan er aangenomen worden dat contracten tussen staten en PMSCs samen met beschikbare licentie- en certificatiemechanismen als dergelijke wetgeving fungeren. Om aansprakelijkheid te vermijden moeten staten geen overheidstaken aan PMSCs uitbesteden of ervoor zorgen dat hun activiteiten in overeenstemming met internationaal recht plaatsvinden. De staatsaansprakelijkheid komt ook tot stand als PMSCs onder "overall control" van staten opereren wanneer deze controletest toegepast wordt. Aangezien werknemers van deze ondernemingen als georganiseerde en hiërarchisch gestructureerde groepen functioneren, verdient deze test de voorkeur boven de effectieve controle-theorie.

Zoals gezegd, spelen verschillende positieve staatsverplichtingen een belangrijke rol in het aansprakelijk houden van staten als de toerekening niet slaagt. De bevindingen van dit onderzoek in overweging nemend wordt er opgemerkt dat er een scala aan resultaats- en inspanningsverplichtingen bestaat dat aan alle typen staten toekomt. In het algemeen hebben contracterende staten de meeste positieve verplichtingen. Gastheerstaten worden ook geacht verschillende maatregelen te nemen maar zijn vaak niet in staat om aan hun verplichtingen onder internationaal recht te voldoen. Thuisstaten blijken de minste positieve verplichtingen te hebben. Het behoeft geen betoog dat alle staten op de hoogte van hun rechtsplichten onder internationaal recht dienen te zijn. Een nieuw verdrag dat op PMSCs gericht is, kan ervoor zorgen dat er meer duidelijkheid over positieve verplichtingen van staten en positieve maatregelen die genomen moeten worden, verschaft wordt. Ook al zal dit verdrag waarschijnlijk niet op grote schaal worden geratificeerd, staat het nu vast dat internationaal recht voldoende houvast biedt om staten voor schendingen van internationaal humanitair recht en mensenrechten die door PMSCs en hun personeel begaan worden, aansprakelijk te stellen.

Russian Summary
Краткое Содержание Книги на Русском Языке

В современном мире так называемые частные военные и охранные компании (ЧВОК) уже не вызывают удивления. Они нанимаются частными и юридическими лицами для выполнения определённых задач, таких как, например, охранно-сторожевые функции, во время вооружённых конфликтов и в мирное время. Не только международные организации, негосударственные организации и компании, но и государства используют частные военные и охранные структуры для достижения своих целей. В прошлом данные компании были замешаны в различных нарушениях международного права, включая международное гуманитарное право и права человека. К сожалению, национальное законодательство, межнациональные инструменты и международное право не в силах полностью обеспечить такие аспекты функционирования ЧВОК, как контроль и мониторинг их операций и ответственность за их преступления. Главная цель данной исследовательской работы – установить, возможно ли привлечь к международной ответственности государства за неправомерные действия ЧВОК и их персонала. Вопрос, на который требовалось ответить, звучит следующим образом:

> «Какова степень ответственности, которую государства должны нести и на практике могут нести за неправомерные действия ЧВОК в зонах конфликта, квалифицируемые как нарушения международного гуманитарного права и прав человека?»

Проблема международной ответственности государств была изучена посредством анализа двух путей, ведущих к данной ответственности. В первую очередь, было исследовано, при каких условиях может быть успешно присвоение правонарушений, совершённых ЧВОК, государству, и какова его степень. Также было изучено определённое количество позитивных обязанностей государств в сфере как международного гуманитарного права, так и прав человека, и условия и степень их применения в ответственности государств.

Первая глава книги является введением, в котором раскрываются следующие вопросы: проблема, подверженная анализу, методология исследования, вопросы исследования, гипотеза, цели научной работы, типы правонарушений, совершаемые ЧВОК и их персоналом, морально-этические аспекты использования ЧВОК государствами в вооружённых конфликтах и структура книги. Было отмечено, что на данный момент ЧВОК используются государствами в зонах конфликта для выполнения множества задач. Как оказалось, в прошлом эти компании и их штат были неоднократно замешаны в совершении противозаконных действий. В литературе и прессе были освещены следующие инциденты, в которых они принимали участие: атаки и обстрелы, нацеленные на мирное население, ликвидации людей по заказу правительств, нанесение вреда частной собственности, пытки и другие формы жестокого обращения, торговля людьми, сексуальное насилие, правонарушения в сфере наркотических средств, участие в криминальных организациях, преступления, связанные с незаконным оборотом оружия, и участие в вооружённых конфликтах в качестве наёмников. В связи с тем, что использование ЧВОК в вооружённых конфликтах может быть проблематично с морально-этической точки зрения, и беря во внимание тот факт, что государства могут избежать ответственности, используя ЧВОК, следует подчеркнуть, что они должны нести международную ответственность за правонарушения данных компаний и их персонала.

Вторая глава касается феномена приватизации войны и безопасности и его исторических аспектов. На протяжении многих веков наёмники использовались разными лицами и структурами во всевозможных конфликтах. В данной главе описывается участие наёмников в античности и средневековье. Далее рассказывается о ситуации в семнадцатом, восемнадцатом и девятнадцатом веках. После заключения Вестфальского мира в 1648-ом году и появления современного понятия «государство» наёмники перестали быть нормой, и их использование значительно снизилось. В шестидесятых годах прошлого века они снова появились на мировой арене. С окончанием холодной войны в начале девяностых годов зародилась индустрия ЧВОК, какой мы её знаем сейчас. ЧВОК создавались согласно национальному законодательству, имели чёткую структуру и культуру предприятий и часто функционировали как международные компании. Данные аспекты подчёркивают важные различия между ЧВОК и наёмниками, но между ними есть и определённые сходства, например, финансовая составляющая их мотивации для предоставления услуг. С одной стороны ЧВОК являются частными предприятиями в соответствии с национальным законодательством, но с другой стороны они могут рассматриваться в качестве транснациональных корпораций согласно международному праву. В целом, ЧВОК попадают под определение компаний, создающихся на территории государств в соответствии с их законодательством, которые имеют структуру частных предприятий, предоставляют военные и охранные услуги, часто оперируют как международные компании и могут прибежать к использованию военной силы.

На сегодняшний день ЧВОК нанимаются не только компаниями, международными и общественными организациями, но и государствами. Можно выделить три вида государств в использовании ЧВОК: государства-контрагенты, которые заключают контракты с данными компаниями, государства территориальной юрисдикции, на территории которых они действуют, и государства происхождения, где они зарегистрированы или инкорпорированы. Разделение государств на вышеуказанные типы играет непосредственную роль при анализе международно-правовых обязательств государств и степени их ответственности. Как транспарентность, так и легитимность ЧВОК могут вызывать сомнения, но индустрия данных предприятий находится в постоянном развитии, и каждый год в мире заключается большое количество контрактов на предоставление их услуг. К сожалению, данные контракты не являются доступными для широкой публики или спрятаны за цепью субконтрактов.

Неверно предположение, что ЧВОК оперируют в правовом вакууме. Существуют определённые механизмы регулирования, которые нацелены на деятельность данных компаний, пребывающих в зонах конфликта, и в третьей главе книги обсуждаются данные механизмы. В первую очередь следует указать на наличие правил и принципов международного публичного права, международного гуманитарного права и прав человека. Они могут быть найдены в Статьях об ответственности государств, Женевских конвенциях и различных международных договорах по правам человека. В особенности экстратерриториальное действие договоров в сфере прав человека создаёт препятствия для применения данных инструментов по отношению к государствам-контрагентам и государствам происхождения. Был сделан вывод о том, что большинство договоров имеют экстратерриториальное действие, если государство, которое заключило договор, осуществляет эффективный контроль над территорией другого государства или имеет власть и полномочия над индивидами. Также существуют определённые механизмы регулирования международного характера, которые имеют непосредственное отношение к самим ЧВОК и другим компаниям: например, Документ Монтрё и Руководящие принципы ООН. Естественно, национальное законодательство государств регулирует ЧВОК и их деятельность, в то время, как контракты составляют правовую базу для использования данных компаний. В дополнение стоит отметить стандартные операционные процедуры, правила ведения боевых действий и соглашения о статусе сил, которые имеют отношение к операциям ЧВОК. В заключение существуют инструменты транснационального регулирования и саморегулирования, такие как Международный кодекс поведения частных охранных компаний, которые не несут в себе высокого уровня эффективности из-за частого отсутствия механизмов мониторинга и инфорсмента.

Когда ЧВОК принимают участие в вооружённых конфликтах, их правовой статус часто неясен. В четвёртой главе представлен анализ данного правового статуса персонала ЧВОК, который влияет не только на их права и обязанности, но и на обязанности и ответственность государств по отношению к определённым группам лиц. Персонал ЧВОК нанимается напрямую или посредством контракта. В международных вооружённых конфликтах только в ограниченном количестве случаев они могут быть квалифицированы как личный состав вооружённых сил стороны, находящейся в конфликте, а также как личный состав ополчения и добровольческих отрядов, входящих в состав этих вооружённых сил. Это демонстрируется рассмотрением статьи 43 Первого Дополнительного Протокола к Женевским Конвенциям и статьи 4(A)(1) Третьей Женевской Конвенции. В определённых случаях эти индивиды могут иметь статус личного состава других ополчений и добровольческих отрядов, согласно статье 4(A)(2) Третьей Женевской Конвенции. Вероятнее всего они должны быть рассмотрены в качестве лиц, которые следуют за вооружёнными силами, но не входят в их состав непосредственно, как упомянуто в статье 4(A)(4) Третьей Женевской Конвенции. Данные лица защищены Женевскими Конвенциями и могут иметь статус военнопленных, когда они принимают участие в международных вооружённых конфликтах. Также персонал ЧВОК может быть расценен как гражданские лица, попадающие под сферу действия Четвёртой Женевской Конвенции. Стоит заметить, что существует вероятность их квалификации в качестве наёмников, согласно статье 47 Первого Дополнительного Протокола к Женевским Конвенциям, а также положениям Африканской Конвенции и Конвенции ООН о наёмниках. В конфликтах, не несущих международный характер, члены персонала ЧВОК только в ограниченном количестве сценариев могут иметь правовой статус вооружённых сил государства и антиправительственных вооружённых сил или других организованных вооружённых групп, которые указаны в статье 1(1) Второго Дополнительного Протокола к Женевским Конвенциям. К тому же маловероятно, что они попадают под определение наёмников, содержащееся в Африканской Конвенции и Конвенции ООН о наёмниках, применимых в немеждународных военных конфликтах. Беря во внимание специфику обстоятельств, при которых ЧВОК участвуют в вооружённых конфликтах данного типа, в большинстве случаев они принадлежат к категории гражданских лиц.

Пятая глава затрагивает проблему международной ответственности государств и посвящена призванию государств к ответственности за действие и бездействие ЧВОК и их персонала, которые могут быть расценены как нарушения международного гуманитарного права и прав человека. Государства могут нести ответственность за нарушения международного права, но степень из ответственности зависит от определённых обстоятельств каждого отдельного случая. Помимо государств, международные организации также могут быть ответственны за данные нарушения, особенно в контексте миротворческих операций. Эта глава касается так называемой прямой ответственности государств, связанной с концептом присвоения поведения государству. Статья 1 Статей об ответственности государств гласит, что каждое международно-противоправное деяние, совершённое государством, влечёт за собой международную ответственность данного государства. Эта ответственность порождает определённые юридические последствия: согласно статьям 30 и 31 Статей об ответственности государств, ответственное государство должно прекратить международно-противоправное деяние и осуществить возмещение нанесённого вреда.

В соответствии со статьёй 2 международно-противоправное деяние государства материализуется, когда определённое поведение – действие или же бездействие – присваивается государству по международному праву и является нарушением международно-правового обязательства этого государства. В первую очередь в книге раскрывается вопрос, могут ли действия ЧВОК и их персонала быть присвоены государствам на основе Статей об ответственности государств. Посредством данного присвоения определяется, возможно ли приравнять это поведение к деяниям самих государств по международному праву. Отмечаются четыре вида присвоения, которые имеют большое значение в контексте найма ЧВОК. Они определены в статьях 4, 5, 8 и 11 Статей об ответственности государств. Статья 4 применяется в тех случаях, когда ЧВОК выполняют свои функции в качестве органов государства и имеют данный статус по внутригосударственному праву. Такая возможность существует, если они сделаны частью вооружённых сил государства и имеют правовой статус органа данного государства на основе его права. Данная статья имеет ограниченное практическое применение по отношению к ЧВОК в связи с тем, что они могут выступать органами государств лишь в очень небольшом количестве случаев. Применение статьи 11, которая касается признания и принятия определённого поведения государством как его собственное, также маловероятно. В общем, государства не склонны к подобным действиям по отношению к правонарушениям, совершённым ЧВОК и другими частными лицами.

Исследование показало, что самыми важными положениями Статей об ответственности государств, имеющих отношение к присвоению поведения ЧВОК и их персонала государствам, являются статьи 5 и 8. Статья 5 посвящена ситуациям, в которых ЧВОК уполномочены правом государства выполнять элементы государственной власти, при этом не являясь органом данного государства. Что включают в себя данные элементы государственной власти, зависит от определенного общества, его истории и традиций. В случае, если компании используются для принятия прямого участия в вооружённых конфликтах с применением военной силы и уполномочены на данные действия правом определённого государства, то их поведение будет присвоено государству.

Исключительно важным типом присвоения в контексте функционирования ЧВОК в зонах конфликта является статья 8. Она касается ситуаций, в которых лица или группы лиц фактически действуют по указаниям либо под руководством или контролем государства. Указания и руководство должны быть реализованы с целью нарушения международного права. На практике, эти типы вмешательства государства в действия ЧВОК не имеют большого значения, так как часто являются труднодоказуемыми. Контроль государства в свою очередь играет более важную роль и указывает на фактическую связь между данными компаниями и государством. Существует две теории контроля, которые необходимо взять во внимание при рассмотрении вопроса, действуют ли ЧВОК под государственным контролем. Теория эффективного контроля была сформулирована Международным Судом в деле *Никарагуа против США* в 1986-ом году и подтверждена им же в деле *Босния и Герцеговина против Сербии и Черногории* 2007-го года. Согласно данному критерию должно быть установлено, что государство осуществляет эффективный контроль над военными и подобными им операциями ЧВОК, в ходе которых нарушается международное право. При этом следует определить имели ли место практическая поддержка операций и нарушений самим государством. Теория эффективного контроля с высоким пороговым уровнем не может быть использована в большинстве случаев для выявления фактических связей между ЧВОК и государствами и не ведёт к присвоению правонарушений этих компаний и их персонала государствам.

В противовес теории эффективного контроля критерий общего контроля, использованный Международным уголовным трибуналом для бывшей Югославии в деле *Прокурор против Тадича* в 1999-ом году, имеет более низкий пороговый уровень и может с успехом быть использован на практике. В случае операций организованных и иерархически структурированных групп достаточно доказать, что государство играет роль в организации, координации и планировании военных действий данных групп, а не только в их финансировании, тренировке, снабжении и оснащении. Применение данной теории является логичным шагом, нацеленным на реалистичный подход к ответственности государств. Оно может предотвратить использование ЧВОК в зонах конфликта и отказ государств от ответственности при нарушении международного права этими компаниями и их персоналом.

Далее в пятой главе уделяется внимание действиям ЧВОК и их штату, чтобы выяснить, являются ли они нарушениями международного права и, следовательно, нарушениями международно-правовых обязательств государств. В первую очередь должно быть установлено, связаны ли государства обязательствами, вытекающими из международного гуманитарного права и международных договоров по правам человека. При этом важными являются такие аспекты, как прямое участие государств в военных действиях и эффективный контроль над территорией и индивидами. Затем следует ответить на вопрос, было ли совершено нарушение международно-правового обязательства государств. Исследование выявило, что большинство действий ЧВОК и их персонала в описанных инцидентах могут быть расценены как нарушения международно-правовых обязательств государств.

Шестая глава делает упор на второй способ привлечения государств к ответственности посредством позитивных обязательств государств в международном праве, так называемую косвенную ответственность, которая особенно важна в случае, когда присвоение поведения ЧВОК и их штата государству невозможно. Позитивные обязательства предусмотрены не только в международном гуманитарном праве, но и в международных договорах по правам человека и подразделяются на обязательства результата и обязательства поведения. Государства обязаны принять всевозможные меры, чтобы реализовать определённый результат или же предотвратить нарушения международного права и наказать виновных. В первую очередь анализ направлен на спектр позитивных обязательств результата в международном гуманитарном праве, который имеет значение при использовании ЧВОК. К примеру, это касается обязательств держащих в плену держав по отношению к военнопленным. Договоры по правам человека не содержат большое количество положений, связанных с обязательствами результата, которые были бы применимы к ЧВОК. Одним из таких прав может быть право на справедливое судебное разбирательство.

Позитивные обязательства поведения также существуют в международном гуманитарном праве и международных договорах по правам человека и связывают все три типа государств. В международном гуманитарном праве существует обязанность «заставлять соблюдать», которая упоминается в статье 1, общей для Женевских Конвенций. Согласно данной статье, государства-контрагенты, государства территориальной юрисдикции и государства происхождения должны принять определённые меры, такие как регулирование ЧВОК. Они также обязаны вмешаться, если международное гуманитарное право может быть нарушено, и предотвратить возможное нарушение. К тому же государства-контрагенты и государства территориальной юрисдикции, которые нередко принимают участие в вооружённых конфликтах, имеют обязательство защищать определённые группы индивидов, а также предотвращать и пресекать нарушения международного гуманитарного права, когда они выступают в роли оккупирующих держав.

В международном праве по правам человека обязательства поведения государств можно подразделить на обязательства предотвращения и наказания. Первый тип обязательств подразумевает то, что государства должны вмешаться, если существует вероятность нарушения прав человека, и создать регулирование, нацеленное на предотвращение подобных нарушений, которые могут быть совершены ЧВОК и их персоналом. Второй тип касается обязательств государств на расследование, преследование и наказание за нарушения прав человека и создание системы регулирования для адекватной реакции на данные правонарушения. Государства-контрагенты обязаны предотвратить возможные нарушения прав человека, вовремя принимая необходимые действия и контролируя деятельность ЧВОК. Они также должны защищать женщин и детей в зонах конфликта и принимать меры по регулированию, такие как создание инструментов мониторинга. Чтобы наказать лиц, ответственных за нарушения прав человека, им следует иметь необходимое законодательство и принимать другие всевозможные меры. В свою очередь, государства территориальной юрисдикции имеют схожие обязательства в этой области. Государства происхождения ЧВОК также обязаны принять определённые позитивные меры. Чтобы предотвратить нарушения прав человека, им необходимо активно действовать, если данные правонарушения могут иметь место, и контролировать ЧВОК. В дополнение им следует создать регулирование, чтобы эффективно реагировать на такие нарушения прав человека, как пытки. Исследование показало, что некоторые государства готовы действовать согласно передовым практическим методам Документа Монтрё и Руководящим принципам ООН и уже предприняли определённые шаги. В то же время большое количество других государств ещё не приступило к действию в этой области.

Седьмая глава служит заключением книги и содержит краткое изложение результатов исследования, основные выводы и наблюдения автора. Анализ показал, что государства, принимающие участие в использовании ЧВОК в зонах конфликта, во многих случаях могут избежать ответственности за нарушения международного права, совершаемые этими компаниями и их штатом. Соответственно, это значит, что данные действия не влекут за собой юридических последствий, которые особенно важны для пострадавших в результате подобных правонарушений. В двух случаях государства несут международную ответственность на практике. Во первых, это происходит, если ЧВОК выполняют элементы государственной власти и уполномочены на это правом государства. В данном случае можно предположить, что контракты, заключённые между государствами и ЧВОК, в совокупности с механизмами лицензирования и сертификации могут рассматриваться как подобный вид законодательства. Во избежание ответственности государства должны прекратить аутсорсинг государственных функций компаниям, предоставляющим военные и охранные услуги, или обеспечить проведение их операций в соответствии с международным правом. Ответственность государств также имеет место, когда ЧВОК оперируют под общим контролем государств, если данная теория применяется к инцидентам с этими компаниями. Так как персонал ЧВОК функционирует в качестве организованных и иерархически структурированных групп, следует признать практическую ценность применения данной теории, а не теории эффективного контроля.

Как было отмечено выше, различные позитивные обязательства государств играют важную роль в привлечении государств к ответственности, если присвоение поведения государству не может быть успешным. Беря во внимание результаты исследования, следует заметить, что существует спектр обязательств результата и поведения, который связывает все три типа государств. В общем, государства-контрагенты имеют наибольшее количество обязательств. Государства территориальной юрисдикции также должны принять различные меры по отношению к ЧВОК, но они часто не в состоянии действовать согласно своим обязательствам, вытекающим из международного права. Государства происхождения ЧВОК имеют наименьшее число обязательств. Несомненно, все государства должны быть в курсе своих международно-правовых обязательств. Новый международный договор, который нацелен на ЧВОК, может внести больше ясности касаемо позитивных обязанностей государств и позитивных мер, которые должны быть ими приняты. Несмотря на то, что данный договор, вероятно, не будет ратифицирован многими государствами, можно сказать с уверенностью, что международное право предоставляет достаточно инструментов для того, чтобы призвать государства к ответственности за нарушения международного гуманитарного права и прав человека, совершённые ЧВОК и их персоналом.

Selected Bibliography

"30 Most Powerful Private Security Companies in the World", *Security Degree Hub*, January 11, 2014, <http://www.securitydegreehub.com/30-most-powerful-private-security-companies-in-the-world/>.

"400 US Commandos Help Kiev in Its Military Offensive in East Ukraine – Reports", *The Voice of Russia*, May 11, 2014, <http://voiceofrussia.com/news/2014_05_11/400-US-commandos-help-Kiev-in-its-military-offensive-in-east-Ukraine-reports-8880/>.

"Abu Ghraib Inmates Sue US Firms", *BBC News*, July 1, 2008, <http://news.bbc.co.uk/2/hi/americas/7482617.stm>.

"Booze, Drugs and Hookers: Hundreds of Offences by Afghan Contractors Uncovered", *The Associated Press*, December 19, 2010, <http://www.rawstory.com/rs/2010/12/hundreds-offenses-afghan-contractors-uncovered>.

"Einsatz Gegen Separatisten: Ukrainische Armee Bekommt Offenbar Unterstützung von US-Söldnern", *Der Spiegel*, May 11, 2014, <http://www.spiegel.de/politik/ausland/ukraine-krise-400-us-soeldner-von-academi-kaempfen-gegen-separatisten-a-968745.html>.

"Kämpfen US-Söldner in der Ukraine?", *Bild*, March 9, 2014, <http://www.bild.de/politik/ausland/ukraine/us-soeldner-von-blackwater-im-einsatz-34992896.bild.html>.

"Licence to Kill for Private Armies", *WarOnWant.org*, October 19, 2009, <http://www.waronwant.org/news/press-releases/16692-licence-to-kill-for-private-armies>.

"NATO Issues New Guidelines for Afghanistan Contracts", *BBC News*, September 13, 2010, <http://www.bbc.co.uk/news/world-south-asia-11287074>.

"Q&A: Arms-to-Africa Scandal", *BBC News*, May 10, 1998, <http://news.bbc.co.uk/2/hi/uk_news/90526.stm>.

"The Story of Mitchell Jessen & Associates: How a Team of Psychologists in Spokane, WA, Helped Develop the CIA's Torture Techniques", *Democracy Now!: The War and Peace Report*, April 21, 2009, <http://www.democracynow.org/2009/4/21/the_story_of_mitchell_jessen_associates>.

"US Mercenaries Pose Threat to Ukrainians Being Ultra-Violent Individuals – US Military Vet", *The Voice of Russia*, April 11, 2014, <http://voiceofrussia.com/2014_04_11/US-mercenaries-pose-threat-to-Ukrainians-being-ultra-violent-individuals-US-military-vet-2839/>.

"US Uses Contractors to Bypass International Humanitarian Laws", *RT.com*, August 28, 2012, <http://rt.com/politics/us-contractors-international-blackwater-754/>.

Abrahamsen, R. and Williams, M.C., *The Globalisation of Private Security, Country Report: Nigeria*, Report, January 2005, Aberystwyth: The University of Wales, <http://users.aber.ac.uk/rbh/privatesecurity/country%20report-nigeria.pdf>.

Alexandra, A., Baker, D.-P., and Caparini, M., "Introduction: The Ethics and Governance of Private Military and Security Companies", in Andrew Alexandra, Deane-Peter Baker, and Marina Caparini (Eds.), (2008) *Private Military and Security Companies: Ethics, Policies and Civil-Military Relations*, London and New York: Routledge.

Alston, P., "The 'Not-a-Cat' Syndrome: Can the International Human Rights Regime Accommodate Non-State Actors", in Philip Alston (Ed.), (2005) *Non-State Actors and Human Rights*, Oxford: Oxford University Press.

Aning, K., Jaye, T., and Atuobi, S., (2008) "The Role of Private Military Companies in US-Africa Policy", *Review of African Political Economy*, Volume 35, Issue 118.

Anzilotti, D., (1929) *Cours de Droit International*, Paris: Editions Panthéon-Assas.

Armstrong, K. and Kilpatrick, C., (2007) "Law, Governance, or New Governance? The Changing Open Method of Coordination", *Columbia Journal of European Law*, Volume 13, Issue 3.

Arts, B., (2003-2004) *Non-State Actors in Global Governance: Three Faces of Power*, Max Planck Project Group on Common Goods, Preprint, Bonn, <http://www.coll.mpg.de/pdf_dat/2003_04online.pdf>.

ASIS International, "ANSI/ASIS PSC Series: Private Security Service Provider Standards", Executive Summary, <https://www.asisonline.org/Standards-Guidelines/Standards/Documents/Executive%20SummaryANSIASIS%20PSC%20series.pdf>.

ASIS International, *Management System for Quality of Private Security Company Operations – Requirements with Guidance: ANSI/ASIS PSC.1-2012*, <http://www.acq.osd.mil/log/PS/p_vault/Item_1997-PSC_1_STD.PDF>.

Asser, M., "Abu Ghraib: Dark Stain on Iraq's Past", *BBC News*, May 25, 2004, <http://news.bbc.co.uk/2/hi/americas/3747005.stm>.

Atteritano, A., "Liability in Tort of Private Military and Security Companies: Jurisdictional Issues and Applicable Law", in Francesco Francioni and Natalino Ronzitti (Eds.), (2011) *War by Contract: Human Rights, Humanitarian Law, and Private Contractors*, Oxford: Oxford University Press.

Augenstein, D. and Kinley, D., "When Human Rights 'Responsibilities' Become 'Duties': The Extra-territorial Obligations of States that Bind Corporations", in Surya Deva and David Bilchitz (Eds.), (2013) *Human Rights Obligations of Business: Beyond the Corporate Responsibility to Respect?*, Cambridge: Cambridge University Press.

Avant, D.D., (2000) "From Mercenary to Citizen Armies: Explaining Change in the Practice of War", *International Organization*, Volume 54, Issue 1.

Avant, D.D., (2004) "The Privatization of Security and Change in the Control of Force", *International Studies Perspectives*, Volume 5.

Avant, D.D., (2005) *The Market for Force: The Consequences of Privatizing Security*, Cambridge: Cambridge University Press.

Avant, D.D. and Sigelman, L., (2008) "What Does Private Security in Iraq Mean for US Democracy at Home?", Research Paper, <http://bc.sas.upenn.edu/system/files/Avant_08.pdf>.

Azarov, V., (2009) *Internationalisation and Attribution: The Tadić Analysis in the Genocide in Bosnia Case*, University of Westminster School of Law Research Paper No. 09-11, <http://ssrn.com/abstract=1470800>.

Baker, L., "British Security Co. Sued over Death of U.S. Soldier", *Reuters.com*, October 26, 2007, <http://www.reuters.com/article/2007/10/26/us-iraq-britain-lawsuit-idUSL2666274820071026>.

Bakker, C. and Sossai. M., (Eds.), (2012) *Multilevel Regulation of Military and Security Contractors: The Interplay Between International, European and Domestic Norms*, Oxford and Portland, Oregon: Hart Publishing.

Bakker, C., "Duties to Prevent, Investigate, and Redress Human Rights Violations by Private Military and Security Companies: The Role of the Host State", in Francesco Francioni and Natalino Ronzitti (Eds.), (2011) *War by Contract: Human Rights, Humanitarian Law, and Private Contractors*, Oxford: Oxford University Press.

Barak-Erez, D., "The Privatization Continuum", in Simon Chesterman and Angelina Fisher (Eds.), (2009) *Private Security, Public Order: The Outsourcing of Public Services and Its Limits*, Oxford: Oxford University Press.

Barker, J.C. and Sivakumaran, S., (2007) "Application of the Convention on the Prevention and Punishment of the Crime of Genocide (Bosnia and Herzegovina v. Serbia and Montenegro)", *International and Comparative Law Quarterly*, Volume 56, Issue 3.

Barnidge, R.P. Jr., (2006) "The Due Diligence Principle Under International Law", *International Community Law Review*, Volume 8.

Bartolini, G., "Private Military and Security Contractors as 'Persons who Accompany the Armed Forces'", in Francesco Francioni and Natalino Ronzitti (Eds.), (2011) *War by Contract: Human Rights, Humanitarian Law, and Private Contractors*, Oxford: Oxford University Press.

Bassiouni, M.C., (2010) "Legal Status of US Forces in Iraq from 2003-2008", *Chicago Journal of International Law*, Volume 11, Issue 1.

Beaucillon, C., Fernandez, J., and Raspail, H., "State Responsibility for Conduct of Private Military and Security Companies Violating Ius ad Bellum", in Francesco Francioni and Natalino Ronzitti (Eds.), (2011) *War by Contract: Human Rights, Humanitarian Law, and Private Contractors*, Oxford: Oxford University Press.

Becker, T., (2008) *Terrorism and the State: Rethinking the Rules of State Responsibility*, Oxford and Portland: Hart Publishing.

Benninger-Budel, C., "Introduction", in Carin Benninger-Budel (Ed.), (2008) *Due Diligence and Its Application to Protect Women From Violence*, Leiden: Koninklijke Brill NV.

Benzing, J., "Rebranded Academi Looks to Shed Blackwater Baggage", *Just Anti-Corruption*, March 27, 2012, <http://www.mainjustice.com/justanticorruption/2012/03/27/rebranded-academi-looks-to-shed-blackwater-baggage>.

Bernaz, N., (2013) "Enhancing Corporate Accountability for Human Rights Violations: Is Extraterritoriality the Magic Potion?", *Journal of Business Ethics*, Volume 117, Issue 3.

Beyani, C. and Lilly, D., "Regulating Private Military Companies: Options for the UK Government", International Alert, August 2001, <http://psm.du.edu/media/documents/reports_and_stats/ngo_reports/intlalert_beyani_regulating_pmcs_2001.pdf>.

Bigo, D., (2003) "Les Entreprises de Coercition Para-Privées: De Nouveaux Mercenaires?", *Cultures & Conflits*, Volume 52, Issue 4.

Birnbaum, M., "Russia to Ukraine: Your Military Actions are 'Criminal'", *The Washington Post*, May 2, 2014, <http://www.washingtonpost.com/world/europe/russia-to-ukraine-your-military-actions-are-criminal/2014/05/02/bd009b57-0ff4-403d-83a3-67f3f9db8908_story.html>.

Boerefijn, I., (2009) "Establishing State Responsibility for Breaching Human Rights Treaty Obligations: Avenues Under UN Human Rights Treaties", *Netherlands International Law Review*, Volume 56, Issue 2.

Bohm, A., Senior, K., and White, A., "The United Kingdom", in Christine Bakker and Mirko Sossai (Eds.), (2012) *Multilevel Regulation of Military and Security Contractors: The Interplay Between International, European and Domestic Norms*, Oxford and Portland, Oregon: Hart Publishing.

Boisson de Chazournes, L. and Condorelli, L., (2000) "Common Article 1 of the Geneva Conventions Revisited: Protecting Collective Interests", *International Review of the Red Cross*, Volume 82, Issue 837.

Boldt, N., (2004) "Outsourcing War – Private Military Companies and International Humanitarian Law", *German Yearbook of International Law*, Volume 47.

Bolkovac, K. and Lynn, C., (2011) *The Whistleblower: Sex Trafficking, Military Contractors, and One Woman's Fight for Justice*, New York: Palgrave Macmillan.

Bosch, S., (2007) "Private Security Contractors and International Humanitarian Law – A Skirmish for Recognition in International Armed Conflicts", *African Security Review*, Volume 16, Issue 4.

Bosch, S., (2008) "Private Security Contractors and State Responsibility: Are States Exempt from Responsibility for Violations of Humanitarian Law Perpetrated by Private Security Contractors?", *The Comparative and International Law Journal of Southern Africa*, Volume 41, Issue 3.

Bourke-Martignoni, J., "The History and Development of the Due Diligence Standard in International Law and Its Role in the Protection of Women against Violence", in Carin Benninger-Budel (Ed.), (2008) *Due Diligence and Its Application to Protect Women From Violence*, Leiden: Koninklijke Brill NV.

Bowett, D.W., (1982) "Jurisdiction: Changing Patterns of Authority over Activities and Resources", *British Yearbook of International Law*, Volume 53, Issue 1.

Brauer, J., (1999) "An Economic Perspective on Mercenaries, Military Companies, and the Privatisation of Force", *Cambridge Review of International Affairs*, Volume 13, Issue 1.

Brinkerhoff, N. and Wallechinsky, D., "Blackwater Mercenaries Rumored to Outnumber Local Police in Pakistani Capital", *AllGov*, February 10, 2010, <http://www.allgov.com/news/us-and-the-world/blackwater-mercenaries-rumored-to-outnumber-local-police-in-pakistani-capital?news=840337>.

Brooks, D. and Solomon, H., (2000) "From the Editor's Desk", *Conflict Trends*, Issue 1, <http://www.accord.org.za/downloads/ct/ct_2000_1.pdf?phpMyAdmin=ceeda2df659e6d3e35a63d69e93228f1>.

Brooks, D., (2000) "Messiahs or Mercenaries? The Future of International Private Military Services", *International Peacekeeping*, Volume 7, Issue 4.

Brooks, D., (2000) "Write a Cheque, End a War: Using Private Military Companies to End African Conflicts", *Conflict Trends*, Issue 1, <http://www.accord.org.za/downloads/ct/ct_2000_1.pdf>.

Brown Weiss, E., (2002) "Invoking State Responsibility in the Twenty-First Century", *The American Journal of International Law*, Volume 96, Issue 4.

Brown, S., (2010) "Bottom-up Lawmaking and the Regulation of Private Military and Security Companies", *Cuadernos de Derecho Transnacional*, Volume 2, Issue 1.

Brownlie, I., "State Responsibility and International Court of Justice", in Malgosia Fitzmaurice and Dan Sarooshi (Eds.), (2004) *Issues of State Responsibility Before International Judicial Institutions*, Oxford and Portland: Hart Publishing.

Brownlie, I., (1983) *System of the Law of Nations: State Responsibility*, Oxford: Oxford University Press.

Bugnion, F., (2001) "Droit de Genève et Droit de la Haye", *International Review of the Red Cross*, Volume 83, Issue 844.

Bulto, T.S., (2011) "Patching the 'Legal Black Hole': the Extraterritorial Reach of States' Human Rights Duties in the African Human Rights System", *South African Journal of Human Rights*, Volume 27, Issue 2.

Buncombe, A., Routledge, P., and Abrams, F., "Inquiry Finds Sandline Did Breach Arms Embargo", *The Independent*, May 17, 1998, <http://www.independent.co.uk/news/inquiry-finds-sandline-did-breach-arms-embargo-1159472.html>.

Business & Human Rights Resourse Center, "Abu Ghraib Lawsuits Against CACI, Titan (now L-3)", <http://business-humanrights.org/en/abu-ghraib-lawsuits-against-caci-titan-now-l-3-0#c17777>.

Cadieux, J., (2008) "Regulating the United States Private Army: Militarizing Security Contractors", *California Western International Law Journal*, Volume 39, Issue 1.

Cameron, L. and Chetail, V., (2013) *Privatizing War: Private Military and Security Companies under Public International Law*, Cambridge: Cambridge University Press.

Cameron, L., (2006) "Private Military Companies: Their Status under International Humanitarian Law and Its Impact on Their Regulation", *International Review of the Red Cross*, Volume 88, Issue 863.

Carmola, K., (2010) *Private Security Contractors and New Wars: Risk, Law, and Ethics*, Oxon and New York: Routledge.

Caron, D.D., (2002) "The ILC Articles on State Responsibility: The Paradoxical Relationship Between Form and Authority", *The American Journal of International Law*, Volume 96.

Carter, P., (2004) "Hired Guns: What to Do About Military Contractors Run Amok", <http://www.slate.com/id/2098571/>.

Cassese, A., "A Judicial Massacre: The International Court Has Set an Unrealistically High Standard of Proof for Finding Serbia Complicit in Genocide", *Guardian Unlimited*, February 27, 2007, <http://www.guardian.co.uk/commentisfree/2007/feb/27/thejudicialmassacreofsrebr>.

Cassese, A., (2005) *International Law*, Second Edition, Oxford: Oxford University Press.

Cassese, A., (2007) "The Nicaragua and Tadić Tests Revisited in Light of the ICJ Judgment on Genocide in Bosnia", *The European Journal of International Law*, Volume 18, Issue 4.

Catan, T., "Private Armies March into a Legal Vacuum", *Financial Times*, February 10, 2005, <http://www.ft.com/cms/s/0/66d5a032-7b0e-11d9-a3ea-00000e2511c8.html#axzz1p06MGP1k>.

Censer, M., "CACI Seeking Payment from Former Abu Ghraib Prisoners", *The Washington Post*, August 12, 2013, <http://www.washingtonpost.com/blogs/capital-business/post/caci-seeking-payment-from-former-abu-ghraib-prisoners/2013/08/12/4c573550-035d-11e3-a07f-49ddc7417125_blog.html>.

Center for Constitutional Rights, "Al Shimari et al v. CACI et al", <http://ccrjustice.org/ourcases/current-cases/al-shimari-v-caci-et-al>.

Center for Constitutional Rights, "Saleh et al v. Titan et al", <http://ccrjustice.org/Saleh-v-Titan>.

Chakrabarti, S., (2009) *Privatisation of Security in the Post Cold-War Period: An Overview of its Nature and Implications*, Institute for Defence Studies and Analyses Monograph Series, No. 2, New Delhi: Magnum Custom Publishing.

Chapman, K.J., (2010) "The Untouchables: Private Military Contractors' Criminal Accountability under the UCMJ", *Vanderbilt Law Review*, Volume 63, Issue 4.

Chatterjee, P., "Iraq War Logs: Military Privatisation Run Amok", *The Guardian*, October 23, 2010, <http://www.guardian.co.uk/commentisfree/cifamerica/2010/oct/23/iraq-war-logs-us-military/>.

Chesterman, S. and Lehnardt, C., "Introduction", in Simon Chesterman and Chia Lehnardt (Eds.), (2007) *From Mercenaries to Market: The Rise and Regulation of Private Military Companies*, Oxford: Oxford University Press.

Chesterman, S., (2008) "We Can't Spy … If We Can't Buy!: The Privatization of Intelligence and the Limits of Outsourcing 'Inherently Governmental Functions'", *The European Journal of International Law*, Volume 19, Issue 5.

Chetail, V., (2003) "The Contribution of the International Court of Justice to International Humanitarian Law", *International Review of the Red Cross*, Volume 85, Issue 850.

Chirwa, D.M., (2004) "The Doctrine of State Responsibility as a Potential Means of Holding Private Actors Accountable for Human Rights", *Melbourne Journal of International Law*, Volume 5, Issue 1.

Christenson, G.A., (1991) "Attributing Acts of Omission to the State", *Michigan Journal of International Law*, Volume 12.

Churchill, W.S., (1993) "The Price of Greatness is Responsibility", *Finest Hour*, No. 80, <http://www.winstonchurchill.org/images/finesthour/Vol.01%20No.80.pdf>.

Ciralsky, A., "Tycoon, Contractor, Soldier, Spy", *Vanity Fair*, January 2010, <http://www.vanityfair.com/politics/features/2010/01/blackwater-201001?currentPage=all>.

Clapham, A., "The Drittwirkung' of the Convention", in Ronald St. J. Macdonald, Franz Matscher and Herbert Petzold (Eds.), (1993) *The European System for the Protection of Human Rights*, Dordrecht, Boston, London: Martinus Nijhoff Publishers.

Clapham, A., (2006) *Human Rights Obligations of Non-State Actors*, Oxford: Oxford University Press.

Cockayne, J. and Others, (2009) *Beyond Market Forces: Regulating the Global Security Industry*, New York: International Peace Institute.

Cockayne, J., (2008) "Regulating Private Military and Security Companies: The Content, Negotiation, Weaknesses and Promise of the Montreux Document", *Journal of Conflict & Security Law*, Volume 13.

Coker, C., (1999) "Outsourcing War", *Cambridge Review of International Affairs*, Volume 13, Issue 1.

Committee on Oversight and Government Reform, Congress of the United States, House of Representatives, Memorandum, October 1, 2007.

Condorelli, L. and Kress, C., "The Rules of Attribution: General Considerations", in James Crawford, Alain Pellet, and Simon Olleson (Eds.), (2010) *The Law of International Responsibility*, Oxford: Oxford University Press.

Coomans, F. and Kamminga, M.T., "Comparative Introductory Comments on the Extraterritorial Application of Human Rights Treaties", in Fons Coomans and Menno T. Kamminga (Eds.), (2004) *Extraterritorial Application of Human Rights Treaties*, Antwerp and Oxford: Intersentia.

Cottier, M., (2006) "Elements for Contracting and Regulating Private Security and Military Companies", *International Review of the Red Cross*, Volume 88, Issue 863.

Council of the European Union, EU Strategic Framework and Action Plan on Human Rights and Democracy, June 25, 2012, <http://www.consilium.europa.eu/uedocs/cms_data/docs/pressdata/EN/foraff/131181.pdf>.

Crawford, J. and Olleson, S., "The Nature and Forms of International Responsibility", in Malcolm D. Evans (Ed.), (2010) *International Law*, Third Edition, Oxford: Oxford University Press.

Crawford, J. and Olleson, S., (2005) "The Continuing Debate on a UN Convention on State Responsibility", *International and Comparative Law Quarterly*, Volume 54, Issue 4.

Crawford, J., "The System of International Responsibility", in James Crawford, Alain Pellet, and Simon Olleson (Eds.), (2010) *The Law of International Responsibility*, Oxford: Oxford University Press.

Crawford, J., (2002) "The ILC's Articles on Responsibility of States for Internationally Wrongful Acts: A Retrospect", *The American Journal of International Law*, Volume 96, Issue 4.

Crawford, J., (2002) *The International Law Commission's Articles on State Responsibility: Introduction, Text and Commentaries*, Cambridge: Cambridge University Press.

Crawford, J., (2013) *State Responsibility: The General Part*, Cambridge Studies in International and Comparative Law, No. 100, Cambridge: Cambridge University Press.

Crnić-Grotić, V., "Non-State Actors and International Humanitarian Law: The Status of Private Military Companies", in Budislav Vukas and Trpimir M. Šošić (Eds.), (2010) *International Law: New Actors, New Concepts – Continuing Dilemmas; Liber Amicorum Božidar Bakotić*, Leiden [etc.]: Brill.

Cullen, P., (2000) "Keeping the New Dog of War on a Tight Leash: Assessing Means of Accountability for Private Military Companies", *Conflict Trends*, Issue 1, <http://www.accord.org.za/publications/conflict-trends/downloads.html?start=40>.

Cusumano, E., "Policy Prospects for Regulating Private Military and Security Companies", in Francesco Francioni and Natalino Ronzitti (Eds.), (2011) *War by Contract: Human Rights, Humanitarian Law, and Private Contractors*, Oxford: Oxford University Press.

Davidson, M.J., (2000) "Ruck Up: An Introduction to the Legal Issues Associated with Civilian Contractors on the Battlefield", *Public Contract Law Journal*, Volume 29, Issue 2.

DCAF, (2006) *Private Military Companies*, DCAF Backgrounder, <http://www.dcaf.ch/publications/kms/details.cfm?lng=en&id=18412&nav1=5>.

De Frouville, O., "Attribution of Conduct to the State: Private Individuals", in James Crawford, Alain Pellet, and Simon Olleson (Eds.), (2010) *The Law of International Responsibility*, Oxford: Oxford University Press.

De Schutter, O., "The Challenge of Imposing Human Rights Norms on Corporate Actors", in Olivier De Schutter (Ed.), (2006) *Transnational Corporations and Human Rights*, Oxford and Portland, Oregon: Hart Publishing.

De Schutter, O., "The Responsibility of States", in Simon Chesterman and Angelina Fisher (Eds.), (2009) *Private Security, Public Order: The Outsourcing of Public Services and Its Limits*, Oxford: Oxford University Press.

De Schutter, O., (2006) *Extraterritorial Jurisdiction as a Tool for Improving the Human Rights Accountability of Transnational Corporations*, Report and Background Paper to the seminar organized in collaboration with the Office of the UN High Commissioner for Human Rights in Brussels on November 3-4, 2006, <http://www.reports-and-materials.org/Olivier-de-Schutter-report-for-SRSG-re-extraterritorial-jurisdiction-Dec-2006.pdf>.

De Schutter, O., (2010) *International Human Rights Law*, Cambridge: Cambridge University Press.

Del Mar, K., (2010) "The Requirement of 'Belonging' under International Humanitarian Law", *The European Journal of International Law*, Volume 21, Issue 1.

Del Prado, J.L.G., (2009) "Private Military and Security Companies and the UN Working Group on the Use of Mercenaries", *Journal of Conflict & Security Law*, Volume 13, Issue 3.

Delbrück, H., (1975) *History of the Art of War: Within the Framework of Political History*, Westport: Greenwood Press, Volume II: The Germans, Walter J. Renfroe Jr. tr., Westport and London: Greenwood Press.

Den Dekker, G., "The Netherlands", in Christine Bakker and Mirko Sossai (Eds.), (2012) *Multilevel Regulation of Military and Security Contractors: The Interplay Between International, European and Domestic Norms*, Oxford and Portland, Oregon: Hart Publishing.

Den Heijer, M. and Lawson, R., "Extraterritorial Human Rights and the Concept of 'Jurisdiction'", in Malcolm Langford, Wouter Vandenhole, Martin Scheinin, and Willem van Genugten, (2014) *Global Justice, State Duties: The Extraterritorial Scope of Economic, Social and Cultural Rights in International Law*, Cambridge: Cambridge University Press.

Deva, S., "The Human Rights Obligations of Business: Reimagining the Treaty Business", an extended version of the paper presented at the Workshop on Human Rights and Transnational Corporations: Paving the Way for a Legally Binding Instrument, organized by the Permanent Missions of Ecuador and South Africa to the United Nations, Geneva, March 11-12, 2014, <http://business-humanrights.org/media/documents/reimagine_int_law_for_bhr.pdf>.

DeWinter-Schmitt, R., "Human Rights and Self-Regulation in the Apparel Industry", in Simon Chesterman and Angelina Fisher (Eds.), (2009) *Private Security, Public Order: The Outsourcing of Public Services and Its Limits*, Oxford: Oxford University Press.

DeWinter-Schmitt, R., (Ed.), *Montreux Five Years On: An Analysis of State Efforts to Implement Montreux Document Legal Obligations and Good Practices*, Report, 2013.

Dickinson, L.A., "Contract as a Tool for Regulating Private Military Companies", in Simon Chesterman and Chia Lehnardt (Eds.), (2007) *From Mercenaries to Market: The Rise and Regulation of Private Military Companies*, Oxford: Oxford University Press.

Dickinson, L.A., (2006) "Public Law Values in a Privatized World", *The Yale Journal of International Law*, Volume 31, Issue 2.

Dixon, M., (2007) *International Law*, Sixth Edition, Oxford: Oxford University Press.

Dörmann, K., (2003) "The Legal Situation of 'Unlawful/Unprivileged Combatants'", *International Review of the Red Cross*, Volume 85, Issue 849.

Dorn, N. and Levi, M., (2007) "European Private Security, Corporate Investigation and Military Services: Collective Security, Market Regulation and Structuring the Public Sphere", *Policing and Society*, Volume 17, Issue 3.

Doswald-Beck, L., "Private Military Companies under International Humanitarian Law", in Simon Chesterman and Chia Lehnardt (Eds.), (2007) *From Mercenaries to Market: The Rise and Regulation of Private Military Companies*, Oxford: Oxford University Press.

Dupuy, P.-M., (1992) *Droit International Public*, Paris: Dalloz-Sirey.

Dupuy, P.-M., (1999) "Reviewing the Difficulties of Codification: On Ago's Classification of Obligations of Means and Obligations of Result in Relation to State Responsibility", *The European Journal of International Law*.

Dutch Advisory Council on International Affairs, *Employing Private Military Companies: A Question of Responsibility* (Dutch: *De Inhuur van Private Militaire Bedrijven: Een Kwestie van Verantwoordelijkheid*), No. 59, December 2007, The Hague.

Eagleton, C., (1950) "International Organization and the Law of Responsibility", *Collected Courses of the Hague Academy of International Law (Recueil des Cours)*, Volume 76.

Ebrahim, A., (2010) "Going to War with the Army You Can Afford: The United States, International Law and the Private Military Industry", *Boston University International Law Journal*, Volume 28.

European Commission, A Renewed EU Strategy 2011-2014 for Corporate Social Responsibility, Communication from the Commission to the European Parliament, the Council, the European Economic and Social Committee and the Committee of the

Regions, October 25, 2011, COM(2011) 681, <http://eur-lex.europa.eu/LexUriServ/LexUriServ.do?uri=COM:2011:0681:FIN:EN:PDF>.

Fainaru, S., "Cutting Costs, Bending Rules, and a Trail of Broken Lives", *The Washington Post*, July 29, 2007, <http://www.pulitzer.org/archives/7688>.

Fainaru, S., "Guards in Iraq Cite Frequent Shootings", *The Washington Post*, October 3, 2007, <http://www.washingtonpost.com/wp-dyn/content/article/2007/10/02/AR2007100202456.html>.

Fallah, K., (2006) "Corporate Actors: The Legal Status of Mercenaries in Armed Conflict", *International Review of the Red Cross*, Volume 88, Issue 863.

Farrior, S., (1998) "State Responsibility for Human Rights Abuses by Non-state Actors", *American Society of International Law Proceedings*, Volume 92.

Fisher, A., "Accountability to Whom?", in Simon Chesterman and Angelina Fisher (Eds.), (2009) *Private Security, Public Order: The Outsourcing of Public Services and Its Limits*, Oxford: Oxford University Press.

Fleck, D. (Ed.), (2001) *The Handbook of the Law of Visiting Forces*, Oxford: Oxford University Press.

Fleck, D., "Status of Forces in Peace Operations", in Terry D. Gill and Dieter Fleck (Eds.), (2010) *The Handbook of the International Law of Military Operations*, Oxford: Oxford University Press.

Fleck, D., (2007) "Individual and State Responsibility for Intelligence Gathering", *Michigan Journal of International Law*, Volume 28.

Francioni, F. and Ronzitti, N., "Introduction", in Francesco Francioni and Natalino Ronzitti (Eds.), (2011) *War by Contract: Human Rights, Humanitarian Law, and Private Contractors*, Oxford: Oxford University Press.

Francioni, F., "The Role of the Home State in Ensuring Compliance with Human Rights by Private Military Contractors", in Francesco Francioni and Natalino Ronzitti (Eds.), (2011) *War by Contract: Human Rights, Humanitarian Law, and Private Contractors*, Oxford: Oxford University Press.

Frost, M., "The Ethics of PMCs in Global Civil Society", in Andrew Alexandra, Deane-Peter Baker, and Marina Caparini (Eds.), (2008) *Private Military and Security Companies: Ethics, Policies and Civil-Military Relations*, London: Routledge.

Frulli, M., "Immunity for Private Military Contractors: Legal Hurdles or Political Snags?", in Francesco Francioni and Natalino Ronzitti (Eds.), (2011) *War by Contract: Human Rights, Humanitarian Law, and Private Contractors*, Oxford: Oxford University Press.

G4S plc, *Annual Report and Accounts 2012*, <http://www.g4s.com/~/media/C15A0466A2F842258247347391C48379.ashx>.

Gathii, J.T., (2009-2010) *Commercializing War: Private Military and Security Companies, Mercenaries and International Law*, Legal Studies Research Paper Series, No. 49, Albany Law School.

Gayle, D., "Has Blackwater Been Deployed to Ukraine? Notorious U.S. Mercenaries 'Seen on the Streets of Flashpoint City' as Russia Claims 300 Hired Guns Have Arrived in Country", *MailOnline*, March 8, 2014, <http://www.dailymail.co.uk/news/article-2576490/Are-Blackwater-active-Ukraine-Videos-spark-talk-U-S-mercenary-outfit-deployed-Donetsk.html>.

Gill, T.D. and Fleck, D., "Private Contractors and Security Companies", in Terry D. Gill and Dieter Fleck (Eds.), (2010) *The Handbook of the International Law of Military Operations*, Oxford: Oxford University Press.

Gillard, E.-C., "International Humanitarian Law and Extraterritorial State Conduct", in Fons Coomans and Menno T. Kamminga (Eds.), (2004) *Extraterritorial Application of Human Rights Treaties*, Antwerp and Oxford: Intersentia.

Gillard, E.-C., "Private Military/Security Companies: The Status of Their Staff and Their Obligations under International Humanitarian Law and The Responsibilities of States in Relation to Their Operation", in Andrew Alexandra, Deane-Peter Baker, and Marina Caparini (Eds.), (2008) *Private Military and Security Companies: Ethics, Policies and Civil-Military Relations*, London and New York: Routledge.

Gillard, E.-C., (2006) "Business Goes to War: Private Military / Security Companies and International Humanitarian Law", *International Review of the Red Cross*, Volume 88, Issue 863.

Goldenberg, S., "US Soldier's Family Brings Legal Action Against British Private Security Firm", *The Guardian*, October 30, 2007, <http://www.guardian.co.uk/world/2007/oct/30/iraq.usa1>.

Gondek, M., (2009) *The Reach of Human Rights in a Globalising World: Extraterritorial Application of Human Rights Treaties*, School of Human Rights Research Series, Volume 32, Antwerp and Portland: Intersentia.

Gorman, S., "CIA Had Secret Al Qaeda Plan", *The Wall Street Journal*, July 13, 2009, <http://online.wsj.com/article/SB124736381913627661.html>.

Gorman, S., "US: CIA Likely Let Contractors Perform Waterboarding", *The Wall Street Journal*, February 8, 2008, <http://www.corpwatch.org/article.php?id=14926>.

Graduate Institute of International and Development Studies, *Small Arms Survey 2011: States of Security*, Geneva, 2011, <http://www.smallarmssurvey.org/publications/by-type/yearbook/small-arms-survey-2011.html>.

Graeber, D., "U.N. Report Cites War Crimes by Private Contractors", October 12, 2007, <http://warcrimes.foreignpolicyblogs.com/2007/10/12/un-reports-cites-war-crimes-by-private-contractors/>.

Green, L.C., (2008) *The Contemporary Law of Armed Conflict*, Melland Schill Studies in International Law, Third Edition, Manchester: Manchester University Press.

Greenwood, C., (2006) *Essays on War in International Law*, London: Cameron May.

Griffith, G.T., (1968) *The Mercenaries of the Hellenistic World*, Groningen: Boom's Boekhuis B.V.

Grofe, J., "Human Rights and Private Military Companies: A Double-Edged Sword Too Dangerous to Use?", in Thomas Jäger and Gerhard Kümmel (Eds.), (2007) *Private Military and Security Companies: Chances, Problems, Pitfalls and Prospects*, Wiesbaden: VS Verlag für Sozialwissenschaften.

Grotius, H., (1625) *The Rights of War and Peace*, Book III, Edited and with an Introduction by Richard Tuck, in Jean Barbeyrac (Ed.), (2005) *Major Legal and Political Works of Hugo Grotius*, Indianapolis: Liberty Fund.

Guillaume, G., "Overview of Part One of the Articles on State Responsibility", in James Crawford, Alain Pellet, and Simon Olleson (Eds.), (2010) *The Law of International Responsibility*, Oxford: Oxford University Press.

Hakimi, M., (2010) "State Bystander Responsibility", *The European Journal of International Law*, Volume 21, Issue 2.

Hallo de Wolf, A., (2006) "Modern Condottieri in Iraq: Privatizing War from the Perspective of International Human Rights Law", *Indiana Journal of Global Legal Studies*, Volume 13, Issue 2.

Hallo de Wolf, A., (2012) *Reconciling Privatization with Human Rights*, School of Human Rights Research Series, Volume 49, Cambridge, Antwerp and Portland: Intersentia.

Hallo de Wolf, A., (2013) "Human Rights and the Regulation of Privatized Essential Services", *Netherlands International Law Review*, Volume 60, Issue 2.

Hearing before the House Committee on Oversight and Government Reform, Blackwater USA, October 2, 2007, <http://www.gpo.gov/fdsys/pkg/CHRG-110hhrg45219/html/CHRG-110hhrg45219.htm>.

Heintze, H.-J., (2004) "On the Relationship Between Human Rights Law Protection and International Humanitarian Law", *International Review of the Red Cross*, Volume 86, Issue 856.

Henckaerts, J.-M. and Doswald-Beck, L., (2005) *Customary International Humanitarian Law, Volume I: Rules*, ICRC, Cambridge: Cambridge University Press.

Herbst, K., "Private Security Companies and Civil-Military Cooperation", in Thomas Jäger and Gerhard Kümmel (Eds.), (2007) *Private Military and Security Companies: Chances, Problems, Pitfalls and Prospects*, Wiesbaden: VS Verlag für Sozialwissenschaften.

Hersh, S.M., "Torture at Abu Ghraib: American Soldiers Brutalized Iraqis. How Far up Does the Responsibility Go?", *The New Yorker*, May 10, 2004, <http://www.newyorker.com/archive/2004/05/10/040510fa_fact>.

Hessbruegge, J.A., (2004) "The Historical Development of the Doctrines of Attribution and Due Diligence in International Law", *New York University Journal of International Law and Politics*, Volume 36.

Hillier, T., (1999) *Principles of Public International Law*, Second Edition, London and Sydney: Cavendish Publishing Limited.

Hobbes, T., (1651) *Leviathan*, Revised Edition, Edited by A.P. Martinich, and Brian Battiste, 2011, Peterborough [etc.]: Broadview Press.

Hobér, K., "State Responsibility and Attribution", in Peter Muchlinski, Federico Ortino, and Christoph Schreuer (Eds.), (2008) *The Oxford Handbook of International Investment Law*, Oxford: Oxford University Press.

Hoppe, C., (2008) "Passing the Buck: State Responsibility for Private Military Companies", *The European Journal of International Law*, Volume 19, Issue 5.

Hoppe, C., "Positive Human Rights Obligations of the Hiring State in Connection with the Provision of 'Coercive Services' by a Private Military or Security Company", in Francesco Francioni and Natalino Ronzitti (Eds.), (2011) *War by Contract: Human Rights, Humanitarian Law, and Private Contractors*, Oxford: Oxford University Press.

Horton, S., "Security Contractors Immune from Torture Charges, Judges Rule", *Harpers*, September 14, 2009, <http://harpers.org/blog/2009/09/security-contractors-immune-from-torture-charges-judges-rule/>.

Hough, L., (2007) "A Study of Peacekeeping, Peace-enforcement and Private Military Companies in Sierra Leone", *African Security Review*, Volume 16, Issue 4.

Human Rights Advocates, *Holding Private Military and Security Companies and Mercenaries Accountable for Human Rights Violations*, Report, <http://www.humanrightsadvocates.org/wp-content/uploads/2010/05/Arellano2008-Mercenaries-and-Corporate-Accountability-Long.pdf>.

Human Rights First, *Private Security Contractors at War: Ending the Culture of Impunity*, Report, 2008, New York.

Human Rights Watch, *The Road to Abu Ghraib*, Report, June 9, 2004, <http://www.hrw.org/reports/2004/06/08/road-abu-ghraib>.

Human Rights Watch, *Hopes Betrayed: Trafficking of Women and Girls to Post-Conflict Bosnia and Herzegovina for Forced Prostitution*, Report, November 2002, <http://www.hrw.org/reports/2002/bosnia/Bosnia1102.pdf>.

Huskey, K. and Sullivan, S., "United States: Law and Policy Governing Private Military Contractors After 9/11", in Christine Bakker and Mirko Sossai (Eds.), (2012) *Multilevel Regulation of Military and Security Contractors: The Interplay Between International, European and Domestic Norms*, Oxford and Portland, Oregon: Hart Publishing.

ICRC, "How is the Term 'Armed Conflict' Defined in International Humanitarian Law?", Opinion Paper, March 2008, <http://www.icrc.org/eng/assets/files/other/opinion-paper-armed-conflict.pdf>.

ICRC, *Interpretive Guidance on the Notion of Direct Participation in Hostilities under International Humanitarian Law*, Prepated by Nils Melzer, May 2009, <https://www.icrc.org/eng/assets/files/other/icrc-002-0990.pdf>.

ICRC, Report of the International Committee of the Red Cross (ICRC) on the Treatment by the Coalition Forces of Prisoners of War and Other Protected Persons by the Geneva Conventions in Iraq During Arrest, Internment and Interrogation, February 2004, <http://www.informationclearinghouse.info/pdf/icrc_iraq.pdf>.

International Service for Human Rights, "Human Rights Council: Little Support for Proposed Convention Regulating Private Security Companies", September 23, 2010, <http://www.ishr.ch/news/human-rights-council-little-support-proposed-convention-regulating-private-security-companies>.

Ipsen, K., "Combatants and Non-Combatants", in Dieter Fleck (Ed.), (2008) *The Handbook of International Humanitarian Law*, Oxford: Oxford University Press.

"Iraq Shooting Firm Loses Licence", *BBC*, September 17, 2007, <http://news.bbc.co.uk/2/hi/middle_east/6998788.stm>.

Isenberg, D., "A Government in Search of Cover: Private Military Companies in Iraq", in Simon Chesterman and Chia Lehnardt (Eds.), (2007) *From Mercenaries to Market: The Rise and Regulation of Private Military Companies*, Oxford: Oxford University Press.

Isenberg, D., "How May We Service You?", *The Huffington Post*, August 22, 2012, <http://www.huffingtonpost.com/david-isenberg/how-may-we-service-you_b_1814311.html?utm_source=Alert-blogger>.

Isenberg, D., "Pedagogic Military Contractors", *The Huffington Post*, December 18, 2011, <http://www.huffingtonpost.com/david-isenberg/military-contractors-blackwater_b_1156080.html>.

Isenberg, D., "PMC Sexual Violence: It's Still a Problem", *The Huffington Post*, January 30, 2012, <http://www.huffingtonpost.com/david-isenberg/pmc-sexual-violence-its-s_b_1240751.html?view=screen>.

Isenberg, D., "The State Department Tells Us How They Really Felt: Part 1", *The Huffington Post*, February 17, 2012, <http://www.huffingtonpost.com/david-isenberg/state-department-blackwater_b_1280847.html>.

Isenberg, D., (2009) *Shadow Force: Private Security Contractors in Iraq*, Westport and London: Praeger Security International.

Jägers, N., (2002) *Corporate Human Rights Obligations: in Search of Accountability*, School of Human Rights Research Series, Volume 17, Antwerpen, Oxford, New York: Intersentia.

Jägers, N., (2012) "Regulating the Private Security Industry: Connecting the Public and the Private Through Transnational Private Regulation", *Human Rights and International Legal Discourse*, Volume, Issue 1.

Jägers, N., Moyakine, E., and Van Genugten, W., "Transnational Private Regulation, Private Security Companies and Human Rights: Promise of Effectiveness?", Case Study of the Center for Transboundary Legal Development (CTLD), Tilburg University, for the Hague Institute for the Internationalisation of Law (HiiL), September 2012 (on file with the author). Details on the case study and broader research project "Transnational Private Regulation: Constitutional Foundations and Governance Design" are available at www.privateregulation.eu.

Jennings, K.M., (2006) *Armed Security: Regulating the Private Military Industry*, New Security Programme, Fafo-report 532, <http://www.fafo.no/pub/rapp/532/532.pdf>.

Johnston, D., "Immunity Deals Offered to Blackwater Guards", *The New York Times*, October 30, 2007, <http://www.nytimes.com/2007/10/30/washington/30blackwater.html>.

Jones, O.R., (2009) "Implausible Deniability: State Responsibility for the Actions of Private Military Firms", *Connecticut Journal of International Law*, Volume 24, Issue 2.

Kalshoven, F., (1991) "State Responsibility for Warlike Acts of the Armed Forces", *International and Comparative Law Quarterly*, Volume 40, Issue 4.

Keegan, J., (1993) *A History of Warfare*, London: Hutchinson.

Kessler, B., (2001) "The Duty to 'Ensure Respect' under Common Article 1 of the Geneva Conventions: Its Implications on International and Non-International Armed Conflicts", *German Yearbook of International Law*, Volume 44.

Kidane, W., (2010) "The Status of Private Military Contractors under International Humanitarian Law", *Denver Journal of International Law and Policy*, Volume 38, Issue 3.

Kinsey, C., (2006) *Corporate Soldiers and International Security: The Rise of Private Military Companies*, London: Routledge.

Kontos, A.P., (2004) "'Private' Security Guards: Privatized Force and State Responsibility under International Human Rights Law", *Non-State Actors and International Law*, Volume 4.

Krahmann, E., "The Rise of Non-State Actors in Security Governance", in Patricia Kennett (Ed.), (2008) *Governance, Globalization and Public Policy*, Cheltenham and Northampton: Edward Elgar Publishing.

Krahmann, E., (2012) "From 'Mercenaries' to 'Private Security Contractors': The (Re) Construction of Armed Security Providers in International Legal Discourses", *Millennium: Journal of International Studies*, Volume 40, Issue 2.

Kratovac, K., "U.N.: Shootings by Guards Might Well Be War Crimes", *The Star-Ledger*, October 12, 2007.

Lam, J.S., (2009) "Accountability for Private Military Contractors Under the Alien Tort Statute", *California Law Review*, Volume 97, Issue 5.

Lambooy, T., (2010) "Corporate Due Diligence As a Tool to Respect Human Rights", *Netherlands Quarterly of Human Rights*, Volume 28, Issue 3.

Lamp, N., (2011) "Conceptions of War and Paradigms of Compliance: The 'New War' Challenge to International Humanitarian Law", *Journal of Conflict & Security Law*, Volume 16, Issue 2.

Lane, F.C., (1979) *Profits from Power: Readings in Protection Rent and Violence-Controlling Enterprises*, Albany: SUNY Press.

Langford, M., Vandenhole, W., Scheinin, M., and Van Genugten, W., "Introduction: An Emerging Field", in Malcolm Langford, Wouter Vandenhole, Martin Scheinin, and Willem van Genugten (Eds.), (2013) *Global Justice, State Duties: The Extraterritorial Scope of Economic, Social and Cultural Rights in International Law*, Cambridge: Cambridge University Press.

Lardner, R., "DynCorp, Other Private Contractors in Afghanistan Behaving Badly", *The Huffington Post*, December 19, 2010, <http://www.huffingtonpost.com/2010/12/19/dyncorp-afghanistan-private-contractors_n_798753.html>.

Lawson, R., "DynCorp: Beyond the Rule of Law", *Colombia Journal*, August 27, 2001, <http://colombiajournal.org/colombia78.htm>.

Lawson, R., "Life after Bankovic: On the Extraterritorial Application of the European Convention on Human Rights", in Fons Coomans and Menno T. Kamminga (Eds.), (2004) *Extraterritorial Application of Human Rights Treaties*, Antwerp and Oxford: Intersentia.

Lawson, R., "Out of Control. State Responsibility and Human Rights: Will the ILC's Definition of the 'Act of State' Meet the Challenges of the 21st Century?", in Monique Castermans-Holleman, Fried van Hoof, and Jacqueline Smith (Eds.), (1998) *The Role of the Nation-State in the 21st Century: Human Rights, International Organisations and Foreign Policy. Essays in Honour of Peter Baehr*, The Hague: Kluwer Law International.

Leander, A., "Regulating the Role of Private Military Companies in Shaping Security and Politics", in Simon Chesterman and Chia Lehnardt (Eds.), (2007) *From Mercenaries to Market: The Rise and Regulation of Private Military Companies*, Oxford: Oxford University Press.

Leander, A., (2005) "The Power to Construct International Security: On the Significance of Private Military Companies", *Millennium: Journal of International Studies*, Volume 33, Issue 3.

Lehnardt, C., "Peacekeeping", in Simon Chesterman and Angelina Fisher (Eds.), (2009) *Private Security, Public Order: The Outsourcing of Public Services and Its Limits*, Oxford: Oxford University Press.

Lehnardt, C., "Private Military Companies and State Responsibility", in Simon Chesterman and Chia Lehnardt (Eds.), (2007) *From Mercenaries to Market: The Rise and Regulation of Private Military Companies*, Oxford: Oxford University Press.

Lendman, S., "Blackwater Killers in Eastern Ukraine", *thepeoplesvoice.org*, May 12, 2014, <http://www.thepeoplesvoice.org/TPV3/Voices.php/2014/05/12/blackwater-killers-in-eastern-ukraine>.

Lendman, S., "Outsourcing War: The Rise of Private Military Contractors", *The Intelligence Daily*, January 20, 2010, <http://dissidentvoice.org/2010/01/outsourcing-war-the-rise-of-private-military-contractors-pmcs/>.

Lewis, P., "It's Not Just Governments That Make War and Peace Now", *The New York Times*, November 28, 1998, <http://www.nytimes.com/1998/11/28/arts/not-just-governments-make-war-or-peace.html>.

Lieutenant General Anthony R. Jones, Article 15-6 Investigation of the Abu Ghraib Prison and 205th Military Intelligence Brigade, August 2004, <news.findlaw.com/hdocs/docs/dod/fay82504rpt.pdf>.

Lieutenant-Colonel Tim Spicer Obe, (2000) *An Unorthodox Soldier: Peace and War and the Sandline Affair*, Edinburgh and London: Mainstream Publishing.

Likosky, M., "The Privatization of Violence", in Simon Chesterman and Angelina Fisher (Eds.), (2009) *Private Security, Public Order: The Outsourcing of Public Services and Its Limits*, Oxford: Oxford University Press.

Lilly, D., (2000) "The Privatization of Peacekeeping: Prospects and Realities", *Disarmament Forum*, Issue 3.

Lilly, D., (2000) *The Privatization of Security and Peacebuilding: A Framework for Action*, International Alert.

Liu, H.-Y., (2010) "Leashing the Corporate Dogs of War: The Legal Implications of the Modern Private Military Company", *Journal of Conflict & Security Law*, Volume 15, Issue 1.

Lococo, E., "Titan Competing with Northrop, L-3 to Keep Its Largest Contract", *Bloomberg*, June 7, 2004, <http://www.bloomberg.com/apps/news?pid=newsarchive&sid=afR4zWgUnpDo>.

Lowe, V., (2007) *International* Law, Oxford: Oxford University Press.

Luckey, J.R., Bailey Grasso, V., and Manuel, K.M., "Inherently Governmental Functions and Department of Defense Operations: Background, Issues, and Options for Congress", CRS Report for Congress, February 1, 2010, <http://www.fas.org/sgp/crs/natsec/R40641.pdf>.

Lynch, C., "Misconduct, Corruption by U.S. Police Mar Bosnia Mission", *The Washington Post*, May 29, 2001, <http://dailyuw.com/archive/2001/05/29/imported/misconduct-corruption-us-police-mar-bosnia-mission#.VBlpSfmSyiw>.

Lynch, T. and Walsh, A.J., (2000) "The Good Mercenary?", *The Journal of Political Philosophy*, Volume 8, Issue 2.

Machairas, D., (2014) "The Ethical Implications of the Use of Private Military Force: Regulatable or Irreconcilable?", *Journal of Military Ethics*, Volume 13, Issue 1.

Machiavelli, N., (1513) *The Prince*, Translated by W.K. Marriott, in Robert Maynard Hutchins (Ed. in Chief), (1952) *Great Books of the Western World: Machiavelli, Hobbes*, Volume 23, Chicago [etc.]: Encyclopaedia Britannica, Inc.

MacLeod, S., "The Role of International Regulatory Initiatives on Business and Human Rights for Holding Private Military and Security Contractors to Account", in Francesco Francioni and Natalino Ronzitti (Eds.), (2011) *War by Contract: Human Rights, Humanitarian Law, and Private Contractors*, Oxford: Oxford University Press.

Major General Antonio N. Taguba, Report of the Article 15-6 Investigation of the 800th Military Police Brigade, March 2004, <http://www.dod.mil/pubs/foi/operation_and_plans/Detainee/taguba/TAGUBA_REPORT_CERTIFICATIONS.pdf>.

Major General George R. Fay, Investigating Officer, Article 15-6 Investigation of the Abu Ghraib Prison and 205th Military Intelligence Brigade, August 2004, <news.findlaw.com/hdocs/docs/dod/fay82504rpt.pdf>.

Mallaby, S., "Paid to Make Peace: Mercenaries Are No Altruists, But They Can Do Good", *The Washington Post*, June 4, 2001, <http://www.sandline.com/hotlinks/article9.html>.

Mandel, R., (2002) *Armies Without States: the Privatization of Security*, Boulder: Lynne Rienner Publishers.

Mann, F.A., (1984) "The Doctrine of International Jurisdiction Revisited After Twenty Years", *Collected Courses of the Hague Academy of International Law (Recueil des Cours)*, Volume 186.

Maogoto, J.N. and Sheehy, B., (2009) "Private Military Companies & International Law: Building New Ladders of Legal Accountability & Responsibility", *Cardozo Journal of Conflict Resolution*, Volume 11.

Martin-Ortega, O., (2013) "Human Rights Due Diligence for Corporations: From Voluntary Standards to Hard Law at Last?", *Netherlands Quarterly of Human Rights*, Volume 31, Issue 4.

Mayer, D., (2010) "Peaceful Warriors: Private Military Security Companies and the Quest for Stable Societies", *Journal of Business Ethics*, Volume 89.

Mazzetti, M., "C.I.A. Sought Blackwater's Help to Kill Jihadists", *The New York Times*, August 19, 2009, <http://www.nytimes.com/2009/08/20/us/20intel.html?_r=1&>.

McCarthy, E., "CACI Contract: From Supplies to Interrogation", *The Washington Post*, May 17, 2004, <http://www.washingtonpost.com/wp-dyn/articles/A31611-2004May16.html>.

McCormack, J., (1993) *On Million Mercenaries: Swiss Soldiers in the Armies of the World*, London: Leo Cooper.

McCorquodale, R. and Simons, P., (2007) "Responsibility Beyond Borders: State Responsibility for Extraterritorial Violations by Corporations of International Human Rights Law", *Modern Law Review*, Volume 70, Issue 4.

McGoldrick, D., "Extraterritorial Application of the International Covenant on Civil and Political Rights", in Fons Coomans and Menno T. Kamminga (Eds.), (2004) *Extraterritorial Application of Human Rights Treaties*, Antwerp and Oxford: Intersentia.

McIntyre, A. and Weiss, T., "Weak Governments in Search of Strength: Africa's Experience of Mercenaries and Private Military Companies", in Simon Chesterman and Chia Lehnardt (Eds.), (2007) *From Mercenaries to Market: The Rise and Regulation of Private Military Companies*, Oxford: Oxford University Press.

McKenna L. and Johnson, R., "A Look at the Largest Private Armies in the World", *Business Insider*, June 3, 2012, <http://www.businessinsider.com/worlds-most-powerful-mercenary-armies-2012-06?op=1>.

Mehra, A., (2010) "Bridging Accountability Gaps – The Proliferation of Private Military and Security Companies and Ensuring Accountability for Human Rights Violations", *Pacific McGeorge Global Business & Development Legal Journal*, Volume 22, Issue 2.

Melis, G. and Persiani, A., "The Regulation of Private Military and Security Companies: Tax Aspects", in Francesco Francioni and Natalino Ronzitti (Eds.), (2011) *War by Contract: Human Rights, Humanitarian Law, and Private Contractors*, Oxford: Oxford University Press.

Memorandum from Majority Staff to Members of the Committee on Oversight and Government Reform Re: Additional Information About Blackwater USA, October 1, 2007.

Merle, R. and McCarthy, E., "6 Employees From CACI International, Titan Referred for Prosecution", *The Washington Post*, August 26, 2004, <http://www.washingtonpost.com/wp-dyn/articles/A33834-2004Aug25.html>.

Milanović, M., (2009) "State Responsibility for Acts of Non-state Actors: A Comment on Griebel and Plücken", *Leiden Journal of International Law*, Volume 22, Issue 2.

Milbank, D., "Bush Seeks to Reassure Nation on Iraq", *The Washington Post*, May 25, 2004, <http://www.washingtonpost.com/wp-dyn/articles/A52711-2004May24.html>.

Miller, R.L. and Jentz, G.A., (2008) *Business Law Today: The Essentials*, 8th Edition, Mason: Thomson West.

Miller, T.C., "Private Contractors Outnumber U.S. Troops in Iraq", *The Los Angeles Times*, July 4, 2007, <http://articles.latimes.com/2007/jul/04/nation/na-private4>.

Milliard, T.S., (2003) "Overcoming Post-Colonial Myopia. A Call to Recognize and Regulate PMCs", *Military Law Review*, Volume 176, Issue 1.

Miluna, I., "Jurisdictional Competence and Applicable Law with Regard to Private Military and Security Companies", in Christine Bakker and Mirko Sossai (Eds.), (2012) *Multilevel Regulation of Military and Security Contractors: The Interplay Between International, European and Domestic Norms*, Oxford and Portland, Oregon: Hart Publishing.

Minow, M., (2005) "Outsourcing Power: How Privatizing Military Efforts Challenges Accountability, Professionalism, and Democracy", *Boston College Law Review*, Volume 46, Issue 5.

Miretski, P.P. and Bachmann, S.-D., (2012) "The UN 'Norms on the Responsibility of Transnational Corporations and Other Business Enterprises with Regard to Human Rights': A Requiem", *Deakin Law Review*, Volume 17, Issue 1.

Mitchell, D.S., (2005) "The Prohibition of Rape in International Humanitarian Law as a Norm of Jus Cogens: Clarifying the Doctrine", *Duke Journal of Comparative and International Law*, Volume 15.

Mockler, A., (1985) *The New Mercenaries: The History of the Mercenary from the Congo to the Seychelles*, Letchworth: The Garden City Press.

Momtaz, D., "Attribution of Conduct to the State: State Organs and Entities Empowered to Exercise Elements of Governmental Authority", in James Crawford, Alain Pellet, and Simon Olleson (Eds.), (2010) *The Law of International Responsibility*, Oxford: Oxford University Press.

Moyakine, E., (2015) "From National and International Frustrations to Transnational Triumph? Hybrid Transnational Private Regulatory Regimes in the Private Military and Security Companies' Industry and Their Effectiveness in Ensuring Compliance with Human Rights", *Pacific McGeorge Global Business & Development Law Journal*, Volume 28, Issue 1 (*forthcoming*).

Muchilinski, P.T., (2007) *Multinational Enterprises & The Law*, Oxford: Oxford University Press.

Narula, S., "International Financial Institutions, Transnational Corporations and Duties of States", in Malcolm Langford, Wouter Vandenhole, Martin Scheinin, and Willem van Genugten, (2014) *Global Justice, State Duties: The Extraterritorial Scope of Economic, Social and Cultural Rights in International Law*, Cambridge: Cambridge University Press.

NATO/ISAF, *COMISAF's Counterinsurgency (COIN) Contracting Guidance*, September 8, 2010, <http://www.isaf.nato.int/images/stories/File/100908-NUI-COMISAF%20 COIN%20GUIDANCE.pdf>.

Netherlands Institute for Human Rights, *Response to the National Action Plan on Business and Human Rights "Knowing and Showing"* (Dutch: *Reactie op het Nationaal Actieplan Bedrijfsleven en Mensenrechten "Knowing and Showing")*, Advisory Report, February 2014, <http://zoekservice.mensenrechten.nl/StippWebDLL/Resources/Handlers/DownloadBestand.ashx?id=2236>.

Nieminen, K., (2004) "The Rules of Attribution and the Private Military Contractors at Abu Ghraib: Private Acts or Public Wrongs?", *Finnish Yearbook of International Law*, Volume 15.

Nollkaemper, A., (2003) "Concurrence Between Individual Responsibility and State Responsibility in International Law", *International & Comparative Law Quarterly*, Volume 52, Issue 3.

Nossal, K.R., (1998) "Roland Goes Corporate: Mercenaries and Transnational Security Corporations in the Post-Cold War Era", *Civil Wars*, Volume 1, Issue 1.

Nowak, M., "Obligations of States to Prevent and Prohibit Torture in an Extraterritorial Perspective", in Mark Gibney and Sigrun Skogly (Eds.), (2010) *Universal Human Rights and Extraterritorial Obligations*, Philadelphia: University of Pennsylvania Press.

Nurick, L., (1945) "The Distinction Between Combatant and Noncombatant in the Law of War", *The American Journal of International Law*, Volume 39.

O'Brien, K.A., "What Should and What Should Not Be Regulated", in Simon Chesterman and Chia Lehnardt, (Eds.), (2007) *From Mercenaries to Market: The Rise and Regulation of Private Military Companies*, Oxford: Oxford University Press.

O'Loughlin, E., "Sandline Scandal Arms Shipment Reaches Forces", *The Independent*, May 22, 2000.

O'Meara, K.P., "US: DynCorp Disgrace", *Insight*, January 14, 2002, <http://www.corpwatch.org/article.php?id=11119>.

Oppenheim, L., (1920) *International Law: A Treatise*, Volume I. – Peace, Third Edition Edited by Ronald F. Roxburgh, Clark, New Jersey: The Lawbook Exchange Ltd.

Orakhelashvili, A., (2003) "Restrictive Interpretation of Human Rights Treaties in the Recent Jurisprudence of the European Court of Human Rights", *The European Journal of International Law*, Volume 14, Issue 3.

Ortiz, C., "The Private Military Company: An Entity at the Center of Overlapping of Overlapping Spheres of Commercial Activity and Responsibility", in Thomas Jäger and Gerhard Kümmel (Eds.), (2007) *Private Military and Security Companies: Chances, Problems, Pitfalls and Prospects*, Wiesbaden: VS Verlag für Sozialwissenschaften.

Ortiz, C., (2009) "Private Military and Șecurity Companies: Misconceptions, Reconceptualisations, and Regulation", Working Paper No. 9, The Centre for Global Political Economy, University of Sussex.

Ortiz, C., (2010) "The New Public Management of Security: the Contracting and Managerial State and the Private Military Industry", *Public Money & Management*, Volume 30, Issue 1.

Østensen, Å.G., (2011) *UN Use of Private Military and Security Companies: Practices and Policies*, SSR Paper 3, DCAF.

Palou-Loverdos, J. and Armendáriz, L., *The Privatization of Warfare, Violence and Private Military & Security Companies: A Factual and Legal Approach to Human Rights Abuses by PMSC in Iraq*, Report, 2011, Nova-Institute for Active Non-violence, <http://nova.cat/wp-content/uploads/2011/12/Informe_PMSC_Iraq_Nova_ok.pdf>.

Pariotti, E., (2009) "International Soft Law, Human Rights and Non-State Actors: Towards the Accountability of Transnational Corporations?", *Human Rights Review*, Volume 10, Issue 2.

Partlow, J. and Raghavan, S., "Iraq Probe of U.S. Security Firm Grows: Blackwater, Accused of Killing 11 on Sunday, Cited in Earlier Deaths", *The Washington Post*, September 22, 2007, <http://www.washingtonpost.com/wp-dyn/content/article/2007/09/21/AR2007092102271.html?sid=ST2007092200176>.

Pellet, A., "The ILC's Articles on State Responsibility for Internationally Wrongful Acts and Related Texts", in James Crawford, Alain Pellet, and Simon Olleson (Eds.), (2010) *The Law of International Responsibility*, Oxford: Oxford University Press.

Pelton, R.Y., "Crescent Goes Down Hard in Washington Post: Violations and Kidnapping Take Down Media-Friendly Security Company", *Privateforces.com*, July 29, 2007, <http://privateforces.com/index.php?option=com_content&task=view&id=1894&Itemid=99999999>.

Percy, S., "Morality and Regulation", in Simon Chesterman and Chia Lehnardt (Eds.), (2007) *From Mercenaries to Market: The Rise and Regulation of Private Military Companies*, Oxford: Oxford University Press.

Percy, S., (2006) *Regulating the Private Security Industry*, Adelphi Series, Volume 46, Issue 384, Oxon: Routledge.

Percy, S., (2007) *Mercenaries: The History of a Norm in International Relations*, Oxford: Oxford University Press.

Percy, S., (2009) "Private Security Companies and Civil Wars", *Civil Wars*, Volume 11, Issue 1.

Perez, E., "Appeals Court Revives Blackwater Case", *The Wall Street Journal*, April 22, 2011, <http://online.wsj.com/news/articles/SB100014240527487035213045762789216175099O8>.

Peterson, M.J., (1992) "Transnational Activity, International Society and World Politics", *Millennium: Journal of International Studies*, Volume 21, Issue 3.

Pictet, J.S., (1952) *Commentary, I Geneva Convention for the Amelioration of the Condition of the Wounded and Sick in Armed Forces in the Field*, Geneva: International Committee of the Red Cross.

Pictet, J.S., (1958) *Commentary, IV Geneva Convention Relative to the Protection of Civilian Persons in Time of War*, Geneva: International Committee of the Red Cross.

Pictet, J.S., (1960) *Commentary, II Geneva Convention for the Amelioration of the Condition of the Wounded, Sick and Shipwrecked Members of Armed Forces at Sea*, Geneva: International Committee of the Red Cross.

Pictet, J.S., (1960) *Commentary, III Geneva Convention Relative to the Treatment of Prisoners of War*, Geneva: International Committee of the Red Cross.

Pillai, P., (2007) "The Genocide Convention Case: Confusion or Clarity?", *Australian International Law Journal*, Volume 14.

Pilloud, C. and Others, (1987) *Commentary on the Additional Protocols of 8 June 1977 to the Geneva Conventions of 12 August 1949*, International Committee of the Red Cross, Geneva: Martinus Nijhoff Publishers.

Pingeot, L., (2012) *Dangerous Partnership: Private Military & Security Companies and the UN*, Report, June 2012, Global Policy Forum and Rosa Luxemburg Foundation: New York and Berlin.

Pisillo-Mazzeschi, R., (1992) "The Due Diligence Rule and the Nature of the International Responsibility of States", *German Yearbook of International Law*, Volume 35.

Plato, *The Republic*, as found in (1973) *The Republic of Plato*, Translated with Introduction and Notes by Francis MacDonald Cornford, London, Oxford and New York: Oxford University Press.

Poole, R.W. Jr., "Privatization", in D.R. Henderson (Ed.), *The Concise Encyclopedia of Economics*, <http://www.econlib.org/library/Enc/Privatization.html>.

Prado, M.M., "Regulatory Choices in the Privatization of Infrastructure", in Simon Chesterman and Angelina Fisher (Eds.), (2009) *Private Security, Public Order: The Outsourcing of Public Services and Its Limits*, Oxford: Oxford University Press.

Priest, D. and Arkin, W.M., "Top Secret America: A Washington Post Investigation", *The Washington Post*, July 20, 2010.

Quirico, O., "A Comparative Overview of European and Extra-European National Regulation of Private Military and Security Services", in Christine Bakker and Mirko Sossai (Eds.), (2012) *Multilevel Regulation of Military and Security Contractors: The Interplay Between International, European and Domestic Norms*, Oxford and Portland, Oregon: Hart Publishing.

Raghavan, S. and Ricks, T.E., "Private Security Puts Diplomats, Military at Odds", *The Washington Post*, September 26, 2007, <http://www.washingtonpost.com/wp-dyn/content/article/2007/09/25/AR2007092502675.html>.

Raghavan, S., Partlow, J., and DeYoung, K., "Blackwater Faulted in Military Reports From Shooting Scene", *The Washington Post*, October 5, 2007, <http://www.washingtonpost.com/wp-dyn/content/article/2007/10/04/AR2007100402654.html>.

Ranganathan, S., (2010) "Between Complicity and Irrelevance? Industry Associations and the Challenge of Regulating Private Security Contractors", *Georgetown Journal of International Law*, Volume 41, Issue 2.

Ratner, S.R., (2001) "Corporations and Human Rights: A Theory of Legal Responsibility", *Yale Law Journal*, Volume 111.

Raymond, L., "Abu Ghraib Torture Survivors Fight on Against Military Contractors", *The Huffington Post*, November 6, 2013, <http://www.huffingtonpost.com/the-center-for-constitutional-rights/abu-ghraib-torture-surviv_b_4227182.html>.

Republic of Ecuador, "Statement on Behalf of a Group of Countries at the 24rd Session of the Human Rights Council", including among others Ecuador, Cuba, Nicaragua, Pakistan, and Sri Lanka, September 2013, <http://business-humanrights.org/media/documents/statement-unhrc-legally-binding.pdf>.

Riechmann, D., "Petraeus Issues Afghan Contracting Guidance", *The Associated Press*, September 13, 2010, <http://www.airforcetimes.com/article/20100913/NEWS/9130308/Petraeus-issues-Afghan-contracting-guidance>.

Risen, J. and Mazzetti, M., "Blackwater Guards Tied to Secret C.I.A. Raids", *The New York Times*, December 10, 2009, <http://www.nytimes.com/2009/12/11/us/politics/11blackwater.html>.

Risen, J., "Ex-Blackwater Guards Face Renewed Charges", *The New York Times*, April 22, 2011, <http://www.nytimes.com/2011/04/23/us/23blackwater.html>.

Romany, C., "State Responsibility Goes Private: A Feminist Critique of the Public/Private Distinction in International Human Rights Law", in Rebecca J. Cook (Ed.), (1994) *Human Rights of Women: National and International Perspectives*, Philadelphia: University of Pennsylvania Press.

Rosemann, N., (2005) "The Privatization of Human Rights Violations – Business' Impunity or Corporate Responsibility? The Case of Human Rights Abuses and Torture in Iraq", *Non-State Actors and International Law*, Volume 5.

Rosemann, N., (2008) "Code of Conduct: Tool for Self-Regulation for Private Military and Security Companies", DCAF.

Rosenstock, R., (2002) "The ILC and State Responsibility", *The American Journal of International Law*, Volume 96, Issue 4.

Ruggie, J.G., (2007) "Business and Human Rights: The Evolving International Agenda", *The American Journal of International Law*, Volume 101, Issue 4.

Ruggie, J.G., (2013) *Just Business: Multinational Corporations and Human Rights*, New York & London: W. W. Norton & Company.

Ryngaert, C.M.J., (2008) "Litigating Abuses Committed by Private Military Companies", *The European Journal of International Law*, Volume 19, Issue 5.

Ryngaert, C.M.J., (2008) *Jurisdiction in International Law*, Oxford: Oxford University Press.

Ryngaert, C.M.J. and D'Aspermont, J., *Third Report on the Obligations and Responsibility of Non-State Actors in International Law*, April 2014, International Law Association, Committee on Non-State Actors.

Ryngaert, C.M.J., *Private Military Contractors: Regulatory Options*, Policy Brief No. 8, May 2008, <https://ghum.kuleuven.be/ggs/publications/policy_briefs/pb05.pdf>.

Salzman, Z., (2007-2008) "Private Military Contractors and a Taint of a Mercenary Reputation", *New York University Journal of International Law and Politics*, Volume 40, Issue 3.

Scheimer, M., (2009) "Separating Private Military Companies from Illegal Mercenaries in International Law: Proposing an International Convention for Legitimate Military and Security Support that Reflects Customary International Law", *American University International Law Review*, Volume 24, Issue 3.

Scheinin, M., "Just Another Word? Jurisdiction in the Roadmaps of State Responsibility", in Malcolm Langford, Wouter Vandenhole, Martin Scheinin, and Willem van Genugten (Eds.), (2013) *Global Justice, State Duties: The Extraterritorial Scope of Economic, Social and Cultural Rights in International Law*, Cambridge: Cambridge University Press.

Schmitt, M.N., (2005) "Humanitarian Law and Direct Participation in Hostilities by Private Contractors or Civilian Employees", *Chicago Journal of International Law*, Volume 5, Issue 2.

Schreier, F. and Caparini, M., (2005) *Privatising Security: Law, Practice and Governance of Private Military and Security Companies*, Occasional Paper No. 6, DCAF.

Schwartz, M., *Department of Defense Contractors in Iraq and Afghanistan: Background and Analysis*, Congressional Research Service Report for Congress, July 2, 2010, <http://www.fas.org/sgp/crs/natsec/R40764.pdf>.

Seiberth, C., (2013) *Private Military and Security Companies in International Law – A Challenge for Non-binding Norms: the Montreux Document and the International Code of Conduct for Private Security Service Providers*, Cambridge, Antwerp, Portland: Intersentia.

Shakespeare, W., (1602) *The History of Troilus and Cressida*, Stephen Orgel and A.R. Braunmuller (Eds.), (2002) *William Shakespeare: The Complete Works*, New York: Penguin Group.

Shane, S., "2 U.S. Architects of Harsh Tactics in 9/11's Wake", *The New York Times*, August 11, 2009, <http://www.nytimes.com/2009/08/12/us/12psychs.html?pagewanted=all>.

Shaw, M.N., (2008) *International Law*, Sixth Edition, Cambridge: Cambridge University Press.

Shearer, D., (1999) "Private Military Force and Challenges for the Future", *Cambridge Review of International Affairs*, Volume 13, Issue 1.

Sheehy, B. and Maogoto, J.N., (2008) "The Private Military Company – Unravelling the Theoretical, Legal & Regulatory Mosaic", *ILSA Journal of International & Comparative Law*, Volume 15, Issue 1.

Shorrock, T., (2008) *Spies for Hire: The Secret World of Intelligence Outsourcing*, New York: Simon & Schuster.

Silverstein, K., (2000) *Private Warriors*, New York and London: Verso.

Singer, P.W., (2003) "Peacekeepers, Inc.", *Policy Review*, Volume 119.

Singer, P.W., (2004) "War, Profits and the Vacuum of Law: Privatized Military Firms and International Law", *Columbia Journal of International Law*, Volume 42, Issue 2.

Singer, P.W., (2004) *The Private Military Industry and Iraq: What Have We Learned and Where to Next*, Policy Paper, DCAF.

Singer, P.W., (2005) "Outsourcing War", *Foreign Affairs*, Volume 84, Issue 2.

Singer, P.W., (2008) *Corporate Warriors: The Rise of the Privatized Military Industry*, Ithaca and London: Cornell University Press.

Singer, P.W., "Humanitarian Principles, Private Military Agents: Some Implications of the Privatised Military Industry for the Humanitarian Community", in Victoria Wheeler and Adele Harmer (Eds.), *Resetting the Rules of Engagement: Trends and Issues in Military Humanitarian Relations*, Humanitarian Policy Group Report 21, March 2006, London: Overseas Development Institute.

Skogly, S.I., "Causality and Extraterritorial Human Rights Obligations", in Malcolm Langford, Wouter Vandenhole, Martin Scheinin, and Willem van Genugten (Eds.), (2013) *Global Justice, State Duties: The Extraterritorial Scope of Economic, Social and Cultural Rights in International Law*, Cambridge: Cambridge University Press.

Slaughter, A.-M., "International Law and International Relations Theory: Twenty Years Later", in Jeffrey L. Dunoff and Mark A. Pollack (Eds.), (2012) *Interdisciplinary Perspectives on International Law and International Relations: The State of the Art*, Cambridge: Cambridge University Press.

Smith, A., (1776) *An Inquiry into the Nature and Causes of the Wealth of Nations*, the Edition Published in 2007, Petersfield: Harriman House.

Smutny, A.C., "State Responsibility and Attribution: When is a State Responsible for the Acts of State Enterprises? Emilio Agustín Maffezini v. The Kingdom of Spain", in Todd Weiler (Ed.), (2005) *International Investment Law and Arbitration*, London: Cameron May.

Sontag, S., "Regarding the Torture of Others", *The New York Times*, May 23, 2004, <http://www.nytimes.com/2004/05/23/magazine/regarding-the-torture-of-others.html>.

Sossai, M., "Status of Private Military and Security Company Personnel in the Law of International Armed Conflict", in Francesco Francioni and Natalino Ronzitti (Eds.), (2011) *War by Contract: Human Rights, Humanitarian Law, and Private Contractors*, Oxford: Oxford University Press.

Spaight, J.M., (1911) *War Rights on Land*, London: Macmillan and Co.

Spinedi, M., (2005) "Private Contractors: Responsabilité Internationale des Entreprises ou Attribution à l'Etat de la Conduite des Personnes Privées?", *International Law FORUM du Droit International*, Volume 7.

Staino, E., (2010) "Suing Private Military Contractors for Torture: How to Use the Alien Tort Statute Without Granting Sovereign Immunity-Related Defenses", *Santa Clara Law Review*, Volume 50, Issue 4.

Statement of Erik D. Prince, Chairman and CEO, Blackwater, for the House Committee on Oversight and Government Reform, October 2, 2007, <http://oversight-archive.waxman.house.gov/documents/20071003153621.pdf>.

Steinhoff, U., "What are Mercenaries?", in Andrew Alexandra, Deane-Peter Baker, and Marina Caparini (Eds.), (2008) *Private Military and Security Companies: Ethics, Policies and Civil-Military Relations*, London: Routledge.

Stern, B., "The Elements of An Internationally Wrongful Act", in James Crawford, Alain Pellet, and Simon Olleson (Eds.), (2010) *The Law of International Responsibility*, Oxford: Oxford University Press.

Stern, B., "The Obligation to Make Reparation", in James Crawford, Alain Pellet, and Simon Olleson (Eds.), (2010) *The Law of International Responsibility*, Oxford: Oxford University Press.

Stewart, J.G., (2003) "Towards a Single Definition of Armed Conflict in International Humanitarian Law: A Critique of Internationalized Armed Conflict", *International Review of the Red Cross*, Volume 85, Issue 850.

Stöber, J., "Contracting in the Fog of War... Private Security Providers in Iraq: A Principal-Agent Analysis", in Thomas Jäger and Gerhard Kümmel (Eds.), (2007) *Private Military and Security Companies: Chances, Problems, Pitfalls and Prospects*, Wiesbaden: VS Verlag für Sozialwissenschaften.

Stohr, G., "Abu Ghraib Contractor Suit Draws U.S. Supreme Court Interest", *Bloomberg*, October 4, 2010, <http://www.bloomberg.com/news/2010-10-04/abu-ghraib-case-involving-private-contractors-draws-top-court-s-interest.html>.

Sturzaker, D. and Cawood, C., (1999) "The Sandline Affair: Illegality and International Law", *Australian International Law Journal*.

Swiss Federal Department of Foreign Affairs, "Participating States of the Montreux Document", February 21, 2014, <http://www.eda.admin.ch/eda/en/home/topics/intla/humlaw/pse/parsta.html>.

Tan, L., "Private Military and Security Companies Wanted for Hire by CIA as 'Hitmen'", *The Huffington Post*, August 27, 2009, <http://www.huffingtonpost.com/amnesty-international/private-military-and-secu_b_270734.html>.

Tarzwell, A., (2009) "In Search of Accountability: Attributing the Conduct of Private Security Contractors to the United States Under the Doctrine of State Responsibility", *Oregon Review of International Law*, Volume 11, Issue 1.

Taulbee, J.L., (1998) "Reflections on the Mercenary Option", *Small Wars and Insurgencies*, Volume 9, Issue 2.

Thompson, G., "2 Say They Reported Abuses at Embassy", *The New York Times*, September 10, 2009, <http://www.nytimes.com/2009/09/11/world/asia/11contractor.html>.

Thomson, J.E., (1994) *Mercenaries, Pirates and Sovereigns: State-Building and Extraterritorial Violence in Early Modern Europe*, Princeton and Chichester: Princeton University Press.

Thurnher, J.S., (2008) "Drowning in Blackwater: How Weak Accountability over Private Security Contractors Significantly Undermines Counterinsurgency Efforts", *Army Lawyer*, Volume 64.

Tickler, P., (1987) *The Modern Mercenary: Dog of War, or Soldier of Honour?*, Wellingborough: Stephens.

Tomuschat, C., (2008) *Human Rights: Between Idealism and Realism*, Second Edition, Oxford: Oxford University Press.

Tomuschat, C., (2010) "Human Rights and International Humanitarian Law", *The European Journal of International Law*, Volume 21, Issue 1.

Tomuschat, C., "Specificities of Human Rights Law and International Humanitarian Law Regarding State Responsibility", in Robert Kolb and Gloria Gaggioli (Eds.), (2013) *Research Handbook on Human Rights and Humanitarian Law*, Cheltenham and Northampton: Edward Elgar Publishing.

Tonkin, H., (2009) "Common Article 1: A Minimum Yardstick for Regulating Private Military and Security Companies", *Leiden Journal of International Law*, Volume 22, Issue 4.

Tonkin, H., (2011) *State Control over Private Military and Security Companies in Armed Conflict*, Cambridge: Cambridge University Press.

Trebilcock, M.J. and Iacobucci, E.M., (2003) "Privatization and Accountability", *Harvard Law Review*, Volume 116.

Trubek, D.M. and Trubek, L.G., (2007) "New Governance & Legal Regulation: Complementarity, Rivalry, and Transformation", *Columbia Journal of European Law*, Volume 13, Issue 3.

Turns, D., "The Law of Armed Conflict (International Humanitarian Law)", in Malcolm D. Evans (Ed.), (2010) *International Law*, Third Edition, Oxford: Oxford University Press.

UK Foreign and Commonwealth Office, *Private Military Companies: Options for Regulation 2001-02*, February 2002, HCP 577, London: The Stationary Office, <http://www.official-documents.gov.uk/document/hc0102/hc05/0577/0577.pdf>.

UK Foreign and Commonwealth Office, "Promoting High Standards in the Private Military and Security Company Industry", June 21, 2011, <https://www.gov.uk/government/news/promoting-high-standards-in-the-private-military-and-security-company-industry--2>.

UK Parliament, Select Committee on Foreign Affairs, Ninth Report, Session 2001-2002, <http://www.parliament.the-stationery-office.co.uk/pa/cm200102/cmselect/cmfaff/922/92209.htm>.

University Centre for International Humanitarian Law, *Expert Meeting on Private Military Contractors: Status and State Responsibility for Their Actions*, Report, Geneva, August 29-30, 2005, <http://www.geneva-academy.ch/docs/expert-meetings/2005/2rapport_compagnies_privees.pdf>.

US Congressional Research Service, *Private Security Contractors in Iraq: Background, Legal Status, and Other Issues*, Report for Congress, August 25, 2008.

US Defense Department Operational Update Briefing, May 4, 2004, <http://www.defense.gov/transcripts/transcript.aspx?transcriptid=2973>.

US Department of Justice, United States Attorney's Office, Eastern District of North Carolina, "David Passaro Sentenced to 100 Months Imprisonment: First American Civilian Convicted of Detainee Abuse During the Wars in Iraq and Afghanistan", Press Release, February 13, 2007, <http://www.justice.gov/usao/nce/press/2007/2007-Feb-13.html>.

US Department of State, Bureau of International Information Programs, *Abraham Lincoln: A Legacy of Freedom*, <http://www.america.gov/media/pdf/books/lincoln.pdf>.

US Government Accountability Office, *Rebuilding Iraq: Actions Needed to Improve Use of Private Security Providers*, Report to Congressional Committees, July 2005, <http://www.gao.gov/assets/250/247252.pdf>.

US House of Representatives, Committee on Oversight and Government Reform, "Additional Information About Blackwater USA", Memorandum, October 1, 2007, <http://graphics8.nytimes.com/packages/pdf/national/20071001121609.pdf>.

US House of Representatives, Committee on Oversight and Government Reform, "Blackwater USA. Hearing before the House Committee on Oversight and Government Reform", Serial No. 110-89, October 2, 2007, <http://www.gpo.gov/fdsys/pkg/CHRG-110hhrg45219/html/CHRG-110hhrg45219.htm>.

US Senate, Committee on Armed Services, *Report of the Inquiry into the Role and Oversight of Private Security Contractors in Afghanistan*, September 28, 2010, <http://www.wbir.com/pdf/SASC.PSCReport.100710.pdf>.

Van der Maas, T., (2010) "Private Military Companies in het Amerikaanse en Britse defensiebeleid", *Vrede en Veiligheid*, Volume 39, Issue 4.

Van Genugten, W., Homan, K., and Others, (2006) *The United Nations of the Future: Globalization with a Human Face*, Amsterdam: KIT Publishers.

Van Genugten, W., Jägers, N., and Van der Heijden, M.-J., "Protecting the Victims of the Privatization of War", in Rianne Letschert and Jan Van Dijk (Eds.), (2011) *The New Faces of Victimhood*, Dordrecht: Springer.

Vandenhole, W., (2011) "Beyond Territoriality: The Maastricht Principles on Extra-Territorial Obligations in the Area of Economic, Social and Cultural Rights", *Netherlands Quarterly of Human Rights*, Volume 29, Issue 4.

Vasquez, J.P. III, (2010) "Blackwater's Rise and the Draft's Demise", *Yale Journal of International Affairs*, Volume 5, Issue 2.

Verkuil, P.R., (2007) *Outsourcing Sovereignty: Why Privatization of Government Functions Threatens Democracy and What We Can Do About It*, Cambridge: Cambridge University Press.

Vernon, R.R., (2003-2004) "Battlefield Contractors: Facing the Tough Issues", *Public Contract Law Journal*, Volume 33, Issue 2.

Vierucci, L., "Private Military and Security Companies in Non-International Armed Conflicts: Ius ad Bellum and Ius in Bello Issues", in Francesco Francioni and Natalino Ronzitti (Eds.), (2011) *War by Contract: Human Rights, Humanitarian Law, and Private Contractors*, Oxford: Oxford University Press.

Voillat, C., (2004) "Private Military Companies: A Word of Caution", *Humanitarian Exchange Magazine*, Issue 28, <http://www.odihpn.org/report.asp?id=2675>.

Volkov, V., (2002) *Violent Entrepreneurs: the Use of Force in the Making of Russian Capitalism*, Ithaca: Cornell University Press.

VPs, Fact Sheet – General, <http://www.voluntaryprinciples.org/wp-content/uploads/2014/05/VPs_Fact_Sheet_-_May_2014.pdf>.

Walzer, M., "Mercenary Impulse: Is There an Ethics That Justifies Blackwater?", *The New Republic*, March 12, 2008, <http://www.tnr.com/article/mercenary-impulse>.

Wayne, L., "America's For-Profit Secret Army", *The New York Times*, October 13, 2002, <http://www.nytimes.com/2002/10/13/business/america-s-for-profit-secret-army.html?pagewanted=all&src=pm>.

Wedgwood, R., "Slobodan Milosevic's Last Waltz", *The New York Times*, March 12, 2007, <http://www.nytimes.com/2007/03/12/opinion/12wedgwood.html?pagewanted=1&_r=1>.

Weiner, R., "Private Military Contractors Come With Strings Attached", *Belfer Center Newsletter*, Winter 2005-2006, Belfer Center for Science and International Affairs, Harvard Kennedy School, <http://belfercenter.ksg.harvard.edu/publication/19402/private_military_contractors_come_with_strings_attached.html>.

White, A., (2010) *The Politics of Private Security: Regulation, Reform and Re-Legitimation*, Houndmills, Basingstoke, New Hampshire: Palgrave Macmillan.

White, J. and Witte, G., "Tension, Confusion Between Troops, Contractors in Iraq", *The Washington Post*, July 10, 2005, <http://www.washingtonpost.com/wp-dyn/content/article/2005/07/09/AR2005070901175.html>.

White, N.D., "Institutional Responsibility for Private Military and Security Companies", in Francesco Francioni and Natalino Ronzitti (Eds.), (2011) *War by Contract: Human Rights, Humanitarian Law, and Private Contractors*, Oxford: Oxford University Press.

White, N.D., "Regulatory Initiatives at the International Level", in Christine Bakker and Mirko Sossai (Eds.), (2012) *Multilevel Regulation of Military and Security Contractors: The Interplay Between International, European and Domestic Norms*, Oxford and Portland, Oregon: Hart Publishing.

Wilber, D.Q. and Hsu, S.S., "Judge Ordered to Revisit Blackwater Guard Shooting Case", *The Washington Post*, April 22, 2011, <http://www.washingtonpost.com/local/judge-ordered-to-revisit-blackwater-guard-shooting-case/2011/04/22/AF4flCRE_story.html>.

Wilber, D.Q., "Judge Dismisses Charges Against Blackwater Guards in Iraq Deaths", *The Washington Post*, January 1, 2010, <http://www.washingtonpost.com/wp-dyn/content/article/2010/01/01/AR2010010100562.html>.

Williamson, O.E., (1999) "Public and Private Bureaucracies: A Transaction Cost Economics Perspective", *Journal of Law, Economics and Organization*, Volume 15, Issue 1.

Wilson, J.Q., (1989) *Bureaucracy*, New York: Basic Books.

Wittich, S., (2013) "Domestic Courts and the Content and Implementation of State Responsibility", *Leiden Journal of International Law*, Volume 26, Issue 3.

Written Statement of Dr. T.X. Hammes, Senior Research Fellow, Provided to the House Committee on Oversight and Government Reform, Subcommittee on National Security and Foreign Affairs, Hearing on June 22, 2010, <http://oversight.house.gov/wp-content/uploads/2012/01/20100622Hammes.pdf>.

Wulf, H., (2005) *Internationalizing and Privatizing War and Peace*, Basingstoke: Palgrave Macmillan.

Xenophon, *Anabasis*, translated by Henry G. Dakyns, in Henry G. Dakyns, (1897) *The Works of Xenophon*, London, New York: Macmillan, <http://www.gutenberg.org/files/1170/1170-h/1170-h.htm>.

Yeoman, B., "Need an Army? Just Pick Up the Phone", *The New York Times*, April 2, 2004, <http://www.nytimes.com/2004/04/02/opinion/02YEOM.html?ex=1081912176&ei=1&en=bce2ecd0c21b8bb4>.

Zarate, J.C., (1998) "The Emergence of a New Dog of War: Private International Security Companies, International Law, and the New World Disorder", *Stanford Journal of International Law*, Volume 34, Issue 1.

Zucchino, D., "Deadly Contractor Incident Sours Afghans", *The Los Angeles Times*, August 13, 2009, <http://www.latimes.com/la-fg-afghan-contractors13-2009aug13-story.html#page=1>.

TABLE OF CASES

INTERNATIONAL COURT OF JUSTICE

Case Concerning Armed Activities on the Territory of the Congo (Democratic Republic of the Congo v. Uganda), Judgment (Merits), December 19, 2005, ICJ Reports 2005.

Case Concerning Military and Paramilitary Activities in and against Nicaragua (Nicaragua v. United States of America), Judgment (Merits), June 27, 1986, ICJ Reports 1986.

Case Concerning the Application of the Convention on the Prevention and Punishment of the Crime of Genocide (Bosnia and Herzegovina v. Serbia and Montenegro), Judgment (Merits), February 26, 2007, ICJ Reports 2007.

Corfu Channel (United Kingdom v. Albania), Judgment (Merits), April 9, 1949, ICJ Reports 1949.

Difference Relating to Immunity from Legal Process of a Special Rapporteur of the Commission on Human Rights, Advisory Opinion, April 29, 1999, ICJ Reports 1999.

Gabčíkovo-Nagymaros Project (Hungary v. Slovakia), Judgment, September 25, 1997, ICJ Reports 1997.

Interpretation of the Agreement of 25 March 1951 Between the WHO and Egypt, Advisory Opinion, December 20, 1980, ICJ Reports 1980.

Legal Consequences of the Construction of a Wall in the Occupied Palestinian Territory, Advisory Opinion, July 9, 2004, ICJ Reports 2004.

Legality of the Threat or Use of Nuclear Weapons, Advisory Opinion, July 8, 1996, ICJ Reports 1996.

North Sea Continental Shelf Case (Federal Republic of Germany v. Netherlands), Judgment, February 20, 1969, ICJ Reports 1969.

Reparation for Injuries Suffered in the Service of the United Nations, Advisory Opinion, April 11, 1949, ICJ Reports 1949.

United States Diplomatic and Consular Staff in Tehran (United States of America v. Iran), Judgment, May 24, 1980, ICJ Reports 1980.

Dissenting Opinion of the Vice-President of the International Court of Justice A.S. Al-Khasawneh in International Court of Justice, *Case Concerning the Application of the Convention on the Prevention and Punishment of the Crime of Genocide (Bosnia and Herzegovina v. Serbia and Montenegro)*, Judgment (Merits), February 26, 2007, ICJ Reports 2007.

Separate Opinion of Judge M. Ago in International Court of Justice, *Case Concerning Military and Paramilitary Activities in and against Nicaragua (Nicaragua v. United States of America)*, Judgment (Merits), June 27, 1986, ICJ Reports 1986.

INTERNATIONAL CRIMINAL TRIBUNAL FOR THE FORMER YUGOSLAVIA

Prosecutor v. Delalic and Others, Trial Chamber, Judgment, November 16, 1998, ICTY-IT-96-21-T.

Prosecutor v. Delalic, Appeals Chamber, Judgment, February 20, 2001, ICTY-IT-96-21-A.

Prosecutor v. Furundžija, Trial Chamber, Judgment, December 10, 1998, ICTY-IT-95-17/1-T.

Prosecutor v. Kunarac and Others, Trial Chamber, Judgment, February 22, 2001, ICTY-IT-96-23-T & ICTY-IT-96-23/1-T.

Prosecutor v. Tadić, Appeals Chamber, Judgment, July 15, 1999, ICTY-IT-94-1-A.

PERMANENT COURT OF INTERNATIONAL JUSTICE

Case of Certain Questions Relating to Settlers of German Origin in the Territory Ceded by Germany to Poland, Advisory Opinion, September 10, 1923, PCIJ Series B 1923, No. 6.

Factory at Chorzów (Germany v. Poland), Jurisdiction, Judgment, July 26, 1927, PCIJ Reports, Series A, No. 9.

Phosphates in Morocco case (Italy v. France), Judgment of June 14, 1938, PCIJ Series A/B, No. 74.

The Case of the S.S. "Lotus" (France v. Turkey), Judgment, September 7, 1927, PCIJ Reports, Series A, No. 10.

UN HUMAN RIGHTS COMMITTEE

Ibrahim Gueye v. France, April 6, 1989, UN Doc. CCPR/C/35/D/196/1985 (1989).

Joaquín David Herrera Rubio et al. v. Colombia, November 2, 1987, Communication No. 161/1983, UN Doc. CCPR/C/OP/2 at 192 (1990).

Lilian Celiberti de Casariego v. Uruguay, July 29, 1981, Communication No. R.13/56 (July 17, 1979), UN Doc. Supp. No. 40 (A/36/40) at 185 (1981).

Mr. Carlos Cabal and Mr. Marco Pasini Bertran v. Australia, September 19, 2003, Communication No. 1020/2001, UN Doc. CCPR/C/78/D/1020/2001 (2003).

Nunez v. Uruguay, July 22, 1983, UN Doc. Supp. No. 40 (A/38/40).

Sergio Euben Lopez Burgos v. Uruguay, July 29, 1981, Communication No. R.12/52, UN Doc. Supp. No. 40 (A/36/40) at 176 (1981).

Suarez de Guerrero v. Colombia, March 31, 1982, UN Doc. Supp. No. 40 (A/37/40).

William Eduardo Delgado Páez v. Colombia, July 12, 1990, Communication No. 195/1985, UN Doc. CCPR/C/39/D/195/1985 (1990).

African System

African Commission on Human and Peoples' Rights

Association pour la Sauvegarde de la Paix au Burundi v. Kenya, Uganda, Rwanda, Tanzania, Zaire (DRC), Zambia, Communication No. 157/96 (2003).
Commission Nationale des Droits de l'Homme et des Libertes v. Chad, Communication No. 74/92 (1995).
D.R. Congo v. Burundi, Rwanda and Uganda, Communication No. 227/99 (2003).
The Social and Economic Rights Action Center and the Center for Economic and Social Rights v. Nigeria, Communication No. 155/96 (2001).
Zimbabwe Human Rights NGO Forum v. Zimbabwe, May 15, 2006, Communication No. 245/02 (2006).

European System

European Commission of Human Rights

Cyprus v. Turkey, Decision, May 26, 1975, Application Nos. 6780/74 and 6950/75.
W v. United Kingdom, Judgment, February 28, 1983, Application No. 9348/81, Series A, No. 12.

European Court of Human Rights

Al-Jedda v. the United Kingdom, Grand Chamber, Judgment, July 7, 2011, Application No. 27021/08.
Andronicou and Constantinou v. Cyprus, Judgment, October 9, 1997, Application No. 25052/94, ECHR 1997-VI.
Assenov and Other v. Bulgaria, Judgment, October 28, 1998, Application No. 24760/94, Reports 1998-VIII.
Avsar v. Turkey, Judgment, July 10, 2001, Application No. 25657/94.
Banković and Others v. Belgium and Others, Decision, December 12, 2001, Application No. 52207/99, Reports 2001-XII.
Behrami and Behrami v. France and Saramati v. France, Germany and Norway, Grand Chamber, Decision as to the Admissibility of Application No. 71412/01 and Application No. 78166/01, May 2, 2007.
Bevacqua and S. v. Bulgaria, Judgment, June 12, 2008, Application No. 71127/01.
Cyprus v. Turkey, Judgment, May 10, 2001, Application No. 25781/94, Reports 2001-IV.
De Cubber v. Belgium, Judgment (Merits), October 26, 1984, Application No. 9186/80.
Drozd and Janousek v. France and Spain, Judgment, June 26, 1992, Application No. 12747/87.
Ergi v. Turkey, Judgment, July 28, 1998, Application No. 23818/94, ECHR 1998-IV, No. 81.

Gezici v. Turkey, Judgment, March 17, 2005, Application No. 34594/97.
Ilaşcu v. Moldova and Russia, Judgment, July 8, 2004, Application No. 48787/99, Reports 2004-VII.
Ilhan v. Turkey, Judgment, June 27, 2000, Application No. 22277/93, Reports 2000-VII.
Issa and Others v. Turkey, Judgment, November 16, 2004, Application No. 31821/96, ECHR 629.
Kaya v. Turkey, Judgment, February 19, 1998, Application No. 22729/93, Report 1998-I.
Kiliç v. Turkey, Judgment, March 28, 2000, Application No. 22492/93, ECHR Reports 2000-III.
Loizidou v. Turkey, Judgment, December 18, 1996, Application No. 15318/89, Reports 1996-VI.
Loizidou v. Turkey, Preliminary Objections, March 23, 1995, Application No. 15318/89, Series A, No. 310.
Mahmut Kaya v. Turkey, Judgment, March 28, 2000, Application No. 22535/93, ECHR Reports 2000-III.
Makaratzis v. Greece, Judgment, December 20, 2004, Application No. 50385/99, Reports 2004-XI.
Mastromatteo v. Italy, Judgment, October 24, 2002, Application No. 37703/97.
McCann and Others v. United Kingdom, Judgment, September 27, 1995, Application No. 18984/91, Series A, No. 324.
McKerr v. United Kingdom, Judgment, May 4, 2001, Application No. 28883/95, Reports 2001-III.
Nachova and Others v. Bulgaria, Judgment, July 6, 2005, Application Nos. 43577/98 and 43579/98, Reports 2005-VII.
Öcalan v. Turkey, Judgment, May 12, 2005, Application No. 46221/99, Reports 2005-IV.
Ogur v. Turkey, Judgment, May 20, 1999, Application No. 21954/93, Reports 1999-III.
Osman v. United Kingdom, Judgment, October 28, 1998, Application No. 23452/94, ECHR Reports 1998-VIII.
X and Y v. The Netherlands, Judgment, March 26, 1985, Application No. 8978/80, ECHR Series A.

INTER-AMERICAN SYSTEM

Inter-American Commission on Human Rights

Alejandre v. Cuba, Report No. 86/99, Case No. 11.589, September 29, 1999.
Coard v. United States, Report No. 109/99, Case No. 10.951, September 29, 1999.
Saldano v. Argentina, Petition, Report No. 38/99, March 11, 1999.

Inter-American Court of Human Rights

Blanco-Romero et al. v. Venezuela, Judgment, November 28, 2005, Series C, No. 138 (2005).

Bulacio v. Argentina, Judgment, September 18, 2003, Series C, No. 100 (2003).
Godínez Cruz v. Honduras, Judgment, January 20, 1989, Series C, No. 5 (1989).
Ituango Massacres v. Colombia, Judgment, July 1, 2006, Series C, No. 148 (2006).
Juan Humberto Sánchez v. Honduras, Judgment, June 7, 2003, Series C, No. 99 (2003).
Montero-Aranguren et al. v. Venezuela, Judgment, July 5, 2006, Series C, No. 150 (2006).
Neira Alegria et al. v. Peru, Judgment, January 19, 1995, Series C, No. 21 (1995).
Paniagua Morales et al. v. Guatemala, Judgment, March 8, 1998, Series C, No. 37 (1998).
Pueblo Bello Massacre v. Colombia, Judgment, January 31, 2006, Series C, No. 140 (2006).
Servellón-García et al. v. Honduras, Judgment, September 21, 2006, Series C, No. 150 (2006).
Valle Jaramillo et al. v. Colombia, Judgment, November 27, 2008, Series C, No. 192 (2008).
Velasquez Rodriguez v. Honduras, Judgment, July 29, 1988, Series C, No. 4 (1988).
Velasquez Rodriguez v. Honduras, Preliminary Objections, Judgment, June 26, 1989, Series C, No. 1 (1994).
Villagran-Morales et al. v. Guatemala, Judgment, November 19, 1999, Series C, No. 63 (1999).

OTHER INTERNATIONAL

France-New Zealand Arbitration Tribunal, *Rainbow Warrior (New Zealand v. France)*, Decision, April 30, 1990, 20 RIAA 215.
Gertrude Parker Massey (United States of America v. United Mexican States), April 15, 1927, Reports of International Arbitral Awards, Volume IV.
International Centre for Settlement of Investment Disputes, *Emilio Agustín Maffezini v. The Kingdom of Spain*, Decision on Jurisdiction (English translation), January 25, 2000, ICSID Case No. ARB/97/7.
Iran – United States Claims Tribunal, *Computer Sciences Corporation v. Government of the Islamic Republic of Iran*, (1986) 10 Iran – United States Claims Tribunal Reports.
Iran – United States Claims Tribunal, *Daley v. Government of the Islamic Republic of Iran*, (1988) 18 Iran – United States Claims Tribunal Reports.
Iran – United States Claims Tribunal, *Hyatt International Corporation v. Government of the Islamic Republic of Iran*, (1985) 9 Iran – United States Claims Tribunal Reports.
Iran – United States Claims Tribunal, *Oil Field of Texas, Inc. v. Government of the Islamic Republic of Iran*, (1982) 1 Iran – United States Claims Tribunal Reports.
Iran – United States Claims Tribunal, *William L. Pereira Associates, Iran v. Government of the Islamic Republic of Iran*, (1984) 5 Iran – United States Claims Tribunal Reports.
Iran – United States Claims Tribunal, *Yeager v. Government of the Islamic Republic of Iran*, (1987) 17 Iran – United States Claims Tribunal Reports.
Laura M. B. Janes et al. (United States of America v. United Mexican States), Reports of International Arbitral Awards, Volume IV.
Thomas H. Youmans (United States of America v. United Mexican States), Reports of International Arbitral Awards, Volume IV.

Belgium

The Netherlands

United States of America

SELECTED UNITED NATIONS DOCUMENTS

COMMISSION ON HUMAN RIGHTS

UN Doc. E/CN.4/1997/24, February 20, 1997, *Report on the Question of the Use of Mercenaries as a Means of Violating Human Rights and Impeding the Exercise of the Right of Peoples to Self-Determination, Submitted by Mr. Enrique Bernales Ballesteros, Special Rapporteur.*

UN Doc. E/CN.4/2005/7, December 22, 2004, *Summary or Arbitrary Executions, Report of the Special Rapporteur, Philip Alston.*

COMMITTEE AGAINST TORTURE

UN Doc. CAT/C/GC/2, January 24, 2008, *General Comment No. 2, Implementation of Article 2 by States Parties.*

COMMITTEE ON ECONOMIC, SOCIAL AND CULTURAL RIGHTS

UN Doc. E/C.12/2000/13, *Maastricht Guidelines on Violations of Economic, Social and Cultural Rights*, Twenty-Fourth Session, November 13 – December 1, 2000.

COMMITTEE ON THE ELIMINATION OF DISCRIMINATION AGAINST WOMEN

UN Doc. A/47/38, January 30, 1992, *General Recommendation 19.*

COMMITTEE ON THE RIGHTS OF THE CHILD

UN Doc. CRC/C/15/Add.195, October 9, 2002, *Concluding Observations: Israel.*

UN Doc. CRC/C/GC/16, April 17, 2013, *General Comment No. 16 on State Obligations Regarding the Impact of the Business Sector on Children's Rights.*

GENERAL ASSEMBLY

UN Doc. A/56/83, January 28, 2002, Resolution 56/83, *Responsibility of States for Internationally Wrongful Acts.*

UN Doc. A/63/467-S/2008/636, October 8, 2008, *Montreux Document on Pertinent International Legal Obligations and Good Practices for States Related to Operations of Private Military and Security Companies During Armed Conflict.*

UN Doc A/RES/174(II), November 21, 1947, Resolution 174(II), *Establishment of an International Law Commission.*

HUMAN RIGHTS COMMITTEE

Concluding Observations on the Second and Third U.S. Reports to the Committee, Eighty-Seventh Session, July 10-28, 2006.

UN Doc. CCPR/C/21/Rev.1/Add.13, May 26, 2004, *General Comment 31.*

UN Doc. CCPR/C/79/Add.50, April 6, 1995, *Concluding Observations of the Human Rights Committee: United States of America.*

UN Doc. CCPR/C/79/Add.93, August 18, 1998, *Concluding Observations of the Human Rights Committee: Israel.*

UN Doc. CCPR/C/81/Add.4, August 24, 1994, *Initial Reports of State Parties due in 1993: United States of America.*

UN Doc. CCPR/C/SR.1405, April 24, 1995, *Summary Record of the 1405th Meeting: United States of America.*

UN Doc. CCPR/C/USA/3, Annex I, November 28, 2005, *Consideration of Reports Submitted by States Parties under Article 40 of the Covenant, Third Periodic Reports of States Parties due in 2003, United States of America.*

UN Doc. CCPR/C/USA/CO/3, September 15, 2006, *Consideration of Reports Submitted by States Parties under Article 40 of the Covenant, Concluding Observations of the Human Rights Committee, United States of America.*

UN Doc. CCPR/CO/78/ISR, August 5, 2003, *Concluding Observations of the Human Rights Committee: Second Periodic Report of Israel.*

UN Doc. HRI/GEN/1/Rev.1, *General Comment 20 Replacing General Comment 07.*

HUMAN RIGHTS COUNCIL

UN Doc. A/HRC/17/31, March 21, 2011, *Report of the Special Representative of the Secretary-General on the Issue of Human Rights and Transnational Corporations and Other Business Enterprises, John Ruggie, Guiding Principles on Business and Human Rights: Implementing the United Nations 'Protect, Respect and Remedy' Framework*, Seventeenth Session.

UN Doc. A/HRC/WG.10/1/2, May 13, 2011, *Draft of a Possible Convention on Private Military and Security Companies (PMSCs) for Consideration and Action by the Human Rights Council*, Prepared by the Working Group on the use of mercenaries as a means of violating human rights and impeding the exercise of the right of peoples to self-determination, First Session, Geneva, May 23-27, 2011.

UN Doc. A/HCR/7/7, January 9, 2008, Report of the Working Group on the Use of Mercenaries as a Means of Violating Human Rights and Impeding the Exercise of the Right of People to Self-Determination.

UN Doc. A/HRC/8/5/Add.2, May 23, 2008, *Corporations and Human Rights: A Survey of the Scope and Patterns of Alleged Corporate-related Human Rights Abuse*, Report of the Special Representative of the Secretary-General on the Issue of Human Rights and Transnational Corporations and Other Business Enterprises.

UN Doc. A/HRC/10/14, January 21, 2009, *Promotion and Protection of All Human Rights, Civil, Political, Economic, Social and Cultural Rights, Including the Right to Development*, Report of the Working Group on the use of mercenaries as a means of violating human rights and impeding the exercise of the right of peoples to self-determination.

UN Doc. A/HRC/11/13, April 22, 2009, *Report of the Special Representative of the Secretary-General on the Issue of Human Rights and Transnational Corporations and Other Business Enterprises, John Ruggie, Business and Human Rights: Towards Operationalizing the 'Protect, Respect and Remedy' Framework,* Eleventh Session.

International Law Commission

Draft Articles on Responsibility of States for Internationally Wrongful Acts with Commentaries, in Report of the International Law Commission on the Work of Its Fifty-Third Session, UN General Assembly Official Records, Fifty-Sixth Session, Supplement No. 10, UN Doc. A/56/10 (2001), reprinted in (2001) Yearbook of the International Law Commission, Volume II(2), UN Doc A/CN.4/SER.A/2001/Add.1 (Part 2); also reprinted in James Crawford (Ed.), (2002) *The International Law Commission's Articles on State Responsibility: Introduction, Text and Commentaries*, Cambridge: Cambridge University Press.

Draft Articles on State Responsibility with Commentaries Thereto Adopted by the International Law Commission on First Reading, as adopted in 1996, January 1997, <http://legal.un.org/ilc/texts/instruments/english/commentaries/9_6_1996.pdf>.

Draft Articles on the Responsibility of International Organizations, With Commentaries, as Adopted by the International Law Commission at its Sixty-Third Session, (2011) Yearbook of the International Law Commission, Volume II(2), UN Doc. A/66/10.

Statute of the International Law Commission as adopted by the UN General Assembly in Resolution 174(II) of November 21, 1947 and as amended by Resolutions 485 (V) of December 12, 1950; 984 (X) of December 3, 1955; 985 (X) of December 3, 1955 and 36/39 of November 18, 1981, <http://untreaty.un.org/ilc/texts/instruments/english/statute/statute_e.pdf>.

UN Doc A/CN.4/233, *Second Report on State Responsibility, by Roberto Ago, Special Rapporteur – the Origin International Responsibility*, (1970) Yearbook of the International Law Commission, Volume II.

UN Doc. A/10010/Rev.1, *Report of the International Law Commission on the Work of Its Twenty-Seventh Session, 5 May – 25 July 1975*, Official Records of the General Assembly, Thirtieth Session, Supplement No. 10, (1975) Yearbook of the International Law Commission, Volume II.

UN Doc. A/9610/Rev.1, *Report of the International Law Commission on the Work of Its Twenty-Sixth Session, 6 May – 26 July 1974*, (1974) Yearbook of the International Law Commission, Volume II(1).

UN Doc. A/CN.4/13 and Corr.1-3, *Report of the International Law Commission on the Work of Its First Session, 12 April 1949*, Official Records of the General Assembly, Fourth Session, Supplement No. 10, (1949) Yearbook of the International Law Commission, Volume 1.

UN Doc. A/CN.4/264 and Add.1, *Fourth Report on State Responsibility, by Roberto Ago, Special Rapporteur – the Internationally Wrongful Act of the State, Source of International Responsibility (Continued)*, (1972) Yearbook of the International Law Commission, Volume II.

UN Doc. A/CN.4/264 and Add.1-3, *Third Report on State Responsibility, by Mr. Roberto Ago, Special Rapporteur, the Internationally Wrongful Act of the State, Source of International Responsibility*, (1971) Yearbook of the International Law Commission, Volume II(1).

UN Doc. A/CN.4/490 and Add 1-7, *First Report on State Responsibility by Mr. James Crawford, Special Rapporteur,* (1998) Yearbook of the International Law Commission, Volume II(2).

UN Doc. A/CN.4/515 and Add.1-3, *Comments and Observations Received from Governments*, (2001) Yearbook of the International Law Commission, Volume II(1).

OFFICE OF THE UNITED NATIONS HIGH COMMISSIONER FOR HUMAN RIGHTS

"State National Action Plans", <http://www.ohchr.org/EN/Issues/Business/Pages/NationalActionPlans.aspx>.

"New Guiding Principles on Business and Human Rights Endorsed by the UN Human Rights Council", Press Release, June 16, 2011, <http://www.business-humanrights.org/media/documents/ruggie/ruggie-guiding-principles-endorsed-16-jun-2011.pdf>.

"Special Rapporteur of the Commission on Human Rights on the Use of Mercenaries as a Means of Impeding the Exercise of the Right of Peoples to Self-Determination", <http://www2.ohchr.org/english/issues/mercenaries/specialrap.htm>.

Summary Report of the Expert Panel on the Use of Private Military and Security Companies by the United Nations, July 31, 2013, New York, <http://www.ohchr.org/Documents/Issues/Mercenaries/WG/StudyPMSC/EventSummary.pdf>.

"It's High Time to Close the Legal Gap for Private Military and Security Contractors", <http://www.ohchr.org/EN/NewsEvents/Pages/DisplayNews.aspx?NewsID=10000&LangID=E>.

Working Group on the Use of Mercenaries as a Means of Violating Human Rights and Impeding the Exercise of the Right of Peoples to Self-Determination, "UN Expert Group Calls for Robust International Regulation of Private Military and Security Companies", November 4, 2013, <http://www.ohchr.org/EN/NewsEvents/Pages/DisplayNews.aspx?NewsID=13940&LangID=E>.

"Guns for Hire", April 29, 2010, <http://www.ohchr.org/EN/NewsEvents/Pages/Gunsforhire.aspx>.

"Mercenaries Still Pose a Serious Threat to Human Rights", April 8, 2011, <http://www.ohchr.org/EN/NewsEvents/Pages/DisplayNews.aspx?NewsID=10923&LangID=E>.

SECURITY COUNCIL

UN Doc. S/Res/1132, October 8, 1997, Resolution 1132.
UN Doc. S/RES/1540, April 28, 2004, Resolution 1540.

INDEX

D

E

G

H

I

CURRICULUM VITAE

Evgeni Moyakine obtained his first Master's degree in International and European Law *cum laude* at Radboud University Nijmegen in 2009 after having studied at the same university and at Manchester Metropolitan University, the United Kingdom. In 2010, he successfully completed the Research Master in Law program at Tilburg University and Leuven University, Belgium, with *the same grade point average*. During his study, he was appointed as an academy assistant with financial support of the Royal Netherlands Academy of Arts and Sciences (KNAW), participated in a research project funded by the Hague Institute for the Internationalization of Law (HiiL) on transnational private regulation, and contributed to the case study "Transnational Private Regulation, Private Security Companies and Human Rights: Promise of Effectiveness?". In 2010, he was awarded a Mosaic grant by the Netherlands Organization for Scientific Research (NWO) for his research project and, from 2010 to 2014, worked as a doctoral candidate at the Department of European and International Public Law, Tilburg Law School, Tilburg University. In addition to his main responsibility with regard to the PhD thesis, he took part in a number of other research projects, such as the above-mentioned HiiL research on transnational private regulation and the report prepared for ACCESS Facility, The Hague, and the UN Working Group on Business and Human Rights on company-community conflicts and the effectiveness of outcomes of non-judicial grievance mechanisms. He also gave lectures and seminars on the issue of State responsibility and private military and security contractors, supervised Bachelor's and Master's students with their theses and the Moot Courts, and obtained his university teaching qualification in 2014. Additionally, throughout the years he functioned as – *inter alia* – an active member of the Center for Transboundary Legal Development at Tilburg University, the Graduate Program Committee of the Utrecht School of Human Rights Research and the Editorial Board of its Newsletter, and a senior editor at the *Tilburg Law Review: Journal of International and European Law*. Finally, separately and jointly with other researchers he published several articles and participated in various research activities and conferences.

School of Human Rights Research Series

The School of Human Rights Research is a joint effort by human rights researchers in the Netherlands. Its central research theme is the nature and meaning of international standards in the field of human rights, their application and promotion in the national legal order, their interplay with national standards, and the international supervision of such application. The School of Human Rights Research Series only includes English titles that contribute to a better understanding of the different aspects of human rights.

For previous volumes in the series, please visit http://shr.intersentia.com.

Published titles within the Series:

62. Ramona Biholar, *Transforming Discriminatory Sex Roles and Gender Stereotyping: The implementation of Article 5(a) CEDAW for the realisation of women's right to be free from gender-based violence in Jamaica*
ISBN 978-1-78068-167-2

63. Jasper Krommendijk, *The Domestic Impact and Effectiveness of the Process of State Reporting under UN Human Rights Treaties in the Netherlands, New Zealand and Finland. Paper-pushing or policy prompting?*
ISBN 978-1-78068-244-0

64. Jennifer Anna Sellin, *Access to Medicines. The Interface between Patents and Human Rights. Does one size fit all?*
ISBN 978-1-78068-247-1

65. Gustavo Arosemena, *Rights, Scarcity and Justice. An Analytical Inquiry into the Adjudication of the Welfare Aspects of Human Rights*
ISBN 978-1-78068-275-4

66. Etienne Ruvebana, *Prevention of Genocide under International Law. An analysis of the obligations of states and the United Nations to prevent genocide at the primary, secondary and tertiary levels*
ISBN 978-1-78068-273-0

www.ingramcontent.com/pod-product-compliance
Ingram Content Group UK Ltd.
Pitfield, Milton Keynes, MK11 3LW, UK
UKHW051918100926
8556IPUK00021B/1247

9 781780 68281